Transduction in
Biological Systems

Series of the Centro de Estudios Científicos de Santiago

Series Editor: Claudio Teitelboim
Centro de Estudios Científicos de Santiago,
Santiago, Chile
and Institute for Advanced Study
Princeton, New Jersey, USA

IONIC CHANNELS IN CELLS AND MODEL SYSTEMS
Edited by Ramon Latorre

PHYSICAL PROPERTIES OF BIOLOGICAL MEMBRANES AND
THEIR FUNCTIONAL IMPLICATIONS
Edited by Cecilia Hidalgo

PRINCIPLES OF STRING THEORY
Lars Brink and Marc Henneaux

QUANTUM MECHANICS OF FUNDAMENTAL SYSTEMS 1
Edited by Claudio Teitelboim

QUANTUM MECHANICS OF FUNDAMENTAL SYSTEMS 2
Edited by Claudio Teitelboim and Jorge Zanelli

TRANSDUCTION IN BIOLOGICAL SYSTEMS
Edited by Cecilia Hidalgo, Juan Bacigalupo, Enrique Jaimovich, and Julio Vergara

Transduction in Biological Systems

Edited by
Cecilia Hidalgo
*Centro de Estudios Científicos de Santiago
and Universidad de Chile
Santiago, Chile*

Juan Bacigalupo
*Universidad de Chile
Santiago, Chile*

Enrique Jaimovich
*Centro de Estudios Científicos de Santiago
and Universidad de Chile
Santiago, Chile*

and
Julio Vergara
*University of California at Los Angeles
Los Angeles, California*

PLENUM PRESS • NEW YORK AND LONDON

Library of Congress Cataloging-in-Publication Data

Transduction in biological systems / edited by Cecilia Hidalgo ... [et
al.].
        p.   cm. -- (Series of the Centro des Estudios Científicos de
Santiago)
    Originated from the Workshop "Transduction in Biological Systems",
held May 23-30, 1988, at the Marine Biological Station of the
Universidad de Valparaiso, Montemar, Chile, and contains
contributions from most of the participants.
    Includes bibliographical references.
    ISBN-13: 978-1-4684-5738-4   e-ISBN-13: 978-1-4684-5736-0
    DOI: 10.1007/978-1-4684-5736-0

    1. Cellular signal transduction--Congresses.   I. Hidalgo,
Cecilia. II. Workshop "Transduction in Biological Systems" (1988 :
Marine Biological Station of the Universidad de Valparaiso)
III. Series.
QP517.C45T7   1990
591.1'8--dc20   90-35639
                                                                CIP

© 1990 Plenum Press, New York
Softcover reprint of the hardcover 1st edition 1990
A Division of Plenum Publishing Corporation
233 Spring Street, New York, N.Y. 10013

Contributors

Barbara H. Alderson, Department of Physiology and Biophysics, The University of Texas Medical Branch, Galveston, Texas 77550

Bojena Antoniu, Department of Muscle Research, Boston Biomedical Research Institute, Boston, Massachusetts 02114

Illani Atwater, Laboratory of Cell Biology and Genetics, National Institute of Diabetes and Digestive and Kidney Diseases, National Institutes of Health, Bethesda, Maryland 20892

Juan Bacigalupo, Departamento de Biologia, Facultad de Ciencias, Universidad de Chile, Santiago, Chile

Nelly Bennett, Laboratoire Biophysique Moléculaire et Cellulaire, Centre d'Etudes Nucléaires de Grenoble, 38041 Grenoble Cedex, France

Michael V. L. Bennett, Department of Neuroscience, Albert Einstein College of Medicine, Bronx, New York 10461

Barbara Block, Departments of Biology and Anatomy, University of Pennsylvania, Philadelphia, Pennsylvania 19104-6018.   *Present address*: Department of Organismal Biology and Anatomy, University of Chicago, Chicago, Illinois 60637

Neil R. Brandt, Department of Pharmacology, University of Miami School of Medicine, Miami, Florida 33101

v

Mariangela Bravin, Centro di Studio per la Fisiologia dei Mitocondri del CNR, Istituto di Patologia Generale dell'Università di Padova, 35131 Padua, Italy

Gustavo Brum, Departamento de Biofísica, Facultad de Medicina, Universidad de la República, Montevideo, Uruguay

Donald Brunder, Department of Physiology and Biophysics, The University of Texas Medical Branch, Galveston, Texas 77550

Ricardo Bull, Departamento de Fisiología y Biofísica, Facultad de Medicina, Universidad de Chile, Santiago, Chile

A. Lee Burns, Laboratory of Cell Biology and Genetics, National Institute of Diabetes and Digestive and Kidney Diseases, National Institutes of Health, Bethesda, Maryland 20892

M. Angélica Carrasco, Departamento de Fisiología y Biofísica, Facultad de Medicina, Universidad de Chile, Santiago, Chile

Patricia Carroll, Laboratory of Cell Biology and Genetics, National Institute of Diabetes and Digestive and Kidney Diseases, National Institutes of Health, Bethesda, Maryland 20892

Anthony H. Caswell, Department of Pharmacology, University of Miami School of Medicine, Miami, Florida 33101

Kevin J. Catt, Endocrinology and Reproduction Research Branch, National Institute of Child and Human Development, National Institutes of Health, Bethesda, Maryland 20892

Armel Clerc, Laboratoire Biophysique Moléculaire et Cellulaire, Centre d'Etudes Nucléaires de Grenoble, 38041 Grenoble Cedex, France

Deida Compagnon, Department of Physiology, University of California–Los Angeles, Los Angeles, California 90024

Serge Crouzy, Laboratoire Biophysique Moléculaire et Cellulaire, Centre d'Etudes Nucléaires de Grenoble, 38041 Grenoble Cedex, France

Yves Chapron, Laboratoire Biophysique Moléculaire et Cellulaire, Centre d'Etudes Nucléaires de Grenoble, 38041 Grenoble Cedex, France

Pierre Deslongchamps, Laboratory of Organic Synthesis, Department of Chemistry, Faculty of Science, University of Sherbrooke, Sherbrooke, Quebec J1K 2R1, Canada

Christine Dettbarn, Department of Physiology and Biophysics, The University of Texas Medical Branch, Galveston, Texas 77550

Francesco Di Virgilio, Centro di Studio per la Fisiologia dei Mitocondri del CNR, Istituto di Patologia Generale dell'Università di Padova, 35131 Padua, Italy

Donald G. Ferguson, Department of Physiology and Biophysics, University of Cincinnati, Cincinnati, Ohio 45267

Robert Fitts, Department of Physiology, Rush University, Chicago, Illinois 60612. Permanent address: Departmennt of Biology, Marquette University, Milwaukee, Wisconsin 53233.

Clara Franzini-Armstrong, Department of Anatomy, University of Pennsylvania, Philadelphia, Pennsylvania 19104-6018

W. Hanke, Universität Osnabrück, FB Biologie, Biophysik, D-4500 Osnabrück, Federal Republic of Germany

Lawrence W. Haynes, Howard Hughes Medical Institute, and Department of Neuroscience, The Johns Hopkins University School of Medicine, Baltimore, Maryland 21205. *Present address:* Department of Medical Physiology, University of Calgary, Alberta T2N 4N1, Canada

Cecilia Hidalgo, Departamento de Fisiología y Biofísica, Facultad de Medicina, Universidad de Chile, and Centro de Estudios Científicos de Santiago, Santiago, Chile

Noriaki Ikemoto, Department of Muscle Research, Boston Biomedical Research Institute, Boston, Massachusetts, 02114, and Department of Neurology, Harvard Medical School, Boston, Massachusetts 02115

Michele Ildefonse, Laboratoire Biophysique Moléculaire et Cellulaire, Centre d'Etudes Nucléaires de Grenoble, 38041 Grenoble Cedex, France

Veronica Irribarra, Departamento de Fisiología y Biofísica, Facultad de Medicina, Universidad de Chile, Santiago, Chile

Vincent Jacquemond, Laboratoire de Physiologie des Elements Excitables, Université Claude Bernard, F-69622 Villeurbanne Cedex, France

Enrique Jaimovich, Departamento de Fisiología y Biofísica, Facultad de Medicina, Universidad de Chile, Santiago, Chile

Edwin Johnson, Marshall University School of Medicine, Huntington, West Virginia 25704

Kyung Sook Kim, Department of Pharmacology, University of Miami School of Medicine, Miami, Florida 33101

Nestor Lagos, Department of Physiology, University of California–Los Angeles, Los Angeles, California 90024

Frank A. Lattanzio, Jr., Department of Pharmacology, University of Nevada School of Medicine, Reno, Nevada 89557

Daniel P. Lew, Division des Maladies Infectieuses, Hôpital Cantonal Universitaire, 1211 Geneva 4, Switzerland

John E. Lisman, Department of Biology, Brandeis University, Waltham, Massachusetts 02254

Maria V. Lobo, Departamento de Fisiología y Biofísica, Facultad de Medicina, Universidad de Chile, Santiago, Chile

Mario Luxoro, Laboratorio de Fisiología Celular, Facultad de Ciencias y Facultad de Medicina, Universidad de Chile, Santiago, Chile

Juan José Marengo, Departamento de Fisiología y Biofísica, Facultad de Medicina, Universidad de Chile, Santiago, Chile

Elisa T. Marusic, Departamento de Fisiología y Biofísica, Facultad de Medicina, Universidad de Chile, Santiago, Chile

Gerhard Meissner, Departments of Biochemistry and Physiology, School of Medicine, University of North Carolina, Chapel Hill, North Carolina 27599-7260

Jacopo Meldolesi, Dipartimento di Farmacologia dell'Universita di Milano, Centro di Studio, di Cito Farmacologia del CNR, Istituto Scientifico San Raffaele, 20132 Milan, Italy

Verónica Nassar-Gentina, Laboratorio de Fisiología Celular, Facultad de Ciencias y Facultad de Medicina, Universidad de Chile, Santiago, Chile

Philip Palade, Department of Physiology and Biophysics, The University of Texas Medical Branch, Galveston, Texas 77550

Richard Payne, Department of Zoology, University of Maryland, College Park, Maryland 20742

Gonzalo Pérez, Unidad de Inmunología Celular, Instituto de Nutrición y Tecnología de los Alimentos, Universidad de Chile, Santiago, Chile

Gonzalo Pizarro, Department of Physiology, Rush University, Chicago, Illinois 60612

Harvey B. Pollard Laboratory of Cell Biology and Genetics, National Institute of Diabetes and Digestive and Kidney Diseases, National Institutes of Health, Bethesda, Maryland 20892

Tullio Pozzan, Centro di Studio per la Fisiologia dei Mitocondri del CNR, Istituto di Patologia Generale dell'Università di Padova, 35131 Padua, Italy

Evelyn Reilley, Department of Muscle Research, Boston Biomedical Research Institute, Boston, Massachusetts 02114

Eduardo Ríos, Department of Physiology, Rush University, Chicago, Illinois 60612

Esther Robinson, Department of Pharmacology, University of Nevada School of Medicine, Reno, Nevada 89557

Phyllis Robinson, Department of Biology, Brandeis University, Waltham, Massachusetts 02254

Luis Robles, Departamento de Fisiología y Biofísica, Facultad de Medicina, Universidad de Chile, Santiago, Chile

Cecilia Rojas, Departamento de Fisiología y Biofísica, Facultad de Medicina, Universidad de Chile, Santiago, Chile

Eduardo Rojas, Laboratory of Cell Biology and Genetics, National Institute of Diabetes and Digestive and Kidney Diseases, National Institutes of Health, Bethesda, Maryland 20892

Mario S. Rosemblatt, Unidad de Immunología Celular, Instituto de Nutrición y Tecnología de los Alimentos, Universidad de Chile, Santiago, Chile, and Department of Muscle Research, Boston Biomedical Research Institute, Boston, Massachusetts 02114

Oger Rougier, Laboratoire de Physiologie des Elements Excitables, Université Claude Bernard, F-69622 Villeurbanne Cedex, France

Luc Ruest, Laboratory of Organic Synthesis, Department of Chemistry, Faculty of Science, University of Sherbrooke, Sherbrooke, Quebec J1K 2R1, Canada

Juan C. Sáez, Department of Neuroscience, Albert Einstein College of Medicine, Bronx, New York 10461

Ximena Sánchez, Departamento de Fisiología y Biofísica, Facultad de Medicina, Universidad de Chile, Santiago, Chile

Rosa Santos, Laboratory of Cell Biology and Genetics, National Institute of Diabetes and Digestive and Kidney Diseases, National Institutes of Health, Bethesda, Maryland 20892

Robert G. Schlatterer, Department of Pharmacology, University of Nevada School of Medicine, Reno, Nevada 89557

Arthur Sherman, Mathematics Research Branch, National Institute of Diabetes and Digestive and Kidney Diseases, National Institutes of Health, Bethesda, Maryland 20892

R. Simmoteit, Universität Osnabrück, FB Biologie, Biophysik, D-4500 Osnabrück, Federal Republic of Germany

David C. Spray, Department of Neuroscience, Albert Einstein College of Medicine, Bronx, New York 10461

Philip Stein, Department of Physiology and Biophysics, The University of Texas Medical Branch, Galveston, Texas 77550

Stanko S. Stojilkovic, Endocrinology and Reproduction Research Branch, National Institute of Child and Human Development, National Institutes of Health, Bethesda, Maryland 20892

Andres Stutzin, Laboratory of Cell Biology and Genetics, National Institute of Diabetes and Digestive and Kidney Diseases, National Institutes of Health, Bethesda, Maryland 20892. *Present address:* Laboratorio de Fisiopatología Molecular, Departamento de Medicina Experimental, Facultad de Medicina, Universidad de Chile, Santiago, Chile

Benjamín A. Suárez-Isla, Departamento de Fisiología y Biofísica, Facultad de Medicina, Universidad de Chile, and Centro de Estudios Científicos de Santiago, Santiago, Chile

John L. Sutko, Department of Pharmacology, University of Nevada School of Medicine, Reno, Nevada 89557

Jane A. Talvenheimo, Department of Pharmacology, University of Miami School of Medicine, Miami, Florida 33101

Ismael Uribe, Department of Physiology, Rush University, Chicago, Illinois 60612. *Present address:* Centro de Investigación del Instituto Politécnico Nacional, Mexico City, Mexico

R. A. Venosa, Cátedra de Fisiología y Biofísica, Facultad de Ciencias Médicas, Universidad Nacional de La Plata, 1900 La Plata, Argentina

Pedro Verdugo, Center for Bioengineering, University of Washington, Seattle, Washington 98195

Julio Vergara, Department of Physiology, University of California–Los Angeles, Los Angeles, California 90024

Manuel Villalon, Center for Bioengineering, University of Washington, Seattle, Washington 98195

Pompeo Volpe, Department of Physiology and Biophysics, The University of Texas Medical Branch, Galveston, Texas 77550

Richard E. Weiss, Department of Pediatrics, Division of Cardiology, University of California–Los Angeles, Los Angeles, California 90024–1743

Shu-Rong Wen, Department of Pharmacology, University of Miami School of Medicine, Miami, Florida 33101; and Department of Pharmacology, Beijing Medical University, Beijing, People's Republic of China

King-Wai Yau, Howard Hughes Medical Institute, and Department of Neuroscience, The Johns Hopkins University School of Medicine, Baltimore, Maryland 21205

Patricio Zapata, Laboratorio de Neurobiología, Universidad Católica de Chile, Santiago, Chile

Preface

The present volume originated from the workshop "Transduction in Biological Systems," held at the Marine Biological Station of the Universidad de Valparaiso, Montemar, Chile, May 23–30, 1988, and contains contributions from most of the participants in the workshop.

The title of both the workshop and the book reflects accurately the central theme discussed during several days of intense debate and profound intellectual exchange in the peaceful environment offered by the central coast of Chile. It was apparent that the workshop was a great success—a sentiment expressed by many seasoned attendees, some of whom dared opinions as strong as "It was the best ever."

There is no single reason to explain why this workshop was so successful. Certainly instrumental was the incredible effort displayed by the Chilean Organizing Committee in selecting adequate facilities and in organizing social events that supplemented the scientific sessions and provided an authentic fraternal environment for the participants. Equally important were the foreign participants, who enthusiastically gave of their time to take part in the event, and the students, who came from Chile as well as from several other Latin American countries, and who applied the necessary pressure in their repeated demands for scientific clarity, accuracy, and sincerity.

Transduction processes are of fundamental importance in biological systems. In spite of their diversity, several similarities can be distinguished among them. Thus, the discussions in the chapters that compose this book are oriented toward the definition of common features in this field. We believe that this heterogeneous group of contributors, with widely different backgrounds ranging from classical electrophysiology to highly specialized biochemistry, and working in dissimiˌar biological systems, constituted an excellent group to discuss the general principles underlying biological transduction mechanisms. It is our hope that the outcome oˉ these interdisciplinary interactions, crystallized in this book, will provide the reader ⱳith a unifying perspective on the role of second messengers in biological systems, and that this book will transmit to the scientific community the unique spirit of integration attained during the workshop.

We would like to thank the National Science Foundation, U.S.A., for their grant NSF INT 87-15069, which provided travel suppcrt for most of the scientists from the U.S. who participated in the workshop. We also thank the Chilean sponsors, Fundación Andes, CONICYT, and the Universidad de Chileˌ for providing funds for the organization of the workshop. Our special thanks go to the Chilean Organizing Committee, the tireless efforts of whose members (I. Behrens, R. Bull, A. Carrasco, P. Donoso, M. E. Fernandez, V. Nassar-Gentina, C. Rojas, and C. Vergara) made this workshop, and hence this book, possible.

The Editors

Santiago and Los Angeles

Contents

B. Excitation–Secretion Coupling

B. Calcium Channels in T-Tubule

C. Molecular Architecture of the Triad

D. Transduction at the Triad

Introduction

This book is intended to discuss the most recent advances in the field of transduction in biological systems and to pursue an integrative view of three specific and related areas: sensory transduction, excitation–secretion coupling, and excitation–contraction coupling in striated muscle.

Part I is subdivided into three sections: "Transduction in Sensory Cells," "Excitation–Secretion Coupling," and "Other Transduction Mechanisms." We grouped into the first section the chapters dealing with various aspects of transduction in different sensory receptor cells. Among invertebrate photoreceptors, those present in the ventral eye of *Limulus* constitute the favorite preparation and the best understood. The chapters by Payne and by Bacigalupo et al. discuss the role of Ca^{2+}, IP_3, and cGMP in invertebrate phototransduction, a central point. In the first of these chapters, the fine control of Ca^{2+} release by IP_3 is discussed, with particular attention to the localization of the process within the cell, its timing, and the effect of cytoplasmic free Ca^{2+} on the process.

In the second contribution, Bacigalupo et al. evaluate the evidence for Ca^{2+} and cGMP, two second messengers that have been proposed as chemical messengers that interact directly with the light-activated channels, causing them to open during exposure to light. Three chapters on vertebrate phototransduction follow. Haynes and Yau describe the basic properties of the light-dependent ion channel in amphibian rods and cones and the role of cGMP in the activation of that channel in both receptor cells. They studied the channel in the native membrane using the patch-clamp technique in the cell-attached and excised configurations.

Electrophysiological and biochemical experiments performed on the purified

1

cGMP-activated channel obtained from bovine rod outer segments are described in the next chapter by Hanke et al. The authors worked with the channel purified to homogeneity, but they explain how membrane preparations containing the channel at different degrees of purity can be used in bilayer experiments. They characterized the channel, and from their data they propose a kinetic model for channel gating. Further studies on the reconstituted cGMP-activated channel, dealing now with the regulation of channel activity by different molecules that are participants in the phototransduction cascade, are presented by Ildefonse et al. in the following chapter. These authors found that particular combinations of such components can enhance the effect of cGMP on the channel or even, in some cases, activate the channel in the absence of cGMP. The section ends with a contribution dealing with the properties of chemoreceptors, another class of sensory receptors on which substantial progress has been made in recent years. In this chapter, Zapata compares transduction and signal encoding in two composite chemoreceptors: gustatory and arterial receptors.

The second section of Part I includes discussions of excitation–secretion coupling. This field has seen rapid progress in the last few years, and various types of secretory processes have been described. A number of cellular components have been identified as relevant in secretion, and the participation of second messengers controlling the activity of stimulus-dependent ion channels has been documented in several experimental systems. While a thorough review of the subject is beyond the scope of this book, we do include six chapters in which some aspects of secretion are discussed in detail.

In the first, Santos et al. introduce the subject of analyzing, in a rather general way, diverse aspects of secretion. The authors focus their attention on two of the most studied secretory processes: insulin secretion by pancreatic β-cells, and release of catecholamines by medullary chromaffin cells. The following chapter, by Sherman et al., describes more specific problems in insulin secretion by pancreatic β-cells, namely, the regulation of the bursting electrical activity evoked by glucose in these cells and the possible mechanisms involved in the control of the graded response to glucose. Secretion in mouse and human medullary chromaffin cells is the subject of the next contribution by Nassar-Gentina et al. These authors characterize the electrophysiological properties of both preparations with regard to their response to acetylcholine. These studies were done using cells *in situ*. In the chapter that follows, Stutzin et al. report the presence of two types of Ca^{2+} channels in the plasma membrane of rat pituitary gonadotrophs. In these hormone-secreting cells, Ca^{2+} influx through these channels is primarily responsible for the rise in the intracellular concentration of the free form of the ion, which is necessary for the response to occur. Rojas et al. then give a detailed characterization of the recently discovered protein, synexin, first isolated from bovine adrenal gland. The evidence provided by the authors suggests that synexin is involved in fusion of secretory granules, a Ca^{2+}-mediated process. Experiments dealing with the mechanism of action of synexin are described, and a model of the possible mechanism is proposed. The section ends with a discussion by Marusic and Lobo of secretion of aldosterone by adrenal glomerulosa cells in response to angiotensin II. In their work, potassium permeability was examined in relation to secretion, using flux measurements. A model is proposed in which K^+ permeability is related to changes in membrane potential and intracellular Ca^{2+} concentration that occur during secretory response of these cells.

In the final section of Part I, entitled "Other Transduction Mechanisms," we have grouped together four chapters discussing diverse aspects of transduction mechanisms. In the first one, Volpe et al. propose the existence of the calciosome, an organelle that has the ability to accumulate Ca^{2+}. This organelle, which would be distinct from the endoplasmic reticulum, would release Ca^{2+} in nonmuscle cells in response to IP_3. The calciosomes would be ubiquitous among the different cell types and would consist of vesicles and small vacuoles distributed in the cytoplasm. In the following chapter, Villalon and Verdugo report studies on the mechanism underlying the stimulation of ciliary movement by prostaglandin. The authors report the direct measurements, using a specific fluorescent probe, of fluctuations in intracellular Ca^{2+} associated with ciliary movement in ciliated cells from mammalian oviducts. The evidence presented in this chapter thus supports previous claims for a role of this ion in mediating this transduction process. In the next chapter, by Saez et al., the properties of gap junction channels in hepatocytes are reviewed, with particular attention to their short- and long-term regulation and their permeability to molecules with second messenger properties, as Ca^{2+} and IP_3. In the process of short-term regulation, a raise in cAMP evoked by glucagon would stimulate phosphorylation of a 27-kDa protein, which is part of the purified junction, thereby causing an increase in junctional conductance. cAMP would also participate in long-term regulation of gap juctions by delaying the removal of the channels from the plasma membrane. In the last chapter of this section, Robles reviews the recent evidence for the existence of two different mechanisms of frequency tuning in the inner ear of vertebrates. In amphibia and reptiles, electrical resonance at the hair cells accounts for the tuning of their auditory organs, whereas mechanical tuning is the mechanism involved in frequency discrimination in the ear of mammals.

Part II of the book is devoted to excitation–contraction coupling in striated muscle, with a strong emphasis on skeletal muscle. Excitation–contraction coupling in the past few years has been a field of intense research, marked by exciting new findings as well as by constructive controversy. Different models have been proposed to explain this process, and most of the chapters in this section deal with them in one way or another. The direct coupling hypothesis favors a mechanical link between the transverse tubule and the junctional SR membrane. The chemical transduction hypothesis has two variations: one that postulates calcium ions as the messenger, and the other that proposes IP_3 as the chemical link in the excitation–contraction process. A clear conclusion is that all of these models have been extremely useful in providing working hypotheses for most of the recent work in the field. It is also clear that more experiments are needed before a complete picture of the mechanism of excitation–contraction coupling emerges.

The first two chapters of Part II deal with sodium channels and the sodium pump in skeletal muscle. Sodium channels are responsible for the first step in excitation–contraction coupling, that is, the propagation of an action potential from the surface membrane into the T-tubule. The updated discussion by Weiss on the different types of sodium channels present in muscle and nerve explores the possible functions of the various sodium channel families in different cell types.

The chapter by Venosa puts an end to an existing controversy by giving a reasonable explanation for the discrepancies found in the literature regarding the density of sodium pumps present in T-tubules, and thus establishes that these membranes indeed have a similar density of sodium pumps to that in the surface membranes.

The three following chapters deal with voltage-dependent Ca^{2+} channels; in skeletal muscle these channels are located in the T-tubule. Luxoro et al. examine calcium currents and calcium transients in barnacle muscle, once more addressing the question of whether calcium entry triggers contraction. The biochemical structure of the dihydropyridine receptor as a marker for Ca^{2+}-channel proteins in skeletal muscle is discussed in great detail by Talvenheimo et al. in Chapter 20. The possible role of slow inward calcium currents in excitation–contraction coupling is analyzed by Jacquemond and Rougier by measuring tension transients in the presence of dihydropyridines and inorganic Ca^{2+}-channel blockers.

The next section is entitled "Molecular Architecture of the Triad." In the first chapter, Franzini-Armstrong et al. review the morphology of the T-SR junction and present extensive evidence for a molecular complex that links the junctional SR membrane to the opossing T-tubule membrane, in different muscle fiber types. The authors argue that the morphology of the T-SR junction favors a direct coupling model for excitation–contraction coupling. A description of some of the proteins present in the triads is presented by Caswell et al., and Rosemblatt et al. describe the properties of some monoclonal antibodies that react with triadic antigens.

The mechanism of transduction is extensively discussed in the next section; it begins with a chapter by Ríos et al. in which they propose the existence of a receptor for calcium as well as for Ca^{2+}-channel blockers in the T-tubule membrane. They also discuss its possible role in excitation–contraction coupling. The use of pharmacological agents that interfere with calcium release as a strategy to explore the calcium-induced calcium release hypothesis is extensely discussed by Palade et al. in the next chapter.

The evidence in favor of a chemical mechanism for excitation–contraction coupling in skeletal muscle is discussed in the next chapter by Vergara et al. These authors emphasize the experimental findings that allowed them to propose inositol trisphosphate as the internal messenger in this process. In the following chapter, Volpe et al. offer a critical review of the positive and negative results regarding a role for IP_3 in skeletal muscle. Some of these negative results are contested by Jaimovich et al. in the following chapter, in which they describe measurements of calcium transients in skinned fibers via detection of aequorin light signals.

The metabolism of the phosphoinositides in skeletal muscle membranes, described by Hidalgo et al., underlines the fact that all the enzymes required to produce and metabolize IP_3 are present in the T-tubule membranes.

The three final chapters of this book deal directly with the calcium release process in isolated SR membranes. Sutko et al., while reviewing the pharmacology of *Ryania* alkaloids, describe the effect of an analogue of ryanodine, the molecule that allowed the purification of the Ca^{2+} channel from SR. In the next chapter, Meissner describes a study of the regulation of the SR Ca^{2+} channel carried out by measuring calcium fluxes in membrane vesicles. Finally, Suárez-Isla et al. describe the properties of the Ca^{2+} channels present in SR vesicles fused into lipid bilayers, and the activation of a large-conductance channel by IP_3.

I. STIMULUS–RESPONSE COUPLING

A. Transduction in Sensory Cells

Chapter 1

Dynamics of the Release of Calcium by Light and Inositol 1,4,5-Trisphosphate in *Limulus* Ventral Photoreceptors

Richard Payne

1. INTRODUCTION

Inositol 1,4,5-trisphosphate (IP$_3$) is thought to mediate the release of calcium from endoplasmic reticulum (ER).[1] Cells in many types of tissue have the biochemical machinery required to produce IP$_3$ when an appropriate receptor is activated in the plasma membrane, as well as intracellular stores for calcium that are sensitive to release by IP$_3$.

Interest has also grown in the mechanism of calcium release by IP$_3$. Experiments using permeabilized cell preparations or microsomal fractions of cells have suggested that calcium release *in vitro* may be modulated by many factors, including calcium, pH, and adenosine and guanosine nucleotides,[2] and that release may require the cooperative action of several IP$_3$ molecules.[3] Relatively little is known, however, about the importance of these factors in regulating the actions of IP$_3$ in a living cell. Nor is much known about the dynamics of the action of IP$_3$ in living cells.

Our lack of knowledge about the dynamics and modulation of the mechanism by which IP$_3$ releases calcium is largely because of the difficulty of rapidly introducing IP$_3$ into living cells. Because of their giant size, *Limulus* ventral photoreceptors are ideal for studies of the dynamics and modulation of calcium release. This chapter will describe experiments performed on *Limulus* ventral photoreceptors and draw some conclusions

RICHARD PAYNE • Department of Zoology, University of Maryland, College Park, Maryland 20742.

that may apply to other cells. First, we shall briefly review the characteristics of calcium release by light in *Limulus* ventral photoreceptors and the role of calcium in phototransduction. Then we shall discuss three characteristics of calcium release by IP_3 that may be of general interest: (1) its localization within ventral photoreceptors, (2) its timing, and (3) its modulation by the concentration of intracellular free calcium ions (Ca_i).

2. THE ACTIONS OF INTRACELLULAR CALCIUM IN *LIMULUS* VENTRAL PHOTORECEPTORS AND ITS RELEASE BY LIGHT

The ventral photoreceptors of *Limulus* are clearly segmented into two lobes (Fig. 1A), only one of which (the rhabdomeral or R-lobe) bears microvilli encrusted with rhodopsin, the visual pigment, and is therefore light sensitive.[4,5] The arhabdomeral (A) lobe contains no microvilli and is insensitive to light. The photoreceptor cell bodies are typically 100–200 μm in length and 40—80 μm in diameter.

Illumination increases the cationic permeability of the R-lobe, resulting in a flow of current into the cell.[6–9] This light-activated current (photocurrent), which is carried mainly by sodium ions,[10] depolarizes the membrane of the entire cell body. At very low light intensities each photon opens many thousand sodium channels,[11] resulting in depolarizing events ("quantum bumps") of amplitude 1–10 mV.[12,13]

Prolonged illumination adapts the photoreceptor by rapidly reducing the ability of light to activate the photocurrent.[7] The consequent rapid (< 2 sec) decline of the depolarization during a bright step of light avoids saturation of the response of the photoreceptor. Adaptation is triggered by light stimuli that deliver more than about 100 effective photons within 1 sec.[14] Thus there are two processes initiated by light: *excitation*, the generation of the photocurrent, and *adaptation*, the reduction—by prolonged or bright illumination—of the ability of light to excite the cell. Calcium has been proposed to play a role in both processes.

The evidence in favor of calcium as a messenger for adaptation is strong. The reduction of the sensitivity of the photocurrent to light during adaptation can be mimicked by intracellular injection of calcium[15,16] and opposed by the injection of calcium chelators.[17] In addition, the rise in calcium is rapid enough to mediate the 1–2 sec time course of the onset of adaptation. Thus, calcium released by light is an important factor in mediating light adaptation. Steady elevations in Ca_i of between 0.1 and 10 μM are believed to be sufficient to cause adaptation.[18]

A role for calcium in mediating excitation is more problematic because the intracellular injection of calcium chelators, which block adaptation, only slows down excitation by light and cannot block it.[17] However, pressure injection of calcium into the R-lobe of ventral photoreceptors does activate a conductance with similar properties to that opened by light, showing that a rise in Ca_i is a sufficient signal to excite the photoreceptor under some conditions[19] (Fig. 2). Calcium injections are only effective in depolarizing the photoreceptor when they are made into the R-lobe, implying a specific action on the phototransduction machinery (compare Figs. 2A and B). The depolarization caused by pressure injection of calcium into the R-lobe closely follows the rise in Ca_i as indicated by the photoprotein aequorin (Fig. 2D). The depolarization takes the form of a

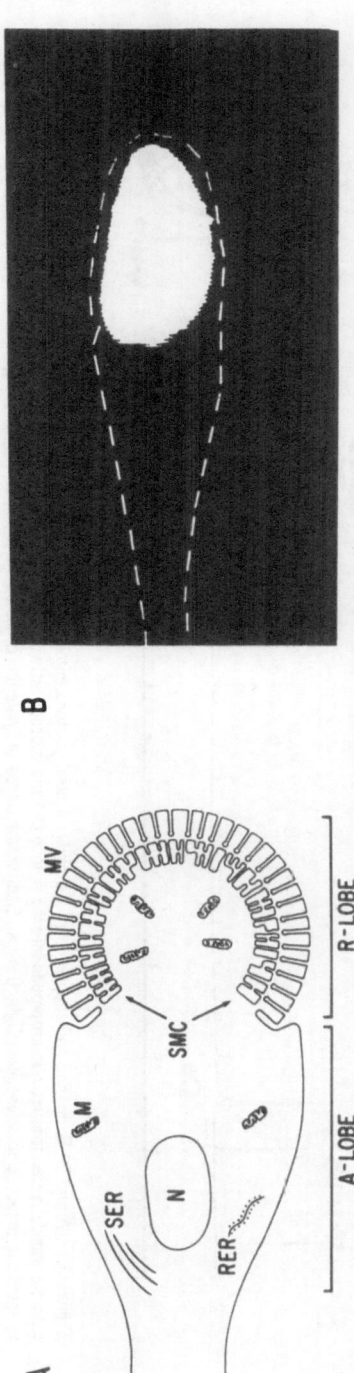

Figure 1. (A) Diagram of a cross section through a *Limulus* ventral photoreceptor, showing the structures in the A- and R-lobes of the photoreceptor, as described by Calman and Chamberlain.[4] The A-lobe contains the nucleus (N), smooth endoplasmic reticulum (SER), rough endoplasmic reticulum (RER), and mitochondria (M). The R-lobe contains microvilli (MV), submicrovillar cisternae of smooth ER (SMC), and mitochondria. (B) Light-induced aequorin luminescence recorded from a *Limulus* ventral photoreceptor following the delivery of a 10-msec flash of light that uniformly illuminated the entire cell. The dotted line outlines the cell body, with the axon exiting from the A-lobe to the left. The luminescence, which signals the light-induced rise in Ca$_i$, is confined to the distal end of the cell body, which comprises the light-sensitive R-lobe, as shown in part (A). For experimental details see Payne and Fein.[24]

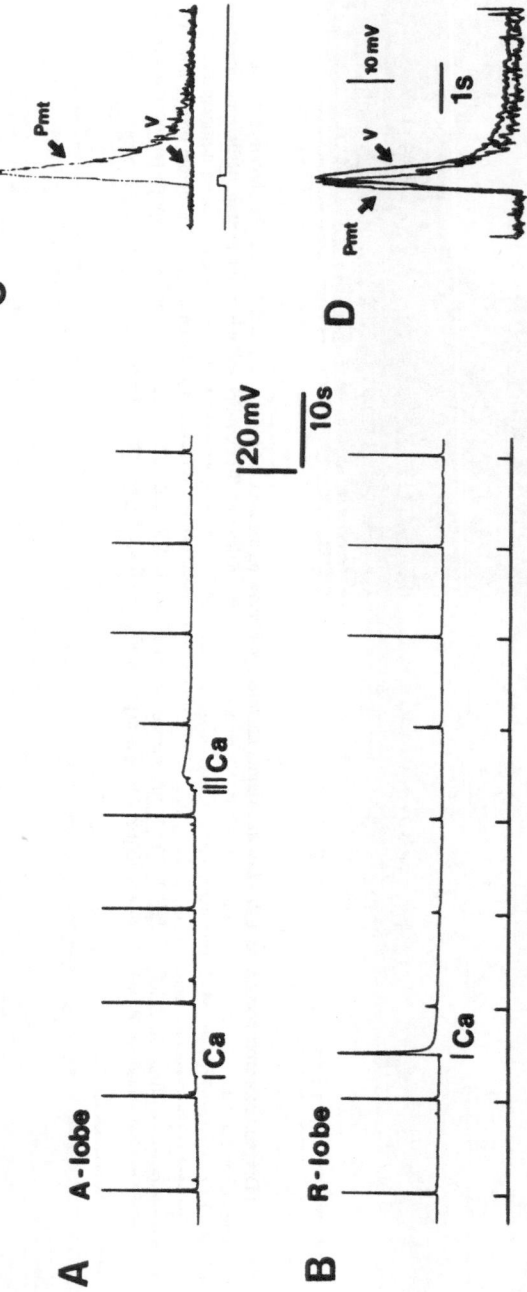

Figure 2. (A and B) Excitation and adaptation of a *Limulus* ventral photoreceptor by pressure injection of calcium into the R-lobe. A brief pulse of 2 m*M* calcium was delivered through a micropipette into the A-lobe (A) and subsequently into the R-lobe (B). Only injection into the R-lobe depolarized the cell and adapted the response to subsequent light flashes. Bars below traces indicate the times of the injections. Diffuse light flashes were delivered at the times indicated by the bottom trace. (C and D) Use of aequorin to confirm calcium injections into the A-lobe (C) and R-lobe (D) of two other cells. The cells were injected with aequorin, and the calcium-induced luminescence was monitored with a photomultiplier. The aequorin traces have been arbitrarily normalized. The peak luminescence was 320 counts per 10 msec bin in trace C and 120 counts per 10 msec bin in trace D. (C) Injection of a pulse of 2 m*M* calcium into the A-lobe produced a rapid transient rise in luminescence (PMT trade) but no depolarization. (D) A similar injection of calcium into the R-lobe of another cell produced a similar aequorin luminescence and an accompanying depolarization. The bottom traces in (C) and (D) indicate the time of the pressure injections. Reproduced with permission from Payne[64].

single, smooth transient lasting a few seconds and is followed by a strong inhibition of the cell's response to subsequent light flashes. It is likely that the rapid elevations in Ca_i that excite the cell are larger than those required for adaptation. Gradual elevation of Ca_i by iontophoretic injection of calcium is able to adapt the photoreceptor but does not depolarize it.[20]

In neither the case of adaptation nor of excitation do we know whether calcium excites the cell by direct interaction with ionic channels or by modulating the level of another messenger.

3. LIGHT-INDUCED CALCIUM RELEASE IS LOCALIZED WITHIN *LIMULUS* VENTRAL PHOTORECEPTORS

Bright flashes of light delivered to *Limulus* ventral photoreceptors produce a rise in Ca_i that is localized to the light-sensitive R-lobe (Fig. 1B).[18,21–24] The rise in Ca_i after a brief, bright flash reaches its peak amplitude of more than 40 μM in less than 300 msec and declines to half-maximal amplitude within 2 sec. A substantial fraction of the rise in Ca_i is due to the release of calcium from internal stores rather than entrance of calcium from the bathing medium.[18,21,25] If bright illumination is sustained for several seconds, the large initial rise in Ca_i is followed by a rapid collapse to a much lower level.[18,21]

The large magnitude of the transient rise in Ca_i in *Limulus* ventral photoreceptors demands some explanation. In the experiments of Levy and Fein,[18] a flash delivering 10^5 photons produced a rise in Ca_i of at least 40 μM. On the assumption that the elevation of Ca_i is uniform in the R-lobe (where the volume is approximately 80 pl) but does not spread to the A-lobe during the transient, then this concentration corresponds to about 10^4 calcium ions per photon. Thus, each photon releases at least that many calcium ions during the flash. This amplification is needed to produce a detectable rise in Ca_i at much lower light intensities, where adaptation begins. Levy and Fein[18] estimate that at the threshold of adaptation, when about 100 photons are effective each second, each photon must release at least 10^3 calcium ions into the R-lobe within the first second of illumination, thus producing a detectable rise in Ca_i.

Calcium release by light is rapid. The approximately 50 msec delay between the absorption of light and the beginning of the rise in Ca_i[21] allows little time for diffusion of a messenger, released from the microvilli, to a distant site within the photoreceptor. This consideration makes the submicrovillar cisternae (SMC) of smooth ER that lie close to the bases of the microvilli a good candidate for the light-sensitive calcium store.[4,26–28] The cisternae, which actively accumulate calcium,[28,29] are ideally localized very close to the bases of the microvilli. The gap between the base of the microvilli and the SMC is less than 100 nm. (See Refs. 29, 30 for a discussion of similar structures in leech photoreceptors.)

4. ACTION OF IP₃ IN *LIMULUS* VENTRAL PHOTORECEPTORS

Inositol 1,4,5-trisphosphate has been proposed as a messenger that is released by rhodopsin from the plasma membrane of the microvilli and that can then cross the gap to the ER so as to mobilize calcium stores.[29,31,32] Injection of IP_3 into *Limulus* ventral photoreceptors releases calcium from internal stores, resulting in a rise in intracellular calcium, similar in amplitude and timecourse to that produced by a flash of light (Fig. 3).[19,33]

As might be expected from the above discussion of the actions of calcium, the rise in Ca_i caused by pressure injection of 2–100 μM IP_3 into *Limulus* ventral photoreceptors both depolarizes the photoreceptor and adapts the response to subsequent light flashes.[31,32] Subsequent responses to IP_3 are, like the light response, suppressed by the prior injection of IP_3. Thus, the response to IP_3 may contain an "adaptation" mechanism, which is discussed in detail below. Prior injection of calcium chelators abolishes the depolarization and adaptation caused by injections of IP_3,[19,34] suggesting that IP_3 acts solely by releasing calcium and that it has no direct action on the light-sensitive conductance.

5. EVIDENCE FOR LIGHT-INDUCED PRODUCTION OF IP_3

Biochemical studies support the proposal that IP_3 or a closely related compound may be the natural mediator of calcium release in *Limulus* and other invertebrate photoreceptors, according to the scheme of Fig. 4.[31,32,35] Light-induced increases in IP_3 content have been measured in *Limulus* ventral eyes[31] and squid retinae,[36,37] and light-induced IP_3 production has been observed in membrane preparations from squid and housefly eyes.[38–40] These recent experiments expand on earlier reports of phosphoinositide turnover in cephalopod retinae.[41,42] A current focus of interest[35,39,40,43] is the possibility that photoactivated rhodopsin is coupled to the activation of phospholipase-C by a GTP-binding protein as has been proposed for the activation of phospholipase-C by other receptor molecules in other cells.[44]

6. LOCALIZATION OF THE ACTIONS OF IP_3: IS ALL ENDOPLASMIC RETICULUM SENSITIVE TO IP_3?

Several studies of permeabilized cells and preparations of ER have concluded that while IP_3 acts only on ER to release calcium, not all ER is sensitive to IP_3.[45–47] The

Figure 3. Aequorin luminescence and membrane depolarization elicited by injection of IP_3 into a ventral photoreceptor. Injection of 1 mM IP_3 (bar) elicited both depolarization of the cell membrane (trace labelled V) and aequorin luminescence (trace labelled PMT), which indicates a rise in Ca_i. The bar to the right indicates a scale of 20 mV for the voltage traces. The PMT trace is arbitrarily normalized to the peak of the voltage trace. (After Ref. 19.)

PMT
V
200 msec

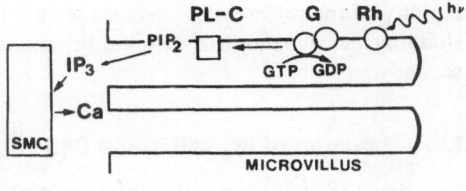

Figure 4. Cascade proposed by Fein[35] for the coupling of photoactivated rhodopsin to the release of inositol 1,4,5-trisphosphate. After absorption of a photon, photoactivated rhodopsin (Rh) catalyzes the exchange of GTP for GDP on a GTP-binding protein (G). The G-protein, with GTP bound, activates a phospholipase-C (PL-C) which cleaves IP$_3$ from phosphatidylinositol bisphosphate (PIP$_2$). Inositol 1,4,5-trisphosphate then releases calcium from the submicrovillar cisternae (SMC).

ER of the *Limulus* ventral photoreceptor may represent a specialized example of differential sensitivity to IP$_3$ amongst ER types.

Given the extensive network of smooth ER that makes up the SMC under the microvilli[4] it comes as no surprise to find that, like light, IP$_3$ releases calcium from the R-lobe of ventral photoreceptors.[24,33] Within the R-lobe, the rise in Ca$_i$ is localized to within a radius of about 25 µm centered on the site of an injection of IP$_3$. However, using an image intensifier to view aequorin luminescence, injections of IP$_3$ into the light-insensitive A-lobe do not release detectable calcium from that lobe.[24] This result is surprising since the A-lobe also contains ER[4] which can utilize ATP to accumulate calcium (B. Walz, personal communication). Thus, while all of the cell's ER may accumulate calcium, only ER in the R-lobe releases calcium in response to injections of IP$_3$. It is possible that the ER in the A-lobe lacks receptor sites for IP$_3$ which are present on the SMC. A more careful, quantitative examination of the density of ER in the two lobes and of the amount of calcium released (if any) in the A-lobe is required.

7. DYNAMICS OF THE ACTIONS OF IP$_3$

7.1. Rapidity of the Action of IP$_3$ in the R-Lobe

When care is taken to inject IP$_3$ directly into the R-lobe, where it is effective, the delay between the delivery of a pressure pulse to the injecting pipette and the onset of calcium release and consequent depolarization is less than 200 msec (Fig. 3) and can be as little as 50 msec. Given the uncertainty in estimating the time between the application of pressure to the pipette and the ejection of IP$_3$, these data represent upper estimates to the delay. At 20°C and in the presence of 10 m*M* extracellular calcium, the latency before depolarization by injection of IP$_3$ into the R-lobe is similar to the latency of depolarization by a flash of light. Preliminary experiments indicate, however, that low temperature or removal of extracellular calcium prolongs the latency of the light response[48] much more than that of the response to IP$_3$. Therefore, at 12°C, or in seawater containing no external calcium, the depolarization caused by a pulse of IP$_3$ into the R-lobe rises faster than that following a light flash. This suggests that IP$_3$ acts on a stage of transduction that follows those responsible for the calcium- and temperature-sensitive latent period of the light response.

The rapidity of the effects of IP$_3$ in the R-lobe of *Limulus* photoreceptors is consistent with the idea that IP$_3$ acts directly on a calcium channel in the ER. The rapidity of the release also underscores the limitations of permeabilized preparations in studying the dynamics of calcium release, where time resolution is usually limited to seconds.

7.2. Diffusion of IP_3 within the Cell

Since IP_3 preferentially releases calcium in the R-lobe[24] and only elevations of Ca_i in the R-lobe cause excitation of the photoreceptor,[49] it follows that any effect of injections of IP_3 into the A-lobe probably results from the diffusion of IP_3 into the R-lobe from the injection site.[32] Injections of IP_3 into the A-lobe of ventral photoreceptors are not always effective. In those cases where IP_3 does depolarize the cell, the depolarization is typically delayed by many seconds when compared with the much greater depolarization of the same cell caused by injection into the R-lobe.[32] Thus, IP_3 can diffuse through a cell so as to release calcium from distant sites, overcoming the very limited diffusion of calcium itself. This ability of IP_3 to diffuse within a cell is unlikely to be of much significance in *Limulus* photoreceptors, because the calcium stores of the SMC are very close to the microvilli where IP_3 is likely to be produced. However, it may be of some importance in other cell types where receptors on the plasma membrane are no so closely apposed to calcium stores.

7.3. Bumps, Bursts, and Oscillations of Calcium Release by IP_3

Depolarization by IP_3 in ventral photoreceptors usually consists of one or more transients. Brief injections (less than 200 msec duration) of 100 μM or more IP_3 delivered directly into the R-lobe usually produce single transients (Fig. 5A). Longer injections of lesser concentrations of IP_3 into the R-lobe, or injections into the A-lobe, generally produce a train of irregular transient bursts of depolarization. The smallest of these bursts resemble the quantum bumps produced by single photons. After long injections of high concentrations of IP_3 into the R-lobe, a large initial transient is sometimes followed by damped oscillations of depolarization lasting for several seconds, which eventually break up into an irregular pattern of transient bursts (Fig. 5B–D).

Present evidence favors the idea that these transient bursts of depolarizations are due to transient bursts of calcium release. First, measurement of Ca_i during transient depolarizations shows that each transient is accompanied by a transient rise in Ca_i, even in the absence of extracellular calcium (Fig. 6).[50] Second, injections of calcium never produce a train of irregular bursts of depolarization, nor oscillations of membrane potential, but simply an immediate, smooth depolarization (Fig. 2).[49] This implies that the site producing the irregular bursts of depolarization is not the one responding to the elevation in Ca_i, but rather the one producing it.

Transient bursts of calcium release and oscillations of Ca_i are common responses to IP_3 in other cells, having been observed following injection of IP_3 into *Xenopus* oocytes[51,52] and during intracellular dialysis of hepatocytes and chromaffin cells with solutions containing IP_3.[2,53] These oscillations may constitute part of the normal response to hormonal stimuli in hepatocytes.[54] In discussing the cause of these transients, we now look in detail at the regulation of calcium release by IP_3 in ventral photoreceptors.

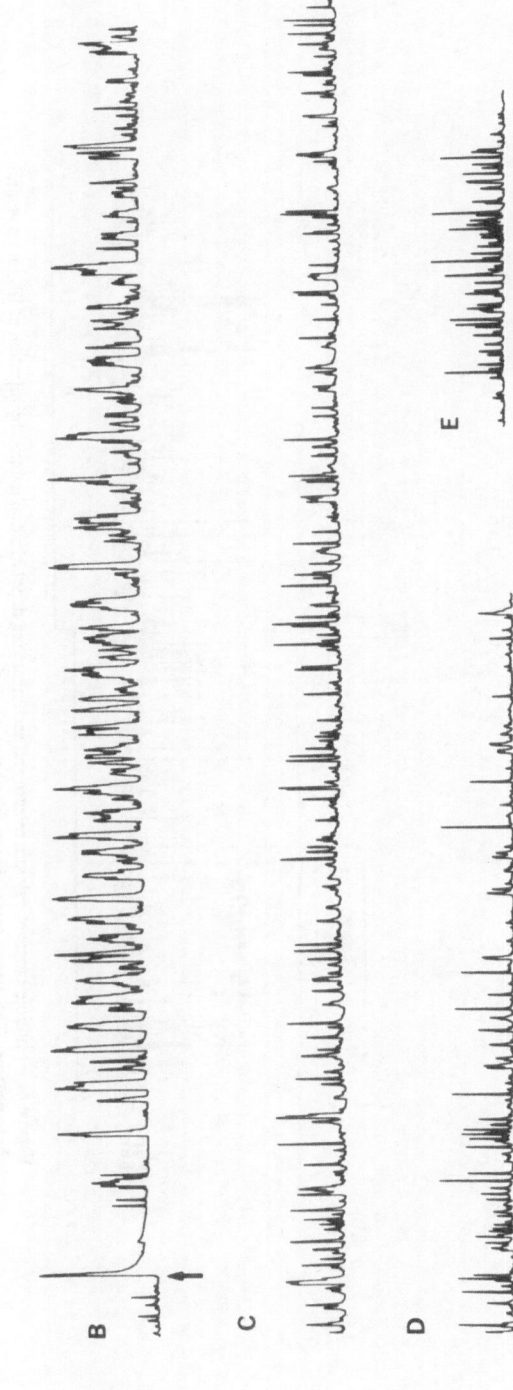

Figure 5. (A) A single 100-msec pressure injection of 100 μM IP₃ into a *Limulus* ventral photoreceptor produces a single smooth transient depolarization. (B) A subsequent 300-msec injection of 100 μM IP₃ causes a single smooth transient depolarization, followed by a "silent period" of desensitization and then a series of oscillatory bursts of transient depolarization that eventually break up (C and D) into events resembling those produced by single quanta, as shown in (E). (E) Individual depolarizing events ("quantum bumps") produced by very dim illumination (bar) of the same photoreceptor.

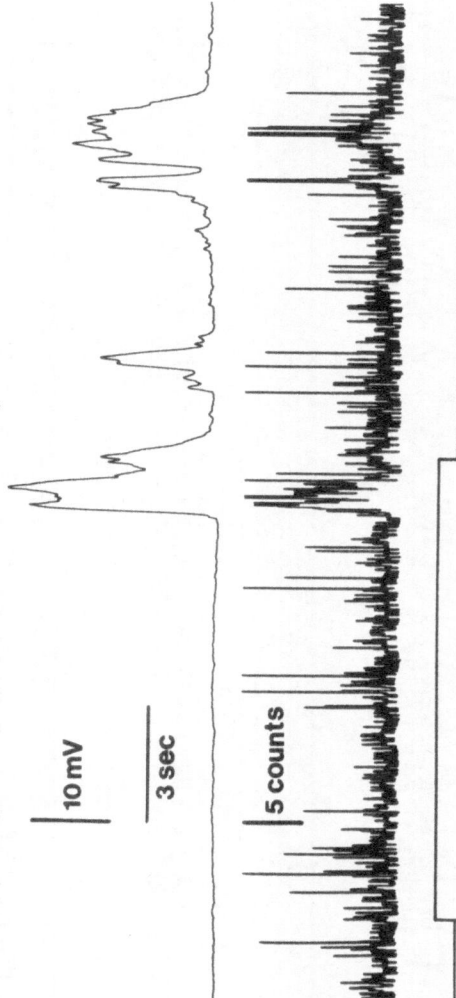

Figure 6. Multiple bursts of calcium release and accompanying depolarization induced by injection of IP$_3$ into a ventral photoreceptor. The upper trace shows membrane potential, the middle trace indicates photon counts from a photomultiplier that recorded the luminescence of the calcium-sensitive photoprotein aequorin, and the lower trace monitors the injection pressure. (Reproduced, with permission, from Ref. 50.)

8. FEEDBACK CONTROL OF IP$_3$-INDUCED CALCIUM RELEASE BY ELEVATIONS IN INTRACELLULAR CALCIUM

The ability of IP$_3$ to depolarize *Limulus* ventral photoreceptors is subject to reversible inhibition by prior bright illumination or by prior injections of IP$_3$.[31,32] The second of a pair of pressure injections of IP$_3$ will not depolarize the photoreceptor if it is delivered within 2 sec of the first injection. Recovery of the response to the second injection takes 10–30 sec (Fig.7).[29] Since both light and the injection of IP$_3$ have a common ability to raise Ca$_i$, a possible explanation for the transient inhibition of the response to IP$_3$ is that elevated Ca$_i$ inhibits the ability of IP$_3$ to depolarize the photoreceptor. To test this possibility, a prior pulsed pressure injection of calcium was delivered before an injection of IP$_3$ and a light flash (Fig. 8A–C). This injection of calcium reversibly suppressed the ionic current activated by subsequent pulses of IP$_3$ and by the light flash. Thus, the activation of ion channels in the plasma membrane by both IP$_3$ and light is inhibited by raising Ca$_i$.

In contrast to the response to IP$_3$ and light, depolarization by injection of calcium does not appear to be inhibited by prior elevation of Ca$_i$. The second injection in a pair of pulsed pressure injections of calcium depolarizes the cell to nearly the same extent as the first, even when the two injections are spaced less than 2 sec apart.[29] Nor does adaptation by prolonged illumination substantially inhibit the depolarization caused by subsequent injections of calcium.[49] It seems likely, therefore, that the inhibition of the response to IP$_3$ by prior injection of IP$_3$ or calcium is largely due to an inhibition of IP$_3$-induced calcium release, rather than an inhibition of the ability of calcium to activate channels in the plasma membrane. This interpretation is supported by direct observation of the greatly reduced rise in Ca$_i$ that accompanies the reduced depolarization caused by the second in a pair of injections of IP$_3$.[19] This mechanism of inhibition of calcium release by elevated Ca$_i$ constitutes a feedback pathway that prevents complete discharge of the calcium stores during sustained elevations of IP$_3$. Figure 8D illustrates the control of the release of calcium by IP$_3$.[29]

Recent studies of permeabilized cells show that micromolar free calcium can inhibit calcium release from intracellular stores by IP$_3$.[55–57] Also, binding of IP$_3$ to homogenized neural tissue appears to be inhibited by micromolar free calcium.[58] Feedback inhibition of calcium release in living *Limulus* ventral photoreceptors may therefore

Figure 7. The peak amplitude of the response to a second pressure injection of 100 μM IP$_3$ into a *Limulus* ventral photoreceptor is plotted against the time elapsed after a similar prior injection. The inset shows an example of the photoreceptor's transmembrane potential during one of the paired injections of IP$_3$. Injection duration was 50 msec. Injections were made into the R-lobe of the photoreceptor. (After Ref. 29.)

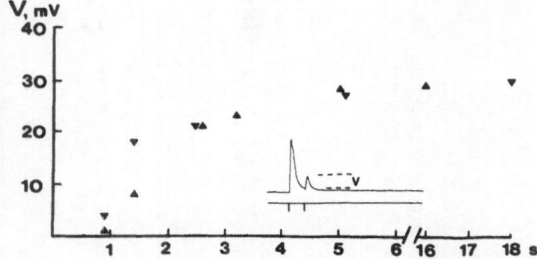

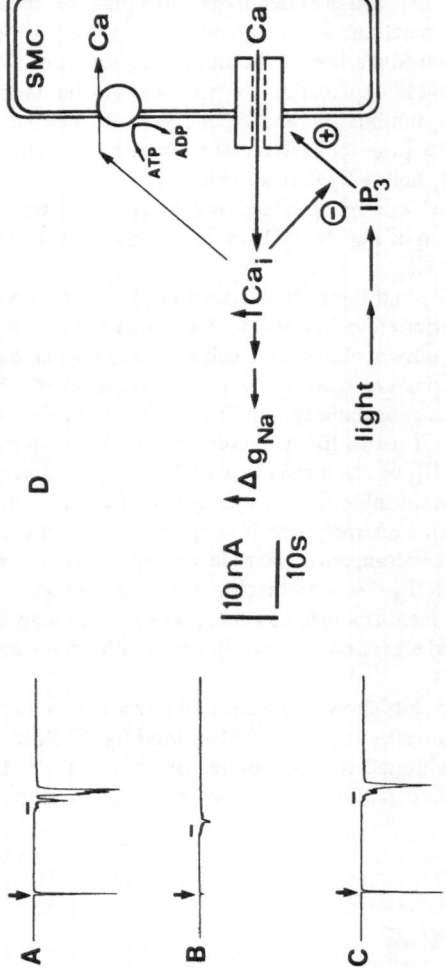

Figure 8. (A–C) Reversible desensitization by an injection of calcium of responses to light and IP$_3$. Trans-membrane current, recorded using a voltage-clamp, during 10-msec flashes of light (arrows) and 1.5-sec intracellular pressure injections of 100 μM IP$_3$ (bars). Downward deflections indicate that current is flowing into the photoreceptor. The control inward currents elicited by light and IP$_3$ in (A) were strongly inhibited by an injection of 1 mM calcium aspartate delivered 30 sec before the record in (B). The currents elicited by light and IP$_3$ recovered to control values 3 min later (C). Both IP$_3$ and calcium aspartate were injected in a carrier solution containing 100 mM potassium aspartate, 10 mM HEPES, pH 7.0. (After Ref. 49.) (D) A model of the release of calcium from the SMC by IP$_3$ in *Limulus* ventral photoreceptors. The membrane of the SMC contains a calcium pump that utilizes ATP to sequester calcium and a calcium pore that is opened by IP$_3$ to release calcium into the cytosol. The released calcium feeds back to inhibit the action of IP$_3$ on the calcium pore in the SMC. (After Ref. 29.)

reflect a general attribute of the mechanism that responds to IP_3 in some other types of cell.

9. THE ROLE OF FEEDBACK IN CREATING OSCILLATORY AND TRANSIENT ELEVATIONS IN Ca$_i$

The introduction of feedback into the control of calcium release by IP_3 may explain why the elevations in Ca_i caused by injection of IP_3 into *Limulus* ventral photoreceptors are usually limited to brief transient events. After an injection, released calcium may rapidly inhibit the actions of IP_3. Following a brief injection, the concentration of IP_3 would decline below that required for further calcium release following the 2–10 sec period of inhibition. However, following a longer injection, when sufficient IP_3 is still present after the period of inhibition, a new cycle of calcium release, depolarization, and desensitization may be triggered, resulting in continued bursts and oscillations of Ca_i. In accord with this model, preliminary experiments show that if an additional injection of IP_3 is delivered at the foot of the declining phase of an oscillatory burst of depolarization, then it is ineffective in causing a depolarization. However, if delivered just before the beginning of the next wave of depolarization, it causes a large depolarization. This suggests that the sensitivity of the cell to IP_3 is indeed oscillating, as required by the model.

Although the model of Fig. 8D provides an initial guess at the explanation of the oscillations in Ca_i, other refinements may be needed. Models of the control of Ca_i that contain a negative feedback loop will generate sustained oscillations of Ca_i only if calcium acts in a cooperative manner to inhibit the reaction that raises Ca_i.[59] The reaction that raises Ca_i must display the characteristics of an "allosteric enzyme" as defined by Monod et al..[60] The application of the definitions of Monod et al.[60] to the model of Fig. 8B would imply a calcium channel having interacting subunits that bind IP_3 and calcium, the affinity for IP_3 binding at one site being inhibited by binding of calcium at the other.

Experimental evidence supporting a cooperative model for the calcium-release mechanism is provided by the bursts of release produced by IP_3 in *Limulus* ventral photoreceptors.

These bursts suggest that calcium channels are opening in the ER in a concerted manner. More direct evidence is provided by Meyer et al.,[3] who have demonstrated cooperative release of calcium by IP_3 within permeabilized basophilic leukemia cells. Another recent study suggests a multimeric structure for an IP_3-binding site in rat cerebellar membranes.[61] The suggestion that the bursts and oscillations of calcium release are produced by a cooperative action of IP_3 on an allosteric protein that gates calcium movements across the ER and is subject to feedback inhibition by calcium deserves careful testing in living cells.

Finally, one must consider that the bursts and oscillations of Ca_i are localized within the cell to small regions due to the limited diffusion of calcium. Thus, the oscillations and bursts of calcium release are, in reality, complex spatial and temporal waves that may propagate across the cell.

10. SUMMARY

Calcium release by IP_3 within *Limulus* ventral photoreceptors is rapid and shows evidence of the concerted release of thousands of calcium ions in bursts. Desensitization of calcium release by feedback may play a critical role in shaping the dynamics of the response. All of these features are also required of the mechanism that responds to light. Single photons appear to release thousands of calcium ions from internal stores, and feedback inhibition of release may contribute to the rapid fall of Ca_i during prolonged illumination, thus protecting the calcium stores from depletion. These similarities between light- and IP_3-induced calcium release strengthen the argument that IP_3, or a closely related compund, is the natural mediator of light-induced calcium release. Discrepancies, of course, remain. For instance, light never induces oscillations in Ca_i under normal conditions. The close apposition of the microvilli producing IP_3 to the calcium stores may limit the spread of IP_3 beyond the site where a photon is effective so that overloading of the cell with IP_3, and consequent oscillation in Ca_i, is prevented. Alternatively, light may release other factors that damp the oscillations.

IP_3 mimics most of the actions of light in depolarizing the cell under physiological conditions. Depolarization by IP_3 shows adaptation and can take the form of events that closely resemble quantum bumps. Although calcium might not be the only messenger mediating excitation by light[62,63] and the ability of calcium to open light-sensitive channels directly has not so far been demonstrated, the striking similarities in excitation by light and by IP_3 suggest that part of phototransduction in invertebrate photoreceptors lies in the phosphoinositide pathway.

ACKNOWLEDGMENTS.   Many of the experiments reviewed in this paper were performed in the laboratory of Dr. Alan Fein at the Marine Biological Laboratory, Woods Hole, and were supported by NIH grant EY 03793 to A.F. Later work was supported by grant EY-07743 to R. Payne. R. Payne is an Alfred P. Sloan Foundation Fellow.

REFERENCES

1. Berridge, M. J., and Irvine, R. F., 1984, Inositol trisphosphate: a novel second messenger in cellular signal transduction, *Nature* **312:** 315–321.
2. Berridge, M. J., 1987, Inositol 1,4,5-trisphosphate and diacylglycerol: two interacting second messengers, *Ann. Rev. Biochem.* **56:** 159–195.
3. Meyer, T., Holowka, D., and Stryer, L., 1988, Highly cooperative opening of calcium channels by inositol 1,4,5-trisphosphate, *Science* **240:** 653–656.
4. Calman, B. G., and Chamberlain, S. C., 1982, Distinct lobes of *Limulus* photoreceptors. II. Structure and ultrastructure, *J. Gen. Physiol.* **80:** 839–862.
5. Stern, J., Chinn, K., Bacigalupo, J., and Lisman, J. E., 1982, Distinct lobes of *Limulus* ventral photoreceptors. I. Functional and anatomical properties of lobes revealed by removal of glial cells, *J. Gen. Physiol.* **80:** 825–837.
6. Hagins, W. A., Zonana, H. V., and Adams, R. G., 1962, Local membrane current in the outer segments of squid photoreceptors, *Nature. (London)* **194:** 844–847.
7. Millecchia, R., and Mauro, A., 1969, The ventral photoreceptor cells of *Limulus*. III. A voltage-clamp study, *J. Gen. Physiol.* **54:** 331–351.

8. Lasansky, A., and Fuortes, M. G. F., 1969, The site of origin of electrical responses in visual cells of the leech *Hirudo medicinalis*, *J. Cell Biol.* **42**: 241–252.

9. Payne, R., and Fein, A., 1986, Localization of the photocurrent of *Limulus* ventral photoreceptors using a vibrating probe, *Biophys. J.* **50**: 193–196.

10. Brown, J. E., and Mote, M. I., 1974, Ionic dependence of reversal voltage of the light response in *Limulus* ventral photoreceptors, *J. Gen. Physiol.* **63**: 337–350.

11. Bacigalupo, J., and Lisman, J. E., 1983, Single-channel currents activated by light in *Limulus* ventral photoreceptors, *Nature* **304**: 268–270.

12. Yeandle, S., and Spiegler, J. B., 1973, Light-evoked and spontaneous discrete waves in the ventral eye of *Limulus*, *J. Gen. Physiol.* **61**: 552–571.

13. Stieve, H., 1986, Bumps, the elementary excitatory responses of invertebrates, in: *The Molecular Mechanism of Photoreception* (Stieve, H., ed.), Springer, New York, pp. 199–230.

14. Wong, F., 1978, Nature of light-induced conductance changes in ventral photoreceptors of *Limulus*, *Nature* **276**: 76–79.

15. Lisman, J. E., and Brown, J. E., 1972, The effects of intracellular iontophoretic injection of calcium and sodium ions on the light response of *Limulus* ventral photoreceptors, *J. Gen. Physiol.* **59**: 701–719.

16. Fein, A., and Charlton, J. S., 1977, A quantitative comparison of the effects of intracellular Ca injection and light adaptation on the photoresponse of *Limulus* ventral photoreceptors, *J. Gen. Physiol.* **70**: 591–600.

17. Lisman, J. E., and Brown, J. E., 1975, Effects of intracellular injection of calcium buffers on light adaptation in *Limulus* ventral photoreceptors, *J. Gen. Physiol.* **66**: 489–506.

18. Levy, S., and Fein, A., 1985, Relationship between light sensitivity and intracellular free calcium in *Limulus* ventral photoreceptors, *J. Gen. Physiol.* **85**: 805–841.

19. Payne, R., Corson, D. W., Fein, A., and Berridge, M. J., 1986, Excitation and adaptation of *Limulus* ventral photoreceptors by inositol 1,4,5-trisphosphate result from a rise in intracellular calcium, *J. Gen. Physiol.* **88**: 127–142.

20. Brown, J. E., and Lisman, J. E., 1975, Intracellular calcium modulates the sensitivity and timescale in *Limulus* ventral photoreceptors, *Nature* **258**: 252–253.

21. Brown, J. E., and Blinks, J. R., 1974, Changes in intracellular free calcium during illumination of invertebrate photoreceptors: Detection with aequorin, *J. Gen. Physiol.* **64**: 643–665.

22. Brown, J. E., Brown, P. K., and Pinto, L. H., 1977, Detection of light-induced changes of intracellular ionized calcium concentration in *Limulus* ventral photoreceptors using Arsenazo III, *J. Physiol. Lond.* **267**: 299–320.

23. Nagy, K., and Stieve, H., 1983, Changes in intracellular calcium ion concentration in the course of dark adaptation measured by arsenazo III in the *Limulus* photoreceptor, *Biophys. Struct. Mech.* **9**: 207–223.

24. Payne, R., and Fein, A., 1987, Inositol 1,4,5-trisphosphate releases calcium from specialized sites within *Limulus* photoreceptors, *J. Cell. Biol.* **104**: 933–937.

25. Bolsover, S. R., and Brown, J. E., 1985, Calcium, an intracellular messenger of light adaptation also participates in excitation of *Limulus* ventral photoreceptors, *J. Physiol. Lond.* **364**: 381–393.

26. Clark, A. W., Millecchia, R., and Mauro, A., 1969, The ventral photoreceptors of *Limulus*. I. The microanatomy, *J. Gen. Physiol.* **54**: 289–309.

27. Lisman, J. E., and Strong, J. A., 1979, The initiation of excitation and light adaptation in *Limulus* ventral photoreceptors, *J. Gen. Physiol.* **73**: 219–243.

28. Walz, B., and Fein, A., 1983, Evidence for calcium-sequestering smooth ER in *Limulus* ventral photoreceptors, *Inv. Opthal. Vis. Sci. Suppl.* **24**: 281.

29. Payne, R., Walz, B., Levy, S., and Fein, A., 1988, The localization of calcium release by inositol trisphosphate in *Limulus* photoreceptors and its control by negative feedback, *Phil. Trans. Roy. Soc. Lond. B* **320**: 359–379.

30. Walz, B., 1982, Ca^{2+}-sequestering smooth endoplasmic reticulum in an invertebrate photoreceptor. I. Intracellular topography as revealed by OsFeCN staining and *in situ* Ca accumulation, *J. Cell. Biol.* **93**: 839–848.

31. Brown, J. E., Rubin, L. J., Ghalayini, A. J., Tarver, A. L., Irvine, R. F., Berridge, M. J., and Anderson, R. E., 1984, Myo-inositol polyphosphate may be a messenger for visual excitation in *Limulus* photoreceptors, *Nature* **311**: 160–162.

32. Fein, A., Payne, R., Corson, D. W., Berridge, M. J., and Irvine, R. F., 1984, Photoreceptor excitation and adaptation by inositol 1,4,5-trisphosphate, *Nature* **311**: 157–160.

33. Brown, J. E., and Rubin, L. J., 1984, A direct demonstration that inositol trisphosphate induces an increase in intracellular calcium in *Limulus* photoreceptors, *Biochem. Biophys. Res. Comm.* **125**: 1137–1142.

34. Rubin, L. J., and Brown, J. E., 1985, Intracellular injection of calcium buffers blocks IP_3-induced but not light-induced electrical responses of *Limulus* ventral photoreceptors, *Biophys. J.* **47**: 38a.

35. Fein, A., 1986, Blockade of visual excitation and adaptation in *Limulus* photoreceptors by GDP-βS, *Science* **232**: 1543–1545.

36. Szuts, E. Z., Wood, S. F., Reid, M. A., and Fein, A., 1986, Light stimulates the rapid formation of inositol trisphosphate in squid retinae, *Biochem. J.* **240**: 929–932.

37. Brown, J. E., Watkins, D. C., and Malbon, C. C., 1987, Light-induced changes of inositol phosphates in squid *Loligo pealei* retina, *Biochem. J.* **247**: 293–297.

38. Wood, S. F., Szuts, E. Z., and Fein, A., 1987, Light-induced changes in inositol trisphosphate in distal segments of squid photoreceptors, *Invest. Ophthal. Vis. Sci.* **28**: 96.

39. Baer, K. M., and Saibil, H. R., 1987, Light- and GTP-activated hydrolysis of phosphatidylinositolbisphosphate in squid photoreceptor membranes, *J. Biol. Chem.* **263**: 17–20.

40. Devary, O., Heichal, O., Blumenfeld, A., Cassel, A., Suss, A., Barash, A., Rubinstein, T., Minke, B., and Selinger, Z., 1987, Coupling of photoexcited rhodopsin to phosphoinositide hydrolysis in fly photoreceptors, *Proc. Natl. Acad. Sci.* **84**: 6939–6943.

41. Vandenberg, C. A., and Montal, M., 1984, Light-regulated biochemical events in invertebrate photoreceptors. 2. Light-regulated phosphorylation of rhodopsin and phosphoinositides in squid photoreceptor membranes, *Biochemistry* **23**: 2347–2352.

42. Yoshioka, T., Takagi, M., Hayashi, F., and Amakawa, T., 1983, The effect of isobutylmethylxanthine on the photoresponse and phsophorylation of phosphatidylinositol in squid photoreceptor membranes, *Biochim. Biophys. Acta* **755**: 50–55.

43. Wood, S. F., Szuts, E. Z., and Fein, A., 1987, Aluminium fluoride and GTP increase inositol phosphate production in distal segments of squid photoreceptors, *Biol. Bull.* **173**: 448–449.

44. Cockcroft, S., 1987, Polyphosphoinositide PDE; Regulation by a novel guanine nucleotide binding protein—G_p, *Trends. Biochem.* **12**: 75–79.

45. Dawson, A. P., and Irvine, R. F., 1984, Inositol 1,4,5 trisphosphate-promoted calcium release from a microsomal fraction of rat liver, *Biochem. Biophys. Res. Comm.* **120**: 858–864.

46. Taylor, C. W., and Putney, J. W., Jr., 1985, Size of the inositol 1,4,5 trisphosphate-sensitive calcium pool in guinea pig hepatocytes, *Biochem. J.* **232**: 435–438.

47. Biden, T. J., Wollheim, C. B., and Schlegel, W., 1986, Inositol 1,4,5 trisphosphate and intracellular calcium homeostasis in clonal pituitary cells (GH_3), *J. Biol. Chem.* **261**: 7223–7229.

48. Martinez, J. M., II, and Srebro, S., 1976, Calcium and the control of discrete-wave latency in the ventral photoreceptor of *Limulus*, *J. Physiol.* **261**: 535–562.

49. Payne, R., Corson, D. W., and Fein, A., 1986, Pressure injection of calcium both excites and adapts *Limulus* ventral photoreceptors, *J. Gen. Physiol.* **88**: 107–126.

50. Corson, D. W., and Fein, A., 1987, Inositol 1,4,5-trisphosphate induces bursts of calcium release inside *Limulus* ventral photoreceptors, *Brain Res.* **423**: 343–346.

51. Parker, I., and Miledi, R., 1986, Changes in intracellular calcium and in membrane currents evoked by injection of inositol trisphosphate into *Xenopus* oocytes, *Proc. Roy. Soc. B* **228**: 307–315.

52. Ferguson, J. E., Han, J.-K., and Nuccitelli, R., 1987, The effects of inositol phosphate isomers on Cl^- conductance in *Xenopus laevis* oocytes, *J. Cell. Biol.* **105**: 3a.

53. Capoid, T., Field, A. C., Ogden, D. C., and Sandford, C. A., 1987, Internal perfusion of guinea pig hepatocytes with buffered Ca^{2+} or inositol trisphosphate mimics noradrenaline activation of K^+ and Cl^- conductances, *FEBS. Lett.* **217**: 247–252.

54. Woods, N. M., Cuthbertson, K. S. R., and Cobbold, P. H., 1986, Repetitive transient rises in cytoplasmic free calcium in hormone-stimulated hepatocytes, *Nature* **319**: 600–602.

55. Suematsu, E., Hirata, M., Hashimoto, T., and Kuriyama, H., 1984, Inositol 1,4,5 trisphosphate releases calcium from intracellular store sites in skinned single cells of porcine artery, *Biochem. Biophys. Res. Comm.* **120:** 481–485.

56. Chueh, S. H., and Gill, D. L., 1986, Inositol 1,4,5 trisphosphate and guanine nucleotides activate calcium release from endoplasmic reticulum via distinct mechanisms, *J. Biol. Chem.* **261:** 13883–13886.

57. Thierry, J., and Klee, C. B., 1986, Calcium modulation of inositol 1,4,5 trisphosphate-induced calcium release from neuroblastoma × glioma hybrid NG108-15 microsomes, *J. Biol. Chem.* **261:** 16414–16420.

58. Worley, P. F., Baraban, J. M., Supattapone, S., Wilson, V. S., and Snyder, S. H., 1987, Characterization of inositol trisphosphate receptor binding in brain, *J. Biol. Chem.* **262:** 12132–12136.

59. Rapp, P. E., and Berridge, M. J., 1977, Oscillations in calcium-cAMP control loops form basis of pacemaker activity and other high-frequency biological rhythms, *J. Theor. Biol.* **66:** 497–525.

60. Monod, J., Wyman, J., and Changeux, J. P., 1965, On the nature of allosteric transitions: A plausible model, *J. Mol. Biol.* **12:** 88–118.

61. Supattopone, S., Worley, P. F., Baraban, J. M., and Snyder, S. H., 1988, Solubilization, purification, and characterization of an inositol trisphosphate receptor, *J. Biol. Chem.* **263:** 1530–1534.

62. Saibil, H. R., 1984, A light-stimulated increase in cyclic GMP in squid photoreceptors, *FEBS Lett.* **168:** 213–216.

63. Johnson, E. C., Robinson, P. R., and Lisman, J. E., 1986, cGMP is involved in the excitation of invertebrate photoreceptors, *Nature* **324:** 468–470.

64. Payne, R., 1986, Phototransduction by the microvillar photoreceptors of invertebrates: Mediation of a visual cascade by inositol trisphosphate, *Photobiochem. Photobiophys.* **13:** 373–397.

Second Messengers in Invertebrate Phototransduction

Juan Bacigalupo, Edwin Johnson, Phyllis Robinson,
and John E. Lisman

1. INTRODUCTION

In visual transduction, the absorption of light by specialized photoreceptor cells evokes a change in voltage across the plasma membrane termed the receptor potential. The problem of how this excitation process occurs has fascinated physiologists and biochemists for over 100 years. We now know that the process involves three fundamentally different phases. In the first phase, the absorption of light by the visual pigment, rhodopsin, is transduced into a change in the conformation of the visual pigment. This phase involves no amplification because one photon alters the conformation of only one rhodopsin molecule. In the second phase, chemical amplification produces a large change in the concentration of a second messenger. In the final phase, this concentration change is detected by membrane channels which gate the flow of ions, thereby generating the receptor potential.

In recent years substantial progress has been made in understanding the gain-producing reactions involved in phototransduction in vertebrate photoreceptors. These cells contain an enzyme cascade that is turned on by light and results in the hydrolysis of

JUAN BACIGALUPO • Departamento de Biologia, Facultad de Ciencias, Universidad de Chile, Santiago, Chile. EDWIN JOHNSON • Marshall University School of Medicine, Huntington, West Virginia 25704. PHYLLIS ROBINSON and JOHN E. LISMAN • Department of Biology, Brandeis University, Waltham, Massachusetts 02254.

the second messenger, cGMP[*].[1] This reduction leads to a closure of cGMP-activated channels. These channels are relatively nonselective toward cations; channel closure thus generates a hyperpolarizing receptor potential.[2] Ca^{2+} ions, once thought to mediate excitation in rods, appear instead to be involved in adaptation,[3–6] a light-driven process that reduces the gain of transduction at high light levels. In invertebrate photoreceptors the nature of the gain-producing reactions that produce the receptor potential is less clear. There are two competing theories about which second messenger causes the light-induced opening of the channels that generates the depolarizing receptor potential. One line of evidence favors IP_3[**] and Ca^{2+}, the other line favors cGMP. There is, however, general agreement that Ca^{2+} and IP_3 are central to the light-adaptation process in invertebrates. In this chapter we will first review the evidence that second messengers are involved in phototransduction, and then summarize what is known about the involvement of particular second messengers in invertebrate phototransduction.

2. THE NEED FOR SECOND MESSENGERS IN PHOTOTRANSDUCTION

The first indication that diffusible second messengers are involved in phototransduction followed from consideration of the anatomy of vertebrate rods. In these cells, most of the rhodopsin is contained within the disk membrane (Fig. 1C), a membrane separated physically from the plasma membrane. A second messenger is thus required to transmit information from the disc membrane to the plasma membrane that contains the light-sensitive ion channels. Somewhat different considerations about invertebrate photoreceptors indicate the necessity of a second messenger in these cells. Figures 1A and B show the microvillar plasma membrane specializations typical of invertebrate photoreceptors. The microvilli contain both the rhodopsin and the light-dependent channels, so in principle a soluble second messenger would not be required. A ventral photoreceptor has about a million microvilli over which the light-activated channels are spread, but the number of photons required to saturate these channels is on the order of a thousand.[7] This implies that the current generated by each photon during a saturating stimulus must involve the channels in about 1,000 microvilli. From the packing density of microvilli, it can be calculated that there must be a diffusible messenger that spreads 2 or 3 μm along a plane perpendicular to the long axis of the microvilli.

A second argument for the involvement of second-messenger cascades in transduction is that such cascades can produce the required amplification. The necessity for amplification in phototransduction is demonstrated by comparing the maximum current through a single light-dependent channel with the peak current during the response to a single photon. The single photon response in rods is about 1 pA (Fig. 2A), whereas that in *Limulus* photoreceptors is about 1 nA (Fig. 2B).[8] Noise analysis in rods indicates that the current through a single channel is about 2 fA (Fig. 2A).[9] In contrast, the current carried by a single light-activated channel from a *Limulus* photoreceptor has an amplitude of several picoamperes (Fig. 2B).[10] If the peak current during the single-

*cGMP: 3'5'-cyclic guanosine monophosphate.
**IP_3: Inositol 1,4,5-trisphosphate.

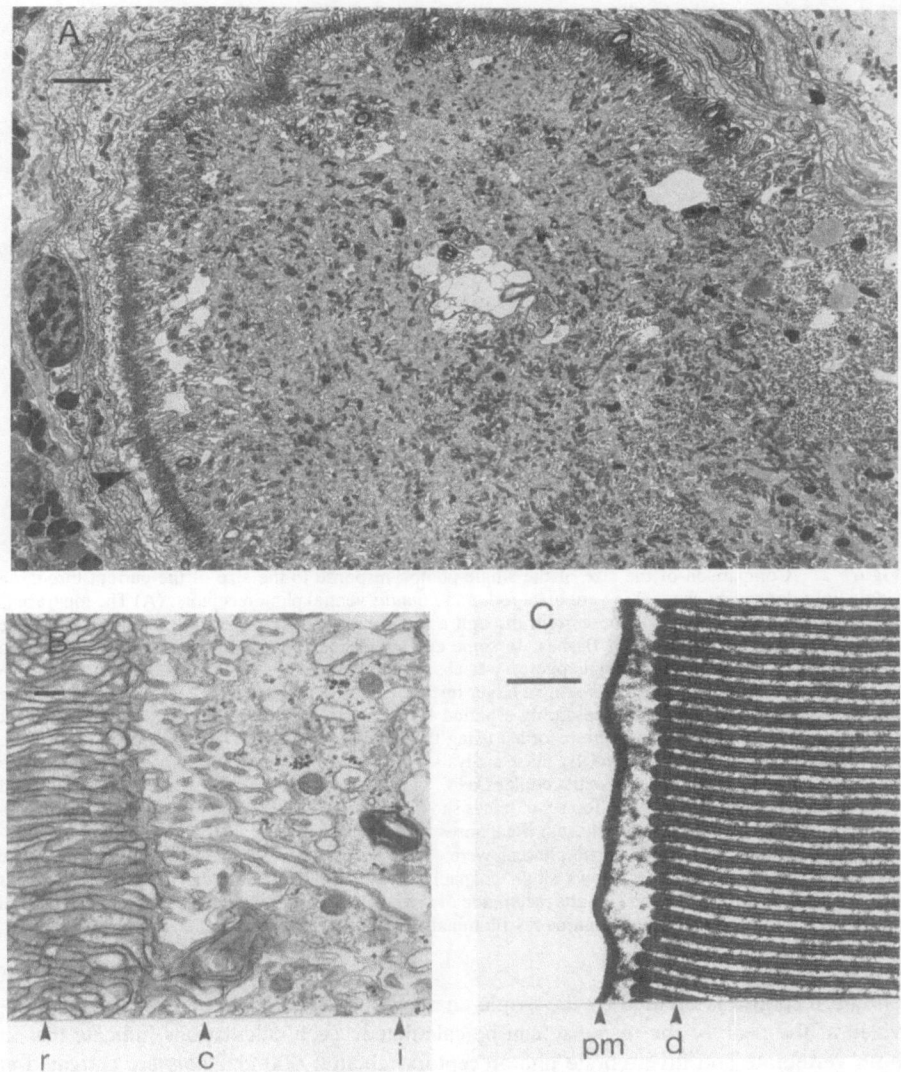

Figure 1. Structure of transducing region of *Limulus* ventral photoreceptor and toad rod. (A) Transducing lobe (R-lobe) of *Limulus* ventral photoreceptor. The arrow points to the rhabdom, the microvillar folding of the plasma membrane. These make an arc along the outer surface of the R-lobe. Calibration bar: $2\mu M$. (B) High-magnification view of rhabdom (r), subrhabdomeric cisternae (c), and interior region of R-lobe (i). Calibration bar: 200 nm. (C) High-magnification view of toad rod outer segment showing plasma membrane (pm) and disks (d). Calibration bar: 100 nm. (Micrographs of *Limulus* courtesy of B. G. Calman and S. C. Chamberlain. Micrograph of rod courtesy of W. H. Schroder and G. L. Fain.)

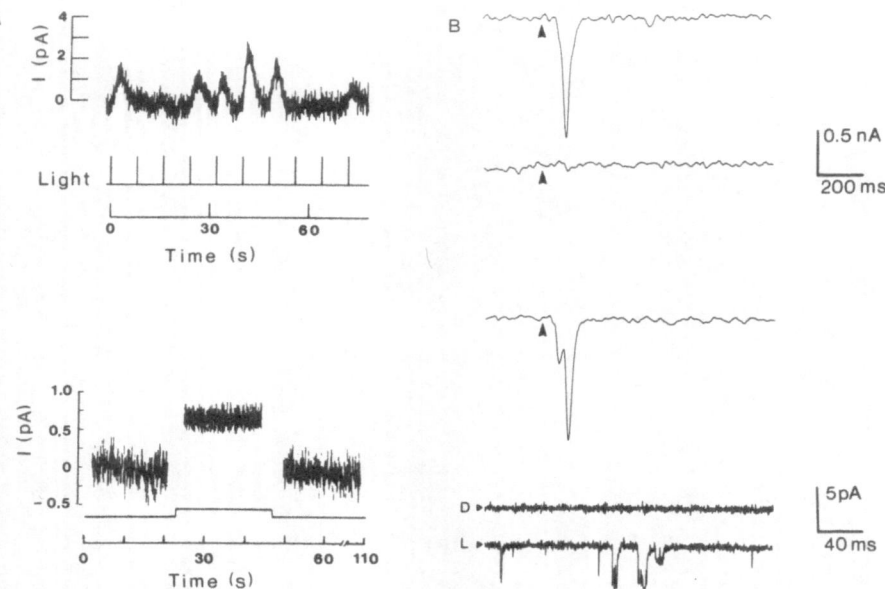

Figure 2. Comparison of the size of the single-photon response to the size of the current through a single light-dependent channel in vertebrate rods and *Limulus* ventral photoreceptors. (A) The top trace is a suction pipette recording of the current through a rod outer segment. The trace shows the variable responses to a series of identical flashes. In some cases no photons were absorbed and there was no response; in some cases only a single photon was absorbed, evoking a response of about 1.5 pA; in two cases two photons were absorbed evoking a larger response (adapted with permission from Ref. 45). The lower trace shows the current noise during a period in darkness (left and right) and a period of bright illumination (center). Currents were recorded using the whole-cell mode of the patch-clamp technique (basal current levels are not shown). By noise analysis it was deduced that the single-channel events have an amplitude of 12 fA in 0.1 mM extracellular Ca^{2+}, the single-channel current would be 2 fA (adapted with permission from Ref. 9). (B) Top set of traces shows responses of a *Limulus* ventral photoreceptor to identical flashes (marked by arrowheads). Responses were measured using the two-microelectrode voltage-clamp method. In some cases no photons were absorbed; in other cases one or more photons were absorbed. Lower set of traces shows single-channel currents recorded from the R-lobe of a *Limulus* ventral photoreceptor using the cell-attached mode of the patch-clamp technique. The first trace (D) is in darkness, the second (L) during continuous illumination. Single-channel currents are about 4 pA.

photon response is divided by the single-channel current, the number of channels activated at the peak of the response can be calculated. Such calculations indicate that in both vertebrate and invertebrate photoreceptors, about 1,000 channels are activated at the peak of the single-photon response. Thus, the chemical gain of phototransduction must be at least a thousand.

3. ANATOMY OF *LIMULUS* VENTRAL PHOTORECEPTORS

Most of the work on the mechanism of phototransduction in invertebrates has been done on the ventral photoreceptors of *Limulus*. The utility of this preparation stems from its large size; its volume is about 1,000 times that of the largest amphibian rods. It is easy therefore to impale these photoreceptors with multiple electrodes and to inject substances of interest. At first it was thought that the transduction machinery was distributed fairly evenly over the whole ventral photoreceptor,[11] but when glial cells that surround the photoreceptors were removed it could be seen easily in the light microscope that the cells were divided into two types of lobes.[12] Scanning electron microscopy[12] and thin-section electron microscopy[13] have revealed that only one type of lobe (termed the R-lobe[*]) contains microvillar specialization (Fig. 3). Furthermore, when the cell was scanned with a small spot of light, it was found that the R-lobe is hundreds of times more sensitive to light than the other lobe (the A-lobe[**]) (Fig. 4).[12]

The existence of these distinct lobes has major consequences for experiments involving the intracellular injection of substances. It has been found that the effect of injecting substances often depends strongly on which lobe the electrode is in. The cross section of the R-lobe shown in Fig. 1A reveals, however, that even the R-lobe is not homogeneous. The microvillar membrane is mostly on the outer surface of the lobe, though some of it is found in large infoldings. Just below the microvilli are cisternae from an internal membrane system termed the subrhabdomeric reticulum. This is a specialized type of endoplasmic reticulum that may be responsible for the IP_3-mediated release of Ca^{2+} into the cytoplasm (see Chapter 1, this volume). The relatively large central core of the R-lobe may not be involved in transduction at all (see the discussion below).

4. THE ROLE OF IP_3 AND Ca^{2+}

The first experiment indicating that intracellular Ca^{2+} is of importance in phototransduction involved the iontophoretic injection of Ca^{2+} into *Limulus* ventral photoreceptors.[14] Ca^{2+} injection had no effect on the resting potential, but dramatically reduced the response to a flash of light (Fig. 5A). Ca^{2+} thus adapted the cell without exciting it. (Later experiments using pressure injection showed that Ca^{2+} could also excite the cell; see below.) The ability of Ca^{2+} to adapt the cell raised the possibility that a light-induced rise in intracellular Ca^{2+} might mediate the normal process of light adaptation. Two lines of evidence strongly support this view. First, light causes a rise in intracellular Ca^{2+} (Fig. 6A).[15,16] Second, the normal light-induced reduction in sensitivity (light adaptation) is blocked by intracellular injection of Ca^{2+} buffer (Fig. 6B).[17]

Brown and Blinks[16] showed that the light-induced rise in Ca^{2+} occurred normally even in the absence of extracellular Ca^{2+}, a finding that implies that Ca^{2+} is released by light from an intracellular compartment. In principle, such an internal compartment could have its own rhodopsin which initiated Ca^{2+} release, but experiments have revealed that the ability of the cell to adapt, a function presumably dependent on this intracellular Ca^{2+} release, is initiated by plasma membrane rhodopsin.[18] Thus, a second messenger is required to couple the rhodopsin in the plasma membrane to the release of Ca^{2+} from internal compartments.

---

*R-lobe: Rhabdomeric lobe.
**A-lobe: Arhabdomeric lobe.

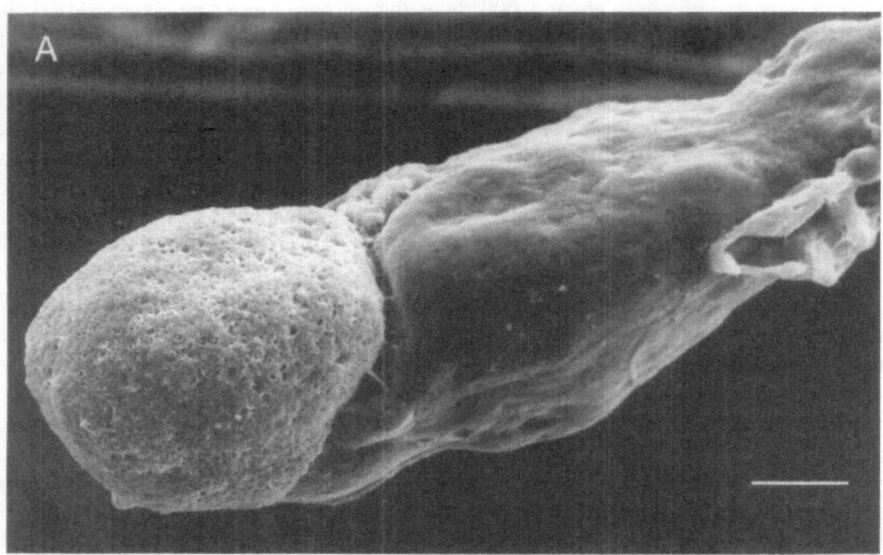

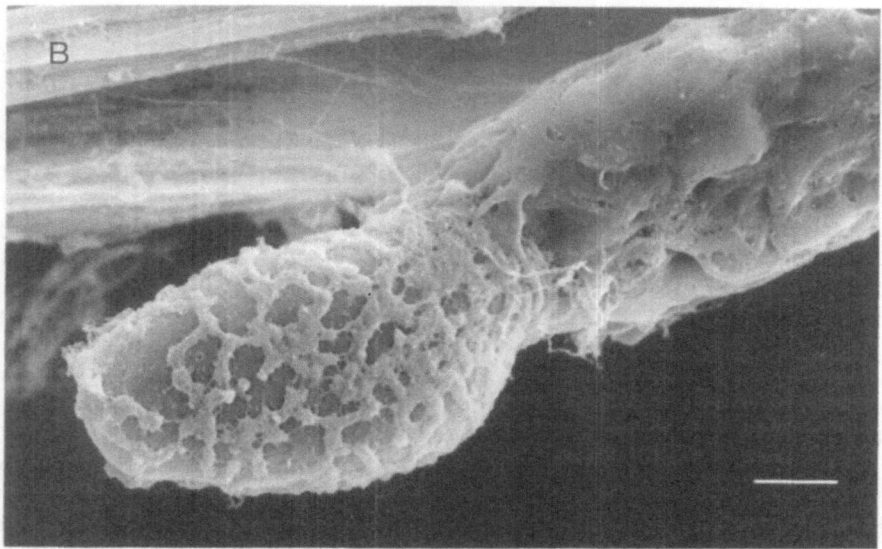

Figure 3. Scanning electron micrographs of a *Limulus* ventral photoreceptor from which glial cells had been "stripped." (A) Entire photoreceptor showing R-lobe (right) and A-lobe (left). (B) Close-up view of R-lobe showing patches of rhabdom. (C) High-resolution view of rhabdom showing microvilli. All photographs from different cells. See Ref. 12 for details of method. Scale bars: A and B: 25 μm; C: 2.2 μm.

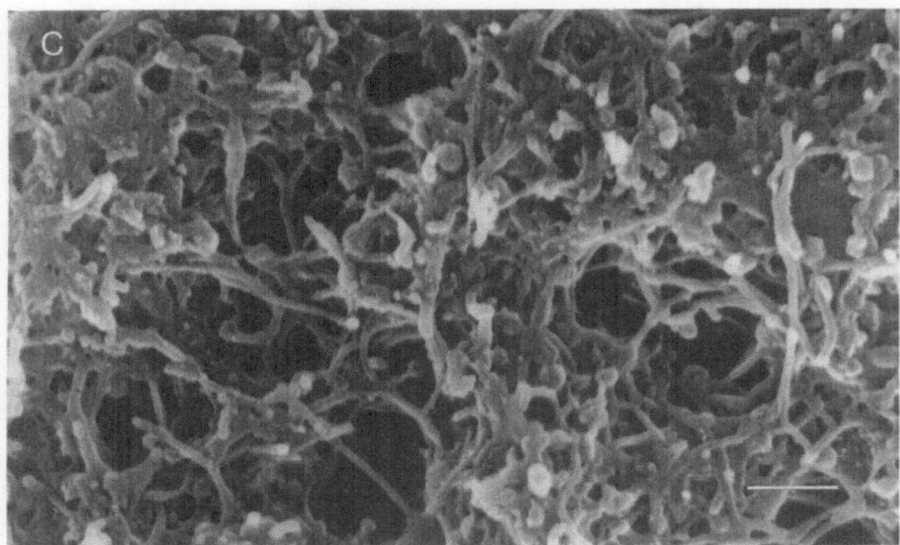

Figure 3. *(Continued)*

Shortly after the discovery that IP_3 could mediate the release of Ca^{2+} from internal stores in permeabilized liver and pancreatic cells, experiments were performed to test the effect of IP_3 injection on *Limulus* ventral photoreceptors. The results were very dramatic (Fig. 7).[19,20] Injection of IP_3 into the R-lobe produced bursts of large, discrete depolarizing waves having kinetics roughly similar to those of the quantum bumps evoked by light. The conductance change that generates these waves is indistinguishable from the conductance activated by light. These results suggest that IP_3 may be involved in the process by which light excites the cell. Figure 7 shows that in addition to exciting the cell, IP_3 adapts the cell; after the injection the cell was less responsive to light. This is consistent with the idea that IP_3 might be releasing Ca^{2+} from internal stores and thereby adapting the cell. To test this idea directly, the Ca^{2+} indicator, aequorin, was used to monitor the effect of IP_3 injection. IP_3 caused a large and rapid rise in Ca^{2+} in the R-lobe.[21,22]

Several lines of biochemical data have strengthened the hypothesis that light-induced IP_3 changes are important in invertebrate phototransduction. A light-induced rise in the concentration of IP_3 has been measured in both *Limulus*[20] and squid.[23,24] The results in squid show that this change is rapid, occurring in less than 1 sec.[23] Furthermore, a light-induced phospholipase-C activity has been measured *in vitro*.[25,26] Most recently a *Drosophila* mutant isolated because it eliminates the receptor potential has been shown to have a defect in the gene coding for phospholipase-C.[27] Work on fly and squid photoreceptors has shown that the phospholipase-C activation is dependent on GTP, suggesting that a G-protein[*] may mediate the effect of light on the phos-

*G-protein: Guanine nucleotide binding protein.

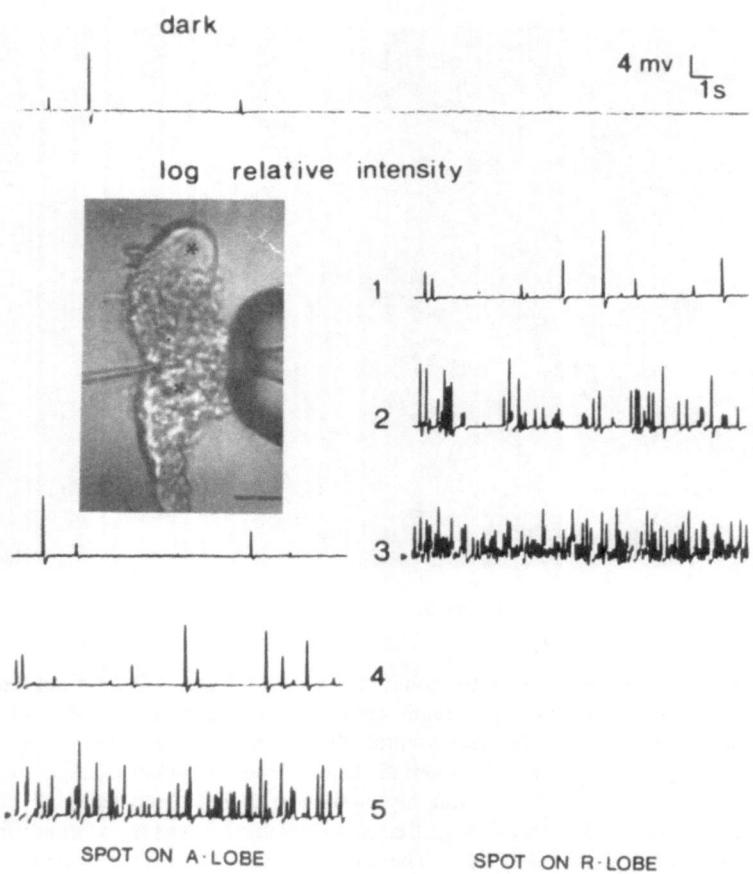

Figure 4. Differential sensitivity to light of different lobes of a *Limulus* ventral photoreceptor. A small spot of light was placed either on the A-lobe or R-lobe (stars in the inset), and the responses to light of various intensities were recorded with an intracellular microelectrode. The numbers represent relative light intensity in 10-fold increments. The inset shows the photoreceptor impaled with a microelectrode and held by a suction pipette (calibration bar: 25 μm). (Reproduced with permission from Ref. 12.)

pholipase.[26,28] Whether this G-protein is the same as that involved in excitation is unclear. Recent work provides evidence for two light-dependent G-proteins[29] (Robinson and Wood, personal communication), raising the possibility that excitation and adaptation might involve different G-proteins.

As described above, IP_3 injection causes both excitation and adaptation. An important question is whether IP_3 has some action other than raising Ca^{2+}, or whether all the effects of IP_3 can be accounted for by its ability to initiate intracellular Ca^{2+} release. The results of Payne et al.[30] support the latter conclusion. They showed that when Ca^{2+} was injected rapidly by pressure, rather than slowly by iontophoresis, Ca^{2+} could cause excitation similar to that caused by IP_3 injection or by light (Fig. 5B).

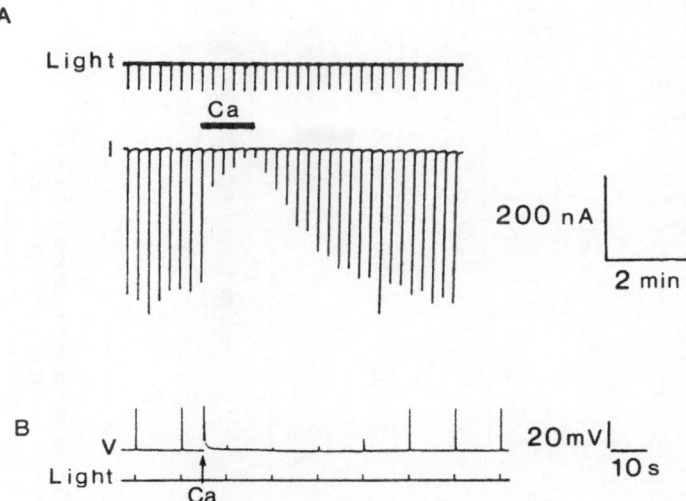

Figure 5. The effect of intracellular injection of calcium in *Limulus* ventral photoreceptors. (A) In this experiment two microelectrodes were used to voltage-clamp the cell and a third for iontophoretic Ca^{2+} injection. Injection caused no excitation, but reversibly reduced the response to flashes (adaptation). (Adapted with permission from Ref. 14.) (B) In this experiment a single electrode was used to measure membrane voltage and pressure inject Ca^{2+}. The injection produced brief excitation (depolarization) and long-lasting adaptation. (Adapted with permission from Ref. 30.)

These results raise the possibility that the pathway (rhodopsin→G-protein→phospholipase-C→IP_3→Ca^{2+}) mediates both the excitation and the adaptation produced by light. A key experiment suggests, however, that this is not the case. As shown in Fig. 8A, injection of the Ca^{2+} chelator, EGTA[*], nearly abolished the excitatory effect of IP_3 injection, but actually enhanced the response to illumination.[22] (The measure of response used is the total charge–flow rather than peak current, a choice dictated by the change in response kinetics produced by EGTA injection.) Indeed previous experiments had shown that injection of Ca^{2+} buffers blocks light adaptation, without dramatically affecting excitation (Fig. 6B).[17] The efficacy of the buffer injection has been demonstrated directly by the ability of the buffer to block the light-induced rise in Ca^{2+} monitored by aequorin.[22]

Experiments with other Ca^{2+} chelators such as Quin2 gave similar results.[22] These experiments with chelators suggest that a rise in Ca^{2+} is not required for excitation to occur. A further argument against the sole involvement of the IP_3→Ca^{2+} pathway is shown in Fig. 8B, from an experiment by Bolsover and Brown.[31] By extensively illuminating the cell that is bathed in solution without Ca^{2+}, a cell can be put into a "Ca^{2+}-depleted" state. In this state, the light-induced aequorin response is reduced by over two orders of magnitude (the reduction in Ca^{2+} is less; see Fig. 8 caption), but there is virtually no reduction in the magnitude of the receptor potential.

---

*EGTA: Ethyleneglycol-bis-(β-amino-ethyl ether)N',N'-tetraacetic acid.

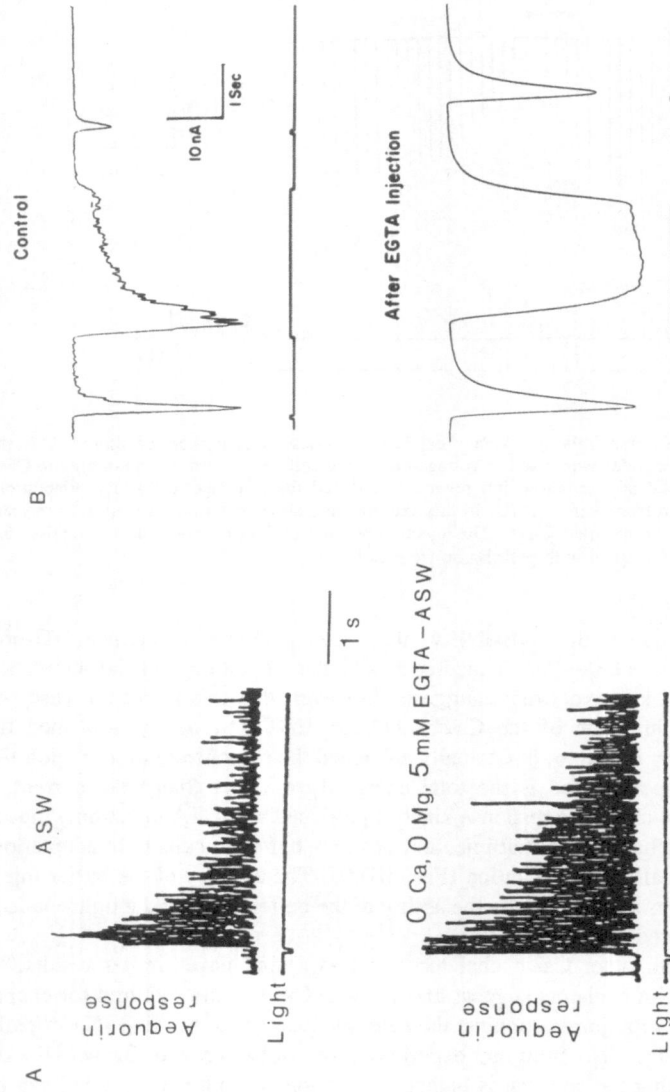

Figure 6. A rise in intracellular Ca^{2+} occurs and is necessary for adaptation in *Limulus* ventral photoreceptors. (A) Light produced a rise in the concentration of intracellular Ca^{2+}, as measured by aequorin. This rise was relatively unaffected by removal of extracellular Ca^{2+}. The output current of a photomultiplier is shown; the current is proportional to the luminescence of the aequorin injected into a single cell. (Adapted with permission from Ref. 16.) (B) Effect of intracellular injection of Ca^{2+} buffer, EGTA, on light adaptation. Before injection, prolonged light stimulus produces reduction (adaptation) of response to brief test flashes. This adaptation was blocked after EGTA injection.

Figure 7. Effect of intracellular injection of IP₃ in a *Limulus* ventral photoreceptor. Top trace is voltage-clamp current. Brief test flashes were given repetitively. At arrow, IP₃ was injected by pressure. Injection caused a large but brief inward current (excitation) and reversible reduction in the responses to subsequent test flashes (adaptation). (Adapted with permission from Ref. 20.)

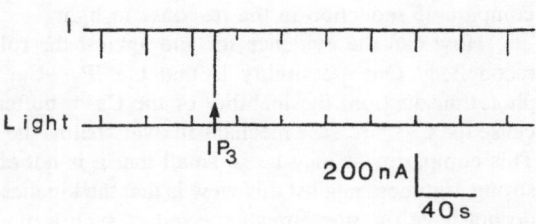

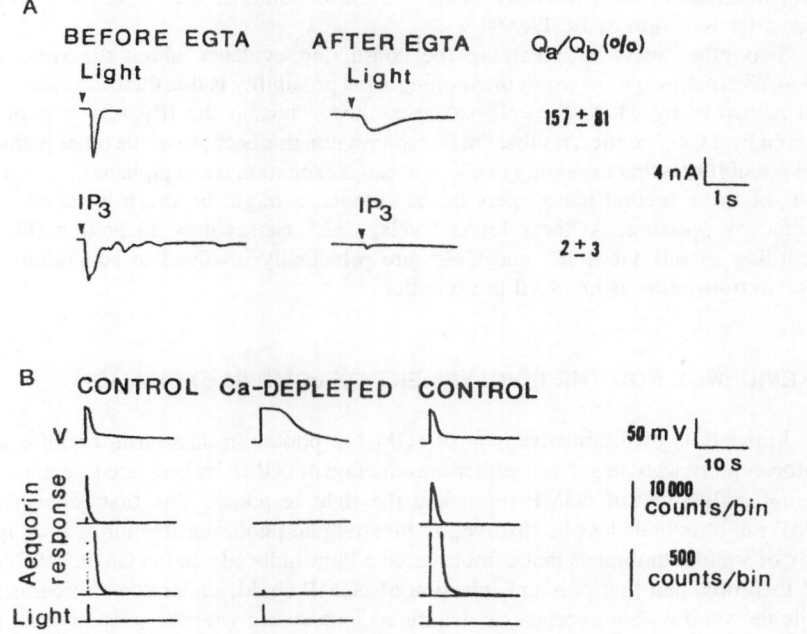

Figure 8. Procedures that dramatically reduce the response to IP₃ injection and the light-induced rise in intracellular Ca^{2+} do not produce a large reduction in the response to light. (A) Top traces: intracellular injection of calcium buffer, EGTA, changed the kinetics of the light-induced current, lowered the peak current, but increased the total charge across the plasma membrane during the response (expressed as the percentage of charge during previous control flashes, Q_a/Q_b, over several cells) by 157%. Lower traces: same EGTA injection reduced the response of IP₃ injection to 2% (percentage of charge during previous control IP₃ injections). (Adapted with permission from Ref. 22.) (B) A "Ca-depleted state" can be induced if the cell is illuminated for long periods of time in the absence of extracellular Ca^{2+} (plus EGTA). In such a state (center traces), the voltage response to a bright light was not reduced, but the aequorin response of the same cell was nearly abolished. All the effects of "Ca-depletion" were reversible. Note that the aequorin responses are presented at two different gains. Note, also, that the differences in the aequorin response may exaggerate the differences in the calcium transients because of the supralinearity of aequorin's dependence on Ca^{2+}. (Adapted with permission from Ref. 31.)

Thus, there are two very different ways of greatly reducing the rise in Ca^{2+} without a comparable reduction in the response to light.

How can the evidence for and against the role of IP_3 and Ca^{2+} in excitation be reconciled? One possibility is that the $IP_3 \rightarrow Ca^{2+}$ pathway is the sole pathway in phototransduction; the inability of the Ca^{2+} buffer to block excitation may occur because the Ca^{2+} release mechanism overwhelms the buffer in a small local compartment. This compartment may be so small that it is not effectively monitored by aequorin. A strong argument against this view is that the kinetics of excitation in EGTA-injected cells do not have the properties expected of such a model. If the occurrence of a response depends on a buffer being overwhelmed, the response should have a nonlinear dependence on light intensity because bright lights should overwhelm the buffer better than dim lights. Just the opposite has been found, however; whereas the response to light varies nonlinearly with intensity under normal conditions, the dependence becomes linear after injection of EGTA.[32]

Two other ways of resolving the conflicting evidence about the roles of the $IP_3 \rightarrow Ca^{2+}$ pathways are worth mentioning. One possibility is that there are two independent pathways by which the cell can be excited. Thus, if the $IP_3 \rightarrow Ca^{2+}$ pathway is blocked by a Ca^{2+} buffer, a substantial response remains because of the other pathway. A final possibility is that the ability of Ca^{2+} to cause excitation is an epiphenomenon; normal levels of these second messengers in the cytoplasm might be much lower than those injected by pressure. At these lower levels, Ca^{2+} causes only adaptation (Fig. 5A). According to this view, IP_3 and Ca^{2+} are principally involved in adaptation, and a different transmitter is involved in excitation.

5. EVIDENCE FOR THE INVOLVEMENT OF cGMP IN EXCITATION

Initial tests of the involvement of cGMP in phototransduction in *Limulus* ventral photoreceptors were negative; neither introduction of cGMP by pressure injection[33] nor internal dialysis[34] of cGMP mimicked the light response. The first indication that cGMP might nonetheless be involved in invertebrate phototransduction was an *in vitro* study of squid membranes that demonstrated a light-induced rise in cGMP.[35] Evidence was then obtained that pressure injection of cGMP could, under some circumstances, excite the ventral photoreceptor, activating a conductance with the same reversal potential and voltage dependence as that opened by light.[36]

As shown in Fig. 9, the response to cGMP injection depends critically on the position of the injection pipette. Not only is it important to have the electrode in the R-lobe, but it is necessary to have the electrode in "hot spots" within the R-lobe. Using a motorized micropositioner capable of making reproducible movements of the electrode along its axis, it was possible to reproducibly move an electrode in and out of such hot spots. The requirement that cGMP reach these hot spots probably accounts for the failure of previous investigators to detect a cGMP-induced depolarization.

The structural basis of these hot spots has not yet been determined, but they are usually found near the edge of the R-lobe. This suggests that the electrode may have to be between the microvillar membrane system and the subrhabdomeric membrane system (Fig. 1B). The subrhabdomeric cisternae cannot be an absolute barrier to diffusion

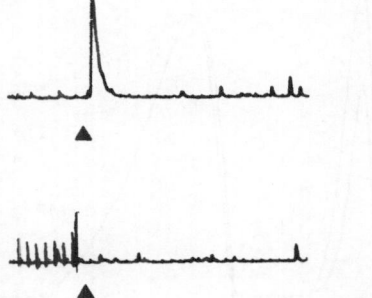

Figure 9. Demonstration of "hot spots" for cGMP injection into the R-lobe of *Limulus* ventral photoreceptors. In top trace, injection into a "hot spot" produced a large depolarization. The electrode was then moved 10 μm along its axis to a different position in R-lobe (note positive-going artifacts produced by stepping motor). The same injection produced no effect (middle trace). After returning the electrode to the original position, the injection again produced a large depolarization (bottom trace). (Reproduced with permission from Ref. 36.)

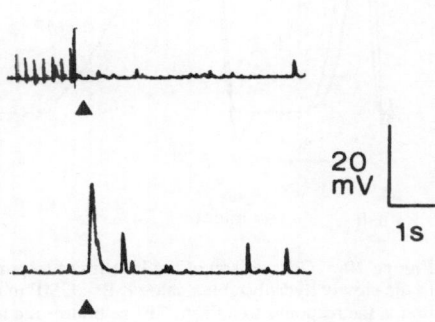

because it is fenestrated.[13] The reticulum could, however, substantially reduce the diffusion of a second messenger away from the microvilli, thereby generating larger concentration changes near the membrane than would occur if diffusion were free. Interestingly, in the regions of the ventral photoreceptor between the microvilli and reticulum, the amount of cytoplasm is small, comparable to that in a vertebrate rod outer segment (Fig. 1).

The strong position dependence of the cGMP effect requires that the comparison of cGMP with other compounds be done with double-barreled electrodes. One barrel is filled with cGMP, the other with a test substance. Once a hot spot is identified using cGMP injection, the other substance can be injected into the same location. This method reveals that cGMP is much more potent than cAMP in exciting the cell. Furthermore, as shown in Fig. 10, injection of 8-Br-cGMP, an analog of cGMP that is more slowly hydrolyzed by phosphodiesterase than cGMP, produces a more prolonged depolarization than a comparable injection of cGMP.[36] These results suggest that the rate of hydrolysis of cGMP determines how fast the depolarization returns to the baseline; it can thus be concluded that cGMP is hydrolyzed in these cells in less than 100 msec. Such rapid hydrolysis will clearly limit the distance over which cGMP can spread. This may provide a further explanation of "hot spots."

The response to cGMP has many properties that would be expected of the transmitter of excitation. The response to cGMP injection is rapid, having an onset less than 100 msec after the beginning of the injection pulse. The response is not blocked by injection of Ca^{2+} buffer and in this respect is like the response to light.[36] In contrast, the response to IP_3 injection is blocked by Ca^{2+} buffer. It would be expected that the response of the channels to the agent that normally activates them would be smoothly graded with concentration. Furthermore, since maintained light can produce a sustained response, the response to the putative transmitter should be sustained. These conditions are met by cGMP. In contrast, IP_3 injection produces a burst of pulsatile responses, the

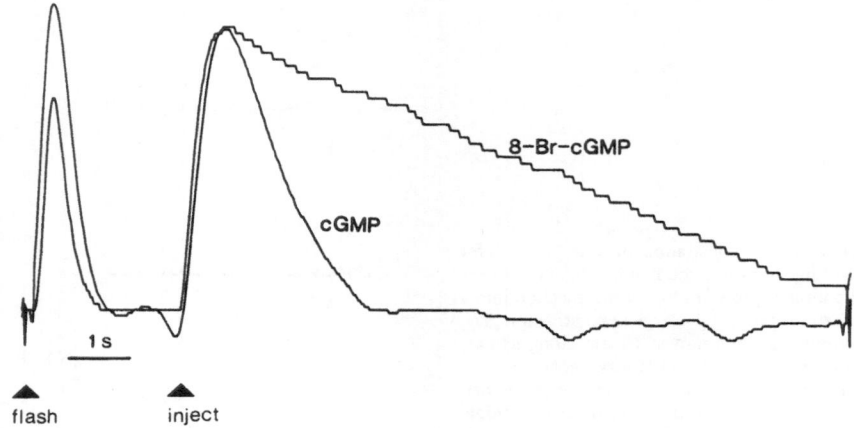

Figure 10. Comparison of the kinetics of response to cGMP injection with the kinetics of the response to the slowly hydrolyzable analog 8-Br-cGMP in *Limulus* ventral photoreceptors. The initial depolarization is the response to a flash. The responses are normalized so that they have the same peak response to injection, but the absolute magnitudes of the peak responses were quite close. The two compounds were pressure-injected out of different barrels of the same electrode. (Reproduced with permission from Ref. 36.)

frequency of which cannot be sustained. This has interesting implications for the Ca^{2+} release mechanism (see Chapter 1, this volume), but makes it hard to see how this pathway could be responsible for the smooth, sustained responses to bright lights.

If cGMP is the excitatory transmitter in invertebrates, its concentration must rise rapidly after illumination. To test whether this occurs, experiments were done on squid retina. The reason for using this preparation is that its large volume makes it much more suitable for biochemical analysis than the much smaller *Limulus* eyes. Squid retina were frozen with a liquid-nitrogen-cooled copper hammer within 100 to 200 msec after illumination. Electrodes on the retina recorded the ERG(*) of the retina, as shown in Fig. 11. One-hundred msec after the flash, the hammer collided with the retina and froze it, creating a large electrical artifact. Radioimmunoassay of the cGMP content of whole squid retina frozen in the dark or 100–200 msec after illumination demonstrates a light-induced rise in cGMP of about 100%.[36]

The enzymatic machinery responsible for this change in cGMP is under active investigation. Squid retina contains guanylate cyclase activity that can be revealed by activation by the detergent Triton, a procedure known to activate this enzyme in a variety of preparations.[37] So far, *in vitro* studies have failed to reveal any light activation of this enzyme. A recent report indicates that cGMP phosphodiesterase activity may be inactivated by light in *Limulus*, a change that could produce the observed rise in cGMP concentration.[38]

One indirect way of testing the role of cGMP in phototransduction is to block its

*ERG: electroretinogram.

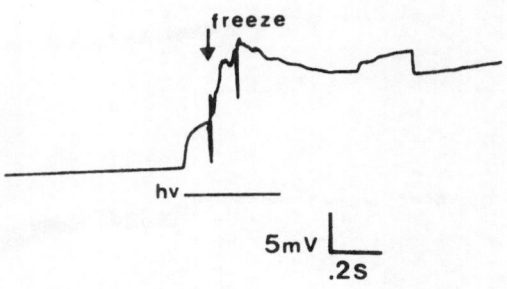

Figure 11. Fast-freezing of squid retina for measurement of light-induced changes in cGMP. electroretinogram of squid retina measured in fast-freezing apparatus. Bar marks onset of light. The initial positive-going deflection of the trace is the electroretinogram, which is followed about 0.1 sec later by the artifact due to impact of the cold copper block.

breakdown by applying phosphodiesterase inhibitors. The most straightforward prediction is that inhibiting the phosphodiesterase should increase cGMP levels and mimic the depolarizing receptor potential. Experiments of this kind have given somewhat conflicting results. Miller et al.[39] and Wulff[40] found that the phosphodiesterase inhibitor aminophylline produced a depolarization of *Limulus* lateral eyes. Corson et al.[41] reported that isobutylmethylxanthine (IBMX), theophylline, and papaverine desensitized the ventral photoreceptor, but did not excite it. Most recently, Faddis and Brown[42] reported that IBMX causes an inward current in the dark, and preloading the cell with EGTA greatly enhanced this current. Clearly, a more direct test of the cGMP hypothesis would be desirable.

In vertebrate rods, the excised patch experiments of Fesenko et al.[43] provided a decisive way of resolving the conflicting claims for the Ca^{2+} and cGMP pathways. Experiments of this kind were done on inside-out patches of *Limulus* ventral photoreceptors and are shown in Fig. 12A,B. The membrane patch potential was held at −50 mV (pipette potential = +50 mV). In part (A), addition of 20 μM cGMP causes an increase in inward current and an increase in noise; individual channel currents are not seen because of the large number of light-activated channels that this patch contained. A further rise in cGMP concentration to 100 μM caused a further increase in the magnitude of the inward current. The current went back to baseline after washing out the cGMP. Application of cGMP was then repeated several times using slightly different protocols (see Fig. 12 caption).

The patch in Fig. 12B contained fewer channels and shows how the activity of such channels is greatly increased by the presence of 50 μM cGMP. These channels have a conductance of 15 pS. The effect of cGMP on these two patches and several other patches studied is consistent with a role of cGMP in opening the light-activated channels in *Limulus*. In the large majority of experiments, however, there was no effect of cGMP on the patches, even though they all contained light-activated channels, as revealed by the tests done prior to excision. It is possible that the channels are very labile after excision or that some cofactor is necessary for the channel to operate normally.

The channels that are activated by cGMP in excised patches are likely to be the same as the light-activated channels because they have similar properties: the cGMP-activated channels have a conductance of 15 pS, the same value as that of the smaller

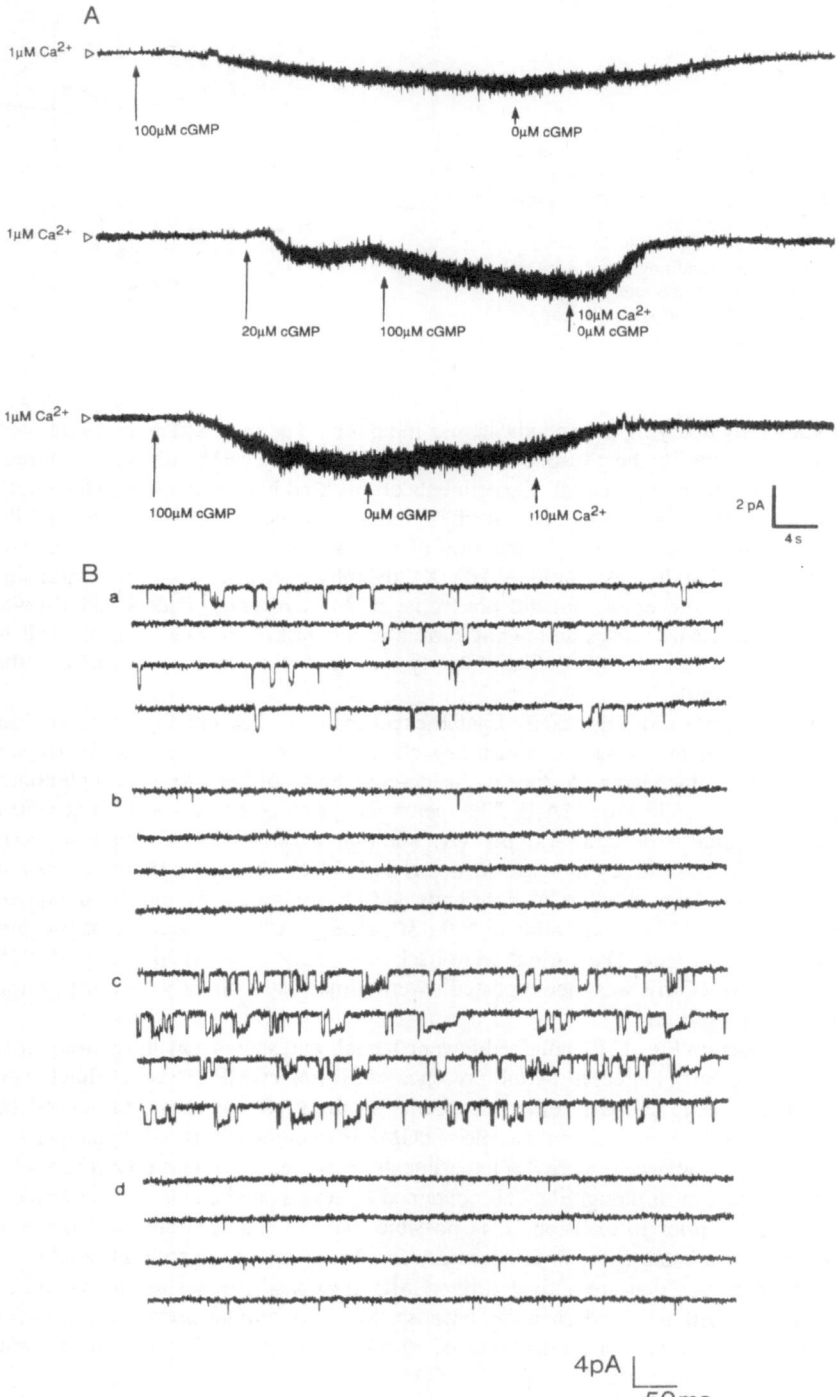

A

1μM Ca²⁺ ▷

↑
100μM cGMP

↑
0μM cGMP

1μM Ca²⁺ ▷

↑
20μM cGMP

↑
100μM cGMP

↑
10μM Ca²⁺
0μM cGMP

1μM Ca²⁺ ▷

↑
100μM cGMP

↑
0μM cGMP

↑
↑10μM Ca²⁺

2 pA

4 s

B

a

b

c

d

4pA

50ms

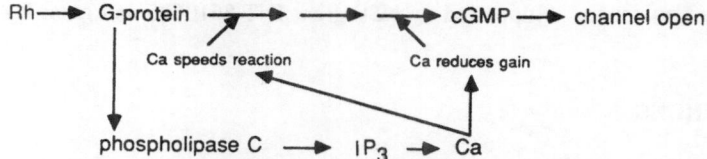

Figure 13. Tentative model of transduction in invertebrate photoreceptors. The excitation pathway involves cGMP, the adaptation pathway IP_3 and Ca^{2+}. In the diagram a single G-protein stimulates both pathways, but this choice is arbitrary. It is possible that rhodopsin activates specialized G-proteins for each pathway. Alternatively, there may be only one G-protein that activates only phospholipase-C. In this case, some produce of phospholipase-C must stimulate the production of cGMP.

conductance state of the light-activated channel[44]; both channels have a reversal potential near 0 mV under similar ionic conditions; and both channels have a mean open time of about 1 msec. Furthermore, while still cell-attached, each patch was checked for voltage-dependent channels; experiments were continued only if patches contained no voltage-dependent channels.

6. CONCLUSIONS

Substantial progress has been made in understanding visual transduction in invertebrate photoreceptors, though the conclusions that can be reached are far less definitive than for vertebrate rods. Figure 13 summarizes a tentative model of invertebrate phototransduction that seems consistent with the known facts. cGMP is the messenger of excitation, IP_3 and Ca^{2+} the messengers of adaptation. Ca^{2+} exerts an adaptational role by reducing the gain of one or more steps in the cGMP cascade. In addition, to account for the slowing of the kinetics of light response that occurs when Ca^{2+} is buffered (Fig. 6B), it must be assumed that Ca^{2+} speeds a step in the cGMP cascade.[32]

Clearly much further work is required to understand how the depolarizing receptor potential of invertebrate photoreceptors is generated. The value of understanding this mechanism is not only for comparing the function of different types of photoreceptors, but also because the mechanisms involved may be related to the membrane transduction initiated by other types of receptors in the rhodopsin family such as the muscarinic and beta-adrenergic receptors.

Figure 12. Response of excised patch of membrane from the R-lobe of *Limulus* ventral photoreceptors to cGMP. (A) In this patch individual channel events could not be resolved. application of cGMP produced inward current and an increase in noise (top trace). The effect was slowly reversible. The effect was graded with the concentration of cGMP (middle trace). Reversal was more rapid than in top trace because of inclusion of 10 μM Ca^{2+}. In bottom trace, application of cGMP was repeated; there was a slow recovery when cGMP was removed, which was speeded by raising Ca^{2+}. (B) Patch in which single channels were resolved. (a) and (c): During application of 50 μM cGMP; (b) and (d): in absence of cGMP.

ACKNOWLEDGMENTS. Supported by NSF grant INT-8610625 and grant EY01496.

REFERENCES

1. Hurley, J. B., 1987, Molecular properties of the cGMP cascade of vertebrate photoreceptors, *Ann. Rev. Physiol.* **49:** 793–812.
2. Pugh, E. N., Jr., 1987, The nature and identity of the internal excitational transmitter of vertebrate phototransduction, *Ann. Rev. Physiol.* **49:** 715–741.
3. McNaughton, P. A., Cervetto, L., and Nunn, B. J., 1986, Measurement of the intracellular free calcium concentration in salamander rods, *Nature* **322:** 261–263.
4. Matthew, H. R., Murphy, R. L. W., Fain, G. L., and Lamb, T. D., 1988, Photoreceptor light adaptation is mediated by cytoplasmic calcium concentration, *Nature* **334:** 67–69.
5. Nakatani, K., and Yau, K.-W., 1988, Calcium and light adaptation in retinal rods and cones, *Nature* **334:** 69–71.
6. Koch, K. W., and Stryer, L., 1988, Highly cooperative feedback control of retinal rod guanylate cyclase by calcium ions, *Nature* **334:** 64–66.
7. Brown, J. E., and Coles, J. A., 1979, Saturation of the response to light in *Limulus* ventral photoreceptor, *J. Physiol.* **296:** 373–392.
8. Fain, G. L., and Lisman, J. E., 1981, Membrane conductances of photoreceptors, *Prog. Biophys. Molec. Biol.* **37:** 91–147.
9. Detwiler, P. B., Conner, J. A., and Bodoia, R. D., 1982, Gigaseal patch clamp recordings from outer segments of intact retinal rods, *Nature* **300:** 59–61.
10. Bacigalupo, J., and Lisman, J. E., 1983, Single-channel currents activated by light in *Limulus* ventral photoreceptors, *Nature* **304:** 268–270.
11. Clark, A. W., Millecchia, R., and Mauro, A., 1969, The ventral photoreceptor of *Limulus*. I. The microanatomy, *J. Gen. Physiol.* **54:** 289–309.
12. Stern, J., Chinn, K., Bacigalupo, J., and Lisman, J. E., 1982, Distinct lobes of *Limulus* ventral photoreceptors. I. functional and anatomical properties of lobes revealed by removal of glial cells, *J. Gen. Physiol.* **80:** 825–837.
13. Calman, B., and Chamberlain, S., 1982, Distinct lobes of *Limulus* ventral photoreceptors. II. Structure and ultrastructure, *J. Gen. Physiol.* **80:** 839–862.
14. Lisman, J. E., and Brown, J. E., 1972, The effects of intracellular iontophoretic injection of calcium and sodium ions on the light response of *Limulus* ventral photoreceptors, *J. Gen. Physiol.* **59:** 701–719.
15. Brown, J. E., Brown, P. K., and Pinto, L. H., 1972, Detection of light-induced changes in intracellular ionized calcium concentration in *Limulus* ventral photoreceptor using arsenazo III, *J. Physiol.* **267:** 299–320.
16. Brown, J. E., and Blinks, J. R., 1974, Changes in intracellular free calcium concentration during illumination of invertebrate photoreceptors, *J. Gen. Physiol.* **64:** 643–665.
17. Lisman, J. E., and Brown, J. E., 1975, Effects of intracellular injection of calcium buffers in *Limulus* ventral photoreceptors, *J. Gen. Physiol.* **66:** 489–506.
18. Lisman, J. E., and Strong, J. A., 1979, The initiation of excitation and adaptation in *Limulus* ventral photoreceptors, *J. Gen. Physiol.* **73:** 219–243.
19. Fein, A., Payne, R., Corson, D. W., Berridge, M. J., and Irvine, R. F., 1984, Photoreceptor excitation and adaptation by inositol 1,4,5-trisphosphate, *Nature* **311:** 157–160.
20. Brown, J. E., Rubin, L. J., Ghalayini, A. J., Tarver, A. P., Irvine, R. F., Berridge, M. J., and Anderson, R. E., 1984, Evidence that myo-inositol polyphosphate may be a messenger for visual excitation in *Limulus* photoreceptors, *Nature* **311:** 160–163.
21. Brown, J. E., and Rubin, L. J., 1984, A direct demonstration that inositol-trisphosphate induces an increase in intracellular calcium in *Limulus* photoreceptors, *Biochem. Biophys. Res. Comm.* **125:** 1137–1142.
22. Payne, R., Corson, D. W., Fein, A., and Berridge, M. J., 1986, Excitation and adaptation of

Limulus ventral photoreceptors by inositol 1,4,5-trisphosphate result from a rise in intracellular calcium, *J. Gen. Physiol.* **88:** 127–142.

23. Szuts, E. Z., Wood, S. F., Reid, M. S., and Fein, A., 1986, Light stimulates the rapid formation if inositol trisphosphate in squid retinas, *Biochem J.* **240:** 929–932.

24. Brown, J. E., Watkins, D. C., and Malbon, C. C., 1987, Light-induced changes in the content of inositol phosphates in squid *Loligo pealei* retina, *Biochem J.* **247:** 293–297.

25. Vandenberg, C. A., and Montal, M., 1984, Light-regulated biochemical events in invertebrate photoreceptors. 2. Light-regulated phosphorylation of rhodopsin and phosphinositides in squid photoreceptor membranes, *Biochemistry* **23:** 2347–2352.

26. Baer, K. M., and Saibil, H. R., 1988, Light- and GTP-activated hydrolysis of phosphatidylinositol bisphosphate in squid photoreceptor membranes, *J. Biol. Chem.* **263:** 17–20.

27. Bloomquist, B. T., Shortridge, R. D., Schneuwly, W., Perdew, M., Montell, C., Steller, H., Rubin, G., and Pak, W. L., 1988, Isolation of a putative phospholipase-C gene of *Drosophila, norpA*, and its role in phototransduction, *Cell* **54:** 723–733.

28. Devary, O., Heichal, O., Blumenfeld, A., Cassel, D., Suss, E., Barash, S., Rubinstein, C. T., Minke, B., and Selinger, Z., 1987, Coupling of photoexcited rhodopsin to inositol phospholipid hydrolysis in fly photoreceptors, *P.N.A.S.* (USA) **84:** 6939–6943.

29. Tsuda, M., 1987, Octopus G-protein: A signal-coupling protein in invertebrate photoreceptor, in: *Proceedings of the International Conference on Retinal Proteins* (V. Ochinnikov, ed.), VNU Science Press, The Netherlands, pp. 393–404.

30. Payne, R., Corson, D. W., and Fein, A., 1986, Pressure injection of calcium both excites and adapts *Limulus* ventral photoreceptors, *J. Gen. Physiol.* **88:** 101–12.

31. Bolsover, S. R., and Brown, J. E., 1985, Calcium ion, an intracellular messenger of light adaptation, also participates in excitation of *Limulus* photoreceptors, *J. Physiol.* **364:** 381–393.

32. Payne, R., and Fein, A., 1986, The initial response of *Limulus* ventral photoreceptors to bright flashes: released calcium as a synergist to excitation, *J. Gen. Physiol.* **87:** 243–269.

33. Bolsover, S. R., and Brown, J. E., 1982, Injection of guanosine and adenosine nucleotides into *Limulus* ventral photoreceptor cells, *J. Physiol.* **332:** 325–342.

34. Stern, J. H., and Lisman, J. E., 1982, Internal dialysis of *Limulus* ventral photoreceptors, *Proc. Natl. Acad. Sci.* **79:** 7580–7584.

35. Saibel, H. R., 1984, A light-stimulated increase in cyclic GMP in squid photoreceptors, *FEBS Lett.* **168:** 213–216.

36. Johnson, E. C., Robinson, P. R., and Lisman, J. E., 1986, Cyclic GMP is involved in the excitation of invertebrate photoreceptors, *Nature* **324:** 468–470.

37. Robinson, P. R., Cote, R. H., and Lisman, J. E., 1987, Guanylate cyclase activity in squid photoreceptor membranes, *Biophys. J.* **51:** 269a.

38. Inoue, M., and Brown, J. E., 1988, Cyclic GMP phosphodiesterase in *Limulus* ventral eye, *ARVO* abstr:218.

39. Miller, W. H., Gorman, R. E., and Bitensky, M. W., 1971, Cyclic adenosine monophosphate: function in photoreceptors, *Science* **174:** 295–297.

40. Wulff, V. J., 1973, The effect of cyclic AMP and aminophylline on *Limulus* lateral eye retinular cells, *Vision Res.* **13:** 2335–2344.

41. Corson, D. W., Fein, A., and Schmidt, J., 1979, Two effects of phosphodiesterase inhibitors in *Limulus* ventral photoreceptors, *Brain Res.* **176:** 365–368.

42. Faddis, M., and Brown, J. E., 1988, Effects of drugs presumed to change intracellular cGMP on voltage-clamp current in *Limulus* ventral photoreceptors, *ARVO* abstr:350.

43. Fesenko, S. S., Kolesnikou, A. L., and Lyubarsky, E. E., 1985, Induction by cyclic GMP of cationic conductance on the plasma membrane of the retinal rod outer segment, *Nature* **313:** 310–313.

44. Bacigalupo, J., Johnson, E., and Lisman, J. E., 1987, A low-conductance light-dependent channel observed in cell-attached and excised patches of *Limulus* ventral photoreceptors, *Biophys. J.* **51:** 15a.

45. Baylor, D. A., Lamb, T. D., and Yau, K.-W., 1979, Responses of retinal rods to single photons, *J. Physiol.* **288:** 613–634.

Chapter 3

The cGMP-Gated Channels of Rod and Cone Photoreceptors

Lawrence W. Haynes and King-Wai Yau

1. INTRODUCTION

Light generates a hyperpolarizing response in retinal photoreceptors by stopping a steady inward current (the dark current) that is present at their outer segments in darkness.[1-3] The mechanism underlying this phototransduction process has been a subject of intense interest and controversy for many years, and is only now becoming fairly clear. The present picture for rods[4-6] is that upon absorbing a photon a visual pigment molecule isomerizes and catalytically activates a GTP-binding protein (also called G-protein) that is bound peripherally to disk membranes within the outer segment. The activated G-protein in turn stimulates a phosphodiesterase that hydrolyzes cGMP, a cyclic nucleotide known to be present at a high concentration in the outer segment. In darkness, this high cGMP level maintains a plasma membrane cationic conductance (the light-regulated conductance) in the open state, thus sustaining the dark current. In the light, the fall in the cGMP level leads to the closure of this conductance, and therefore the cessation of the dark current. This picture for phototransduction in rods now seems to apply to cones as well,[7] despite their different surface membrane geometries.

In this chapter we briefly describe some electrical measurements we have made on

LAWRENCE W. HAYNES and KING-WAI YAU • Howard Hughes Medical Institute, and Department of Neuroscience, The Johns Hopkins University School of Medicine, Baltimore, Maryland 21205. *Present address for L.W.H.:* Department of Medical Physiology, University of Calgary, Calgary, Alberta T2N 4N1, Canada.

the light-regulated conductance in rods and cones with the excised-patch recording method. The properties of this conductance in the two kinds of receptors are similar but not identical. Most of the results described here are already published.[8-13] Besides ourselves, other groups have been engaged in similar experiments on the rod conductance.[14-23]

2. METHODOLOGY

The experiments were done on rods from the toad (*Bufo marinus*) or the larval tiger salamander (*Ambystoma tigrinum*), and on cones from the channel catfish (*Ictalurus punctatus*). A piece of retina was pretreated with enzymes (hyaluronidase and collagenase) and then thoroughly chopped under Ringer's solution to yield a dispersion of isolated cells and broken outer segments for patch-recording. The pipets for recording were pulled from borosilicate glass, coated with Sylgard, and fire polished. Gigaohm seals were obtained by pressing the pipet against the side of a rod outer segment or the tip of a cone outer segment and applying gentle suction. A tap on the micromanipulator holding the pipet jolted the pipet and excised an inside-out membrane patch from the outer segment. The cytoplasmic surface of the membrane was then exposed to bath perfusion with or without cGMP.

Normal Ringer's solution contained (in mM): 110 NaCl, 2.5 KCl, 1.6 MgCl$_2$ and 1.0 CaCl$_2$, buffered to pH 7.6 with 5.0 tetramethylammonium (TMA)-HEPES. The pseudointracellular solution contained (in mM): 100 KGluconate, 12.5 NaCl, 1.6 MgCl$_2$ and 0.1 TMA-EGTA, buffered to pH 7.6 with 5.0 TMA-HEPES. Many experiments were also done with identical NaCl solutions on both sides of the membrane patch, containing (in mM): 118 NaCl, 0.1 NaEGTA, 0.1 NaEDTA, buffered to pH 7.6 with 5.0 NaHEPES.

3. THE CURRENT–VOLTAGE RELATION

The current–voltage relations for the cGMP-induced currents measured in a rod and a cone patch[9,12] are shown in Fig. 1. They are similar to those previously measured for the light-sensitive currents in intact rods and cones.[24-27] In these excised-patch experiments, the extracellular surface of a patch was exposed to normal Ringer's solution, and the cytoplasmic surface was in contact with a pseudointracellular solution containing a saturating concentration of cGMP. The current–voltage relations of both the rod and cone conductances have reversal potentials around 10–15 mV, consistent with the notion that the channel in both cases is not very selective between Na$^+$ and K^{+}[28,29] (L. W. Haynes, K. Nakatani, and K.-W. Yau, unpublished).

The rod conductance shows prominent outward rectification; that is, the cGMP-induced current increases exponentially with increasing depolarization but stays more or less unchanged at hyperpolarizations. The cone conductance, on the other hand, increases exponentially with both depolarizations and hyperpolarizations. The smooth curves in Figs. 1A and 1B are drawn according to the following equation:

$$I(V) = \exp[(1 - \gamma)(V - V_r)/V_0] - \exp[-\gamma(V - V_r)/V_0] \tag{1}$$

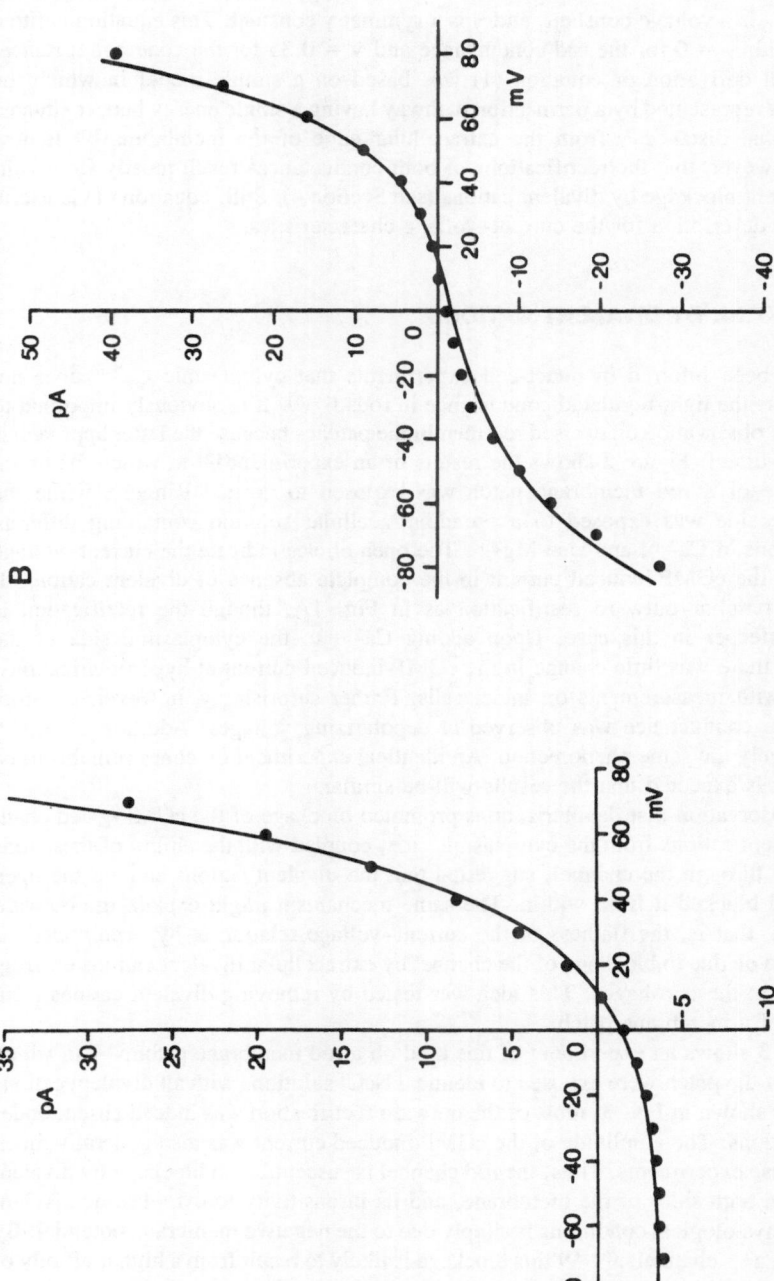

Figure 1. (A) Current–voltage relation for the cGMP-induced current from an excised, inside-out rod membrane patch. (B) The same from a cone patch. In both cases, the patch-pipet contained normal Ringer's solution, while the bath contained a pseudointracellular solution with a saturating concentration of cGMP. Current measurements were taken at the end of a 1 sec voltage step. The smooth curve in (A) is drawn according to equation (1) with $V_r = 12$ mV, $V_0 = 25$ mV, and $\gamma = 0.35$. That in (B) is the same equation but with $V_r = 15$ mV, $V_0 = 12.5$ mV, and $\gamma = 0$. (Redrawn from Ref. 12.)

where $I(V)$ is the transmembrane current at membrane potential V, V_r is the reversal potential, V_0 is a voltage constant, and γ is a symmetry constant. This equation is fitted to Fig. 1 with $\gamma = 0$ for the rod conductance and $\gamma = 0.35$ for the cone conductance. The original derivation of equation (1) was based on a simple model in which the channel was represented by a permeation pathway having a single energy barrier situated at a fractional distance γ from the extracellular edge of the membrane.[30] It now appears, however, that the rectifications of both conductances result mostly from voltage-dependent blockage by divalent cations (see Section 4). Still, equation (1) is useful as a formal description for the current–voltage characteristics.

4. BLOCKAGE BY DIVALENT CATIONS

It has been inferred in intact-cell experiments that cytoplasmic Ca^{2+} does not directly close the light-regulated conductance in rods.[29,31] It is obviously important to confirm this observation on excised rod membrane patches because the latter approach is much more direct. Figure 2 shows the results of an experiment[12] in which the extracellular side of a rod membrane patch was exposed to normal Ringer's while the cytoplasmic side was exposed to a pseudointracellular solution containing different concentrations of Ca^{2+}, and also Mg^{2+}. The open circles indicate the current–voltage relation for the cGMP-induced current in the complete absence of divalent cations. It shows the familiar outward rectification as in Fig. 1A, though the rectification is somewhat steeper in this case. Upon adding Ca^{2+} to the cytoplasmic side of the membrane, there was little change in the cGMP-induced current at hyperpolarizations, consistent with measurements on intact cells. Rather surprisingly, however, a partial block of the conductance was observed at depolarizing voltages. Addition of Mg^{2+} showed largely the same phenomenon. An identical experiment on cones remains to be done, but it is expected that the results will be similar.

The observation that depolarizations promoted blockage of the cGMP-gated channel by divalent cations from the cytoplasmic side, coupled with the ability of these ions to permeate through the channel, suggested that the divalent cations entered the open channel and blocked it from within. The same mechanism might explain the outward rectification; that is, the flatness of the current–voltage relation at hyperpolarizations was likely to be due to blockage of the channel by extracellular divalent cations entering the channel at these voltages. This idea was tested by removing divalent cations from both sides of a membrane patch.

Figure 3 shows an experiment of this kind on a rod membrane patch,[12] in which both sides of the patch were exposed to identical NaCl solutions with all divalent cations omitted. As shown in Fig. 3, most of the outward rectification was indeed absent under these conditions. The amplitude of the cGMP-induced current was also generally much larger in these experiments. Thus, the rod channel is susceptible to blockage by divalent cations from both sides of the membrane, and its insensitivity to cytoplasmic divalent cations in physiological conditions is simply due to the negative membrane potential. By analogy to Ca^{2+} channels,[32,33] this blockage is likely to result from a higher affinity of binding site(s) within the channel for divalent cations over monovalent cations, as a result of which the times required for their transit through the channel are relatively long, thus reducing the overall ion flux rate.

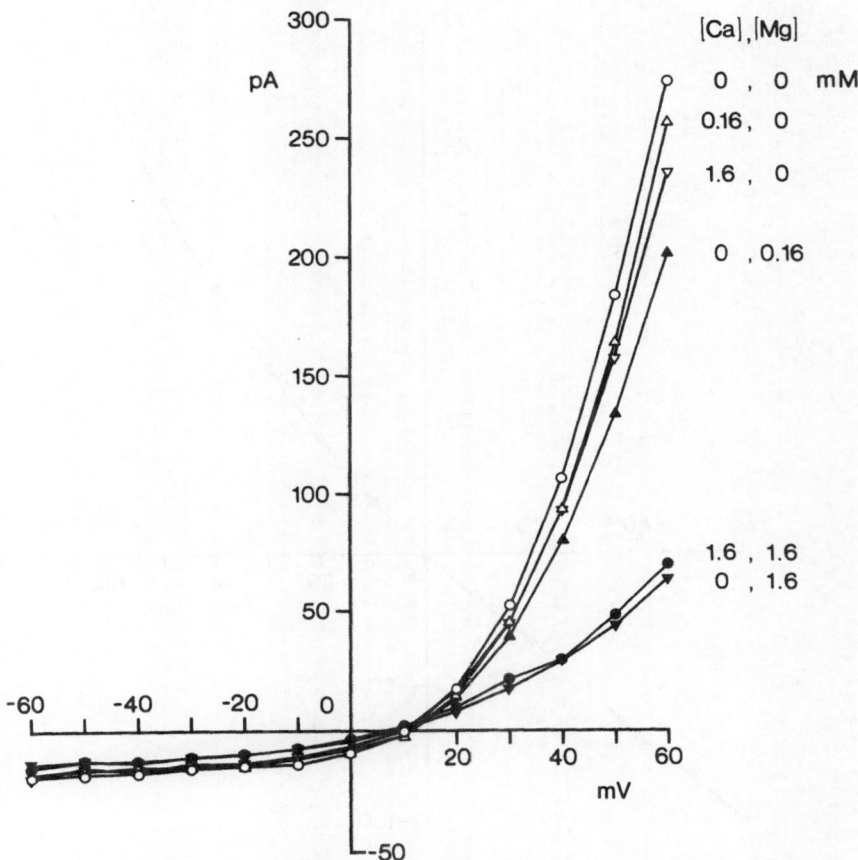

Figure 2. Effect of divalent cations on the cGMP-activated current in a rod patch. The pipette contained normal Ringer's solution, and the bath contained a succession of pseudointracellular solutions with the indicated concentrations of Mg^{2+} and Ca^{2+}. 1000 μM cGMP. Adjacent points are joined by straight lines. (From Ref. 12.)

Recently, Zimmerman and Baylor[34] have developed a quantitative model based on this idea. Similar experiments on the cone channel have suggested a similar picture (L. W. Haynes and K.-W. Yau, unpublished). The difference in rectification between the rod and cone conductances under physiological conditions must reflect a structural difference in their respective pore structures.

5. SINGLE-CHANNEL CHARACTERISTICS

One puzzle about the light-regulated conductance was the very small unit conductance (ca. 0.1 pS) derived from measurements in intact rods or in excised patches of rod membrane with physiological solutions.[25,35–37] The discovery that divalent cations

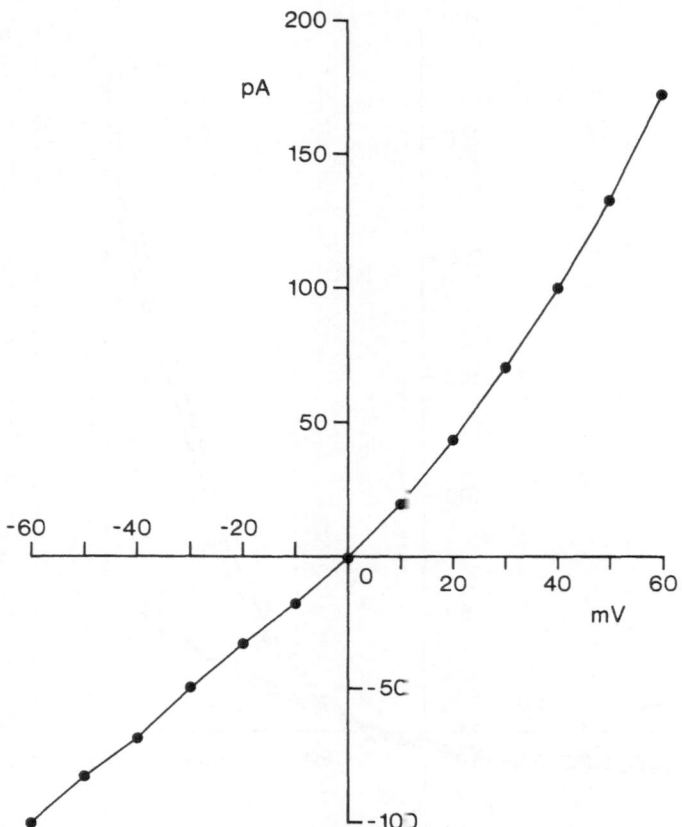

Figure 3. Current–voltage relation for the cGMP-activated current in the absence of divalent cations. Identical, isotonic NaCl solutions were present on both sides of a rod patch. 1000 μM cGMP. Adjacent points are connected by straight lines. (From Ref. 12.)

block the conductance, however, suggested that this blockage may underlie the small apparent unit conductance. This speculation was confirmed by the experiments of Fig. 4, in which an excised rod or cone membrane patch was exposed to a low concentration of cGMP in the complete absence of divalent cations [11,13] Part A shows the results from a rod patch, and part B shows those from a cone patch. In either case the baseline was quiet in the absence of cGMP, but brief steps of current indicative of single-channel activity became apparent in the presence of the ligand. These steps of current could be separated into at least two sizes, with the smaller events about 30% the size of the larger ones. The large events had a conductance of about 25 pS in rods and about 50 pS in cones. These single-channel conductances seemed to stay roughly constant with voltage.

The events, large or small, occurred both singly and in short bursts. With increasing cGMP concentration, the bursts seemed to retain their characteristics but occurred more frequently. This feature suggests that the burst duration represents the time during

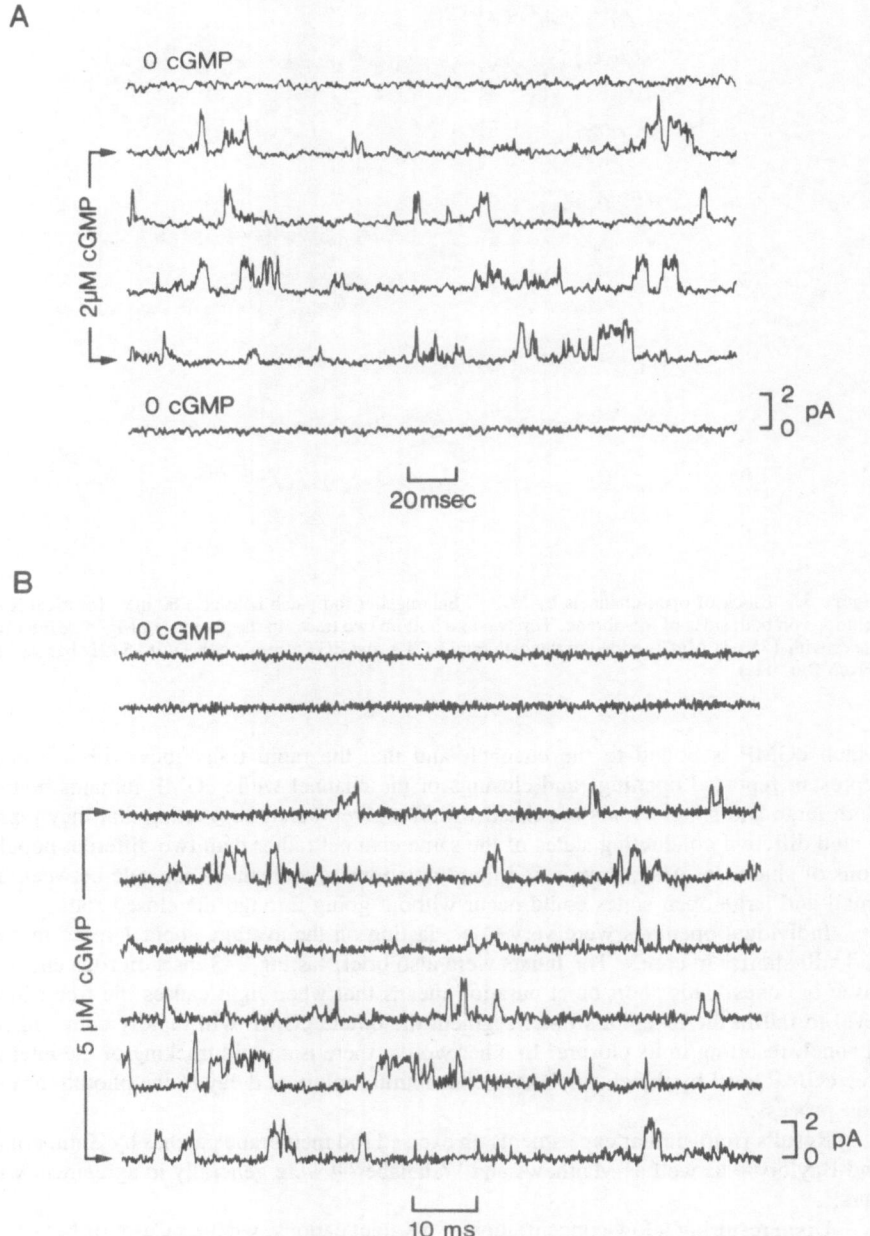

Figure 4. cGMP-induced, single-channel activity recorded from a rod and a cone patch. Identical, isotonic NaCl solutions were present on both sides of each patch. Individual traces are not contiguous. (A) Rod patch held at +60 mV. DC to 2 kHz bandwidth. (From Ref. 11.) (B) Cone patch held at +30 mV. DC to 5 kHz bandwidth.

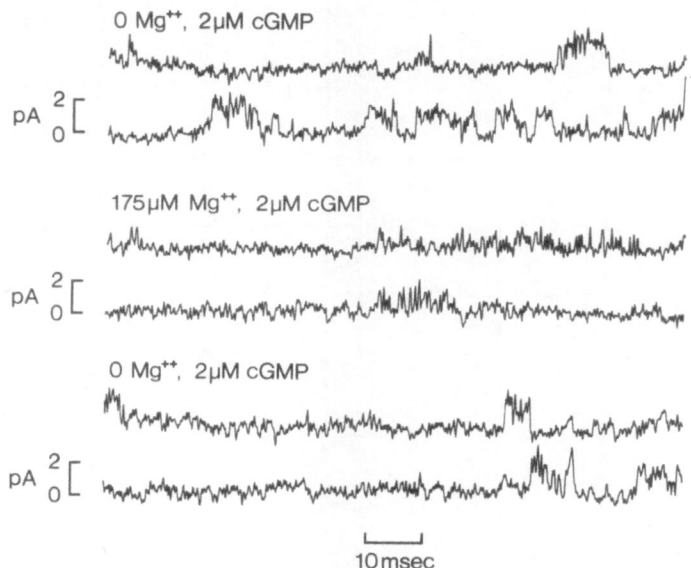

Figure 5. Block of open channels by Mg^{2+}. Salamander rod patch held at $+60$ mV. Identical NaCl solutions on both sides of membrane. Top two and bottom two traces in the absence of Mg^{2+}, center two traces with 175 μM $MgCl_2$ added to the bath (and EDTA and EGTA removed). DC to 5 kHz bandwidth. (From Ref. 11.)

which cGMP is bound to the channel, and that the rapid transitions within a burst represent repeated openings and closings of the channel while cGMP remains bound. Both large and small events coexisted within a given burst, suggesting that they represented different conducting states of the same channel rather than two different populations of channels. Also, it appeared that transitions of a channel molecule between the small and large open states could occur without going through the closed state.

Individual openings were very brief, lasting on the average about 1 msec in rods and still shorter in cones. The bursts were also brief, lasting 2–3 msec in rods and 1–2 msec in cones. This short burst duration means that when light causes the free cGMP level to fall in the receptor's outer segment the bound cGMP will rapidly come off the channel, resulting in its closure. In other words, there is a rapid tracking of the internal free cGMP level by the channel, thus contributing minimal delay to the phototransduction process.

Results from similar experiments on excised rod membrane patches by Zimmerman and Baylor[19] as well as Matthews and Watanabe[22] were generally in agreement with ours.

Upon restoring a low concentration of divalent cations, whether Ca^{2+} or Mg^{2+}, to an excised patch, the single-channel activity was largely suppressed and replaced by bursts of high-frequency flickers (Fig. 5). This phenomenon is consistent with the idea mentioned earlier that divalent cations intermittently enter the open channel and block it by virtue of a lower exit rate.[11] A similar kind of flicker block by divalent cations with probably the same underlying mechanism had previously been described for the Ca^{2+} channel.[38]

6. DEPENDENCE OF CONDUCTANCE ACTIVATION ON cGMP CONCENTRATION

The requirement for cGMP, in order for the channel to open, is stringent, with the probability of opening in the absence of cGMP being near zero.[11,19] Furthermore, in the continuous presence of cGMP there is little sign of desensitization.[11,19] The relation between conductance activation and cGMP concentration is quite steep. Figure 6 shows the relation obtained in an excised rod patch in the absence of divalent cations.[11] The smooth curve is the Hill equation, with a coefficient of 3.0. In the presence of divalent cations the Hill coefficient is somewhat variable, ranging about 1.7 to 3.1.[12] It is possible that divalent cations have an effect on the cooperativity. As for the cone conductance, a Hill coefficient of about 2.5 was obtained in the absence of divalent cations (L. W. Haynes and K.-W. Yau, unpublished); a range similar to that for the rod conductance was otherwise obtained in the presence of divalent cations.[9] At present it is not certain how many cGMP binding sites are on the channel molecule, but the number has to be at least three, possibly four. It is also not certain whether the channel can open without being fully liganded.

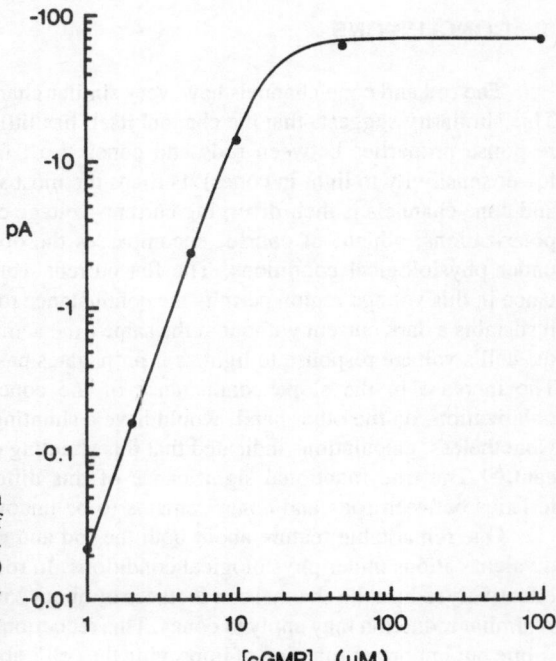

Figure 6. Dependence of current on cGMP concentration. Log–log plot of mean current as a function of cGMP concentration for a rod patch held at -30 mV. The smooth curve is the Hill equation, with $K_{\frac{1}{2}} = 15$ μM, $I_{max} = 67$ pA, and a Hill coefficient of 3.0. (From Ref. 11.)

The half-saturating cGMP concentration, or $K_{\frac{1}{2}}$, ranges from 10 to 50 μM for the rod conductance.[11,12] The value for the cone conductance is slightly higher.[9] The reason underlying the variability in $K_{\frac{1}{2}}$ from patch to patch is unclear at present. We have found that depolarization promotes the activation of the channel by cGMP in both rods and cones, leading to a reduction in the $K_{\frac{1}{2}}$ value.[9,11,12] Karpen et al.[23] have attributed this to a decrease in the closing rate of the liganded channel with depolarization.

All in all, the similarities between the rod and cone channels with respect to their activation characteristics are striking.

7. CHANNEL DENSITY

An estimate of the number of channels on an excised patch could be estimated from the saturating current elicited by cGMP. A rough estimate of the patch area could also be estimated from the size of the lumen at the pipet tip. From these two estimates we could obtain the channel density, which on average is about 1000 μm^{-2} in rods[11] and about 60 μm^{-2} in cones (L. W. Haynes and K.-W. Yau, unpublished data). Our estimate on the rod channel density is in broad agreement with those made in intact rods.[25,36,37] The large difference in channel densities between rods and cones is striking, but, interestingly, their ratio is roughly the inverse of the surface/area ratio for the rod and cone outer segments, suggesting similar total numbers of channels in the two types of receptors.

8. CONCLUSIONS

The rod and cone channels have very similar characteristics of activation by cGMP. This similarity suggests that the channel itself has little to do with the difference in light response properties between rods and cones (i.e., faster response kinetics and much lower sensitivity to light in cones). Perhaps the most striking difference between the rod and cone channels is their diverging current–voltage characteristics at membrane hyper-polarizations, which, of course, encompasses the operating voltage range of the cells under physiological conditions. The flat current–voltage relation for the rod conductance in this voltage region permits the conductance to act as a current generator, that is, it sustains a dark current without at the same time acting as an electrical shunt to degrade the cell's voltage response to light as it propagates passively through the outer segment. The increase in the slope conductance of the cone channel with membrane hyper-polarization, on the other hand, would have a shunting effect on the propagating signal. Nonetheless, calculations indicated that this shunting effect is rather weak and insignificant.[9] The true functional significance of this difference in current–voltage characteristics between rods and cones remains to be uncovered.

One remarkable feature about both the rod and cone channels is their blockage by divalent cations under physiological conditions. In rods, the result of this blockage is to effectively reduce the channel conductance by about 250 times (i.e., from 25 pS to 0.1 pS). A similar reduction may apply to cones. This reduction in the channel conductance has the subtle but important function of improving the cell's ability to detect light. Let us examine this point with rods as an example. At a membrane potential of about -40 mV in the dark, the single-channel currents corresponding to the blocked and unblocked cases would be 4 fA and 1 pA, respectively. For a normal dark current of about 40 pA,[3] the average number of channels open at a given time is therefore 10,000 in the blocked case, versus only 40 in the unblocked case. From the Poisson distribution, the standard deviation of the number of open channels is equal to the square root of the mean. Thus, with 10,000 mean open channels each having a unit current of 4 fA, the standard deviation of the dark current fluctuations will be 100×4 fA, or 0.4 pA. With 40 mean open channels each having a unit current of 1 pA, on the other hand, the standard deviation of the current fluctuations would be 6.3 pA, or more than an order of magnitude higher. In other words, the background electrical noise is substantially lower with smaller (and therefore more numerous) channels, thus improving the signal-to-noise ratio for light detection. Of course, the blockage by divalent cations would also contribute electrical noise to the cell. However, as shown earlier, most of this noise is of relatively high frequency and is filtered out by the membrane time constant of the cell.

REFERENCES

1. Tomita, T., 1970, Electrical activity of vertebrate photoreceptors, *Quart. Rev. Biophys.* **3:** 179–222.
2. Hagins, W. A., Penn, R. D., and Yoshikami, S., 1970, Dark current and photocurrent in retinal rods, *Biophys. J.* **10:** 380–412.
3. Baylor, D. A., Lamb, T. D., and Yau, K.-W., 1979, The membrane current of single rod outer segments, *J. Physiol.* **288:** 589–611.
4. Pugh, Jr., E. N., and Cobbs, W. H., 1986, Visual transduction in vertebrate rods and cones: A tale of two transmitters, calcium and cyclic-GMP, *Vision Res.* **26:** 1613–1643.
5. Stryer, L., 1986, Cyclic GMP cascade in vision, *Ann. Rev. Neurosci.* **9:** 87–119.
6. Yau, K.-W., and Baylor, D. A., 1989, Cyclic-GMP-activated conductance of retinal photoreceptor cells, *Ann. Rev. Neurosci.* **12:** 289–327.
7. Yau, K.-W., Haynes, L. W., and Nakatani, K., 1988, Study of the phototransduction mechanism in rods and cones, in: *Proceedings of the Retina Research Foundation Symposium* (D. M. K. Lam, ed.) Portfolio Publishing, Woodlands, TX, **1:** 41–58.
8. Nakatani, K., and Yau, K.-W., 1985, cGMP opens the light-sensitive conductance in retinal rods, *Biophys. J.* **47:** 356a.
9. Haynes, L. W., and Yau, K.-W., 1985, Cyclic-GMP-sensitive conductance in outer segment membranes of catfish cones, *Nature* **317:** 61–64.
10. Yau, K.-W., and Haynes, L. W., 1986, Effect of divalent cations on the macroscopic cGMP-activated current in excised rod membrane patches, *Biophys. J.* **49:** 33a.
11. Haynes, L. W., Kay, A. R., and Yau, K.-W., 1986, Single cGMP-activated channel activity in excised patches of rod outer segment membranes, *Nature* **321:** 66–70.
12. Yau, K.-W., Haynes, L. W., and Nakatani, K., 1986, Roles of calcium and cyclic-GMP in visual transduction, in: *Membrane Control of Cellular Activity* (H. Ch. Luettgau, ed.), Gustav Fischer, Stuttgart, pp. 343–366.
13. Haynes, L. W., and Yau, K.-W., 1987, Single cGMP-activated channel activity recorded from excised cone membrane patches, *Biophys. J.* **51:** 18a.
14. Fesenko, E. E., Kolesnikov, S. S., and Lyubarsky, A. L., 1985, Induction by cyclic-GMP of cationic conductance in plasma membrane of retinal rod outer segment, *Nature* **313:** 310–313.
15. Zimmerman, A. L., Yamanaka, G., Eckstein, F., Baylor, D. A., and Stryer, L., 1985, Interaction of hydrolysis-resistant analogs of cyclic-GMP with the phosphodiesterase and light-sensitive channel of retinal rod outer segments, *Proc. Natl. Acad. Sci. U.S.A.* **82:** 8813–8817.

16. Fesenko, E. E., Kolesnikov, S. S., and Lyubarsky, A. L., 1986, Direct action of cGMP on the conductance of retinal rods plasma membrane, *Biochem. Biophys. Acta* **856:** 661–671.

17. Matthews, G., 1986, Comparison of the light-sensitive and cyclic-GMP-sensitive conductance of the rod photoreceptor: Noise characteristics, *J. Neurosci.* **6:** 2521–2526.

18. Stern, J. H., Kaupp, U. B., and MacLeish, P. R., 1986, Control of the light-regulated current in rod photoreceptors by cyclic-GMP and L-*cis*-Diltiazem, *Proc. Natl. Acad. Sci., U.S.A.* **83:** 1163–1167.

19. Zimmerman, A. L., and Baylor, D. A., 1986, Cyclic-GMP-sensitive conductance of retinal rods consists of aqueous pores, *Nature* **321:** 70–72.

20. Stern, J. H., Knutsson, H., and MacLeish, P. R., 1987, Divalent cations directly affect the conductance of excised patches of rod photoreceptor membrane, *Science* **236:** 1674–1678.

21. Matthews, G., 1987, Single-channel recordings demonstrate that cGMP opens the light-sensitive ion channel of the rod photoreceptor, *Proc. Natl. Acad. Sci. U.S.A.* **84:** 299–302.

22. Matthews, G., and Watanabe, S.-I., 1987, Properties of ion channels closed by light and opened by guanosine 3′,5′-cyclic monophosphate in toad retinal rods, *J. Physiol.* **389:** 691–716.

23. Karpen, J. W., Zimmerman, A. L. Stryer, L., and Baylor, D. A., 1988, Gating kinetics of the cyclic-GMP-activated channel of retinal rods: Flash photolysis and voltage-jump studies, *Proc. Natl. Acad. Sci., U.S.A.* **85:** 1287–1291.

24. Bader, C. R., MacLeish, P. R., and Schwartz, E. A., 1979, A voltage-clamp study of the light response in solitary rods of the tiger salamander, *J. Physiol.* **296:** 1–26.

25. Bodoia, R. D., and Detwiler, P. B., 1985, Patch-clamp recordings of the light-sensitive dark noise in retinal rods from lizard and frog, *J. Physiol.* **367:** 183–216.

26. Baylor, D. A., and Nunn, B. J., 1986, Electrical properties of the light-sensitive conductance of rods of the salamander *Ambystoma tigrinum.*, *J. Physiol.* **371:** 115–145.

27. Attwell, D., Werblin, F. S., and Wilson, M., 1982, The properties of single cones isolated from the tiger salamander retina, *J. Physiol.* **328:** 259–283.

28. Yau, K.-W., and Nakatani, K., 1984, Cation selectivity of light-sensitive conductance in retinal rods, *Nature* **309:** 352–354.

29. Hodgkin, A. L., McNaughton, P. A., and Nunn, B. J., 1985, The ionic selectivity and calcium dependence of the light-sensitive pathway in toad rods, *J. Physiol.* **358:** 447–468.

30. Jack, J. J. B., Noble, D., and Tsien, R. W., 1975, *Electric Current Flow in Excitable Cells,* Clarenden, Oxford.

31. Yau, K.-W., and Nakatani, K., 1984, Electrogenic Na–Ca exchange in retinal rod outer segment, *Nature* **311:** 661–663.

32. Almers, W., and McCleskey, E. W., 1984, Nonselective conductance in calcium channels of frog muscle: Calcium selectivity in a single-file pore, *J. Physiol* **353:** 585–608.

33. Hess, P., and Tsien, R. W., 1984, Mechanism of ion permeation through calcium channels, *Nature* **309:** 453–456.

34. Zimmerman, A. L., and Baylor, D. A., 1988, Ionic permeation in the cGMP-activated channel of retinal rods, *Biophys. J.* **53:** 472a.

35. Detwiler, P. B., Conner, P. B., and Bodoia, R. D., 1982, Gigaseal patch-clamp recording from outer segments of intact retinal rods, *Nature* **300:** 59–61.

36. Gray, P., and Attwell, D., 1985, Kinetics of light-sensitive channels in vertebrate photoreceptors, *Proc. Roy. Soc. Lond. B.* **223:** 379–388.

37. Zimmerman, A. L., and Baylor, D. A., 1985, Electrical properties of the light-sensitive conductance of salamander retinal rods, *Biophys. J.* **47:** 357a.

38. Lansman, J. B., Hess, P., and Tsien, R. W., 1985, Direct measurement of entry and exit rates for calcium ions in single calcium channels, *Biophys. J.* **47:** 67a.

Chapter 4

Functional Aspects of the cGMP-Activated Channel from Bovine Rod Outer Segments

W. Hanke and R. Simmoteit

1. INTRODUCTION

The cGMP-activated channel of the plasma membrane of bovine rod outer segments is responsible for the change of the electrical potential of this membrane,[1] which occurs after the absorption of light by rhodopsin in the disk membrane. The plasma membrane hyperpolarizes due to the closing of such channels, when the cGMP concentration drops as a consequence of the activation of an enzyme cascade.

The cGMP-activated channel has been characterized previously in some detail by electrophysiological and biochemical experiments and flux studies.[2–5]

As a strategy to obtain the data necessary to create a gating model of channel activity, one of the major goals in investigating ion channels, it is important to have extensive single-channel data available, as well as information regarding the behavior of multichannel systems.[6] Single-channel data provide some basic parameters that are useful in developing a gating model and that cannot be obtained from multichannel recordings. Figure 1 presents a general strategy for investigating ligand-activated ion channels and includes a variety of electrophysiological and biochemical techniques.

Focusing on electrophysiological techniques, the scheme in Fig. 2 illustrates an application to the problem of creating a structure–function model for a particular ion channel, the cGMP-activated channel from rod outer segments. From Fig. 1 it is clear

W. HANKE and R. SIMMOTEIT • Universität Osnabrück, FB Biologie, Biophysik, D-4500 Osnabrück, Federal Republic of Germany.

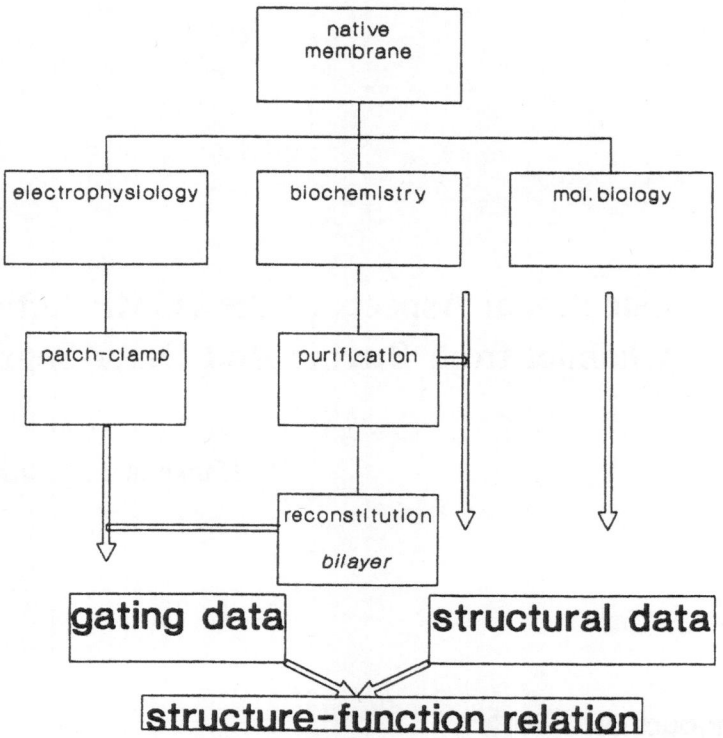

Figure 1.   General strategy for investigating ion channels with different techniques.

that electrophysiological data are insufficient to understand the function of a channel-forming protein, and other experimental techniques appear necessary. In the case of the cGMP-activated channel from vertebrate photoreceptors, SDS-gel electrophoresis—after purification of the protein using biochemical methods—revealed that the channel is most probably composed of one subunit of 65 kDa in the form of a homooligomer.

Flux measurements opened a way for the functional purification of the protein and demonstrated that the channel is permeable to divalent cations. At present, however, no detailed pharmacology of the channel is available. Drugs are usually necessary tools to purify and investigate ion channels, as has been shown for other ion channels (acetyl-choline receptor, AChR).[7]

Following the scheme of Fig. 2, we have performed patch-clamp experiments and studied excised patches from bovine rod outer segment plasma membranes.[5] Similar experiments have been done previously by others in amphibian rods.[2,8,9] Because the channel density in the plasma membrane is very high, about 200 channels μm^{-2},[2,9,10] it is very difficult to obtain single-channel data in patch-clamp experiments. Detailed multichannel data, however, have been obtained with regards to the pharmacology and selectivity of the cGMP-activated channel. In amphibian cones, Haynes and Yau (see

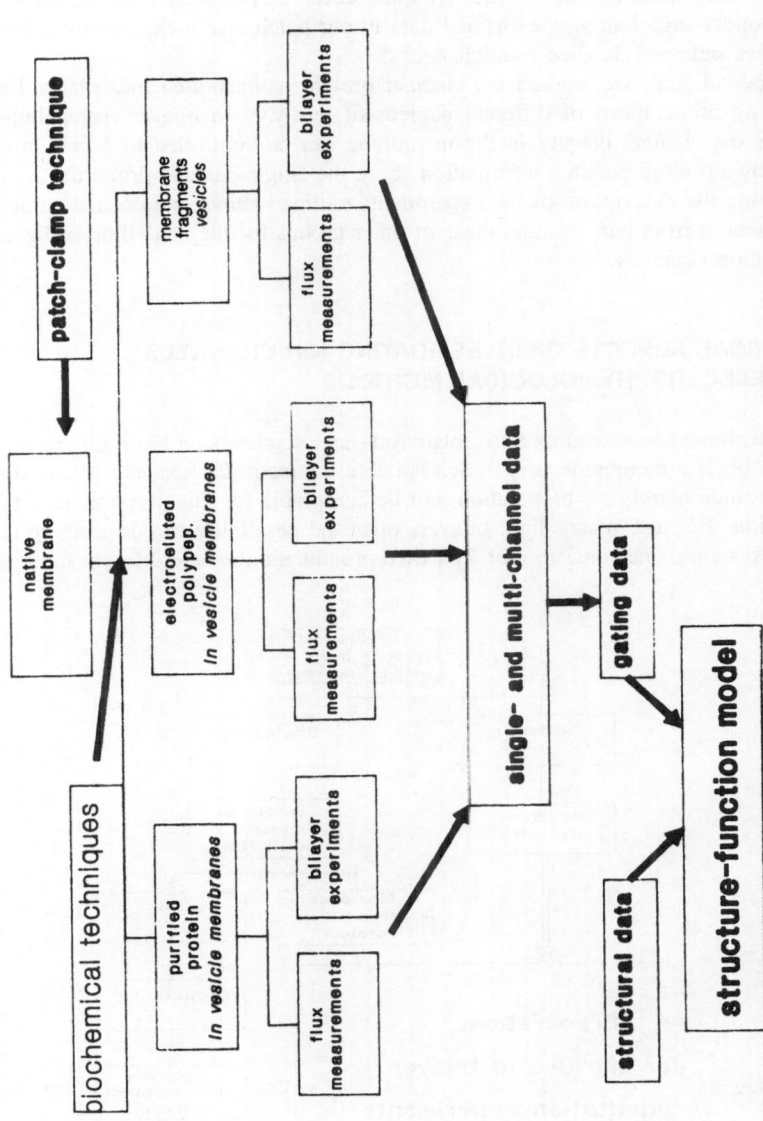

Figure 2. Scheme presenting ways of investigating ligand-activated channels by use of electrophysiological techniques. In addition to these, biochemical and genetic techniques must be included to reach the final goal of creating a structure–function model of such channels. This is indicated in the lower part of the left side of the figure.

Chapter 3, this volume) have been able to perform single-channel patch-clamp experiments, due to the lower density of the cGMP-activated channel in their membranes.

In addition to the high channel density in bovine rod outer segments, the size of these cells is very small and the patches are quite unstable. For these reasons, we were unable to collect sufficient single-channel data in our patch-clamp experiments. However, we have obtained detailed multichannel data.

In a second step, we studied the channel protein reconstituted into planar lipid bilayers using preparations of different degrees of purity.[4] In bilayer reconstitution experiments the channel density in the membrane can be controlled to a certain extent,[11] allowing more detailed information about the single-channel properties.

Following the description of the experiments outlined above, we discuss some of the consequences from our data and those of other groups for the modelling of ligand-activated cation channels.

2. TECHNICAL ASPECTS OF INVESTIGATING ION CHANNELS WITH ELECTROPHYSIOLOGICAL METHODS

Methods have been developed to isolate rod outer segments.[3] From this material one has to obtain a membrane preparation suitable for reconstitution into planar lipid bilayers. A crude membrane preparation will be acceptable for this purpose, as a first approximation. Because planar lipid bilayers open the possibility of constructing two component systems, one consisting of a purified protein and the other of pure lipid, the

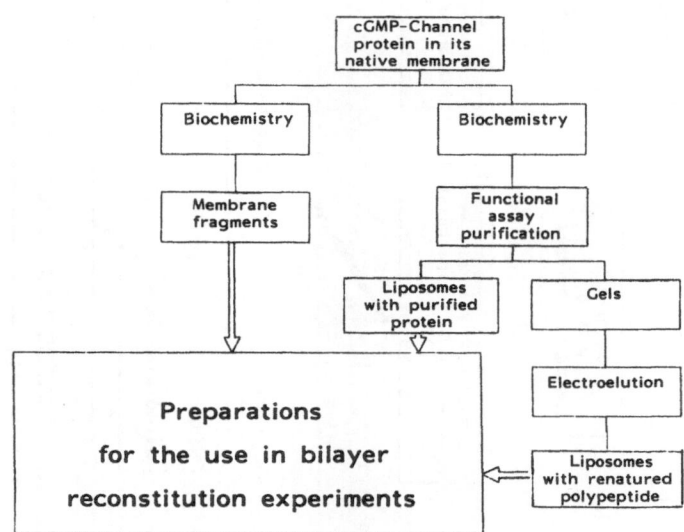

Figure 3. Sketch of preparations to be used in bilayer reconstitution experiments. Starting from the native membrane, the protein may be used at different degrees of purity for the experiment, from crude membrane preparations to the electroeluted and renatured polypeptide.

next step should be the purification of the protein. This was done for the cGMP-activated channel using a functional assay.[3] The resulting protein was represented on SDS-gels by a single polypeptide band (of about 65 kDa). This polypeptide band was electroeluted, in a final step, from SDS-gels, renatured and incorporated into the membranes of liposomes, which were used in the bilayer experiments.

The investigation of a two-component system has the advantage that it may be described theoretically, something that cannot be done for biological membranes because of their complex composition. Thus, planar lipid bilayer experiments using purified proteins are important not only for investigating the channel itself, but also its interactions with the surrounding lipid environment.

Figure 3 gives a sketch of different ways of obtaining, from bovine rod outer segments, preparations containing the cGMP-activated channel to be used in bilayer reconstitution experiments.

3. BIOCHEMISTRY AND PHARMACOLOGY OF THE cGMP-ACTIVATED CHANNEL

In a first series of experiments, we have characterized the channel using biochemical methods. Most probably, the channel is composed of only one polypeptide (65 kDa) in the form of a homooligomer (pentamer). Figure 4 shows a SDS-gel of the purified protein.

Flux measurements showed that the cooperativity of channel activation by cGMP was $n = 2-4$. A number of cGMP analogues and other nucleotides were tested, and it was concluded that the channel can be effectively activated by cGMP and analogues of this molecule modified *only* in the C-8 position.[12] The same was found for the channel in the native membrane. In this latter case, cGMP concentration is controlled by two enzymes: phosphodiesterase and guanylate cyclase. Phosphodiesterase, which hydrolyzes cGMP, is activated via an enzyme cascade after the absorption of a photon by rhodopsin. Guanylate cyclase raised cGMP concentration. The activity of guanylate cyclase is controlled by calcium, as has been shown in electrophysiological experiments.

At present, only one drug has been found to block the channel, I-*cis* Diltiazem, which has an effect at micromolar concentrations.[13] Such a pharmacology is lost upon purification of the channel protein.[3] The loss of I-*cis* Diltiazem sensitivity after purification is only partially understood. The lack of more drugs that could be used as tools to investigate the channel constitutes a major problem when studying the cGMP-activated channel.

4. MULTICHANNEL EXPERIMENTS

Patch-clamp experiments were performed with very small pipettes (of a tip resistance of 25 Mohm) Due to the small size of the cells (5 μm × 1 μm). In the basic experiments, aqueous solutions contaqined 118 mM NaCl, buffered at pH = 7.4. We added a number of pharmacological agents to the bath solution in different concentrations. Current–voltage relations were obtained by giving a voltage ramp to the mem-

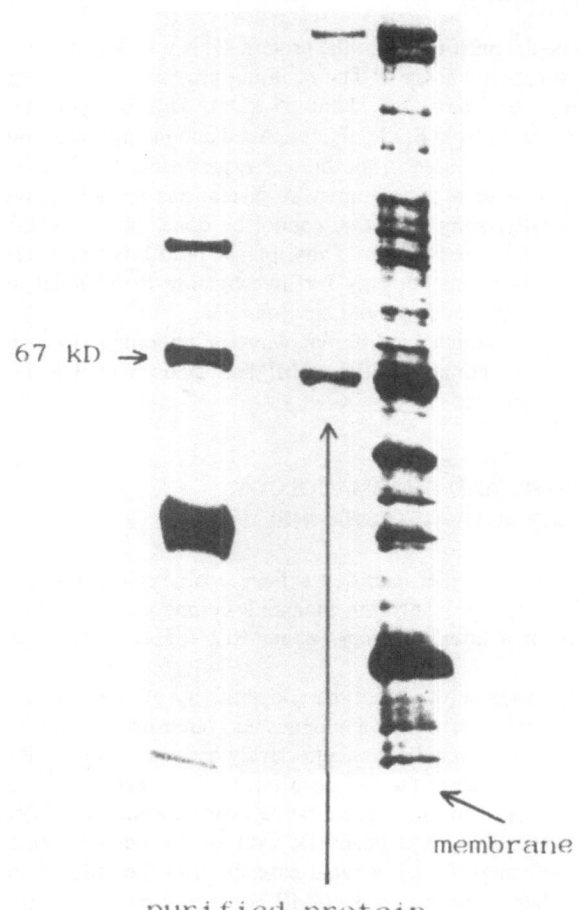

67 kD →

membrane

purified protein

Figure 4. SDS-gel of the purified cGMP-activated channel protein. One band at about 65 kDa represents the channel. In some gels we found an additional band slightly over 200 kDa, which was identified as a spectrinlike protein. The presence or absence of this band did not affect the reconstitution efficiency.

brane and recording current versus voltage in an FM tape recorder. As shown in Fig. 5, a set-up for fast bath solution exchange (less than 100 msec) was used, which allowed us to take complete series of measurements using a variety of different solutions for the same membrane. This was necessary because there was a significant dispersion in the data from patch to patch, especially with regards to the K_D of cGMP binding. In addition, the stability of our patches was usually limited to a few minutes.

We obtained I/V curves from a series of experiments at different 8-Br-cGMP concentrations (a cyclic GMP derivative which is not hydrolyzed and which activates the

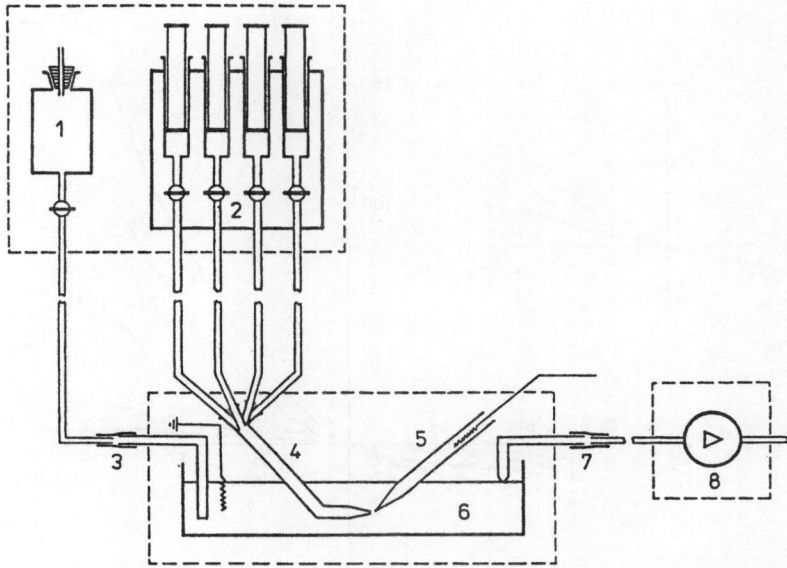

Figure 5. Set-up for fast bath solution exchange in patch-clamp experiments. The excised patch is only shifted from one capillary to the next for solution exchange. This operation can be done in much less than 1 sec.

channel more effectively than cGMP[14]). An example of this is shown in Fig. 6A, in the presence (inset, upper left corner) and in the absence of divalent cations. It can be seen that in both cases the I/V curve slightly rectifies and that divalent cations (1 mM) reduce the current carried by sodium at least threefold at all potentials tested. From the I/V curves, a Hill plot for channel activation was constructed, as shown in Fig. 6B, in which a slope of about 2.5 was obtained. The K_D from this plot is about 2 μM; for cGMP the data were $n = 2.5$ and $K_D = 35$ μM. In other experiments we obtained I/V curves in the presence of different monovalent cations in the bath solution and sodium inside the pipette. From such curves the permeability ratios, according to the Goldmann equation, could be calculated, as well as the current ratios at a fix potential. The same was done for the permeability of the channel for divalent cations. We were able to show that both monovalent and divalent cations permeate the channel.

In pharmacological experiments, we verified that the channel could be blocked by I-*cis* Diltiazem, although much less effectively by *d-cis* Diltiazem and by other drugs. Furthermore, we demonstrated the ability of divalent cations (calcium and magnesium, millimolar concentrations, respectively) to block the current carried by monovalent cations (see Fig. 6A).

The data from these experiments have been listed together with some bilayer and flux data in Table I.

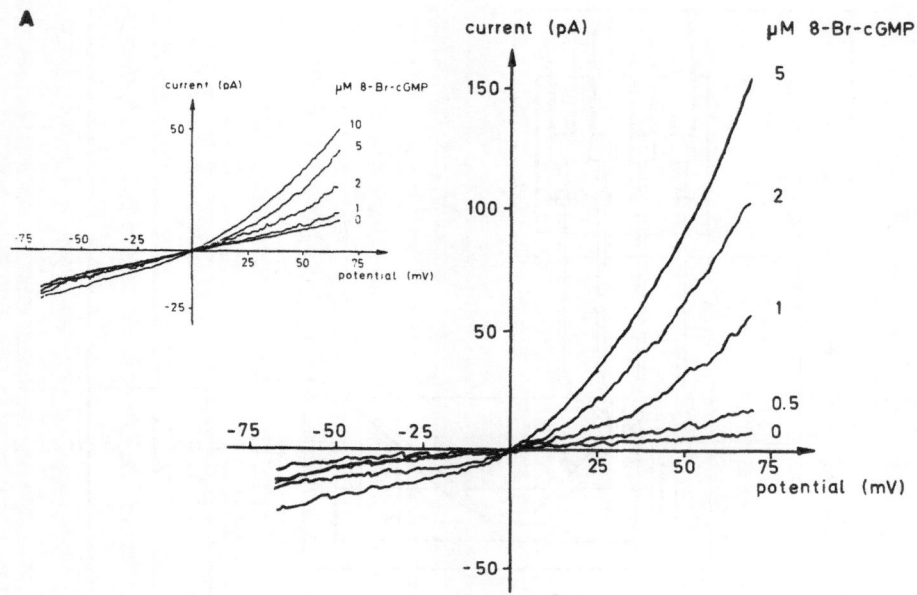

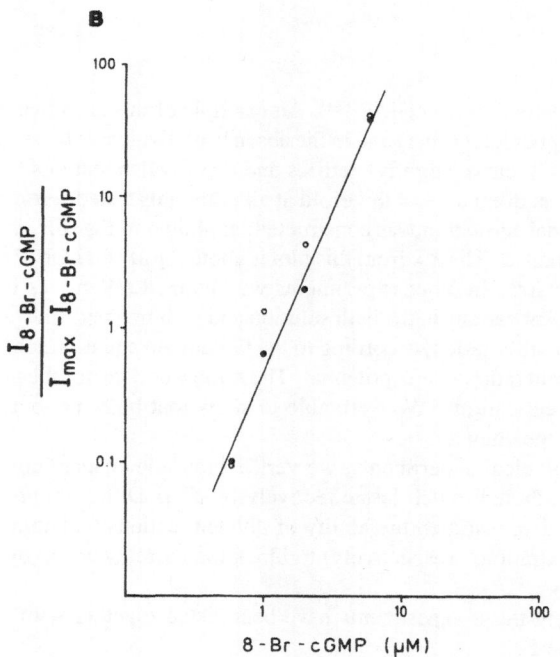

Table I. Properties of the cGMP-Activated Channel from Bovine Rod Outer Segments

			Bilayer		
	Patch-clamp	Flux	Membrane fragment	Purified	Electroeluted
Conductance (S)	25	—	25	25	25
Cooperativity	>2	>3	>2	>3	—
K_D(μM cGMP)	35	35	40	45	—
Open-state lifetime	3 msec	—	5 msec	5 msec	2 msec
Selectivity	Na>K>Rb>Cs	—	—	Na>K>Rb>Cs	—
Ca^{2+} permeability	+	+	—	+	—
Blockage by I-*cis*					
Diltiazem	2+	+	—	+	—
Ca^{2+} blockage	+	—	+	+	—
I/V characteristics with and without Ca^{2+}	Rectifies	—	Rectifies	Rectifies	—

5. SINGLE-CHANNEL STUDIES

Due to the high channel density in the native membrane of the bovine rod outer segments, it was very difficult to obtain single-channel recordings in patch-clamp experiments. Other groups succeeded, however, in recording single-channel currents from amphibian cones (Haynes and Yau, Chapter 3 in this volume). A channel conductance of about 25 pS and open-state lifetimes in the millisecond range were found in the absence of divalent cations.

To create a detailed gating model, as stated above, it is a prerequisite to have detailed single-channel data available. We obtained these data from bilayer reconstitution experiments.[4] The channel protein was used at different degrees of purity, from membrane fragments up to the highly purified protein (see Fig. 3). A scheme of our bilayer experiments is given in Fig. 7.

A channel conductance of 25 pS was found in symmetrical 150 mM NaCl and state-lifetimes in the 5–10 msec range at low agonist concentration. The channel was activated by cGMP in a cooperative way with $n > 2$ and a K_D of about 45 μM. These K_D data were obtained from Hill plots using the following equation:

$$\log[(P_0/(1-P_0)] = f(\log [cGMP])$$

Figure 6. (A) I/V-curves of an excised patch from a bovine rod outer segment plasma membrane at different 8-Br-cGMP concentrations (a cGMP derivative that is more effective that cGMP itself in activating the channel). In the upper left corner the data in the presence of 1 mM Ca^{2+} are shown. The current induced is reduced dramatically by calcium. When no calcium is present in the pipette and in the bath (right part), the currents are significantly larger, but the rectification of the I/V-curves remains. (B) The data from part (A) of this figure (without Ca^{2+}) are replotted in a Hill plot. A slope of $n = 2.5$ and a $K_D = 2\mu M$ were obtained from this plot.

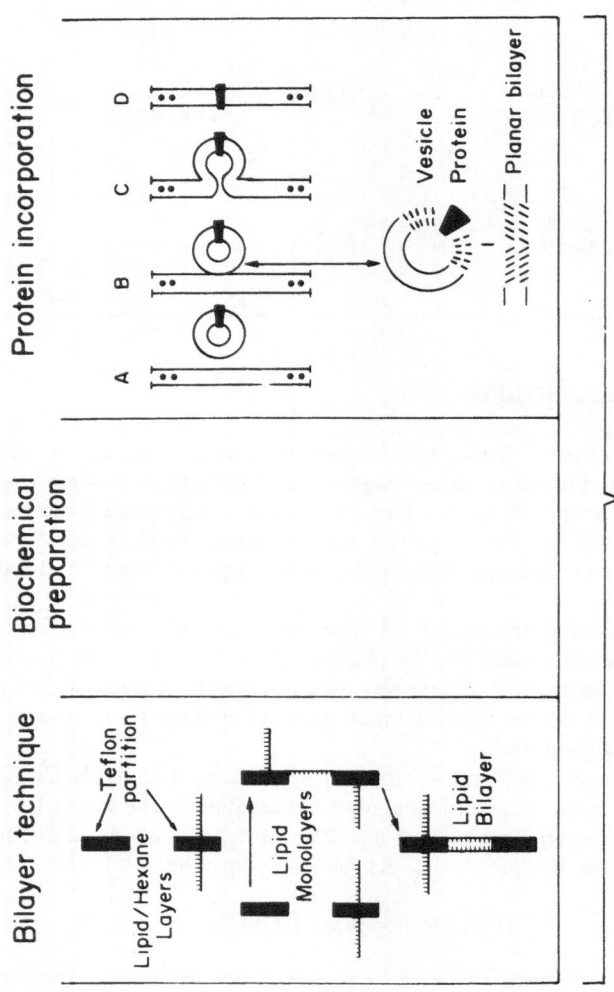

Figure 7. Principal steps of a bilayer reconstitution experiment. The basic set-up and protein incorporation technique are shown. (9,20)

Besides the predominant channel behavior, we found a complex substructure of channel gating and conductance. A long trace of such behavior is shown in Fig. 8A, together with an amplitude histogram (Fig. 8B) and a scatter plot of the conductance of each open-state event plotted as a function of its duration (Fig. 8C). Clearly, the amplitude histogram has more than one open-state peak. Whether the wide peak between 5 and 10 pS is due to one or more open-state levels cannot be decided with our data. However, the predominant open-state peak is clearly located at about 25 pS. Again in the scatter plot at least two channel populations can be resolved, one with lower amplitudes and shorter lifetimes and another with larger amplitudes and longer lifetimes. These data about channel lifetime and channel conductance verify that the cGMP-activated channel has more than one open state. The open-channel states may differ in the number of ligands bound, in addition to the differences in conductance and lifetimes, because we found that the occurrence of sublevels was dependent on agonist concentration. Furthermore, in several experiments we found spontaneous channel activity after incorporation of the protein into planar lipid bilayers (no agonist present). These spontaneously active channels displayed the same pharmacology as the cGMP-activated channels, and therefore the possibility of artifacts could be excluded.

In conclusion, from biochemical experiments it is known that the channel protein is represented by a single polypeptyde band of about 65 kDa on denaturing SDS-gels. This polypeptide band can be electroeluted, partially renatured, and incorporated into the membranes of liposomes. After fusion of such liposomes with planar lipid bilayers, it was found a channel activity which was very similar or even identical to that of the native protein obtained in bilayer and patch-clamp experiments.

6. DISCUSSION

From the data presented in the previous section and in the literature[2–4,8–15] we conclude that the cGMP-activated channel is cooperatively activated by up to four ligand molecules. Furthermore, the data suggest that the channel has several open states, depending on the number of ligands bound, which may have smaller conductances than the fully liganded channel. In addition, these states can be discriminated by their different kinetics. The most general model of such behavior, taking into account the spontaneous activity of the channel, would include at least five open states (with none to four ligand molecules bound to the channel protein) and the same number of closed states; additional states may be necessary to describe a possible desensitization of the channel and other effects.

$$R + 4L \; \rightleftharpoons \; R_{1L} + 3L \; \rightleftharpoons \; R_{2L} + 2L \; \rightleftharpoons \; R_{3L} + 1L \; \rightleftharpoons \; R_{4L}$$
$$\updownarrow \qquad\qquad \updownarrow \qquad\qquad \updownarrow \qquad\qquad \updownarrow \qquad\qquad \updownarrow$$
$$R^* + 4L \rightleftharpoons R_{1L}^* + 3L \; \rightleftharpoons \; R_{2L}^* + 2L \; \rightleftharpoons \; R_{3L}^* + 1L \; \rightleftharpoons \; R_{4L}^*$$

The interpretation of such a complicated gating model is very difficult, due to the large number of free variables. It must be an aim for future investigations to provide a simpler model describing the behavior of the channel under physiological conditions. The question of whether complicated gating models describing a channel in all its details

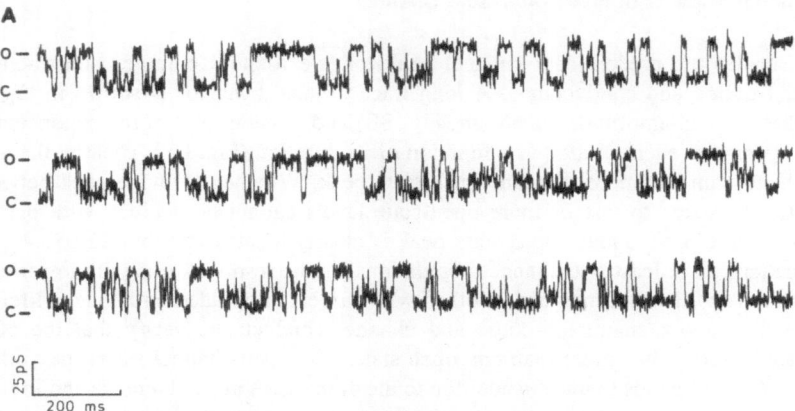

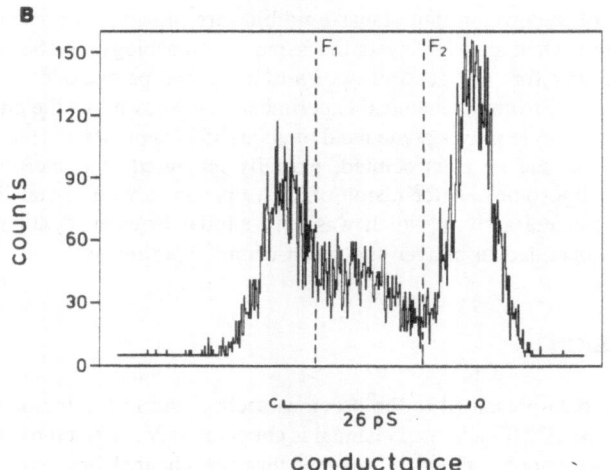

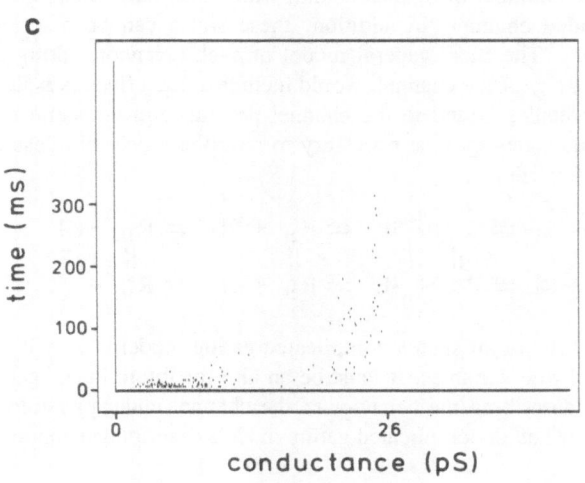

are useful, or whether they should be replaced by somewhat simpler models describing only the predominant channel behavior, should be discussed elsewhere.

Biochemical data have been published,[3] in which evidence is given in support of the notion that the cGMP-activated channel from bovine rod outer segments is most probably composed of four or five identical subunits of about 65 kDa. This ligand-activated channel shows spontaneous activity, as well as open states with one to four ligand molecules bound. It was postulated, as a consequence, that each subunit has one binding site for cGMP. The fully liganded state would have one cGMP molecule bound to each subunit. Perhaps the subunits have different affinities for cGMP for each additional binding of cGMP molecules to the protein.

In this chapter, we have mainly discussed data from our own experiments on the cGMP-activated channel from bovine rods. For this vertebrate channel no additional data have been presented with regards to its electrophysiological properties (bilayer, patch-clamp), except one paper in which bilayer reconstitution was utilized (see Chapter 5, this volume). The data presented are in agreement with our own observations and give additional information about the activation of the channel by the active form of phosphodiesterase. The interpretation of these results is difficult at the present. In addition, this is the only cGMP-activated channel that has been biochemically identified and purified so far.[3,16] The molecular description of this channel is currently under investigation.

However, a substantial amount of electrophysiological data is available about the cGMP-activated channel from amphibian photoreceptors. The channel from that preparation is identical to that from bovine rods in most of its properties, but not in the way in which it is blocked by calcium (see Fig. 6A and Refs. 17, 18).

Very similar behavior has been found for another homooligomeric ligand-activated ion channel: the neuronal (nicotinic) acetylcholine receptor channel from insect nervous tissue.[6,19] This AChR is composed of four (or eventually five) subunits, also of about 65 kDa, which are most probably identical. Some striking similarities between these two channels from totally different sources are the following:

1. both have a homooligomeric structure;
2. both have four (or five) subunits of about 65 kDa;
3. both present spontaneous activity;
4. both bear open states with different conductance and kinetics; and
5. both have open states with no, one, or more ligands bound.

These observations give rise to interesting questions about the principal requirements for ligand-activated ion channels. Thus, as a preliminary working hypothesis one

---

Figure 8. (A) Long recording of a single channel activated by 50 μM cGMP in a planar lipid bilayer. Purified channel protein was used in this experiment (the lipid used was purified soybean lipid, no Ca^{2+} was present). The channel conductance in 150 mM NaCl is about 25 pS; the open-state lifetimes are in the 10-msec range. (B) Amplitude histogram of the above trace, showing one closed-state peak, one major open-state peak at 26 pS, and one wide open-state peak in the range of 5 to 10 pS. (C) Scatter plot of the trace from (A). The lifetime of each open-state event discriminated by findline 1 from (B) of this figure, is plotted as a function of the amplitude of this event. At least two distinct populations can be seen in the plot. If findline 2 from (B) of this figure is used, only one population of channel events is detected (the right one at 26 pS).

can conclude that a homooligomeric structure of four (or five) subunits of 65 kDa is sufficient to do the job of a ligand-activated ion channel. The activation of such a channel by its ligands may be cooperative (cGMP-activated channel, $n \geq 3$) or not (AChR, $n = 1$), as can be observed in the Hill plots. In addition, such channels present spontaneous activity. To find out whether these are general properties of ligand-activated ion channels, especially in neurophysiological tissues, will be an aim of future investigations.

REFERENCES

1. Baylor, D. A., and Fuortes, M. G. F., 1970, Electrical response of single cones in the retina of turtle, *J. Physiol. (London)* **207**: 77–92.
2. Haynes, L. W., Kay, A. R., and Yau, K.-W., 1985, Single cyclic-GMP-activated channel activity in excised patches of rod outer segment membranes, *Nature* **321**: 66–70.
3. Cook, N. J., Hanke, W., and Kaupp, U. B., 1987, Identification, purification, and functional reconstitution of the cyclic-GMP-dependent channel from rod photoreceptors, *Proc. Natl. Acad. Sci. USA.* **84**: 585–589.
4. Hanke, W., Cook, N. J., and Kaupp, U. B., 1988, cGMP-dependent channel protein from photoreceptor membranes: Single-channel activity of the purified and reconstituted protein, *Proc. Natl. Acad. Sci. USA.* **85**: 94–98.
5. Kaupp, U. B., Hanke, W., Simmoteit, R., and Lühring, H., 1988, Electrical and biochemical properties of the cGMP-gated cation channel from rod photoreceptors, Cold Spring Harbor Symposia on Quantitative Biology Vol. LIII, 407–415.
6. Hanke, W., and Breer, H., 1989, Reconstitution of acetylcholine receptors in planar lipid bilayers, in: *Molecular Enzymology* (Harris and Etmadi, eds.), Plenum, New York 339–362.
7. Changeaux, J.-P., Devillers-Thiery, A., and Chemovilli, P., 1984, Acetylcholine receptor: An allosteric protein, *Science* **225**: 1335–1345.
8. Fesenko, E. E., Kolesnikov, S. S., and Lyubarsky, A. L., 1985, Induction by cyclic-GMP of cationic conductance in plasma membrane of retinal rod outer segments, *Nature* **313**: 310–313.
9. Bodoia, R. D., and Detwiler, P. B., 1985, Patch-clamp recordings of the light-sensitive dark noise in retinal rods from the lizard and frog, *J. Physiol. (London)* **367**: 183–216.
10. Zimmerman, A. L., and Baylor, D. A., 1986, Cyclic-GMP-sensitive conductance of retinal rods consist of aqueous pores, *Nature* **321**: 395–398.
11. Hanke, W., 1985, Reconstitution of ion channels, *CRC Crit. Rev. Biochem.* **19**: 1–44.
12. Koch, K.-W., and Kaupp, U. B., 1985, Cyclic-GMP directly regulates a cation conductance in membranes of bovine rods by a cooperative mechanism, *J.B.C.* **260**: 6788–6800.
13. Stern, J. H., Kaupp, U. B., and McLeish, P. R., 1986, Control of the light-regulated current in rod photoreceptors by cyclic-GMP, calcium and 1-*cis* Diltiazem, *Proc. Natl. Acad. Sci. USA.* **83**: 1163–1167.
14. Zimmerman, A. L., Yamanaka, G., Eckstein, F., Baylor, D. A., and Stryer, L., 1985, Interactions of hydrolysis-resistant analogs of cyclic-GMP with the phosphodiesterase and light-sensitive channel of retinal rod outer segments, *Proc. Natl. Acad. Sci. USA.* **82**: 8813–8817.
15. Karpen, J. W., Zimmerman, A. L., Stryer, L., and Balor, D. A., 1986, Gating kinetics of the cGMP-activated channel of retinal rods: Flash photolysis and voltage-jump studies, *Proc. Natl. Acad. Sci. USA.* **85**: 1287–1291.
16. Matesic, D., and Liebmann, P. A., 1987, cGMP-dependent cation channel of retinal rod outer segments, *Nature* **326**: 600–603.
17. Stern, J. H., Knutsson, H., and MacLeish, P. R., 1987, Divalent cations directly affect the conductance of excised patches of rod photoreceptor membranes, *Science* **236**: 1674–1678.
18. Yau, K.-W., Haynes, L. W., and Nakatani, K., 1986, Roles of calcium and cGMP in visual

transduction, in: *Membrane Control of Cellular Activity* (H. C. Lüttgau, ed.), G. Fischer, Stuttgart, pp. 343–366.

19. Hanke, W., and Breer, H., 1987, Characterization of the channel properties of a neuronal acetylcholine receptor reconstituted into planar lipid bilayers, *J. Gen. Physiol.* **90:** 855–879.

20. Hanke, W., 1986, Incorporation by fusion, in: *Ion Channel Reconstitution* (C. Miller, ed.), Plenum, New York, pp. 141–157.

Chapter 5

A Complex Regulation of the cGMP-Dependent Channels of Retinal Rod Membranes by the cGMP Phosphodiesterase

Michele Ildefonse, Nelly Bennett, Serge Crouzy, Yves Chapron, and Armel Clerc

1. INTRODUCTION

Photoexcitation induces a hyperpolarization of the plasma membrane of the rod outer segment, and it is now clearly established that this hyperpolarization results from a decrease of cGMP* concentration which directly regulates the opening of sodium channels of this membrane (see Haynes and Yau, Chapter 3 in this volume, and Hanke and Simmoteit, Chapter 4 in this volume; for a review, see Pugh and Cobbs[1]). cGMP-dependent sodium channels are also present in the membranes of the disks.[2-7]

Both bilayer reconstitution[8,9] and patch-clamp experiments[10,11] have been performed to characterize the cGMP-dependent channel conductance and kinetic properties. They have led to very similar results (see also Chapters 3 and 4 in this volume).

2. cGMP-DEPENDENT ACTIVITY

Vesicles from rod membranes free of peripheral proteins were incorporated into painted decane-lipid bilayers.[12] In the presence of cGMP (from 5 to 100 μM), we

MICHELE ILDEFONSE, NELLY BENNETT, SERGE CROUZY, YVES CHAPRON, and ARMEL CLERC • Laboratoire Biophysique Moléculaire et Cellulaire, Centre d'Etudes Nucléaires de Grenoble, 38041 Grenoble Cedex, France.

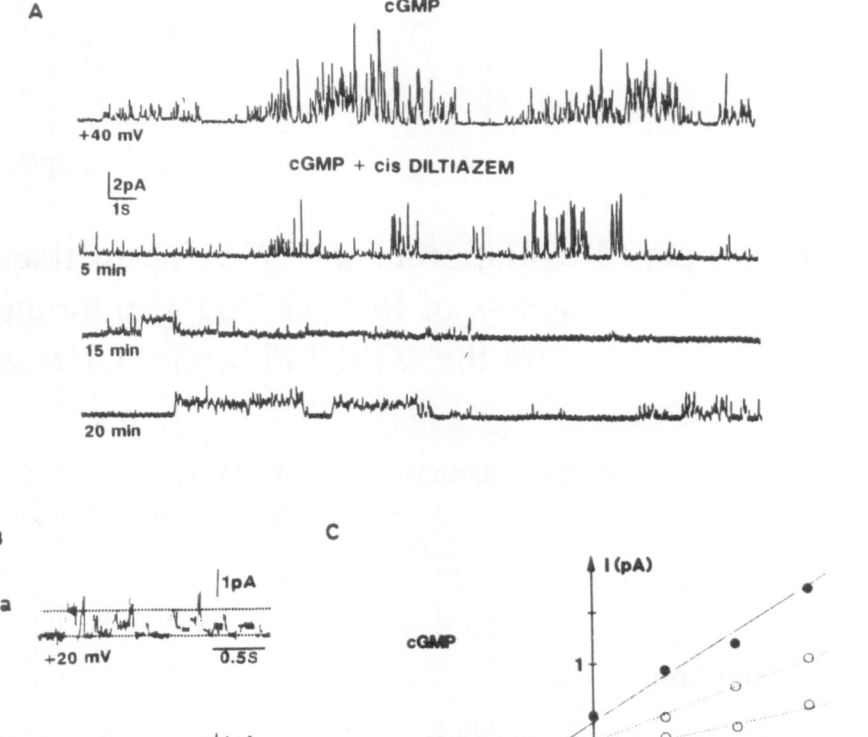

Figure 1. cGMP-activated sodium channels in rod membranes incorporated into planar bilayers (room light; NaCl 500 mM cis/200 mM trans; plus, at both sides, 10 mM HEPES, pH 7.5; 150 μM CaCl$_2$; 100 μM MgCl$_2$; 50 μM EGTA. (A) Multichannel activity (rhodopsin concentration (cis) 5μg/ml; 100 μM cGMP (cis) and inhibition of the activity by 100 μM cis-Diltiazem. (B) Single-channel records [same conditions as in (A)]. Dotted lines indicate closed states and the highest current level in (a), a small level in (b). (C) Current–voltage relations of unitary currents. The three current levels correspond to conductances of 7, 13, and 22 pS, respectively. The most frequent levels (filled symbols) could be resolved by histogram analysis in some experiments; less frequent events are noted with open symbols. Rectification of the current may be due to the sodium gradient between the two chambers. Lipids [PamOle-PdtEtn, 1-palmitoyl-2-oleoylphosphatidyl-ethanolamine/PamOle-PdtCho, 1-palmitoyl-2-phosphatidylcholine (80/20) in decane (30 mg/ml), synthesized by Avanti Polar Lipids] were spread across a hole of 180 or 260 μm diameter to form planar bilayers. Currents were recorded using a Bio-Logic K300 patch-clamp amplifier with a 10-GOhm feedback headstage and a 300 Hz bandwidth. The cis chamber was held at a virtual ground, voltage being applied to the trans chamber. Potentials values were defined as cis minus trans voltages, according to the physiological convention. Chart records were filtered at either 30 or 100 Hz. Histogram analysis was performed after 25 Hz digital filtering using a multigaussian regression method. Data were transferred with a sampling frequency of 700 Hz.

observed sodium currents quite comparable to those described by Tanaka et al.[8] and by Matthews[13]: current signal often consisted of spikes and bursts, as shown on the upper record of Fig. 1A. Their amplitude and duration varied widely from one experiment to the other and they were separated by silent periods from seconds to minutes. During quiet periods, single levels of current were sometimes observed, which correspond to the unitary conductance or to sublevels of conductance.

Cis-Diltiazem, at a concentration of 100 μM, inhibits the sodium activity, as shown in the lower records of Fig. 1A; the bursting events observed in the control progressively disappeared within a few minutes (without stirring) and after 15 to 20 min, clear levels of current were observed. They correspond to elementary conductance and to sublevels of conductance that can be measured easily (Fig. 1A and B). After washing out the *cis*-Diltiazem, addition of cGMP restored the higher bursting currents.

Current–voltage curves (Fig. 1C) have been obtained from measurements of the elementary currents observed in different experiments and done either by amplitude histograms, or by direct estimation of clear jumps of current (Fig. 1B). The steepest curve corresponds to a conductance value of 22 pS, in satisfactory agreement with previous estimates[9–11]; two smaller conductance values, corresponding respectively to 7 and 13 pS, are revealed by these measurements. Note the rectification of the current–voltage relation, which may be accentuated here because of the difference in sodium concentration between the two chambers.

3. POTENTIATION OF cGMP-DEPENDENT ACTIVITY BY INACTIVE PDE

We have attempted to suppress the channel activity by hydrolysis of cGMP. For this purpose, we have used purified preparations of cGMP phosphodiesterase (PDE) and of G-protein (Figs. 2a and b) and reassociated them to stripped rod membranes (Fig. 2c). In

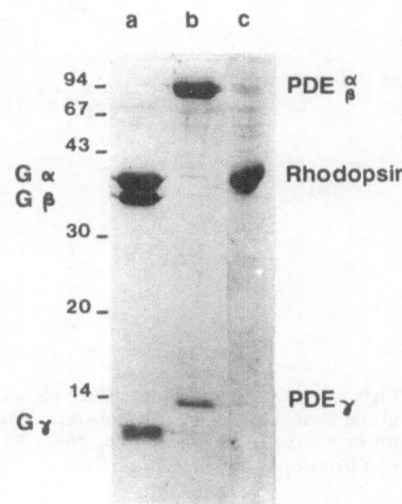

Figure 2. SDS-gel electrophoresis (15% acrylamide; see Ref. 24) of (a) purified G-protein, (b) purified PDE, and (c) stripped rod membranes. G-protein and PDE were prepared from purified rod outer segments [25] according to Baehr et al. [26,27] and kept at −20°C in 50% glycerol for less than 2 weeks. Membranes were washed several times in hypotonic buffer after extraction of PDE and G-protein and kept frozen in liquid nitrogen. PDE and G-protein concentrations were determined by the method of Bradford. [28] The PDE activity of each preparation was measured from the proton release associated with cGMP hydrolysis, [29] in the presence of membranes (room light) and in the absence or presence of G-protein and GTP. The turnover number in the absence of G-protein was less than 0.02 sec^{-1} while it ranged between 1,500 and 3,000 sec^{-1} (20°C, pH 7.5) in the presence of 1 μM G-protein (30 nM PDE), that is, 70–100% of the trypsin-induced activity of the same preparation. Under the ionic conditions used in the experiments described in this work, G-protein and PDE have a strong affinity for the membranes.[25]

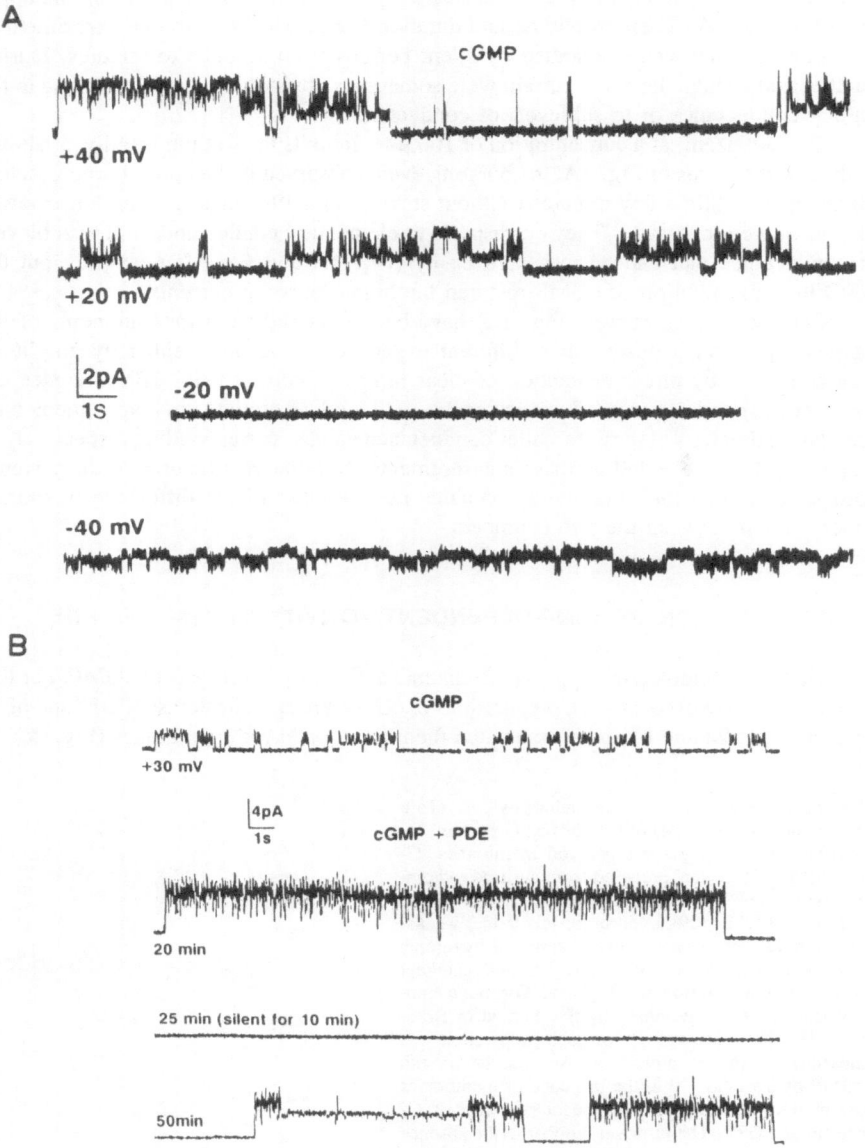

Figure 3. Effect of inhibited cGMP phosphodiesterase (PDE) on the cGMP-activated sodium activity in rod membranes (same solutions as in Fig. 1). (A) cGMP-activated currents at different potentials [rhodopsin concentration (cis) 5μg/ml; 70 μM cGMP]. (B) Effect of inhibited PDE at +30 mV (PDE/rhodopsin molar ratio, *cis:* 2%).

band c of Fig. 2, the 63 kDa protein revealed by Hanke et al.[9] as corresponding to the cGMP-regulated channel is not clearly visible because of its low amount relative to rhodopsin. In the presence of GTP, photoexcited rhodopsin present in the membranes catalyzes the formation of GTP-binding protein (G-αGTP), which is the activator of the PDE.

In the course of these experiments, we have obtained a rather surprising result: after incorporation of vesicles in the presence of cGMP and observation of some activity, the addition of PDE produced a drastic stimulation of the ionic current. In this case, the PDE is inactive, as checked by activity measurements. The stability of the current was verified for 30 to 40 min before the addition of PDE. The increase in amplitude and frequency of the activity occurred within 10 to 20 min after the addition of PDE. The effect of inactive PDE requires the presence of cGMP. After incorporation of a vesicle and washout of cGMP, the bilayer remained silent after the addition of PDE (see Fig. 5C). The absence of channel opening was checked during 1 hr in the control experiment.

The potentiating effect of PDE was particularly impressive in one experiment in which we observed three distinct current levels. In this experiment, for which records at different potentials are shown in Fig. 3A, the amplitude histograms gave a conductance value of 22 pS for the first open state.

The upper record in Fig. 3B shows the control current fluctuations at +30 mV; about 15 min after the addition of PDE, the current amplitude suddenly reached a much higher value. The channels remained open for long periods (tens of seconds) with events bursting around a mean value. Sometimes, the currents stabilized to a smaller value with almost no bursting and, in this case, jumps of current corresponding to the elementary conductance and sublevels of conductance could be measured. The active periods stopped abruptly, sometimes for very long periods. We observed this spectacular activity for about 40 min before the bilayer popped.

These results suggest that inactive PDE interacts directly with the channels, increasing the elementary conductance, or stabilizing the open state, a situation that would lead, in the experiment illustrated in Fig. 3, to a simultaneous and long-lasting opening of all the channels incorporated.

4. ACTIVATED PDE-DEPENDENT ACTIVITY

In the series of experiments reported above, PDE was inactive since the link between excited rhodopsin and PDE in the transduction chain, G-protein, was missing. If G-protein and GTP are also added, the PDE will be activated as long as GTP is present in the chamber. One would therefore expect cGMP to be hydrolyzed and the channel activity to drop progressively. This, however, did not happen. We then decided to incorporate vesicles in the absence of cGMP. We found that channel activity occurred in the presence of purified PDE and G-protein and of GTP.

This rather unexpected situation is illustrated in the records of Fig. 4A: spikes and bursting events occurred in these conditions, too. Increasing the concentration of G-protein seemed to increase the activity. Besides, G-protein preactivated with the non-hydrolyzable GTP analog GTP-γS was equally active.

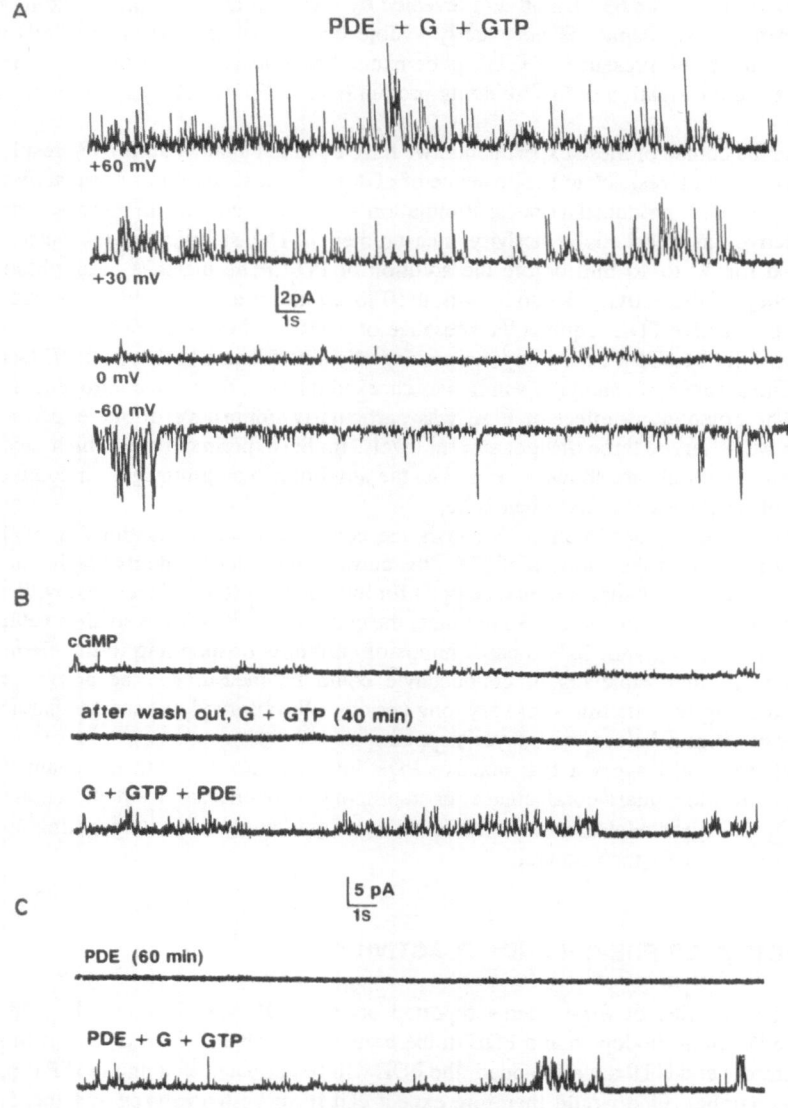

Figure 4. Activation of sodium channels in rod membranes in the absence of cGMP, by addition of G-protein, GTP, and cGMP phosphodiesterase (PDE). (A) Multichannel activity at different potential values [rhodopsin concentration (cis) 50 µg/ml; PDE/rhodopsin molar ratio:5%; G/PDE molar ratio:20; 70 µM GTP; other conditions as in Fig. 1]. (B and C): Control experiments. Vesicles (rhodopsin concentration 50 µg/ml) were incorporated in the presence of cGMP (65 µM). After a stable activity was observed, the rhodopsin and cGMP concentrations were reduced to 0.12 µg/ml and 0.16 µM respectively by washout of the chamber, before addition of either G-protein (200 nM) and GTP (400 µM) in (B) or PDE (25 nM) in (C); continuous recording were carried out for 40 min (B) or 60 min (C) without occurrence of channel opening. Further addition of PDE (25 nM) in (B) or G-protein (270 nM) and GTP (2.2mM) in (C) then induced activity within 10 min.

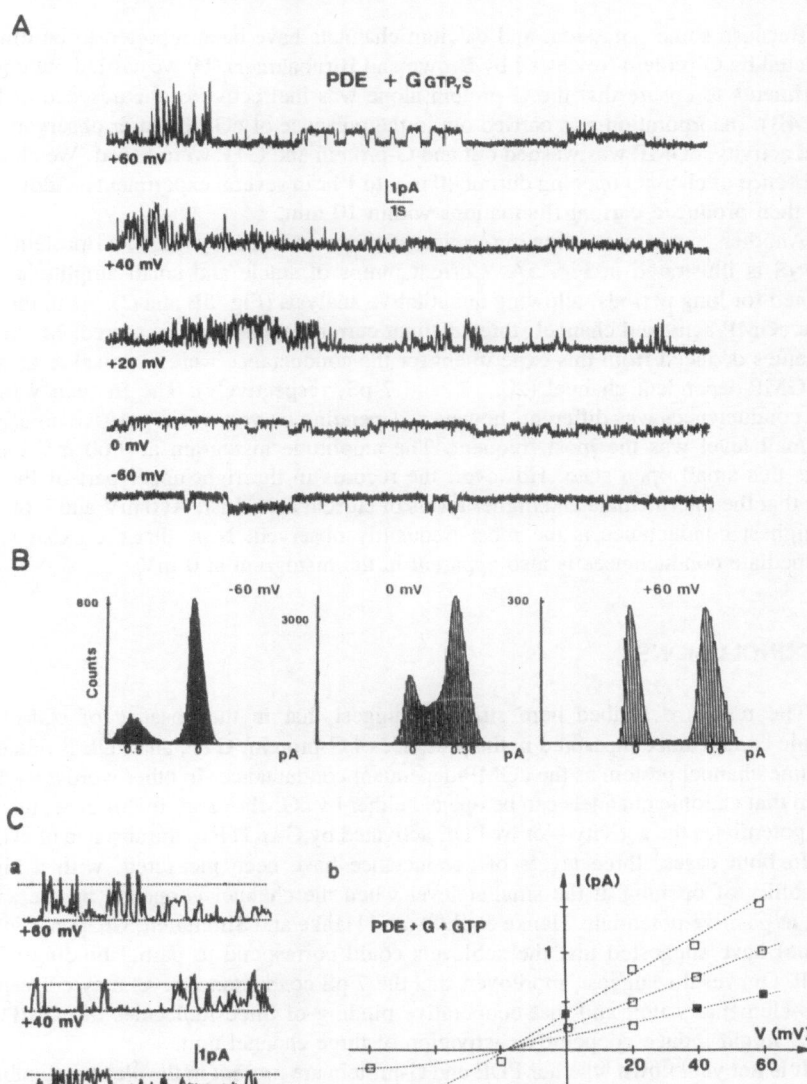

Figure 5.   Activation of sodium channels in rod membranes in the absence of cGMP, by addition of G-protein, GTP, and cGMP phosphodiesterase (PDE). (A and C,a) Single-channel records at different potentials. On the extended records (Ca) of the same experiment as in (A), the dotted lines correspond to the closed level and to the highest level which, in these conditions, is less frequent than the smaller ones [rhodopsin concentration (cis) 5 µg/ml; PDE/rhodopsin molar ratio:8%; 1 µM GTP-γS; other conditions as in Fig. 1]. (B) Amplitude histograms corresponding to current records in (A). Notice the existence of intermediate current level at 0·mV. (Cb) Current–voltage relations corresponding to the unitary currents illustrated in (A) and (Ca) and to measurements from several other experiments. Notice that the three conductance levels are comparable to those in Fig. 1 (here 7, 13, and 20 pS, respectively).

Because some potassium and calcium channels have been reported to be directly regulated by G-protein (reviewed by Brown and Birnbaumer),[14] we carried out control experiments to ensure that the G-protein alone was ineffective in the absence of PDE (Fig. 4B). Incorporation was carried out in the presence of cGMP. After observation of stable activity, cGMP was washed out and G-protein and GTP were added. We checked the absence of channel opening during 40 min to 1 hr in several experiments. Addition of PDE then produced current fluctuations within 10 min.

Another experiment showing cation activity induced by PDE, G-protein, and GTP-γS is illustrated in Fig. 5A. Current jumps of stable and small amplitude were obtained for long periods, allowing quantitative analysis (Fig. 5B and C). As in the case of the cGMP-activated channel, three distinct current levels were measured. Moreover, the values deduced from this experiment for the conductance were identical to those of the cGMP-dependent channel (20, 13, and 7 pS, respectively). The frequency of the three conductances was different, however, depending on voltage: at positive potentials, the small level was the most frequent. The amplitude histogram at +60 mV clearly shows this small open state. However, the records in the right upper part of Fig. 5C show that the intermediate and higher levels of current also exist. At 0 mV and −60 mV, the highest conductance is the most frequently observed. Note that the existence of intermediate conductances is also apparent in the histogram at 0 mV.

5. CONCLUSIONS

The results described here strongly suggest that in the absence of cGMP the cationic conductance measured in the presence of G-protein, GTP, and PDE is related to the same channel protein as the cGMP-dependent conductance. In other words, we have shown that cationic channels can be opened either by cGMP—and, in this case, inactive PDE potentiates the activity—or by PDE activated by G-αGTP in the absence of cGMP.

In both cases, three levels of conductance have been measured, with a higher probability of opening at the smaller level when the channel is opened by the active PDE, at positive potentials. Hanke et al.[9] (and Hanke and Simmoteit, Chapter 4 in this volume) have suggested that the sublevels could correspond to partial binding of the cGMP. Our results suggest, moreover, that the 7 pS conductance level may correspond to the elementary unit and that cooperative binding of three molecules of cGMP[4,6–9,10,11] might induce cooperative activation of three channel units.

It is not yet known whether PDE and G-protein are present in the plasma membrane of the rods. However, they are known to be present in the disk membranes. Our results therefore suggest that channels in the disk membrane are regulated by PDE in a complex manner.

In the dark, the channels are expected to be maximally opened as a result of the concerted action of high cGMP level and inactive PDE; the ionic composition of the intradiskal compartment would reach a steady-state level during dark adaptation. On light excitation, this could be modified by PDE in two ways: (1) directly by positive effect on the disk membrane conductance by activated PDE, and (2) indirectly by negative effect caused by a reduction of the cGMP concentration. The latter not only

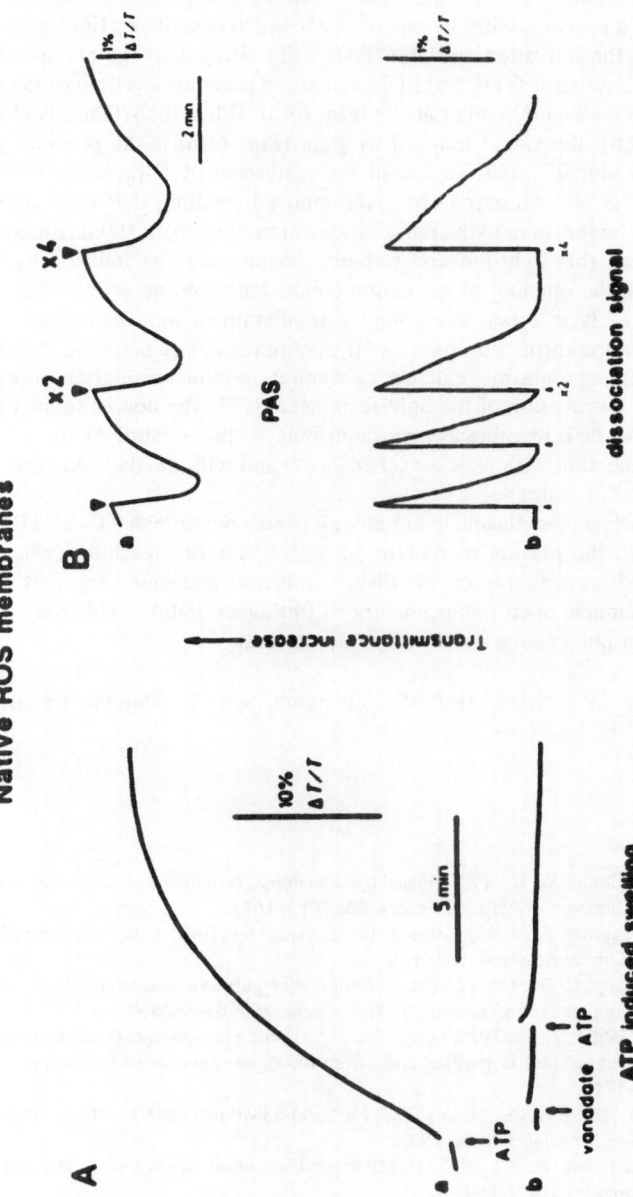

Figure 6. Scattering changes induced (A) by ATP in the dark and (B) by light in the presence of GTP. (A) Scattering change induced by ATP in the dark (a) and after addition of 10 μM orthovanadate when the effect of ATP is inhibited (b). Rhodopsin concentration 9 μM, Mg-ATPase 500 μM; external medium; 120 mM Na acetate, 10 mM MOPS pH 7.0. (B) Scattering change induced by light in the presence of 10 μM GTP after incubation with ATP (shown in A) in the absence (a) or presence (b) of vanadate: a transient scattering increase ("PAS" in a) or decrease ("dissociation signal" in b), respectively, is observed. (Modified from Ref. 15.)

affects the opening of channels in the disk membrane, but also modifies the ionic composition of the cytoplasm by closing channels in the plasma membrane.

However surprising these results might seem, they were not totally unexpected and they can be related to a previous work by one of us[15] which describes a light-scattering signal associated with the activation of PDE ("PAS," Fig. 6Ba). This signal requires not only the presence of G-protein, GTP, and PDE, but also a previous swelling of the disks by the dark ATPase described by Uhl et al.[16] (Fig. 6Aa). When the ATPase is blocked by vanadate (Fig. 6Ab), the signal induced by light (Fig. 6Bb) is the previously described "dissociation signal" associated with the activation of G-protein.[17,18] The "dissociation signal" is also observed after ATP-induced swelling if PDE is omitted. "PAS" was shown to be sensitive to the ionic composition of the bath.[15] It is tempting, therefore, to think that this light-induced turbidity change may be induced by ionic fluxes associated with the opening of the cation conductance by the active PDE.

It is as yet too early to relate this complex regulation of ionic movements to a precise step in the phototransduction process. It may nevertheless be noted that light-induced decrease of the cytoplasmic calcium concentration is increasingly proposed as being involved in the termination of the light-response.[19–23] The double regulation of cation channels of the disk membrane may contribute to this control of the calcium concentration, in connection with Na/Ca exchangers[5] and with the dark ATPase[15,16] also present in the disk membrane.

If there is any PDE on the plasma membrane, it is also possible that G-αGTP could hop from the disks to the plasma membrane at high levels of bleaching (when few inhibited PDEs would be available on the disk membrane) and thus keep part of the plasma membrane channels open independently of further excitation. This may play a role in desensitization phenomena induced by intense light.

ACKNOWLEDGMENTS.   We thank Drs M. Vivaudou and Y. Dupont for helpful discussions.

REFERENCES

1. Pugh, E. N., Jr., and Cobbs, W. H., 1986, Visual transduction in vertebrate rods and cones: a tale of two transmitters: calcium and cGMP, *Vision Res.* **26:** 1613–1643.
2. Caretta, A., and Cavaggioni, A., 1983, Fast ionic flux activated by cGMP in the membrane of cattle rod outer segments, *Eur. J. Biochem.* **132:** 1–8.
3. Koch, K. W., and Kaupp, U. B., 1985, Cyclic-GMP directly regulates a conductance in membranes of bovine rods by a cooperative mechanism, *J. Biol. Chem.* **260:** 6788–6800.
4. Puckett, K. L., and Goldin, S. M., 1986, Guanosine 3′,5′-cyclic-monophosphate stimulates release of actively accumulated calcium in purified disks from rod outer segments of bovine retina, *Biochemistry* **25:** 1739–1746.
5. Schnetkamp, P. P. M., and Bownds, M. D., 1987, Na^+ and cGMP-induced Ca^{2+} fluxes in frog rod photoreceptors, *J. Gen. Physiol.* **89:** 481–500.
6. Matesic, D., and Liebman, P. A., 1987, cGMP-dependent cation channel of retinal rod outer segments, *Nature (London)* **326:** 600–603.
7. Cook, N. J., Hanke, W., and Kaupp, U. B., 1987, Identification, purification, and functional reconstitution of the cGMP-dependent channel from rod photoreceptors, *Proc. Natl. Acad. Sci. USA.* **84:** 585–589.

8. Tanaka, J. C., Furman, R. E., Cobbs, W. H., and Mueller, P., 1987, Incorporation of a retinal rod cGMP-dependent conductance into planar bilayers, *Proc. Natl. Acad. Sci. USA.* **84:** 724–728.

9. Hanke, W., Cook, N. J., and Kaupp, U. B., 1988, cGMP-dependent channel protein from photoreceptor membranes: Single-channel activity of the purified and reconstituted protein, *Proc. Natl. Acad. Sci. USA.* **85:** 94–98.

10. Haynes, L. W., Kay, A. R., and Yau, K.-W., 1986, Single cyclic-GMP-activated channel activity in excised patches of rod outer segment membrane, *Nature (London)* **321:** 66–70.

11. Zimmerman, A. L., and Baylor, D. A., 1986, Cyclic-GMP-sensitive conductance of retinal rods consists of aqueous pores, *Nature* **321:** 70–72.

12. Miller, C., 1978, Voltage-gated cation channel from fragmented sarcoplasmic reticulum: Steady-state electrical properties, *J. Membr. Biol.* **40:** 1–23.

13. Matthews, G., 1987, Single-channel recordings demonstrate that cGMP opens the light-sensitive ion channel of the rod photoreceptor, *Proc. Natl. Acad. Sci. USA.* **84:** 299–302.

14. Brown, A. M., and Birnbaumer, L., 1988, Direct G-protein gating of ion channels, *Am. J. Physiol.* **254:** H401–H410.

15. Bennett, N., 1986, A functional link between the dark Mg-ATPase activity and the light-induced enzymatic cascade in rod outer segment, *Eur. J. Biochem.* **157:** 487–495.

16. Uhl, R., Borys, T., and Abrahamson, E. W., 1979, Evidence for structural changes in the photoreceptor disk membrane enabled by magnesium ATPase activity and triggered by light, *FEBS Lett.* **107:** 317–322.

17. Kuhn, H., Bennett, N., Michel-Villaz, M., and Chabre, M., 1981, Interactions between photoexcited rhodopsin and GTP-binding protein: Kinetic and stoichiometric analyses from light-scattering changes, *Proc. Natl. Acad. Sci. USA* **78:** 6873–6877.

18. Bennett, N., and Dupont, Y., 1985, The G-protein of retinal rod outer segments (transducin): Mechanism of interaction with rhodopsin and nucleotides, *J. Biol. Chem.* **260:** 4156–4168.

19. Yau, K.-W., and Nakatani, K., 1985, Light-induced reduction of cytoplasmic free calcium in retinal rod outer segment, *Nature (London)* **313:** 579–582.

20. Torre, V., Matthews, H. R., and Lamb, T. D., 1986, Role of calcium in regulating the cyclic-GMP cascade of phototransduction in retinal rods, *Proc. Natl. Acad. Sci. USA.* **83:** 7109–7113.

21. McNaughton, P. A., Cervetto, L., and Nunn, B. J., 1986, Measurement of the intracellular free calcium concentration in salamander rods, *Nature (London)* **322:** 261–263.

22. Pepe, I. M., Panafoli, I., and Cugnoli, C., 1986, Guanylate cyclase in rod outer segments of the toad retina: Effect of light and calcium, *FEBS Lett.* **203:** 73–76.

23. Kondo, H., and Miller, W. H., 1988, Rod light adaptation may be mediated by acceleration of the phosphodiesterase-guanylate cyclase cycle, *Proc. Natl. Acad. Sci. USA.* **85:** 1322–1326.

24. Laemmli, U. K., 1970, Cleavage of structural proteins during the assembly of the head of bacteriophage T4, *Nature (London)* **227:** 680–685.

25. Kuhn, H., 1985, Interaction between photoexcited rhodopsin and light-activated enzymes in rods, in: *Progress in Retinal Research, Vol. 3* (N. Osborne and J. Chader, eds.) Pergamon, New York, pp. 123–156.

26. Baehr, W., Devlin, M. J., and Applebury, M. L., 1979, Isolation and characterization of cGMP phosphodiesterase from bovine rod outer segments, *J. Biol. Chem.* **254:** 11669–11677.

27. Baehr, W., Morita, E. A., Swanson, R. J., and Applebury, M. L., 1982, Characterization of bovine rod outer segment G-protein, *J. Biol. Chem.* **257:** 6452–6460.

28. Bradford, M. M., 1976, A rapid and sensitive method for the quantitation of microgram quantities of protein utilizing the principle of protein–dye binding, *Anal. Biochem.* **72:** 248–254.

29. Liebman, P. A., and Evanczuk, A. T., 1982, Real-time assay of rod disk membrane cGMP phosphodiesterase and its controller enzymes, *Methods Enzymol.* **81:** 532–542.

Chapter 6

Transduction, Signal Transference, and Encoding in Composite Chemoreceptors: A Comparison between Gustatory and Arterial Chemoreceptors

Patricio Zapata

1. INTRODUCTION

The chemoreceptor organs are specialized to detect the steady levels and changes in the concentration of chemical constituents in external or internal environments. The receptor sites may be located either on the membrane of neurons themselves (simple or primary receptors) or on specialized receptor cells (composite or secondary receptors). In this last case, stimuli do not act on components of primary sensory neurons, but require the intervention of epithelioid receptor cells as transducer elements and of receptoneural synapses for the signal transference from those cells to the sensory endings of afferent neurons.

The gustatory and arterial chemoreceptors are well-studied composite receptors, which may provide models for a comparison of mechanisms intervening in stimulus transduction and impulse generation.

2. BASIC ARCHITECTURE OF CHEMORECEPTOR ORGANS

2.1. Gustatory Chemoreceptors

All animals possess chemoreceptors to "taste" the chemistry of their environment and react accordingly. While aquatic vertebrates have taste receptors disseminated on their external surface apart from those in the mouth, terrestrial ones have them only

---

PATRICIO ZAPATA • Laboratorio de Neurobiología, Universidad Católica de Chile, Santiago, Chile.

within the mouth and pharynx. Taste receptors in mammals are restricted to the taste buds. Chemical stimulation of a lingual papilla without taste buds does not elicit a taste response. Single taste buds may appear on top of the fungiform papillae of the anterior tongue, but hundreds of them are found in the trenches of the vallate and foliate papillae of the posterior tongue. Few taste buds are located in the palate, pharynx, and epiglottis.

Each taste bud is an ovoid collection of fusiform epithelioid cells. The abundant microvilli of their apices project into a taste pit, filled with a glycoprotein matrix, which provides the site of contact with taste stimuli (tastants or sapid substances). The taste pit is open to the tongue surface through the taste pore, which allows the entrance of liquids and small particles. The rest of the tongue, consisting of stratified squamous epithelial cells, has a thick keratin shelter. Since the contacts between taste bud cells are sealed by tight junctions, their apical surfaces represent the only places where tastants have access to the gustatory apparatus. In fact, radiolabeled tastants applied to the tongue surface were unable to reach the underlying sensory nerve endings. Thus, sensory nerve fibers may be only excited physiologically through transmitters released from receptor cells.

At least three types of cells are found within taste buds: receptor, sustentacular, and basal cells. The extensive synapses between the basal portions of receptor cells and the calyciform sensory nerve endings are characterized by abundant dense-cored granules (containing catecholamines or serotonin) and synaptic-like clear microvesicles crowded toward presynaptic dense projections on the side of receptor cells, and abundant mito-chondria and few synaptic-like microvesicles within nerve terminals. As nerve fibers penetrate taste buds, Schwann cells are replaced by sustentacular cells investments. Basal cells represent undifferentiated precursor cells.

Taste receptor and sustentacular cells have a life span close to 10 days, and they are replaced continuously by basal or perigemmal epithelial cells. This rapid renewal of receptor cells involves a continuous rearrangement of synapses between these ephemeral receptor cells and the nerve terminals of primary sensory neurons during the entire life of adult mammals. Innervation by chemosensory nerve fibers is essential not only for the differentiation of epithelial cells into receptor cells, but for the very existence of taste buds. The integrity of taste buds is dependent on their connection with the perikarya of gustatory nerves (located in sensory ganglia), not necessarily with their connection with the CNS. Thus, after section of the peripheral process of a gustatory nerve (denerva-tion), the taste buds of its territory disappear within 2 weeks, but they persist after section of the central process of the same nerve (decentralization).[1]

Regeneration of chemosensory nerve fibers is followed by the reappearance of taste buds within a few weeks. Cross-innervation with different nerves revealed an important degree of trophic specificity: while foreign chemosensory nerves may restore taste bud differentiation, this cannot be achieved by mechanosensory or motor nerves. Thus, only the perikarya of chemosensory neurons manufacture specific trophic molecules, which move down through the axon toward the periphery.

Each sensory nerve unit may provide innervation to several receptor cells from the same or different taste buds, even those residing in different fungiform papillae. Other-wise, some taste buds and even some receptor cells may be innervated by two nerve fibers. Taste buds in vallate and foliate papillae of the pharyngeal tongue, like those in the palate, are innervated by the glossopharyngeal nerves, containing both chemo- and mechanosensory fibers, with their perikarya in the petrossal ganglia. The central processes of all taste fibers end in the nuclear complex of the solitary tract.

2.2. Arterial Chemoreceptors

The idea that the chemistry of the blood is "tasted" by the arterial glomera, represented mainly by the carotid and aortic bodies, was advanced by the Spanish histologist Fernando de Castro.[2] The essential feature of these organs[3] is the presence of glomus cells, identified by their formaldehyde-induced fluorescence (FIF) and abundance of dense-cored granules, two characteristics common to catecholamine-containing cells. Glomus cells are organized into glomoids around a capillary, provided with a thin fenestrated endothelium and thin basal lamina. One sustentacular cell involves several glomus cells and the nerve terminals apposed to them.

Most studies of peripheral arterial chemoreceptors have been concerned with the carotid body. This organ and its neighboring baroreceptor—the carotid sinus—receive their sensory innervation from the carotid (sinus) branch of the glossopharyngeal nerve, and the perikarya of the primary sensory neurons are located mostly within the petrosal ganglion, with their central processes ending in the nuclear complex of the solitary tract. Within the carotid body, the sensory nerve fibers present extensive calyciform and small boutonlike endings—both rich in mitochondria—in apposition with glomus cells.[4] The abundance of large dense-cored granules within glomus cells and that of synaptic-like microvesicles within nerve endings—a striking similarity with adrenomedullary chromaffin cells innervated by splanchnic nerve fibers—led to the proposal that the carotid nerve endings were the presynaptic element of efferent neurons, while the glomus cells would constitute the postsynaptic elements of an endocrine secretory organ.[5]

Section of the carotid nerve (or extracranial section of the glossopharyngeal nerve) results in rapid degeneration of nerve endings apposed to glomus cells, with some proliferation of foldings by sustentacular cells to occupy empty spaces. Glomus cells are preserved, however, even after long-term denervation. Otherwise, intracranial section of glossopharyngeal nerve roots at their exit from the brain stem (i.e., central to sensory ganglia) did not modify the architecture, innervation, and chemosensory activity of the carotid body.[4] Thus, de Castro's idea that glomus cells were innervated by sensory nerve endings was restored. Further support was provided by the observation that radiolabeled amino acids injected at the site of sensory perikarya in the petrosal ganglion were traced through the carotid nerve to the nerve endings apposed to the glomus cells.[6]

The essential participation of glomus cells in chemosensory transduction is supported by the following observations: (1) after complete removal of the carotid body, the regenerating carotid nerve does not respond to physiological stimuli[7]; (2) after cryocoagulation of the carotid body, the regenerating carotid nerve did not recover its chemosensory activity, except when some fibers reestablished connections with surviving eyelets of glomus cells[8]; (3) after crushing the carotid nerve at different distances from the carotid body, chemosensory activity reappeared at the time of reapposition between nerve fibers and glomus cells[9]; (4) transplantation of the carotid body of the cat to its tenuissimus muscle in the leg resulted in the appearance of sensory discharges in the corresponding nerve which were elicited not only by muscle stretch but also by hypoxic stimuli.[10]

3. TRANSDUCTION AND SIGNAL TRANSFERENCE IN CHEMORECEPTORS

3.1. Gustatory Chemoreceptors

Intracellular recordings have been obtained from taste receptor cells, mostly from amphibians, which possess oral taste disks, devoid of pores and thus in direct contact with oral contents. Beidler, Sato, and coworkers recorded depolarizing potentials in response to common tastants (saltine, sweet, and sour); their amplitudes were directly correlated to the concentrations of the chemical stimuli.[11] Thus, they were assumed to represent "receptor potentials." Most receptor cells respond to more than one of the basic taste submodalities, but with different thresholds.

Intracellular recordings have been obtained from surface cells of the tongue mucosa of the Chilean frog (*Callyptocephalella* or *Caudiverbera*).[12] These cells had resting membrane potentials of -6 to -40 mV (mean -18 mV) when the preparation in vitro was bathed in Ringer's solution. We observed that applying minute amounts of different salts (NaCl, NaF, KCl, Na_2SO_4, $CaCl_2$ and $MgCl_2$) resulted in depolarization, often overshooting the zero membrane potential; repeated application at short intervals showed summating depolarizing potentials. These potentials were increased during hyperpolarization of the membrane by applying inward current and were decreased by current of the opposite sign; they reversed their polarity when salts were applied to cells positively polarized by strong inward current. Interestingly, distilled water provoked hyperpolarizing potentials, which may also summate if given at brief intervals. However, we observed that these responses to tastants were not restricted to the taste disk cells, but were also exhibited by other epithelial cells of the neighborhood. Furthermore, a parallel electron microscopical study of the taste disks of this preparation[13] revealed that the slim rod cells synapsing at their bases with nerve endings (i.e., the presumed receptor cells) were outnumbered by the husky mucose cells. Thus, it seems probable that many of the chemically evoked potential changes recorded from taste disks were originated from mucose cells. Therefore, the difference is that the changes in membrane potential of taste receptor cells lead to synaptic excitation of their adjacent sensory nerve endings.

With regard to the ionic currents underlying receptor potentials, intracellular recordings from frog's taste cells showed that those depolarizing receptor potentials elicited by applying quinine–HCl were increased by reducing the external Cl^- concentration or by injecting Cl^- into the cells, were independent of the external Na^+ concentration, and were reduced when furosemide (an inhibitor of Na^+/Cl^- cotransport) was added to the interstitial fluid.[14] Thus, at least the depolarizing receptor potential evoked by quinine–HCl may be due to active secretion of Cl^- across the receptor membrane.

When the lingual mucosa of the mudpuppy (*Necturus*) was explored in vitro, action potentials (i.e., regenerative impulses) were recorded intracellularly from taste receptor cells (but not from surrounding nongustatory epithelial cells); these responses were elicited by brief depolarizing currents, as by external application of salts and acids.[15] However, action potentials have not been recorded from the same preparation in situ, either in response to depolarizing current or chemical stimuli.[16] An excellent review on the generation of the taste cell potential has been published recently.[17]

The whole-cell configuration of patch-clamp recordings has been applied to isolated lingual cells, dissociated by collagenase and mechanical agitation. Kinnamon and Roper[18,19] observed that the nongustatory surface epithelial cells of the mudpuppy exhibit only passive membrane properties, while basal cells (taste cells precursors) present predominantly outward K^+ currents in response to depolarizing voltage steps. The same procedure evoked the following responses in taste receptor cells:

Current:	Na^+	Ca^{2+}	K^+
Direction:	Inward	Inward	Outward
Activation at:	-40 mV	0 mV	ca. 0 mV
Peak at:	-20 mV	30 mV	Linear ↑ with depolarization
Inactivation:	Rapid	Slow	No
Blockers:	TTX	Cd^{2+}, Mn^{2+}, Co^{2+}	TEA
Substituted by:	Mg^{2+}	Nifedipine	Cs^+

Thus, the sustained K^+ currents are a major contributor of the resting conductance of receptor cells; the transient Na^+ currents participate in their action potentials; and the slow Ca^{2+} currents may mediate their hyperpolarizing after-potentials and transmitter secretion. Furthermore, single-channel recordings in cell-attached and inside-out configurations have shown that a voltage- and Ca^{2+}-dependent K^+ current modulated by sour tastants was highly restricted to the apical membrane of taste cells, providing a basis for sour taste transduction.[20]

Patch-clamp recordings from receptor cells of the frog's taste organ have also revealed action potentials elicited by depolarizing outward current pulses.[21] Reversible depolarization of these cells without action potentials generation was recorded in response to KCl, $BaCl_2$, $CaCl_2$, quinine, cAMP, forskolin, and 3-isobutyl-1-methyl xanthine. The effects of forskolin and methyl xanthine may be attributed to stimulation of intracellular adenylate cyclase and inhibition of intracellular phosphodiesterase, respectively, suggesting that cAMP may operate as a cytosolic messenger in taste receptor cells.

Electrical coupling between lingual epithelial cells has been tested by injecting them with the fluorescent dye Lucifer yellow; nearly 20% in the mudpuppy taste buds were electrically coupled.[22] In other species (bullfrogs and rabbits), translingual potential changes were recorded in response to external application of tastants; their amplitudes were well correlated to the concentrations of substances tested.[23,24]

Extracellular recordings have been obtained from entire gustatory nerves or fine filaments containing a single or few taste fibers. Single taste neurons can also be recorded intracellularly at the level of sensory ganglia. Spontaneous activity, either randomly at low rates (< 4 Hz) or grouped in small bursts, may be observed. Sapid stimulation evokes trains of action potentials, the maximal discharge frequency for which is directly and linearly dependent on the logarithm of the concentration of tastants. Since a single gustatory unit may innervate from 1 to 12 fungiform papillae, its action potentials may arise from activation of multiple generator sites. Thus, an impulse generated from one nerve terminal is conducted orthodromically toward the CNS, but at points of fiber branching also sets impulses conducted antidromically to the other terminals of the same sensory unit. Peripheral interactions do indeed occur: the response to stimulation of one fungiform papilla may be enhanced or suppressed by simultaneous chemical stimulation of other papillae from the same receptor field. Furthermore, the afferent activity of a given gustatory unit may be strongly inhibited by antidromic impulses in another unit innervating the same papilla and set by electrical or chemical stimulation of an adjacent papilla.[25]

The sensitivities of single taste fibers to stimuli of different submodalities have been studied in detail. Taking the rat glossopharyngeal nerve as an example, 60% of its taste fibers respond to 0.3 M NaCl, 60% to 0.01 M HCl, 40% to 0.001 M quinine, and 40% to 0.3 M sucrose. Although the incidence of fibers responding to 1, 2, 3, or 4 of these prototypical stimuli could correspond to a random distribution, other studies indicate that sensitivities are not mutually independent: salty and sweet sensitivities are correlated negatively, while those to sour and bitter tend to occur concomitantly. Experiments using increasing concentrations of tastants revealed that units responding to several submodalities of taste stimuli may present a preferential sensitivity (lowest threshold) and/or better responsiveness (higher frequency of discharges) to a given submodality. Thus, for a given nerve, some gustatory fibers were preferentially sensitive or best responsive to salts, while others were best responsive to sugars.

In Sweden, Borg et al.[26] recorded chorda tympani taste fibers in patients undergoing middle ear surgery. They observed linear correlations between impulse frequency and intensity of taste sensations, as between the logarithms of both of them and that of tastants concentrations. For 0.2 M NaCl solutions flowing continuously over the tongue, the reduction to extinction of taste sensations (adaptation) directly paralleled the decreasing frequency of nerve discharges.

It is assumed that tastants (initially dissolved in saliva and then in the glycoprotein matrix of taste pits) have reactive moieties that are adsorbed to specific affinity sites in the membrane of the microvilli covering the apical surfaces of receptor cells, leading to the ion permeability changes resulting in receptor potentials. These electrical potentials would be coupled to transmitter release from receptor cells. The junctional transmitter will be recognized at postsynaptic receptor sites in sensory nerve endings, leading to the generation of action potentials, conducted along chemosensory nerve fibers. Although taste receptor cells contain acetylcholine, catecholamines (dopamine and noradrenaline), and serotonin,[27,28] further studies are required to identify the transmitter involved in taste chemoreception.

3.2. Arterial Chemoreceptors

Eyzaguirre et al. have performed several studies on whole excised carotid bodies or tissue slices superfused in vitro, to penetrate glomus cells under visual observations utilizing interference contrast (Nomarski) optics; some cells were later identified through dye injections.[29] Glomus cells had resting membrane potentials of about −20 mV and input resistances of 30–40 Mohm. The membrane potential was not significantly modified by changes in external K^+, was increased by removal of external Na^+, and was diminished by reducing external Cl^-.

The use of ion-selective microelectrodes allowed determination of intracellular Cl^- activity of nearly 21 mV. Cells depolarized when the Cl^- equilibrium potential was shifted in a positive (+) direction, an indication that Cl^- ions are not distributed passively across the glomus cell membrane and that they play an important role in determining the resting membrane potential of these cells.[30] The same technique applied to cultured glomus cells from rat carotid bodies revealed a membrane potential of -35.5 mV and a pHi of 6.84 when the external solution was kept at a pH of 7.4 at 31–32°C; acidification of the external medium reduced both the pHi and the membrane potential.[31]

The most consistent change in membrane potential of glomus cells was brought about by modifying the temperature of the bath: cooling induced decreases in both membrane potential and input resistance, while heating had the opposite effects. Depolarizing responses were commonly obtained by decreasing the pH, decreasing the flow, or increasing the osmolarity of the superfusing solution. Variable responses (depolarizations, hyperpolarizations, or no changes in membrane potential) were observed from different glomus cells when increasing the pCO_2 (but maintaining constant pH) or decreasing the pO_2 of the superfusing solution, or applying NaCN (see Ref. 3).

Hayashida and Eyzaguirre[32] have also recorded voltage noise from carotid body glomus cells in vitro. The amplitude of this membrane noise is exceptionally large: frequency histograms show a peak at about 1 mV. Voltage noise is increased markedly after applying acetylcholine, a transmitter known to be present in glomus cells themselves (see Ref. 3), and its analogs nicotine and bethanechol, pointing to the existence of nicotinic and muscarinic autoreceptors in glomus cells.

Very recently, electrotonic coupling between adjacent pairs of glomus cells of the rat carotid body superfused in vitro has been reported.[33] Interestingly, external application of dopamine, another transmitter known to be present in glomus cells (see Ref. 3), produced transient uncoupling of glomus cells.

The whole-cell patch-clamp technique has also been recently employed for recordings of freshly dissociated cells from rabbit carotid bodies.[34–36] Glomus (type I) cells were electrically excitable and generated Na^+-dependent action potentials, while sustentacular (type II) cells did not generate action potentials. Voltage clamp studies revealed at least three voltage-gated currents in glomus cells: (1) an inactivating, TTX-sensitive inward Na^+ current; (2) a high-threshold sustained inward Ca^{2+} current (suppressed by adding Co^{2+} or Cd^{2+}); and (3) outward K^+ currents (abolished by replacement of K^+ with Cs^+ and partially blocked by TEA). Since the outward current was diminished by Co^{2+}, part of this current may be carried by Ca^{2+}-activated K^+ channels. Hypoxia, the classical stimulus for arterial chemoreceptors, did not modify Na^+ and Ca^{2+} currents, but produced a transient decrease in K^+ current, pointing to a key role of K^+ conductance in chemoreceptor transduction. Hypoxia also increased the firing frequency of action potentials in glomus cells. On the other hand, sustentacular cells only exhibited a high-threshold outward K^+ current.

The pronounced dependence of the rates of chemosensory discharges on the temperature and Ca^{2+} concentration of the medium is consistent with the operation of chemical synapses for signal transference between glomus cells and chemosensory nerve terminals. The occurrence of several transmitters (acetylcholine, dopamine, noradrenaline, serotonin, substance P, enkephalins) has been reported in carotid body glomus cells (for review see Ref. 3). At least, the releases of acetylcholine[37] and dopamine[38] have been shown during carotid body stimulation. Recent evidence indicates colocalization of choline acetyl transferase (ChAT), tyrosine hydroxylase (TH), and dopamine-β-hydroxylase (DBH) in most glomus cells of the cat carotid body.[39] Studies with carotid bodies in situ and in vitro indicate that while acetylcholine increases the frequency of impulses recorded from chemosensory nerve fibers, dopamine exerts a modulatory role upon these discharges.[40,41] Thus, substances released by glomus cells may operate as transmitters or modulators on the nerve terminals of chemosensory fibers.

The large size of some afferent nerve endings apposed to glomus cells[42] allowed their penetration for intracellular recordings.[43] They showed spontaneous hypopolarizing potentials (1–10 mV, 5–120 ms), the frequency for which increased during chemoreceptor stimulation. Some action potentials originated from the peak of these spontaneous hypopolarizing potentials, while other spikes were unrelated to the local slow potentials and may correspond to antidromic invasion by action potentials generated in other terminals of the same chemosensory unit. Each carotid nerve chemosensory unit innervates many glomus cells of the same glomoid or distant glomoids.

Simultaneous recordings of the mass receptor potential of the carotid body and the carotid nerve discharges[44] indicate that natural and pharmacological stimuli (hypoxia, asphyxia, NaCN, ACh, nicotine) evoke focal slow potentials and propagated action potentials, the amplitude and frequency for which are respectively related to the intensity of these stimuli. The chemoreceptor depressants (hyperoxia, dopamine) evoked slow potentials of inverse polarity and transient decreases in the frequency of nerve spikes. The focal slow potentials recorded from the carotid body tissue may be the resultant of the receptor potentials of glomus cells and the generator potentials of the chemosensory nerve endings.

Intracellular recordings from petrosal ganglia in vitro have revealed that the perikarya of the chemosensory neurons innervating the carotid body have different electrical properties than those of the mechanosensory neurons innervating the carotid sinus.[45] In response to prolonged depolarizing pulses, chemosensory neurons show usually only one action potential, with a hump on the falling phase, followed by a prolonged afterhyperpolarization. On the other side, barosensory neurons fired repetitively throughout prolonged depolarizations, and the action potentials of the faster ones had no humps and short after-hyperpolarizations. Thus, the properties of primary sensory neurons are partly dependent on the type of receptor that they innervate.

4. NEURAL ENCODING FROM CHEMORECEPTOR ORGANS

4.1. Taste Submodalities

Behavioral and electrophysiological experiments have led to the proposal of six primary taste qualities: sweet, salty, bitter, sour, watery (evoked by distilled water), and umami (evoked by aminoacids, such as glutamate, and nucleotides, such as IMP and

GNP). Whether these categories represent distinctly separate sensory domains or the corners of continuous series of tastes is still open to discussion.

Chemical stimulation of individual fungiform papillae in humans[46] has given conflicting results: when using low concentrations of tastants, most papillae mediated a single taste submodality; but when using high-stimulus concentrations, most papillae mediated tastes of several submodalities. Although the number of taste submodalities (from one to four) elicited by stimulation of fungiform papillae correlated well with the number of taste buds (from one to fifteen) on those papillae, at least one papilla with a single taste bud responded to all four taste submodalities. Thus, few taste buds may exhibit multiple sensitivity.

How taste submodalities are coded along the gustatory pathway has been a subject of much debate[47] in recent years, involving a theoretical problem of general neurobiological importance. Two main theories have been proposed. The labeled-line or channeling theory assumes that each taste submodality is confined to a selective channel or line of neurons, the activity for which is relatively independent of that carried by another line. It assumes a certain degree of specificity in taste receptors for gustatory submodalities. The across-fiber pattern theory proposes that each given taste is read by the simultaneous frequency profile in the gustatory neuron population. It assumes an overlapping in the sensitivity of taste receptors, broadly tuned to sense a gustatory continuum or series of continua. Since experimental data provide partial support for both theories, a taste sensation of a given submodality may be the result of the preponderance in the rate of discharge of a cluster of neurons [not strictly specific but predominantly sensitive ("best responding") to a given submodality], enhanced by the contrast with the concomitant rates of discharge in other clusters.

4.2. Is the Carotid Body a Polymodal Receptor?

A long series of experiments both in situ and in vitro has shown that the rate of discharges recorded from carotid body sensory fibers is increased by several changes in its immediate environment: (1) hypoxia, either by reducing pO_2, or by interference of O_2 utilization by blockers of electron transport (cyanides) and uncouplers of oxidative phosphorylation (2,4-dinitrophenol); (2) hypercapnia; (3) acidosis; (4) hyperthermia; (5) hyperosmolarity; (6) reduced flow (see Ref. 3). Recent work in our laboratory has been concerned with the possibility that the increased frequency of sensory discharges elicited by reduced flow of perfusion in situ or of superfusion in vitro is partly independent from secondary changes in the levels of the classical chemical stimuli.[48,49]

An important point is that evidence presently available indicates that a single chemosensory unit may respond to various chemical, thermal, osmotic, and flow changes, and that different sensory units have only slightly different thresholds for each kind of stimulus. Thus, the discharges carried either by each sensory fiber or by the entire group of chemosensory fibers innervating the carotid body do not apparently provide information with regard to the exact nature of the actual stimulus originating those discharges. It must be noted that the interval analysis of spike trains has shown that while most chemosensory units discharge at random according to the Poisson model, some units depart significantly from this exponential distribution, with an apparent sequential ordering.[50] Since these oscillations have also been recorded in cat's carotid bodies superfused in vitro, they have been attributed to modulation of the spike generator sites of chemosensory nerve endings.

The coding provided by arterial chemoreceptors is therefore entirely different from the one provided by taste chemoreceptors. In the carotid body, not only different submodalities of chemical stimuli (hypoxia, hypercapnia, acidosis) but also different modalities of stimuli (chemical, thermal, osmotic, flow) are capable of increasing the rate of discharges of a given primary sensory neuron to the same levels. Furthermore, the parallel sensory channels formed by the entire population of chemosensory fibers innervating each carotid body do not differ apparently one from another in terms of their sensitivities or frequency profiles. Thus, each carotid body chemosensory unit or the full population may be described as operating as the output of a multimodal receptor.

The above idea that carotid body chemosensory fibers do not differ qualitatively between themselves is in line with our observations on the reflex effects evoked by stimulation of carotid body chemoreceptors: different submodalities or even modalities of stimuli result in similar changes in ventilation, heart rate, and arterial pressure.[51–53] It must be pointed out, however, that the cardiovascular reflexes originated from the aortic bodies are relatively different from those obtained from the carotid bodies.

ACKNOWLEDGMENTS.   Thanks are due to Mrs. Carolina Larrain for her comments on this manuscript. The work of the author is supported by grants from the Catholic University Research Division (DIUC 85/87), the National Fund for Scientific and Technological Development (FONDECYT 716-87), and the Gildemeister Foundation.

REFERENCES

1. Donoso, A., and Zapata, P., 1976, Effects of denervation and decentralization upon taste buds, *Experientia* **32:** 591–592.
2. De Castro, F., 1926, Sur la structure et l'innervation de la glande intercarotidienne (glomus caroticum) de l'homme et des mammifères, et sur un nouveau système d'innervation autonome du nerf glossopharyngien, *Trab. Lab. Invest. Biol. Univ. Madrid* **24:** 365–432.
3. Eyzaguirre, C., and Zapata, P., 1984, Perspectives in carotid body research, *J. Appl. Physiol.* **57:** 931–957.
4. Hess, A., and Zapata, P., 1972, Innervation of the cat carotid body: Normal and experimental studies, *Fed. Proc.* **31:** 1365–1382.
5. Biscoe, T. J., Lall, A., and Sampson, S. R., 1970, Electron microscopic and electrophysiological studies on the carotid body following intracranial section of the glossopharyngeal nerve, *J. Physiol.* **208:** 133–152.
6. Fidone, S. J., Zapata, P., and Stensaas, L. J., 1977, Axonal transport of labeled material into sensory nerve endings of cat carotid body, *Brain Res.* **124:** 9–28.
7. Smith, P. G., and Mills, E., 1979, Physiological and ultrastructural observations on regenerated carotid sinus nerves after removal of the carotid bodies in cats, *Neuroscience* **4:**2009–2020.
8. Verna, A., Roumy, M., and Leitner, L. M., 1975, Loss of chemoreceptive properties of the rabbit carotid body after destruction of the glomus cells, *Brain Res.* **100:** 13–23.
9. Zapata, P., Stensaas, L. J., and Eyzaguirre, C., 1976, Axon regeneration following a lesion of the carotid nerve: electrophysiological and ultrastructural observations, *Brain Res.* **113:** 235–253.
10. Monti-Bloch, L., Stensaas, L. J., and Eyzaguirre, C., 1983, Carotid body grafts induce chemosensitivity in muscle nerve fibers of the cat, *Brain Res.* **270:** 77–92.

11. Sato, T., 1980, Recent advances in the physiology of taste cells, *Progr. Neurobiol.* **14:** 25–67.
12. Eyzaguirre, C., Fidone, S. J., and Zapata, P., 1972, Membrane potentials changes recorded from the mucosa of the toad's tongue during chemical stimulation, *J. Physiol.* **221:** 515–532.
13. Stensaas, L. J., 1971, The fine structure of fungiform papillae and epithelium of the tongue of a South American toad, *Calyptocephalella gayi, Am. J. Anat.* **131:** 443–462.
14. Okada, Y., Miyamoto, T., and Sato, T., 1988, Ionic mechanisms of generation of receptor potential in response to quinine in frog taste cell, *Brain Res.* **450:** 295–302.
15. Roper, S., 1983, Regenerative impulses in taste cells, *Science* **220:** 1311–1312.
16. Teeter, J., 1987, Quasi-regenerative responses to chemical stimuli in in vivo taste cells of the mudpuppy, *Ann. N. Y. Acad. Sci.* **510:** 652–654.
17. Teeter, J., Funakoshi, M., Kurihara, K., Roper, S., Sato, T., and Tonosaki, K., 1987, Generation of the taste cell potential, *Chem. Senses* **12:** 217–234.
18. Kinnamon, S. C., and Roper, S. D., 1987, Voltage-dependent ionic currents in dissociated mudpuppy taste cells, *Ann. N. Y. Acad. Sci.* **510:** 413–416.
19. Kinnamon, S. C., and Roper, S. D., 1988, Membrane properties of isolated mudpuppy taste cells, *J. Gen. Physiol.* **91:** 351–371.
20. Kinnamon, S. C., Dionne, V. E., and Beam, K. G., 1988, Apical localization of K^+ channels in taste cells provides the basis for sour taste transduction, *Proc. Natl. Acad. Sci. USA.* **85:** 7023–7027.
21. Avenet, P., and Lindemann, B., 1987, Patch-clamp study of isolated taste receptor cells of the frog, *J. Membr. Biol.* **97:** 223–240.
22. Yang, J., and Roper, S. D., 1987, Dye-coupling in taste buds in the mudpuppy, *Necturus maculosus, J. Neurosci.* **7:** 3561–3565.
23. Soeda, H., Sakudo, F., and Noda, K., 1985, Relation between translingual potential changes induced by NaCl in the bullfrog tongue and taste nerve activity, *Jap. J. Physiol.* **35:** 1101–1105.
24. Simon, S. A., Robb, R., and Garvin, J. L., 1986, Epithelial responses of rabbit tongues and their involvement in taste transduction, *Am. J. Physiol.* **251:** R598–R608.
25. Murayama, N., 1988, Interaction among different sensory units within a single fungiform papilla in the frog tongue, *J. Gen. Physiol.* **91:** 685–701.
26. Borg, G., Diamant, H., Ström, L., and Zotterman, Y., 1967, The relation between neural and perceptual intensity: A comparative study on the neural and psychophysical responses to taste stimuli, *J. Physiol.* **192:** 13–20.
27. Morimoto, K., and Sato, M., 1982, Role of monoamines in afferent synaptic transmission in frog taste organ, *Jap. J. Physiol.* **32:** 855–871.
28. Nagahama, S., and Kurihara, K., 1985, Norepinephrine as a possible transmitter involved in synaptic transmission in frog taste organs and Ca dependence on its release, *J. Gen. Physiol.* **85:** 431–442.
29. Eyzaguirre, C., Baron, M., Hayashida, Y., Monti-Bloch, L., and Gallego, R., 1980, Effects of different stimuli on the glomus cell membrane, *Adv. Physiol. Sci.* **10:** 399–408.
30. Oyama, Y., Walker, J. L., and Eyzaguirre, C., 1986, The intracellular chloride activity of glomus cells in the isolated rabbit carotid body, *Brain Res.* **368:** 167–169.
31. Hee, S. F., Wei, J. Y., and Eyzaguirre, C., 1988, Intracellular pH and some membrane characteristics of cultured carotid body cells, *9th International Symposium on Arterial Chemoreceptors,* Park City, Abstract, p. 41.
32. Hayashida, Y., and Eyzaguirre, C., 1979, Voltage noise of carotid-body type I cells, *Brain Res.* **167:** 189–194.
33. Monti-Bloch, L., Abudara, V., and Clavijo, J., 1988, Electric communications between glomus cells of the rat carotid body, *9th International Symposium on Arterial Chemoreceptors,* Park City, Abstract, p. 59.
34. Duchen, M. R., Caddy, K. W. T., Kirby, G. C., Patterson, D. L., Ponte, J., and Biscoe, T. J., 1988, Biophysical studies of the cellular elements of the rabbit carotid body, *Neuroscience* **26:** 291–311.
35. Lopez-Barneo, J., Lopez-Lopez, J. R., Ureña, J., and Gonzalez, C., 1988, Chemotransduction in

the carotid body: K^+ current modulated by pO_2 in type I chemoreceptor cells, *Science* **241**: 580–582.

36. Delpiano, M. A., Hescheler, J., and Acker, H., 1988, Evidence for O_2-sensitive ionic channels in carotid body type-I cells, *Physiologist* **31**: A172.
37. Eyzaguirre, C., and Zapata, P., 1968, The release of acetylcholine from carotid body tissues. Further study on the effects of ACh and cholinergic blocking agents on the chemosensory discharge, *J. Physiol.* **195**:589–607.
38. Fidone, S. J., Stensaas, L. J., and Zapata, P., 1983, Sites of synthesis, storage, release and recognition of biogenic amines in carotid bodies, in: *Physiology of the Peripheral Arterial Chemoreceptors* (H. Acker and R. G. O'Regan, eds.), Elsevier/North-Holland, Amsterdam, pp. 21–44.
39. Wang, Z. Z., Dinger, B., Fidone, S., and Stensaas, L. J., 1988, The localization and coexistence of biogenic amines and neuropeptides in carotid body type I cells, *9th International Symposium on Arterial Chemoreceptors*, Park City, Abstract, p. 13.
40. Zapata, P., 1975, Effects of dopamine on carotid chemo- and baroreceptors in vitro, *J. Physiol.* **244**: 235–251.
41. Llados, F., and Zapata, P., 1978, Effects of dopamine analogues and antagonists on carotid body chemosensors in situ, *J. Physiol.* **274**: 487–499.
42. Nishi, K., and Stensaas, L. J., 1974, The ultrastructure and source of nerve endings in the carotid body, *Cell Tissue Res.* **154**: 303–319.
43. Hayashida, Y., Koyano, H., and Eyzaguirre, C., 1980, An intracellular study of chemosensory fibers and endings, *J. Neurophysiol.* **44**: 1077–1088.
44. Zapata, P., and Eyzaguirre, C., 1985, Bioelectric potentials in the carotid body, *Brain Res.* **331**: 39–50.
45. Belmonte, C., and Gallego, R., 1983, Membrane properties of cat sensory neurons with chemoreceptor and baroreceptor endings, *J. Physiol.* **342**: 603–614.
46. Arvidson, K., and Friberg, U., 1980, Human taste: Response and taste bud number in fungiform papillae, *Science* **209**: 807–808.
47. Scott, T. R., and Chang, F. C. T., 1984, The state of gustatory neural coding, *Chem. Senses* **8**: 297–314.
48. Alcayaga, J., Iturriaga, R., and Zapata, P., 1986, Carotid body chemoreceptor excitation produced by carotid occlusion, *Acta Physiol. Pharmacol. Latinoam.* **36**: 199–215.
49. Alcayaga, J., Iturriaga, R., and Zapata, P., 1988, Flow-dependent chemosensory activity in the carotid body superfused in vitro, *Brain Res.* **455**: 31–37.
50. Nolan, W. F., Donnelly, D. F., Smith, E. J., and Dutton, R. E., 1984, Nonrandom chemoreceptor activity during superfusion in vitro, *Brain Res.* **292**: 194–197.
51. Serani, A., and Zapata, P., 1981, Relative contribution of carotid and aortic bodies to cyanide-induced ventilatory responses in the cat, *Arch. Int. Pharmacodyn. Thér.* **252**: 284–297.
52. Serani, A., Lavados, M., and Zapata, P., 1983, Cardiovascular responses to hypoxia in the spontaneously breathing cat: Reflexes originating from carotid and aortic bodies, *Arch. Biol. Med. Exp.* **16**: 29–41.
53. Iturriaga, R., Alcayaga, J., and Zapata, P., 1988, Contribution of carotid body chemoreceptors and carotid sinus baroreceptors to the ventilatory and circulatory reflexes produced by common carotid occlusion, *Acta Physiol. Pharmacol. Latinoam.* **38**: 27–48.

B. Excitation–Secretion Coupling

Chapter 7

Role of Membrane Receptors in Stimulus–Secretion Coupling

Eduardo Rojas, Rosa M. Santos, and Illani Atwater

1. HORMONE–RECEPTOR AND NEUROTRANSMITTER–RECEPTOR INTERACTIONS

In the last two decades there has been an exponential increase in our fundamental knowledge of hormone receptors. The literature accumulated is so vast that it is impossible to cover in any depth, in a short review, all the aspects of hormone–receptor interaction. In this chapter we will consider only two aspects. First, the modulation of the membrane-bound enzyme system adenylate cyclase, and, second, the perturbation of the phosphoinositide cycle caused by neurotransmitters involved in the neural control of catecholamine and insulin secretion. To this end, a mechanism for the generation of intracellular signals resulting from the activation of adrenergic and cholinergic receptors will be also considered, and illustrative examples of receptor-controlled electrical activity and ATP secretion will be presented.

1.1. The GTP-Binding Regulatory Proteins

It is now firmly established on physiological and biochemical grounds that receptors for a large number of hormones and neurotransmitters, as well as for light in the

EDUARDO ROJAS, ROSA M. SANTOS, and ILLANI ATWATER • Laboratory of Cell Biology and Genetics, National Institute of Diabetes and Digestive and Kidney Diseases, National Institutes of Health, Bethesda, Maryland 20892.

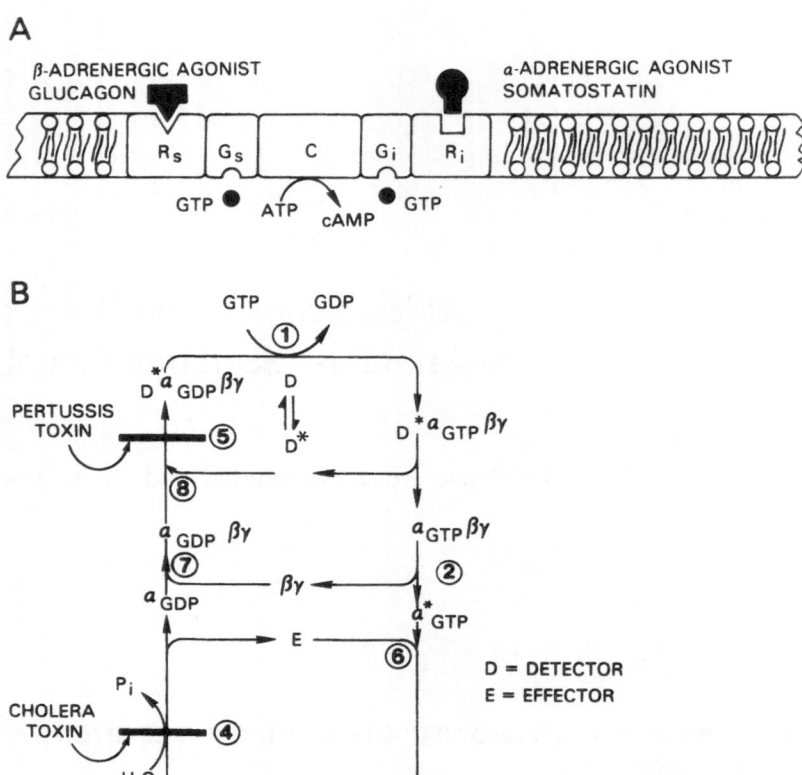

Figure 1. (A) Simplified diagram depicting the receptor proteins (R_s, R_i), GTP regulatory proteins (G_s, G_i), and the catalytic unit (C). (B) Scheme depicting the interactions between subunits of G proteins (see text). (From Ref. 4, with modifications.)

case of rhodopsin, are linked in the plasma membrane to a class of proteins that have in common three fundamental properties (see Fig. 1A):

1. They have high-affinity, GTP-specific binding sites.
2. They degrade GTP to GDP.
3. They are made up of three distinct subunits, one of which is common to all of them.[1]

Three GTP-binding proteins have been identified: G_s, G_i, and retinal transducin. When activated by specific receptors, G_s stimulates, and G_i inhibits, adenylate cyclase activity. Transducin, a protein found in the rod outer segment of vertebrate retina, couples photoexcitation of rhodopsin to stimulation of a specific cGMP phosphodiesterase (PDE).

These proteins have similar subunit structures (each is a heterotrimer of three subunits), similar amino acid compositions of the respective subunits, and a guanine nucleotide binding site in the α-subunit of each protein.

Figure 1B depicts a recently proposed scheme of interactions between subunits of G proteins and detector (D) and effector (E) components of GTP-dependent transmembrane signaling.[2] The three subunits of the coupling G proteins are denoted α, β, and γ.

In the presence of agonist, the coupling protein is activated and the exchange of GTP for GDP at the guanine nucleotide binding site occurs (step 1 in Fig. 1B). This reaction is followed by separation of the α^*_{GTP}- and $\beta\gamma$-subunits (step 2). Termination of the active state is achieved by the GTPase activity present in the α-subunit (step 3). Specific hormones maintain activity of the cognate coupling protein subunit by promoting repeated replacement of GDP by GTP, a process that results in hormone-stimulated GTP hydrolysis (step 3).[3] For transducin and G_s, the active α^*_{GTP}(GTP-bound)-subunit stimulates the effector enzymes, adenylate cyclase and PDE, respectively.

In the model, α^*_{GTP} (the active state of the α-subunit) activates E (step 6); this active conformation decays, however, concomitant with hydrolysis of the γ-phosphoryl bond of GTP (step 3). Free $\beta\gamma$ then binds to α_{GDP} (step 7), forming the α_{GDP}–$\beta\gamma$ complex which has a high affinity for the activated detector component (D^* = hormone–receptor complex; step 8). Guanine nucleotides and D^* displace each other from binding to α_{GDP}–$\beta\gamma$. As a result, GTP can replace GDP on the guanine nucleotide binding site of $D^*\alpha_{GDP}$–$\beta\gamma$ (step 1), with concomitant dissociation of α_{GTP}–$\beta\gamma$ from D^*. A conformational change in α_{GTP} to α^*_{GTP} (step 2), stabilized by GTP, reduces its affinity for $\beta\gamma$, and free α^*_{GTP} is released to activate E once again (step 6).

1.2. The Toxins from *Bordetella pertussis* and *Vibrio cholerae*

Another important homology between the two G proteins, G_s and G_i, is that only the α-subunit of each protein serves as a substrate for ADP-ribosylation by the exotoxins of *Vibrio cholerae* and *Bordetella pertussis*.[3] The toxin-catalyzed covalent modifications have proved useful for identifying the coupling proteins. The action of pertussis toxin led to the discovery of G_i. Although the mechanisms by which G_i inhibits adenylate cyclase are under investigation,[1] the results with pertussis toxin suggested that G_i mediates the effects of certain additional signals that do not depend on alterations of cAMP synthesis. In the case of cholera toxin G_s is the target protein, but the exact details of this interaction are not known.

Pertussis toxin (119 kDa), obtained from the microbial agent responsible for the disease commonly known as whooping cough, is an invaluable tool in exploring the mechanisms for adenylate cyclase regulation. It is well established that the toxin blocks the action of hormones that inhibit adenylate cyclase activity. The target for pertussis toxin appears to be a subunit of G_i of molecular weight of about 40,000 da. Pertussis toxin treatment has been shown to enhance forskolin stimulation of adenylate cyclase.[4,5]

Pertussis toxin is known to enhance insulin secretory responses. This effect was accidentally discovered when pancreases obtained from pertussis vaccine-treated rats showed resistance to the inhibitory effects of epinephrine on insulin release; instead of the α_2-adrenergic receptor mediated blockade of insulin release,[6] a β-adrenergic stimulation was apparent.[7,8] For this reason the toxin is also called islet-activating protein.[9,10] The underlying mechanism of the action of pertussis toxin on islet of Langerhans is the ADP-ribosylation of the α-subunit of G_i by the A-protomer of pertussis toxin, which is transported across the plasma membrane to its target sites within the cell as a result of the binding of the toxin. Since cAMP, the product of adenylate cyclase, is a second messenger of a variety of cell stimuli, it is not surprising to find that the interaction of the A-protomer of pertussis toxin with G_i leads to enhancement of insulin release.

1.3. Structural and Functional Properties of the Regulatory Components G_s and G_i

The regulation of adenylate cyclase by hormones and neurotransmitters requires the simultaneous action of the hormone on a receptor protein and of the guanine nucleotide on a regulatory protein. This was first shown for glucagon, a stimulatory hormone binding to the R_s receptor and acting on the G_s subunit[11] and, later, for inhibitory hormones acting on the G_i subunit.[12] It was clearly established that guanine nucleotide regulation of adenylate cyclase is mediated by proteins (G_s and G_i) distinct from receptor proteins (R_s and R_i) or from the catalytic component of the system. Both G_s and G_i are large proteins, each of 96,000 da; both are trimers with subunits of 42,000 (α_s), 35,000 (β) and 5,000 (γ) for G_s and 40,000 (α_i), 35,000 (β), and approximately 5,000 (γ) da for G_i. It is now clear that there are not only structural similarities between G_s and G_i, but a number of functional similarities as well. The activation of the membrane-bound proteins in the absence of hormonal regulation are qualitatively similar. For both proteins, the nonhydrolyzable GTP analogues, such as GMP-P(NH)P and GTPγS, have greater efficacy for activation than does GTP itself. Both proteins have an absolute requirement for Mg^{2+} for activation, and not only guanine nucleotides but also fluoride will cause activation of both G_s and G_i.[13]

Regulation of G_s and G_i by receptors also resemble each other. The affinity of either stimulatory or inhibitory hormones for their respective receptors is decreased by guanine nucleotides and increased by Mg^{2+}. Both hormonal stimulation and hormonal inhibition of adenylate cyclase are accompanied by an increased release of guanine nucleotides from membranes, which is thought to reflect increased guanine nucleotide exchange and a hormone-dependent increase in GTP hydrolysis. In the case of stimulatory hormones, the G_s itself has been shown to have GTPase activity.

The evidence accumulated in the last five years suggest that G_i and G_s are functionally, as well as structurally, very similar proteins and that they are regulated by an activation/deactivation cycle involving GTP hydrolysis in the deactivation step. Figure 2 summarizes the information available in a kinetic model. The scheme is applicable to either G_s or G_i protein. In both cases there is a guanine nucleotide dependent formation of high-molecular-weight αGTP–$\beta\gamma$ complex which is converted in a Mg^{2+}-dependent

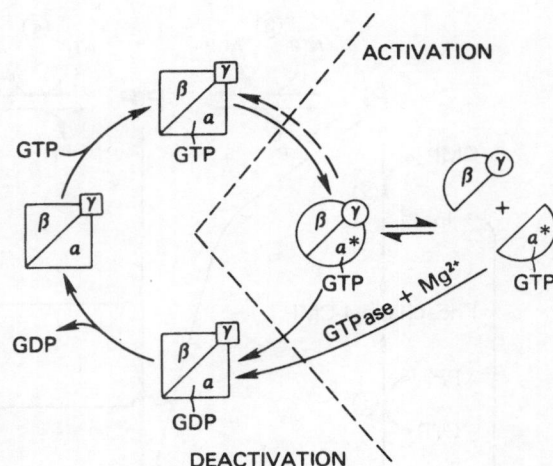

Figure 2. Regulatory cycle of G_s and G_i. (From Ref. 5 with minor changes.)

manner to a high-molecular-weight $\alpha^*GTP-\beta\gamma$ complex capable of dissociating into α^*GTP and $\beta\gamma$ components. It has also been shown that the α-subunits of G_s and G_i are identical.

In spite of the overall similarities in the kinetic cycles for the regulation of the activation of G_s and G_i, there are differences between these two proteins. For example, although both require guanine nucleotide and Mg^{2+} for activation, their dependence on Mg^{2+} is not identical. Another important difference between G_s and G_i is that while pertussis toxin treatment results in inhibition of hormonal regulation of G_i, cholera toxin treatment does not interfere with hormonal regulation of G_s.

1.4. Hormone and Neurotransmitter Stimulation of Membrane Receptors and the Phosphoinositide Cycle

Inositol-containing phospholipids are present as a few percent of the total phospholipids in all eukaryote cells so far examined. Phosphatidylinositol (PtdIns) concentrations in eukaryotes range from 0.5 to 2.5 μmol/g tissue.[14] Measurements in hepatocytes revealed that the total content of phosphatidylinositol 4-phosphate (PtdIns(4)P) and phosphatidylinositol 4,5-bisphosphate (PtdIns(4,5)P$_2$), which represent only a small fraction of the inositol-containing phospholipids, is about 30 nmol/g tissue.[15] If PtdIns(4)P and PtdIns(4,5)P$_2$ were to be confined to the hepatocyte plasma membrane, which probably comprises less than 1% of the tissue, then they would be found there at a local concentration of at least 2 mM.

There is no single cell in which all of the known steps in the metabolic pathways to and from PtdIns and PtdIns(4,5)P$_2$ have been fully characterized. The widely accepted set of metabolic pathways shown in Fig. 3 is based on partial bits of information from many cells.

It is generally agreed that most synthesis of PtdIns from phosphatidate (reactions 1

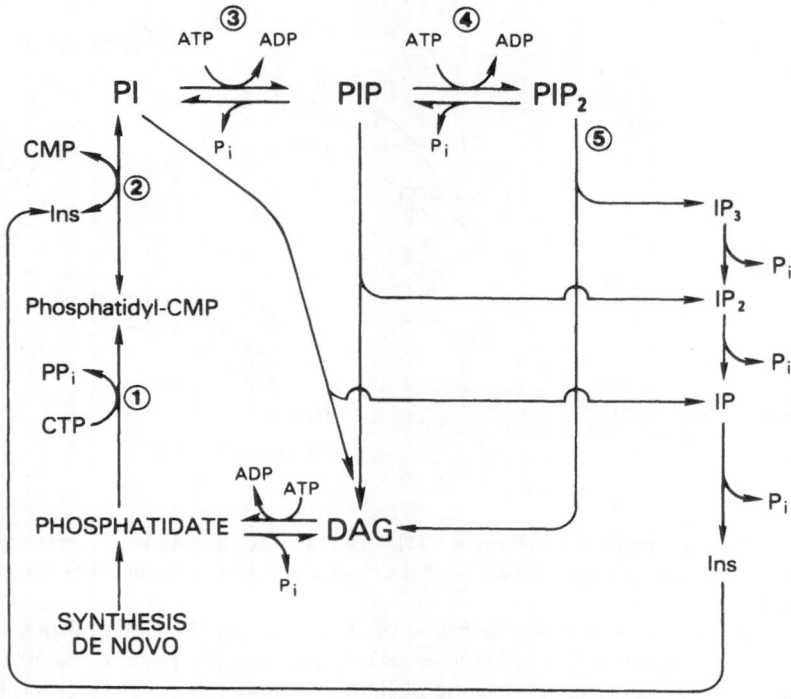

Figure 3. Pathways for the synthesis and degradation of inositol glycerophospholipids.

and 2) occurs at the endoplasmic reticulum. PtdIns(4)P is generated from PtdIns in the reaction step 3. PtdIns(4,5)P_2 is produced by phosphorylation of PtdIns(4)P in reaction 4. 1,2-diacylglycerol (DAG) is one of the degradation products of PtdIns(4,5)P_2 together with Ins(1,4,5)P_3 in reaction 5.

Although the exact details of the turnover of inositol phospholipids during signal transduction remain to be elucidated, it is well known that activation of β-adrenergic and muscarinic receptors in many tissues is accompanied by an increased incorporation of ^{32}P-phosphate into phosphatidic acid and PtdIns.[16]

There are two major mechanisms whereby hormones and neurotransmitters may perturb the phosphoinositide cycle. The first mechanism was discovered nearly 30 years ago.[17–19] The fundamental observation was that ACh stimulation of the pancreas provoked a rapid increase in ^{32}[P]-incorporation into phosphatidic acid and PtdIns. This increase in labeling was associated with a decrease in the mass of PtdIns and an increase in the mass of phosphatidic acid and DAG, and the concept emerged that the primary change was phospholipase-C-mediated phophoinositide hydrolysis. Furthermore, several results indicate that β$_1$-adrenoceptors mediate the effects of catecholamines that perturb the phosphoinositide turnover.[20,21]

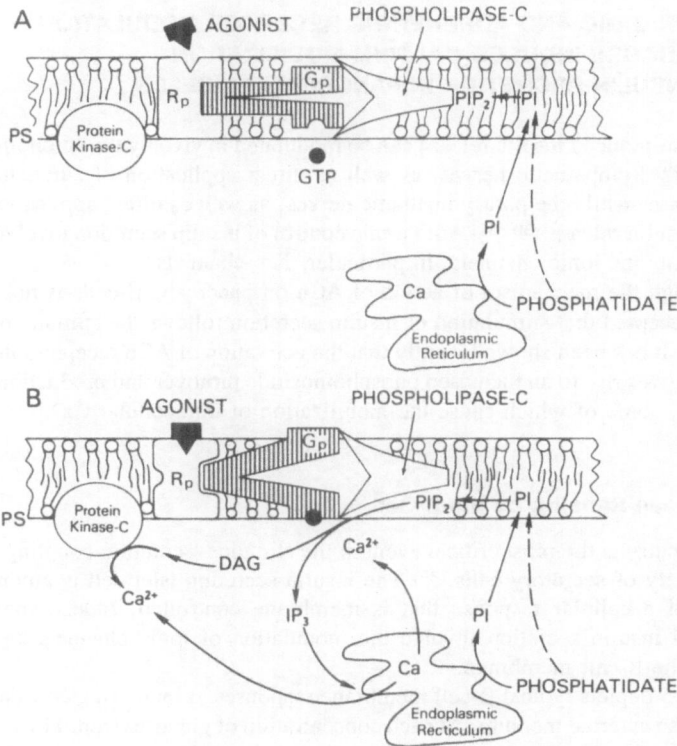

Figure 4. Generation of intracellular signals after receptor stimulation. R_p, receptor protein; G_p, GTP-binding protein; PS, phosphatidylserine.

The second major mechanism for perturbing the cycle, recognized only recently in studies of ACTH action,[22] is to increase *de novo* synthesis of phosphatidic acid.

Figure 4 illustrates a possible sequence of signals resulting from receptor stimulation. Part A illustrates the situation prior to receptor activation by the agonist. When the agonist binds to the high-affinity receptor protein (Part B), another transducer protein, possibly of the G type, stimulates a membrane-bound phospholipase-C which acts on PtdIns(4,5)P_2 to generate Ins(1,4,5)P_3 and DAG.

The consequences of perturbing the PtdIns cycle are multiple. First, it has been clearly established that Ins(1,4,5)P_3 induces Ca^{2+} liberation from intracellular stores, presumably endoplasmic reticulum.[23–25] Second, since phosphatidic acid is a natural Ca^{2+} ionophore,[26] it may facilitate transmembrane Ca^{2+} movements. Third, increases in phosphatidic acid and DAG may provide substrate for phospholipases and lipases to generate arachidonic acid, which is a promoter of membrane fusion. Finally, increases in DAG may activate a Ca^{2+}-dependent protein kinase C, which is also activated by phosphatidylserine (PS) and PtdIns.[27]

2. CHOLINERGIC AND ADRENERGIC RECEPTOR MODULATION OF GLUCOSE-INDUCED CALCIUM MOBILIZATION AND INSULIN SECRETION IN PANCREATIC β-CELLS

Glucose-induced insulin release can be modulated in vivo by the autonomic nervous system.[28,29] Sympathetic nerves, as well as direct application of adrenaline, inhibit insulin release, while the parasympathetic nerves, as well as direct application of ACh, stimulate insulin release.[30-32] Autonomic control of insulin secretion involves modulation of membrane ionic channels, in particular, K^+-channels.

Although the mechanism of action of ACh on pancreatic β-cells is not clear, it is generally accepted that stimulation of insulin secretion follows the stimulation of Ca^{2+} entry.[33,34] It has been shown recently that the activation of ACh receptors in the β-cell membrane gives rise to an increased phosphoinositide turnover and production of second messengers, some of which cause the mobilization of intracellular Ca^{2+}.[35-38]

2.1. Glucose Sensing by the β-Cell

Ca^{2+} entry is the most critical event in the stimulus–secretion coupling process of a wide variety of secretory cells.[39] The insulin-secreting islet cell is among the best examples of a cellular response that is membrane controlled. Indeed, both glucose sensing and insulin secretion involve the modulation of ionic channels known to be present in the β-cell membrane.

Figure 5 depicts typical β-cell membrane responses to increasing concentrations of glucose in the external medium. At each concentration of glucose (from 11 to 21 mM) the membrane potential spontaneously oscillates between two levels: one at around -65 mV (the silent phase) and the other at around -45 mV during which action potentials occur (the active phase). Dean and Matthews[40] coined the term "burst" to describe this electrical activity. Numerous studies have established that the fast depolarization of the membrane during the action potentials is caused by the activation of voltage-gated Ca^{2+}-channels[41-45] and that the rapid repolarization involves the activation of K^+-channels.[46-48]

The ionic mechanisms underlying the slow oscillations have not been established. Atwater et al.[49] explained the slow oscillations in terms of changes in $[Ca^{2+}]_i$ as follows. In absence of glucose, $[Ca^{2+}]_i$ is relatively high, controlled mainly by Na^+–Ca^{2+} exchange. This high level of $[Ca^{2+}]_i$ increases the open probability of a $[Ca^{2+}]_i$-sensitive K^+-channel and membrane potential is at its most negative level while membrane resistance is low. Upon addition and metabolism of glucose, $[Ca^{2+}]_i$ is lowered by the activation of various energy-requiring processes including active Ca^{2+} transport out of the cytosol. As $[Ca^{2+}]_i$ is lowered, the open probability of the K^+-channel decreases and the membrane depolarizes. When $[Ca^{2+}]_i$ is sufficiently reduced, the K^+-channel is deactivated and a rapid depolarization of the membrane to a plateau potential occurs. This depolarization of the membrane reaches the threshold for the activation of voltage-gated Ca^{2+}-channels. Regenerative activation of Ca^{2+}- and K^+-channels gives rise to spike activity. When $[Ca^{2+}]_i$ reaches the level for activation of the large conductance K^+-channel, the membrane hyperpolarizes to the silent phase potential.

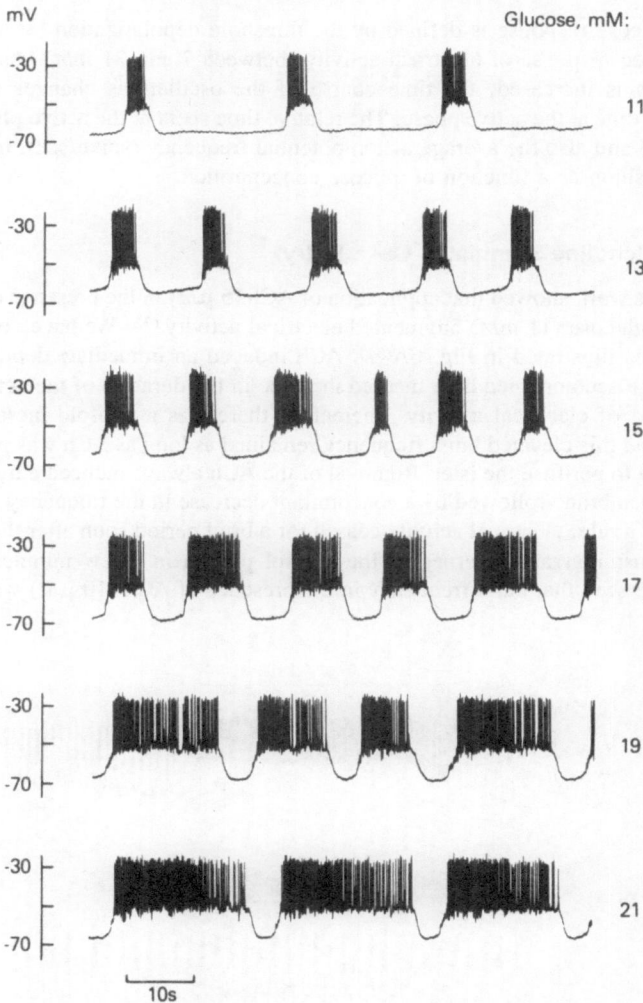

Figure 5. Glucose-evoked electrical activity in the pancreatic β-cell as a function of glucose concentration. (From Ref. 50.)

K$^+$-channel activity of the β-cell membrane plays a major role in glucose sensing.[50] Glucose modulation of K$^+$-channel activity was first determined by observing that glucose inhibited K$^+$ efflux from rat islets in a dose-dependent manner, about 85% inhibition occurring at 7 mM glucose.[50] Membrane potential measurements also showed that the relative permeability to K$^+$ is decreased by glucose.[46] The reduction in K$^+$ permeability obtained by augmenting the concentration of glucose from 0 to 7 mM depolarizes the β-cell membrane from about −70 to ca. −60 mV but does not induce electrical activity or insulin release.

The glucose response is defined by the threshold depolarization (at about 7 mM) and the graded response of electrical activity (between 7 and 21 mM). As the glucose concentration is increased, the time course of the oscillations changes and the cell spends more time at the active phase. The relative time spent at the active phase (percent active phase) and also the average action potential frequency (spike/sec), increases in a sigmoidal fashion as a function of glucose concentration.

2.2. Acetylcholine Stimulates Ca²⁺ Entry

Previous work showed that application of ACh (5 μM) in the presence of stimulatory levels of glucose (11 mM) augmented electrical activity.[33] We have confirmed this observation as illustrated in Fig. 6A.[51] ACh induced an immediate depolarization of the membrane accompanied by a marked increase in the duration of the active phase of the first burst of electrical activity. Thereafter, there was a twofold increase in burst frequency, and this elevated burst frequency remained as long as ACh was present in the solution used to perifuse the islet. Removal of the ACh always induced a hyperpolarization of the membrane followed by a concomitant decrease in the frequency of the bursts ($n = 12$). As a rule, electrical activity ceased for a brief period soon after the removal of the neurotransmitter and returned to the control pattern in a few minutes (Fig. 6A).

We have seen that burst frequency in the presence of ACh (10 μM) was twice that

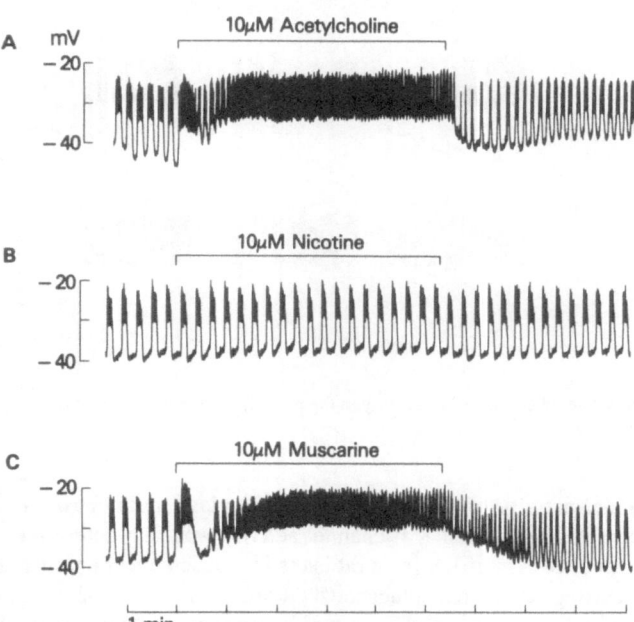

Figure 6. Effects of ACh (A), nicotine (B), and muscarine (C) on glucose-evoked electrical activity. ACh and cholinergic agonists were applied in the presence of glucose (11 mM) as indicated. Records from the same cell. (From Ref. 51 with permission.)

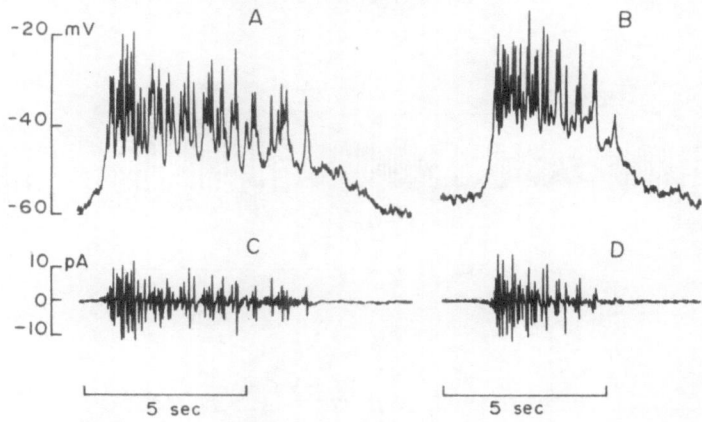

Figure 7.   Effects of ACh on ionic currents. Typical burst recorded (A) prior to the application of ACh and (B) in the presence of ACh. Ionic currents were calculated as $-C_m \, dV/dt$, where C was taken as 6 pF and dV/dt was evaluated from the bursts.

measured in the absence of ACh (Fig. 6A). Thus, it is clear that ACh-evoked additional stimulation of Ca^{2+} entry is related to the depolarization of the membrane and activation of voltage-gated Ca^{2+}-channels rather than receptor-operated Ca^{2+}-channels.

2.3.   Muscarinic Cholinergic Receptor Subtype M1 Is Involved in ACh Action

To determine the cholinergic receptor type involved in the ACh action, we applied to the same cell which had been exposed to ACh (Fig. 6A) nicotine (Fig. 6B) and muscarine (Fig. 6C). Application of muscarine (10 μM) but not nicotine (10 μM) was as effective as ACh in stimulating electrical activity.

The analysis of the electrical activity shown in Fig. 7 suggests that ACh stimulates further Ca^{2+} entry without affecting Ca^{2+} current during the action potentials. Ionic currents derived from two sample bursts recorded before (A) and during (B) the application of ACh are very similar. The amplitude of both inward and outward currents is shown in Fig. 7(C,D). In the β-cell, inward current is carried by Ca^{2+}[50] and outward current is carried by K^+.[50]

To determine the type of muscarinic receptor involved, we used pirenzepine, antagonist at the M1 receptor and gallamine, antagonist at the M2 receptor.[52–54] While pirenzepine impaired the ACh action on glucose-evoked electrical activity (Fig. 8A), gallamine was without effect (Fig. 8B).

Application of pirenzepine (1 μM) in the presence of ACh (1 μM) induced a transient cessation of the electrical activity, mimicking the effects of the removal of ACh (see Fig. 6A). However, while no measurable hyperpolarization of the membrane was observed with pirenzepine (Fig. 8A), removal of the ACh caused a clear hyperpolarization of the membrane (Fig. 6A). We have shown that muscarine caused a rapid depolarization of the β-cell membrane (Fig. 6C). This membrane depolarization was accompanied by a marked increase in electrical activity. Pirenzepine also blocked the

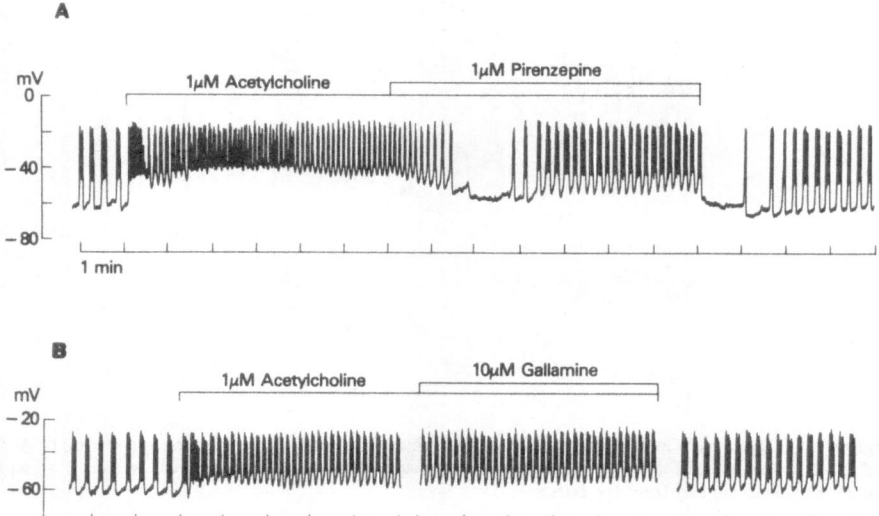

Figure 8. Effects of the muscarinic receptor blockers pirenzepine and gallamine on ACh stimulation. (A) First the islet was exposed to ACh (1 μM) and then to pirenzepine in the presence of glucose (11 mM) as indicated. (B) Effect of gallamine (10 μM) in the presence of ACh (1 μM) and glucose (11 mM). (From Ref. 51 with minor modifications.)

muscarine-evoked depolarization of the membrane and stimulation of electrical activity. This blockade by pirenzepine of the muscarine action was completely reversible (data not shown). Burst frequency measured toward the end of the application of ACH in the presence of pirenzepine showed that the concentration for 50% inhibition was ca. 0.25 μM (data not shown).

We have shown that membrane ionic currents associated with the generation of electrical activity are not affected by muscarinic receptor stimulation (see Fig. 7). It is possible that the K^+-channels controlling the membrane potential could be under muscarinic cholinergic receptor control as in other cells. Indeed, in other cell types, slow depolarization of the membrane in response to topic application of muscarinic agonists is commonly observed.[55–59] This membrane depolarization increases spontaneous electrical activity. It is now clear that cholinergic M1 receptor activation leads to a transient blockade of K^+-channels.[57,59] Thus, the excitatory effects of ACh and muscarine could also result from the blockade of K^+-channels in the β-cell membrane.

2.4. Catecholamines Induce a Transient Inhibition of Glucose-Evoked Calcium Entry and Insulin Release

Adrenaline and noradrenaline (NA) have been observed to induce a sustained inhibition of insulin release in perfused rat pancreas.[34] In contrast, we have observed that adrenaline and NA induced only a transient inhibition of electrical activity (Fig. 9). This might reflect dissociation between electrical activity and insulin release.

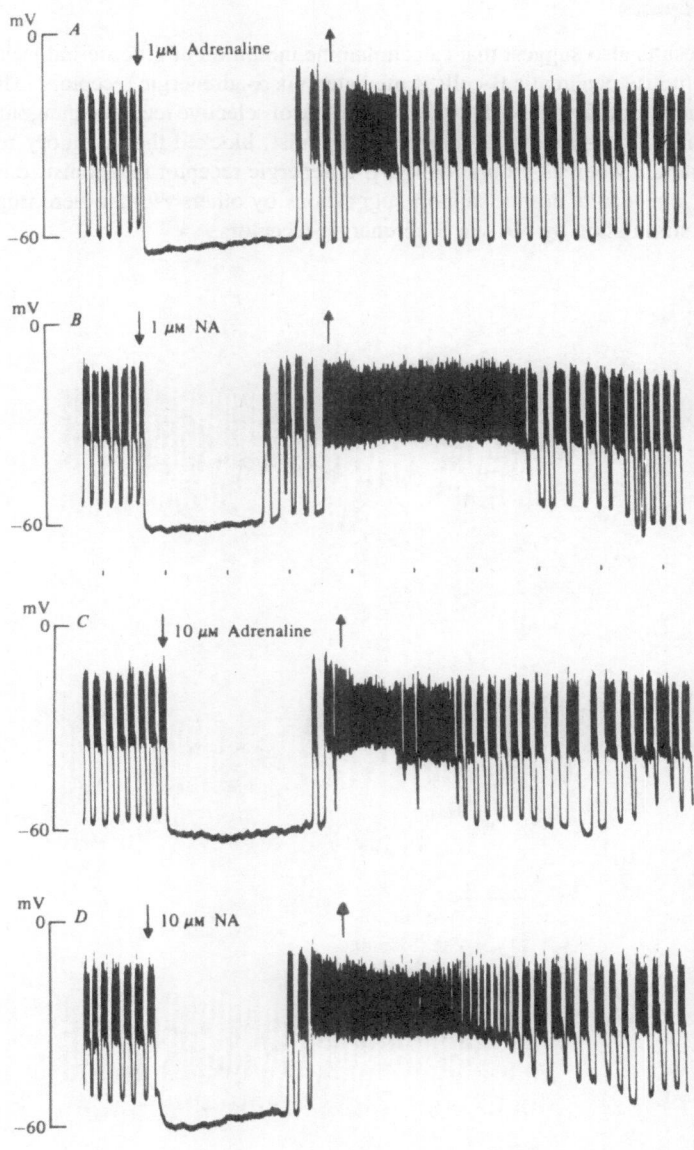

Figure 9. Effects of catecholamines on glucose-induced electrical activity. Adrenaline was applied between the arrow: 1 μ*M* for record (A) and 10 μ*M* for record (C). NA was applied at either 1 μ*M* (B) or 10 μ*M* (D). Time marks every 1 min. Glucose (11 m*M*) throughout. (From Ref. 60 with permission.)

2.5. Catecholamine Effects on the β-Cell Are Mediated by α-Adrenergic Receptors

Our results also suggest that catecholamine inhibition of glucose-induced electrical activity in mouse pancreatic β-cells is mediated via α-adrenergic receptors. The identification of the receptors involved is based on the use of selective receptor antagonists. Thus, phentolamine, an α-adrenergic receptor antagonist, blocked the inhibitory response to catecholamines, whereas propranolol, a β-adrenergic receptor antagonist, did not (Fig. 10). Thus, our results do not support suggestions by others[60] that there might exist a secondary stimulation by NA via β-adrenergic receptors.

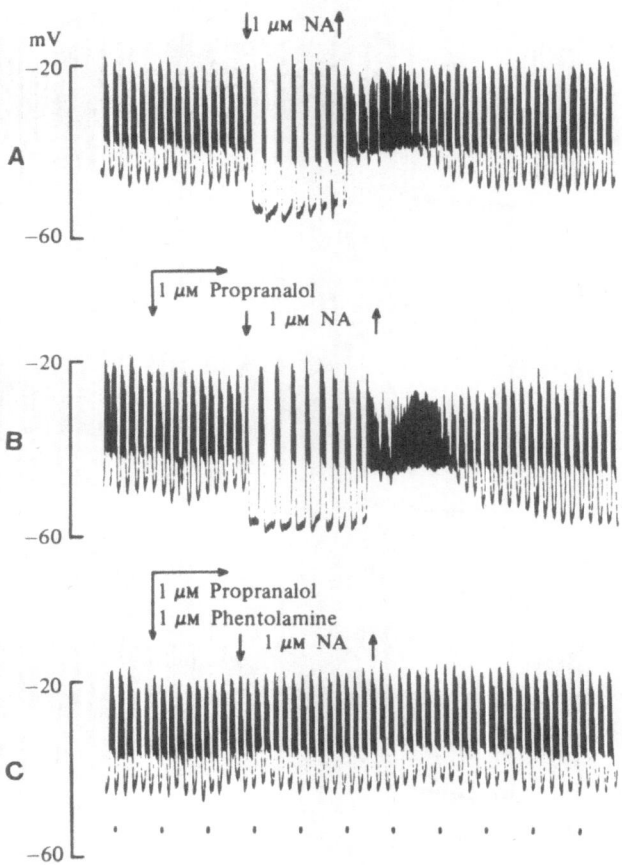

Figure 10. Effects of propranolol and propranolol plus phentolamine on Na-evoked inhibition of electrical activity. Glucose (11 mM) throughout. Time marks every 1 min. (From Ref. 59 with permission.)

2.6. α-Adrenergic Receptor Modulation of K^+-Channels

Our measurements of the effects of catecholamine on input resistance showed that, concomitant with the hyperpolarization of the cell membrane (see Fig. 9), there was a marked decrease in input resistance (data not shown). The measured decrease in input resistance concomitant with hyperpolarization of the cell membrane in response to Ca, indicates an increase in K^+ permeability during the inhibition of the electrical activity. The observation that tetraethylammonium (TEA) had very little effect on the hyperpolarization while quinine[47,48] substantially decreased it[60] suggests that the inhibition of electrical activity could be due stimulation of $[Ca^{2+}]_i$-activated K^+-channel. These observations are in accord with findings in guinea pig *Taenia coli* by Takai and Tomita[61] who also noted that quinine inhibits the adrenaline-induced membrane hyperpolarization.

3. CHOLINERGIC RECEPTOR CONTROLLED CATECHOLAMINE SECRETION FROM MEDULLARY CHROMAFFIN CELLS

Rojas et al. have recently shown[62] that it is possible to obtain real-time measurements of the kinetics of secretion of vesicle contents by monitoring ATP release from stimulated medullary chromaffin cells using luciferin–luciferase. ACh-evoked ATP secretion from freely suspended bovine medullary chromaffin cells was measured on-line by detecting the light emitted by a luciferin–luciferase mixture included in the extracellular medium.

ACh-stimulated ATP release shares many important properties with the process of catecholamine secretion in chromaffin cells. For example, (i) both secretory processes are evoked by activation of acetylcholine receptors; (ii) both processes depend on extracellular Ca^{2+} and are inhibited by Ca^{2+}-channel blockers such as Cd^{2+} and, (iii) both secretory processes are blocked by trifluoroperazine. The conclusion to be drawn from this comparison is that ATP is secreted by exocytosis together with the catecholamines. Thus, the technique can be used to monitor, in real time, the process of secretion of granular content.

We describe here the results of experiments designed to characterize the properties of cholinergic nicotinic and muscarinic receptor stimulation in bovine medullary chromaffin cells.

3.1. Properties of ACh-Evoked ATP Secretion

Figure 11A depicts a typical record of the time integral of the ATP release process evoked by ACh. It may be seen that immediately after the application of the ACh ($10~\mu M$), the signal increased rapidly and reached a steady-state level after 60 s. Because the ATP released by the cells is confined to the reaction chamber, the vertical axis represents the time integral of the ATP released. Thus, the time derivative of the record shown in Fig. 11B (inset) represents the time course of the rate of ATP release. It is apparent that the rate of ATP secretion rose rapidly from its basal level to a peak value and then returned to this basal level. Thus, ATP secretion was almost complete in less than 2 min, consistent with data for the rate of secretion of catecholamines.[63]

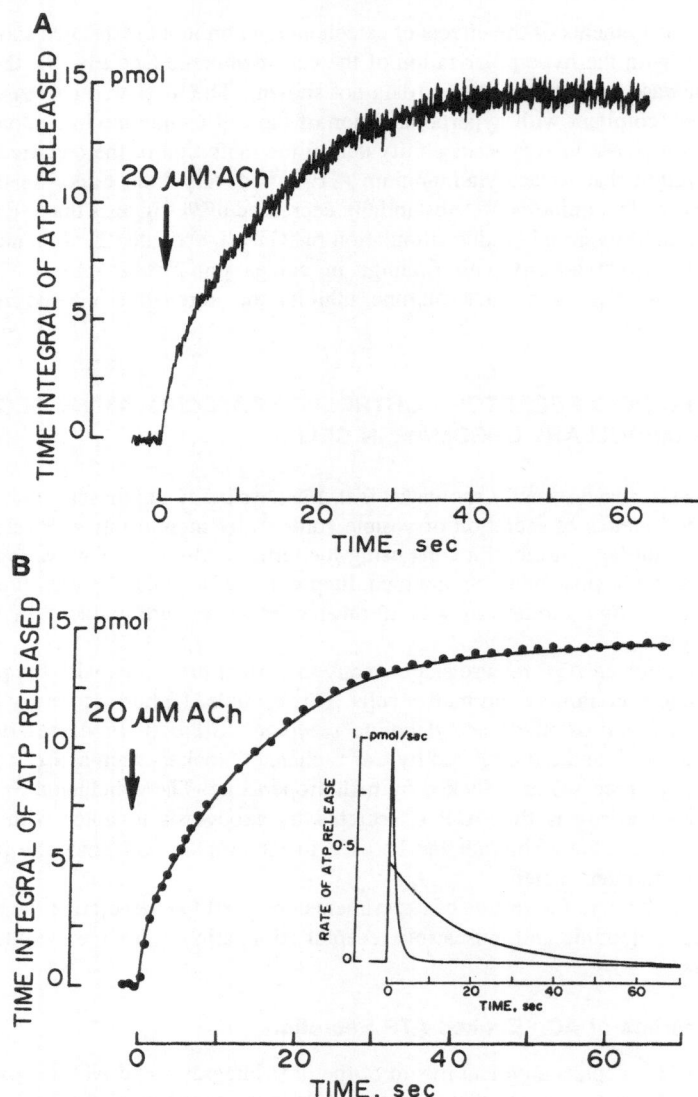

Figure 11. Kinetic properties of ACh (10 μM)-induced ATP secretion from bovine chromaffin cells. (A) The output from the photomultiplier. (B) Filled circles represent measurements made of the light signal shown above; solid line represents the best fit of the points by the double exponential function

$$\text{ATP} (t) = \text{ATP}_{max}[1 - \exp(-t/\tau_1)] + \text{ATP}_{max}[1 - \exp(t/\tau_2)]$$

Inset: Calculated rates of ATP release.

Application of ACh (1–100 μM) to chromaffin cells caused prompt release of ATP (Fig. 11), the extent of which depended on [ACh]. The midpoint of the response was obtained at an ACh concentration of about 15 μM. Because ACh-evoke ATP release from bovine chromaffin cells is blocked by D-tubocurarine (<10 μM) and hexamethonium (<10 μM), specific blockers of the nicotinic receptor channel, ACh-induced ATP secretion appears to be mediated by nicotinic receptor stimulation.[62]

On the other hand, while application of muscarine (1–50 μM) to bovine chromaffin cells alone does not evoke ATP secretion, ACh-evoked ATP release in the presence of atropine, a blocker of the muscarinic acetylcholine receptors, is smaller than in the absence of atropine. These effects of atropine (10 μM) on ATP release are illustrated in Fig. 12. In the absence of atropine, the ATP release evoked by acetylcholine (2.7 μM) amounted to 2% of the ATP in the cells. Further stimulation with the agonist (27.5 μM) elicited further release of ATP, about 16%. The ATP release in the presence of atropine (10 μM) amounted to 1% for the first dose of acetylcholine (2.7 μM) and, 9% for the second dose (27.5 μM), 56% of the release measured in the absence of the blocker of muscarinic receptors.

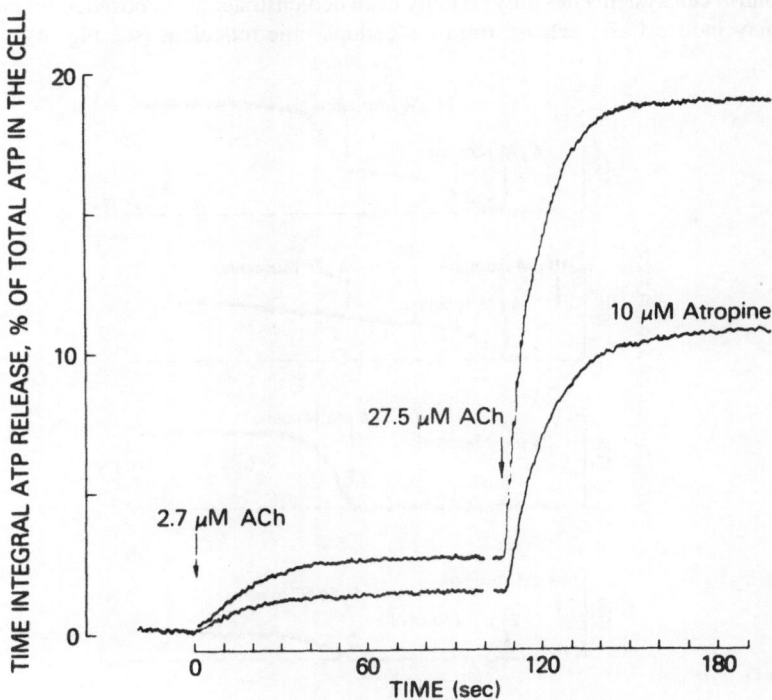

Figure 12. Effects of atropine on ACh-evoked ATP release. About 10,000 freely suspended bovine chromaffin cells in the reaction medium. Upper record: In the absence of atropine two consecutive applications of acetylcholine as indicated by the arrows. Lower record: Protocol as in the upper record but cells were preincubated in the presence of 10 μM atropine. Same calibration for both records. Experiment carried out at 18°C.

Cholinergic receptors in chromaffin cells can be desensitized by prolonged exposure to high concentrations of agonists. This property of the ACh receptors is also expressed in the case of the ATP release. Submaximal doses of nicotine (10 μM) can evoke substantial ATP release (up to 15% of the cellular ATP). Addition of a further identical dose 3 min later (20 μM) resulted in no further ATP release (data not shown). Control experiments showed that the application of a single 20-μM dose elicited more ATP release (up to 25%). Thus, this refractoriness of ATP secretion was not due to the depletion of ATP.

3.3. Synergistic Enhancement of Secretion by Nicotinic and Muscarinic ACh Receptors

Muscarinic receptor activation may directly activate a membrane-bound phospholipase-C, causing PtdIns(4,5)P_2 hydrolysis and generation of DAG and Ins(1,4,5)P_3.[64] Ins(1,4,5)P_3 may then mobilize Ca^{2+} from internal sources and Ca^{2+} may activate a cytosolic phospholipase-C, which enhances phosphatidylinositol hydrolysis in the cell interior (see Fig. 3). The relevance of this hypothetical scheme to the chromaffin cell systems has only recently been demonstrated.[65] Phosphatidyl inositols, too, may induce Ca^{2+} release from the endoplasmic reticulum (see Fig. 4), and this

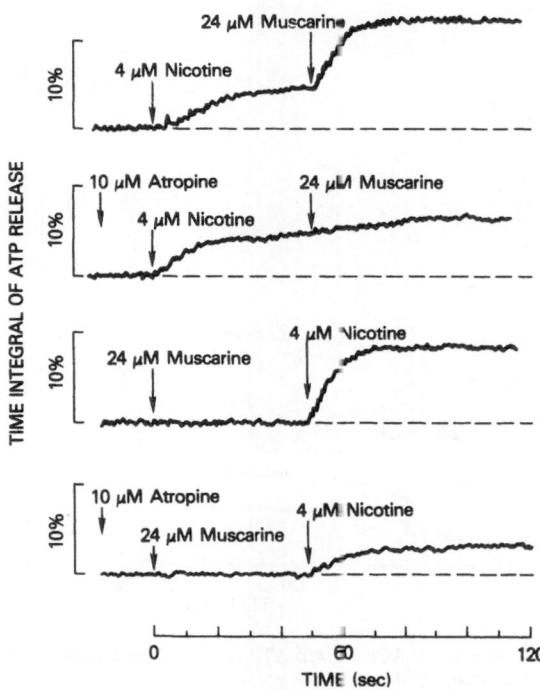

Figure 13. ATP release induced by sequential stimulation of the nicotinic and muscarinic receptors.

additional source of Ca^{2+} may play a fundamental role in stimulus–secretion coupling. Since this sequence of events is extremely rapid, it has been difficult to determine the temporal relationship between the multiple cellular events triggered by the activation of the receptors.

However, our technique can be used to measure, on-line, the secretory response and, thus, the time relationship between receptor activation and ATP release. For example, from sequential stimulation of the cholinergic receptors in chromaffin cells we learned that, provided the time interval between the two stimuli was less than 3 min, sequential nicotinic–muscarinic receptor stimulation always gave an enhanced ATP release. The results of a typical experiment are shown in Fig. 13. Initial stimulation with nicotine (4 μM) elicited a moderate release of ATP (about 2%). Stimulation with muscarine induced an additional ATP release (8%). Reversing the sequence of the stimuli to muscarinic–nicotinic, although the cells responded only to nicotine, the extent of the response was augmented by the pretreatment with muscarine from 2% to 9%.

4. CONCLUDING REMARKS

We have shown that stimulus–secretion coupling may or may not involve membrane receptors. The best example of the first mode of coupling is the pancreatic β-cell and, of the receptor mediated coupling, is the medullary chromaffin cell. However, modulation of the secretory response always involve membrane receptors to specific hormones or neurotransmitters. Indeed, the secretory response of both islet β-cells and adrenal chromaffin cells is modulated by specific receptors.

The GTP-binding regulatory proteins constitute the physical link between the membrane receptor and the effector, which, in many cells, is a membrane enzyme such as the adenylate cyclase and the phospholipase-C. Activation or inhibition of the effector gives rise to either an increase or a decrease in the rate of production of intracellular messengers such as cAMP, Ins(1,4,5)trisphosphate and DAG. These messengers act on intracellular target proteins such as the Ca^{2+} release channel known to be present in the membrane of the reticulum. The messengers may also activate cytosolic enzymes such as the protein kinases.

While in the field of excitation–contraction coupling mechanical activation has been described in molecular terms, in the field of secretion by exocytosis the details have yet to be elucidated.

REFERENCES

1. Gilman, A. G., 1984, G proteins and dual control of adenylate cyclase, *Cell* **36:** 577–579.
2. Smigel, M., Katada, T., Northup, J. K., Bokoch, G. M., Ui, M., and Gilman, A. G., 1984, Mechanisms of guanine nucleotide-mediated regulation of adenylate cyclase activity, *Adv. Cyclic Nucleotide Protein Phosphor. Res.* **17:** 1–18.
3. Smigel, M. D., Northup, J. K., and Gilman, A. G., 1982, Characteristics of the guanine nucleotide-binding regulatory component of adenylate cyclase, *Rec. Proc. Horm. Res.* **38:** 601–626.
4. Bourne, H. R., Medynski, D., Vandop, C., Sullivan, K., and Chang, F. H., 1985, Genetic and

functional studies of pertussis toxin substrates, in: *Pertussis Toxin* (R. D. Sekura, J. Moss, and M. Vaughan, eds.) Academic, Orlando, pp. 167–184.

5. Birnbaumer, L., Codina, J., Sunyer, T., Rosenthal, W., Hilderbrandt, J., Cerione, R. A., Caron, M. G., Lefkowitz, R. J., and Sekura, R. D., 1985, Structural and functional properties of N_s and N_i, the regulatory components of adenyl cyclases, in: *Pertussis Toxin* (R. D. Sekura, J. Moss, and M. Vaughan, eds.), Academic, Orlando, pp. 77–104.

6. Nakadate, T., Nakari, T., Muraki, T., and Kato, R., 1980, Regulation of plasma insulin level by α_2-adrenergic receptors, *Eur. J. Pharmacol.* **65:** 421–424.

7. Katada, T., and Ui, M., 1977, Perfusion of the pancreas isolated from pertussis-sensitized rats: Potentiation of insulin secretory responses due to β-adrenergic stimulation, *Endocrinology* **101:** 1247–1255.

8. Katada, T., and Ui, M., 1979, Effect of in vivo pretreatment of rats with a new protein purified from *Bordetella pertussis* on in vitro secretion of insulin: Role of calcium, *Endocrinology* **104:** 1822–1827.

9. Yajima, M., Hosada, K., Kanbayashi, Y., Nakamura, T., Nogimori, K., Mizushima, Y., and Ui, M., 1978a, Islets-activating protein (IAP) in *Bordetella pertussis* that potentiates insulin secretory responses of rats, *J. Biochem* **83:** 295–303.

10. Yajima, M., Hosoda, K., Kanbayashi, Y., Nakamura, T., Takahashi, I., and Ui, M., 1987b, Biological properties of islets-activating protein (IAP) purified from the culture medium of *Bordetella pertussis*, *J. Biochem.* **83:** 305–312.

11. Rodbell, M., Birnbaumer, L., Pohl, S. L., and Krans, H. M. J., 1971, The glucagon-sensitive adenyl cyclase system in plasma membranes of rat liver, *J. Biol. Chem.* **246:** 1877–1882.

12. Jakobs, K. H., Saur, W., and Schultz, G., 1978, Inhibition of platelet adenylate cyclase by epinephrine requires GTP, *FEBS Lett.* **85:** 167–170.

13. Hildebrandt, J. D., Hanoune, J., and Birnbaumer, L., 1982, Guanine nucleotide inhibition of cyc-S49 mouse lymphoma cell membrane adenylyl cyclase, *J. Biol. Chem.* **257:** 14723–14725.

14. Michell, R. H., 1975, Inositol phospholipids and cell surface receptor function. *Biochem. Biophys. Acta* **415:** 81–147.

15. Creba, J. A., Downes, P., Hawkins, P. T., Brewster, G., Michell, R. H., and Kirk, C. J., 1983, Rapid breakdown of phosphatidylinositol 4-phosphate and phosphatidylinositol 4,5-bisphosphate in rat hepatocytes stimulated by vasopressin and other Ca-mobilizing hormones, *Biochem. J.* **212:** 733–747.

16. Griffin, H. D., Hawthorne, J. N., and Sykes, M., 1979, A calcium requirement for the phosphatidylinositol response following activation of presynaptic muscarinic receptors, *Biochem. Pharmacol.* **28:** 1143–1147.

17. Hokin, L. E., and Hokin, M. R., 1956, The actions of pancreozymin in pancreas slices and the role of phospholipids in enzyme secretion, *J. Physiol.* **132:** 442–453.

18. Hokin, L. E., and Hokin, M. R., 1958a, Phosphoinositides and protein secretion in pancreas slices, *J. Biol. Chem.* **233:** 805–810.

19. Hokin, M. R., and Hokin, L. E., 1958b, Enzyme secretion and the incorporation of ^{32}P into phospholipids of pancreas slices, *J. Biol. Chem.* **233:** 967–977.

20. Fain, J. N., and García-Sainz, J. A., 1980, Role of phosphatidylinositol turnover in α_1 and of adenylate cyclase inhibition in α_2 effects of catecholamines. *Life Sci.* **26:** 1183–1194.

21. Ullrich, S., and Wollheim, C., 1985, Expression of both β_1- and β_2- adrenoceptors in an insulin-secreting cell line: Parallel studies of cytosolic free Ca^{2+} and insulin release, *Mol. Pharmacol.* **28**(2): 100–106.

22. Farese, R. V., Sabir, M. A., and Vandor, S. L., 1979, Adrenocorticotropin acutely increases adrenal phosphoinositides, *J. Biol. Chem.* **254:** 6842–6844.

23. Prentki, M., and Wollheim, C. B., 1984, Cytosolic free Ca^{2+} in insulin-secreting cells and its regulation by isolated organelles, *Experientia* **40**(10): 1052–1060.

24. Vergara, J., Tsien, R., and Delay, M., 1985, Inositol 1,4,5-trisphosphate: A possible chemical link in excitation–contraction coupling in muscle, *Proc. Natl. Acad. Sci. USA* **82**(18): 6352–6356.

25. Berridge, M. J., and Irvine, R. F., 1984, Inositol trisphosphate, a novel second messenger in cellular signal transduction, *Nature* **312:** 315–321.
26. Tyson, C. A., Vande-Zande, H., and Green, D. E., 1976, Phospholipids as ionophores, *J. Biol. Chem.* **251:** 1326–1332.
27. Takai, Y., Kishimoto, A., Kikkawra, U., Mori, T., and Nishizuka, Y., 1979, Unsaturated di-acylglycerol as a possible messenger for the activation of calcium-activated, phospholipid-dependent protein kinase system, *Biochem. Biophys. Res. Commun.* **91:** 1218–1224.
28. Bergman, E. N., and Miller, R. E., 1973, Direct enhancement of insulin secretion by vagal stimulation of the isolated pancreas, *Am. J. Physiol.* **236:** E139–E146.
29. Porte, J., D., Girardier, L., Seydoux, J., Kanazawa, Y., and Posternak, J., 1973, Neural regulation of insulin secretion in the dog, *J. Clinical Investigation* **52:** 210–214.
30. Milner, R. D. G., and Hales, C. N., 1968, The interaction of various inhibitors and stimuli of insulin release studied with rabbit pancreas in vitro, *Biochemical J.* **113:** 472–479.
31. Lerner, R. L., and Porte, Jr., D., 1971, Epinephrine: Selective inhibition of the acute insulin response to glucose, *J. Clinical Investigation* **50:** 2453–2457.
32. Sorenson, R. L., Elde, R. P., and Seybold, V., 1979, Effect of norepinephrine on insulin, glucagon, and somatostatin secretion in isolated perifused rat islets, *Diabetes* **28:** 899–904.
33. Gagerman, E., Idahl, L.-A., Meissner, H. P., and Täljedal, I.-B., 1978, Insulin release, cGMP, cAMP, and membrane potential in acetylcholine-stimulated islets, *Am. J. Physiol.* **4:** E493–E500.
34. Wollheim, C. B., and Sharp, G. W. G., 1981, Regulation of insulin release by calcium, *Physiol. Rev.* **61:** 914–973.
35. Best, L., and Malaisse, W. J., 1983, Stimulation of phosphoinositide breakdown in rat pancreatic islets by glucose and carbamylcholine, *Biochem. Biophys. Res. Commun.* **116**(1): 9–16.
36. Best, L., and Malaisse, W. J., 1984, Nutrient and hormone-neurotransmitter stimuli induce hydrolysis of polyphosphoinositides in rat pancreatic islets, *Endocrinology* **115**(5): 1814–1820.
37. Dunlop, M., Shaw, M., Dimitriadis, E., Gurtler, V., Wark, J., and Larkins, R. G., 1988, Evidence that muscarinic receptors in islet cells are not coupled functionally to adenylate cyclase through the inhibitory guanine nucleotide binding protein (Ni), *Horm. Metab. Res.* **20**(3): 150–153.
38. Mathias, P. C., Best, L., and Malaisse, W. J., 1985. Stimulation by glucose and carbamylcholine of phospholipase-C in pancreatic islets, *Cell. Biochem. Funct.* **3**(3): 173–177.
39. Rubin, R. P., 1982, *Calcium and Cellular Function.* Plenum, New York.
40. Dean, P. M., and Matthews, E. K., 1970, Glucose-induced electrical activity in pancreatic islet cells, *J. Physiol.* **210:** 255–264.
41. Meissner, H. P., and Schmelz, H., 1974, Membrane potential of β-cells in pancreatic islets, *Pfluegers Arch.* **351:** 195–206.
42. Matthews, E. K., and Sakamoto, Y., 1975, Electrical characteristics of pancreatic islet cells, *J. Physiol.* **246:** 421–437.
43. Atwater, I., Dawson, C. M., Eddlestone, G. T., and Rojas, E., 1981, Voltage noise measurements across the pancreatic β-cell membrane: Calcium channel characteristics, *J. Physiol.* **314:** 195–212.
44. Meissner, H. P., and Preissler, M., 1980, Ionic mechanisms of the glucose-induced membrane potential changes in β-cells, *Horm. Metab. Res. (Suppl.)* **10:** 91–99.
45. Ribalet, B., and Beigelman, P. M., 1980, Calcium action potentials and potassium permeability activation in pancreatic β-cells, *Am. J. Physiol.* **239:** C124–C133.
46. Atwater, I., Ribalet, B., and Rojas, E., 1978, Cyclic changes in potential and resistance of the β-cell membrane induced by glucose in islets of Langerhans from mouse, *J. Physiol.* **278:** 117–139.
47. Atwater, I., Ribalet, B., and Rojas, E., 1979, Mouse pancreatic β-cells: Tetraethylammonium blockage of the potassium permeability increase induced by depolarization, *J. Physiol.* **288:** 561–574.
48. Atwater, I., Dawson, C. M., Ribalet, B., and Rojas, E., 1979, Potassium permeability activated by intracellular calcium ion concentration in the pancreatic β-cell, *J. Physiol.* **288:** 575–588.
49. Atwater, I., Dawson, C. M., Scott, A., Eddlestone, G., and Rojas, E., 1980, The nature of the oscillatory behavior in electrical activity from pancreatic β-cell, *Horm. Metab. Res. (Suppl.)* **10:** 100–107.

50. Atwater, I., Carroll, P., and Li, M. X., 1989, Electrophysiology of the pancreatic β-cell, in: *Molecular and Cellular Biology of Diabetes Mellitus* (B. Draznin, S. Melmed, and D. Le Roith, eds.), Volume 1, pp. 49–68.
51. Santos, R. M, and Rojas, E., 1989, Muscarinic receptor modulation of glucose-induced electrical activity in mouse pancreatic β-cells, *FEBS Lett.* **249:** 411–417.
52. Birdsall, N. J. M., Hulme, E. C., and Stockton, J. M., 1983, Muscarinic receptor heterogeneity, in: *Proc. International Symposium on Subtypes of Muscarinic Receptors* (B. I. Hirschowitz, R. Hammer, A. Giachetti, J. K. Keirns, and R. R. Levine, eds.), Supplement to Trends in Pharmacological Sciences, pp. 4–8.
53. Watson, M., Vickrpy, T. W., Roeske, W. R., and Yamamura, H. I., 1983, Subclassification of muscarinic receptors based upon the selective antagonist pirenzepine, in: *Proc. International Symposium on Subtypes of Muscarinic Receptors* (B. I. Hirschowitz, R. Hammer, A. Giachetti, J. K. Keirns, and R. R. Levine, eds.), Supplement to Trends in Pharmacological Sciences, pp. 9–11.
54. Mitchelson, F., 1983, Heterogeneity in muscarinic receptors: Evidence from pharmacological studies with antagonists, in: *Proc. International Symposium on Subtypes of Muscarinic Receptors* (B. I. Hirschowitz, R. Hammer, A. Giachetti, J. K. Keirns, and R. R. Levine, eds.), Supplement to Trends in Pharmacological Sciences, pp. 12–16.
55. Adams, P. R., and Brown, D. A., 1982, Synaptic inhibition of the M-current: Slow excitatory postsynaptic potential mechanism in bullfrog sympathetic neurons, *J. Physiol.* **332:** 263–272.
56. Hashiguchi, T., Kobayashi, H., Tosaka, T., and Libet, B., 1982, Two muscarinic depolarizing mechanisms in mammalian sympathetic neurones, *Brain Res.* **242:** 378–382.
57. Jones, S. W., 1985, Muscarinic and peptidergic excitation of bullfrog sympathetic neurons, *J. Physiol.* **366:** 63–87.
58. Kawatani, M., Rutigliano, M., and Degroat, W. C., 1985, Depolarization and muscarinic excitation induced in a sympathetic ganglion by vasoactive intestinal polypeptide, *Science* **229:** 879–881.
59. Kuffler, S. W., and Selnowski, T. J., 1983, Peptidergic and muscarinic excitation at amphibian synapses, *J. Physiol.* **341:** 257–278.
60. Santana de Sa, S, Ferrer, R., Rojas, E., and Atwater, I., 1983, Effects of adrenaline and noradrenaline on glucose-induced electrical activity of mouse pancreatic β-cell, *Quar. J. Phys.* **68:** 247–258.
61. Takai, A., and Tomita, T., 1980, Effects of quinine on the α-action of adrenaline in the guinea pig *Taenia coli*, *J. Physiol.* **308:** 54–55P.
62. Rojas, E., Pollard, H. B., and Heldman, E., 1985, Real-time measurements of acetylcholine-induced release of ATP from bovine medullary chromaffin cells, *FEBS Lett.* **185:** 323–327.
63. Oka, M., Isosaki, M., and Watanabe, J., 1980, Calcium flux and catecholamine release in isolated bovine adrenal medullary chromaffin cells: Effects of nicotinic and muscarinic stimulation, *Adv. Biosci.* **36:** 29–33.
64. Prentki, M., Biden, T. J., Danjicc, D., Irvine, R. F., Berridge, M. J., and Wollheim, C. B., 1984, Rapid mobilization of Ca from rat insulinoma microsomes by inositol-1,4,5-trisphosphate, *Nature* **309:** 562–565.
65. Forsberg, E. J., Rojas, E., and Pollard, H. B., 1986, Muscarinic receptor enhancement of nicotine-induced catecholamine secretion may be mediated by phosphoinositide metabolism in bovine adrenal chromaffin cells, *J. Biol. Chem.* **261**(11): 4915–4920.

Chapter 8

Glucose Dose Response of Pancreatic β-Cells: Experimental and Theoretical Results

Arthur Sherman, Patricia Carroll,
Rosa M. Santos, and Illani Atwater

1. INTRODUCTION

Islets of Langerhans are functional units of thousands of endocrine cells located within the pancreas. Approximately 85% of the cells within the islet are β-cells. β-cells secrete insulin, the hormone critical in the regulation of fuel metabolism. The major focus of insulin action is the maintenance of glucose homeostasis; however, insulin also regulates storage and breakdown of fats. Through complex interactions between the β-cells and target tissues of insulin action (liver, muscle, and fat cells), fuel homeostasis is closely regulated in both fasting and fed states.

Diabetes mellitus is a disease in which relative (type II) or absolute (type I) insulin deficiency exists. In the non-insulin-dependent type of diabetes (type II), resistance of target tissues to insulin action plays a major role; however, it has recently been appreciated that impaired β-cell sensitivity to glucose is a characteristic of the disease and that the β-cell must also play a role in pathogenesis.[1]

The response of the endocrine pancreas to glucose is due to a combination of a variable threshold for activity from cell to cell and a graded response to glucose of each cell. It is interesting to speculate that this variability in and of itself may have some

ARTHUR SHERMAN • Mathematics Research Branch, National Institutes of Diabetes and Digestive and Kidney Diseases, National Institutes of Health, Bethesda, Maryland 20892.   PATRICIA CARROLL, ROSA M. SANTOS, and ILLANI ATWATER • Laboratory of Cell Biology and Genetics, National Institutes of Diabetes and Digestive and Kidney Diseases, National Institutes of Health, Bethesda, Maryland 20892.

relevance physiologically. It may well be that the threshold to glucose response participates in setting the fasting blood sugar with the graded response, providing a spare capacity within the range of response of the islet. There is some evidence that alterations of a glucose-sensitive K-permeability are present in animal models of diabetes.[2,3] Changes in glucose-sensitive K-permeability may contribute to the development of fasting hyperglycemia, which is the hallmark of diabetes.[4]

This chapter will focus on the glucose-sensitive K-permeability of the β-cell, the regulation of the burst pattern of glucose-evoked electrical activity, and possible mechanisms controlling the threshold and graded response to glucose. Experimental and theoretical modeling approaches, and the profitable interaction between them, will be discussed.

2. ELECTROPHYSIOLOGY

The most striking electrophysiological fact about the β-cell is the bursting activity seen when one records intracellularly from a single cell in an intact islet of Langerhans.[5] In the presence of stimulatory concentrations of glucose there are periods of rapid spiking alternating with silent phases. (See Fig. 2A for examples.) Above the glucose threshold for bursting, there is a graded electrical response to glucose that correlates closely to insulin secretion in these cells.[6] It has been established that K-permeability in the β-cell is modulated by glucose.[7,8] K-channel activity thus governs the β-cell membrane potential, which provides a link between metabolism and electrical activity. Atwater et al.[9] proposed a qualitative model of this behavior based on calcium feedback. Reduction of K-permeability caused by reduction of intracellular calcium and the consequent depolarization lead to calcium entry, which triggers insulin release and restores K-permeability. This negative-feedback system allows the β-cell to sense glucose. Both the burst pattern and the graded electrical response have been modeled mathematically.[10]

2.1. Channels

There are two groups of channels that have been identified in the β-cell: fast, voltage-gated channels which control spiking, and slow, chemically-gated K-channels which have been proposed to play a major role in the control of the threshold and graded response to glucose. Note that it is not the channels themselves that are slow, but the modulation of their total conductance by chemical processes. The first group consists of an inward calcium channel and an outward, delayed rectifying potassium channel, similar to the squid giant axon delayed rectifier. The properties of these channels were measured by Rorsman and Trube.[11]

There are two slow K-channels: the sulfonylurea-sensitive, metabolite-regulated, ATP-blockable K-channel (K-ATP) which in symmetrical K$^+$ solutions has a conductance of 55 pS (inward) and 30 pS (outward)[12–15] (see Ashcroft[16] for a review); and the $[Ca^{2+}]_i$-activated and voltage-dependent (K-Ca) channel which in symmetrical K$^+$ has a conductance of 200 pS (inward and outward).[17] Recently, both of these channels have been described to be glucose sensitive.[18,19] Figure 1 shows the effect of 10 mM glucose to alter the activity of both the K-ATP and K-Ca channels. The K-ATP channel is

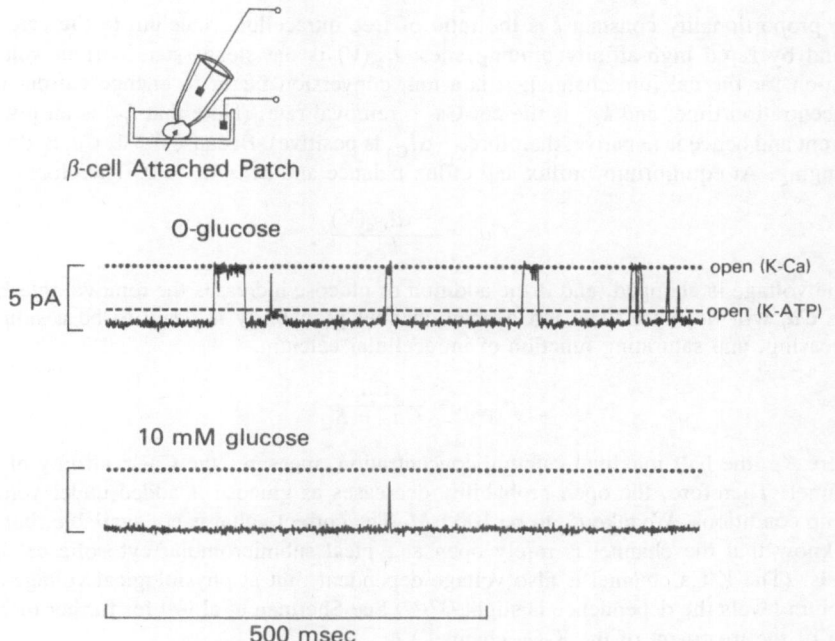

β-cell Attached Patch

Figure 1. 10 m*M* Glucose inhibits K-Ca and K-ATP channel activity. The experiment was obtained from cultured rat β-ells. Cell-attached configuration as shown in the insert. K-current flowing from the cell to the pipet (outward current) is shown by upward deflections. K-ATP and K-Ca openings are as indicated by the lines. (Solutions (m*M:*) pipet 135 NaCl, 5 KCl, 2.6 CaCl$_2$, 10 HEPES pH 7.4, bath 140 KCl, 2.6 CaCl$_2$, 10 HEPES pH 7.4. Pipet potential 0 mV).

maximally blocked by this concentration of glucose.[20] The K-Ca channel is maximally blocked by 20 m*M* glucose.[19]

We attribute the effects of glucose on the two channels to different causes: The K-ATP channel closes because ATP increases, while the K-Ca channel closes because calcium is pumped more rapidly out of the cell and calcium levels decline.

2.2. Calcium Handling

We can quantify the effect of glucose on intracellular calcium levels by examining an equation for calcium handling introduced by Plant,[21] who followed Frankenhaeuser and Hodgkin's[22] treatment of accumulation of potassium in the peri-axonal space. The equation says in words that the rate of change of intracellular calcium is proportional to the difference between calcium influx through the calcium channel and calcium removal by calcium pumps:

$$\frac{dCa_i}{dt} = f(-\alpha I_{Ca}(V) - k_{Ca}Ca_i) \tag{1}$$

The proportionality constant f is the ratio of free intracellular calcium to the calcium bound by rapid high-affinity binding sites; $I_{Ca}(V)$ is the steady-state current voltage relation for the calcium channel; α is a unit conversion factor to change current into concentration/time; and k_{Ca} is the net Ca^{2+} removal rate. (Note that I_{Ca} is an inward current and hence is negative; therefore, $-\alpha I_{Ca}$ is positive). Because $f \ll 1$, Ca_i is slowly changing. At equilibrium, influx and efflux balance and $dCa_i/dt = 0$. Therefore

$$Ca_i = \frac{-\alpha I_{Ca}(V)}{k_{Ca}} \qquad (2)$$

If the voltage is clamped, and if the addition of glucose increases the removal rate k_{Ca}, then Ca_i will decrease. The K-Ca open channel probability is taken to be a simple, increasing, and saturating function of intracellular calcium,

$$p_{open} = \frac{Ca_i}{Ca_i + K_d} \qquad (3)$$

where K_d, the half-maximal calcium concentration, measures the Ca^{2+} affinity of the channel. Therefore, the open probability decreases as glucose is added under voltage clamp conditions. We take K_d to be 100 μM. The correct value is not available, but we do know that the channel is rarely open at typical submicromolar cytosolic calcium levels. (The K-Ca channel is also voltage dependent, but at physiological voltage and calcium levels the dependence is slight.[17,23] See Sherman et al.[24] for further discussion of the treatment of the K-Ca channel.)

3. BURSTING

3.1. Intracellular Recording

Below the threshold for glucose-induced electrical activity, the membrane potential is hyperpolarized and there is no electrical activity (not shown). Figure 2A illustrates the graded oscillatory response of the β-cell membrane potential to superthreshold increments of glucose. Figure 2B illustrates the graded response reproduced by the mathematical model. The experimental records show the steady state during a 6-min exposure to each glucose concentration. As the glucose is increased from 11 to 15 mM, the burst period increases then declines as glucose is increased further. The silent phases become shorter and the time spent in the active phase (depolarized and spiking) increases. Eventually the silent phases disappear and the cell exhibits continuous spike activity (active phase 100%) at a concentration of glucose near 22 mM (not shown). Absolute membrane potentials are not affected by glucose. The relative time spent at the active phase increases in a sigmoidal fashion as a function of glucose concentration (Fig. 3A), as does the average action potential frequency (not shown).

3.2. Modeling Bursting

Our modeling approach follows closely the Chay–Keizer model.[10] Many variants on this basic idea have been proposed. Chay[25] obtains bursting by using calcium

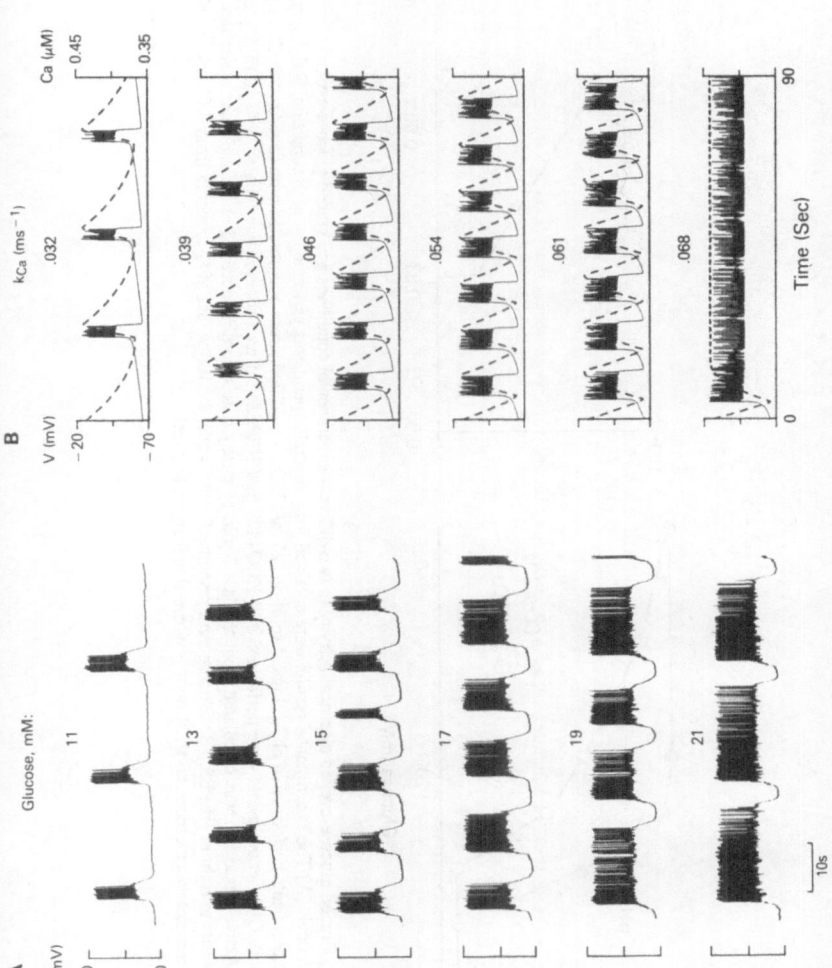

Figure 2. Comparison of effect of graded increments of glucose (experimental) concentration or k_{Ca} (theoretical) on measured or predicted membrane electrical activity. (A) Steady-state portions of a continuous record obtained from a single β-cell with a 6-min exposure to each glucose concentration. (B) Voltage and intracellular calcium time courses obtained by numerical simulation with the model. In each panel the K-ATP conductance, $g_{K\text{-ATP}}$, is 50 pS, while the calcium removal rate, k_{Ca}, is varied to simulate the effect of increasing glucose. Note that minimum and maximum calcium levels are independent of k_{Ca} until continuous spiking is attained.

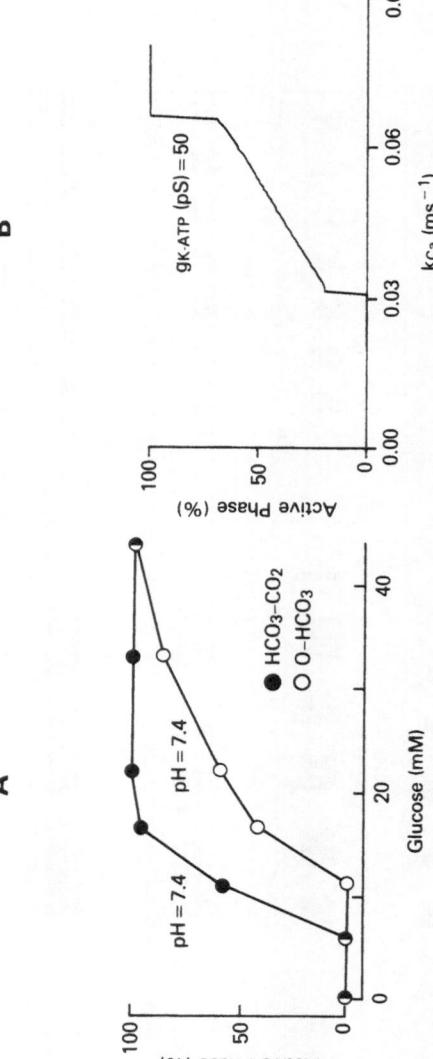

Figure 3. Graded glucose-evoked electrical activity in two different experimental conditions and graded glucose-evoked electrical activity of model. (A) The experimental results were obtained from several experiments performed at each condition. Solid symbols represent experiments performed in physiological solution (25 mM HCO_3) modified Krebs equilibrated with 95% O_2/5% CO_2 to pH 7.4 (external). Other experiments were performed in 0-HCO_3 (20 mM HEPES) buffered solution equilibrated with 100% O_2 to the same pH. Active phase (%) was calculated from the last 3 min of steady-state recording obtained at each glucose concentration. Threshold shifts to right in HEPES. (B) Glucose dose response for the model for 100 values of k_{Ca} ranging from 0.02 ms^{-1}, where the cell is completely quiescent, to 0.08 ms^{-1}, where it spikes continuously.

activation of a voltage-gated, *outward* potassium current, and also[26] by calcium *inactivation* of a voltage-gated *inward* calcium current. Keizer[27] has proposed a model in which the K-ATP channel is indirectly calcium sensitive through mitochondrial interactions. The model presented here follows most closely that of Sherman et al.[24] who adapted measurements by Rorsman and Trube[11] on the voltage-gated calcium and potassium channels. For additional analysis and interpretation of this class of models see Atwater and Rinzel[28] and Rinzel.[29]

As in the standard Hodgkin–Huxley[30] formulation the membrane potential is determined by the equation

$$C_m \frac{dV}{dt} = -I_{ion} \tag{4}$$

where C_m is the membrane capacitance and I_{ion} is composed of the voltage-gated calcium and potassium currents and the slowly varying potassium currents:

$$I_{ion} = I_{Ca} + I_K + g_{K\text{-slow}}(V - V_K) \tag{5}$$

Here $g_{K\text{-slow}}$ is the slow potassium conductance and V_K is the potassium reversal potential. The voltage-gated currents vary on a fast time scale (~ 10 ms[11]) and are responsible for spiking. The slow potassium current modulates the onset and termination of bursting, and varies on the time scale of a burst (~ 10 s; see Fig. 2A).

The slow potassium conductance can be decomposed into the calcium-activated potassium (K-Ca) conductance and the ATP-blockable (K-ATP) conductance:

$$g_{K\text{-slow}} = \bar{g}_{K\text{-Ca}} \frac{Ca_i}{Ca_i + K_d} + g_{K\text{-ATP}} \tag{6}$$

where we have used equation (3) for the fraction of K-Ca channels open, $\bar{g}_{K\text{-Ca}}$ is the maximal K-Ca conductance, K_d is the half-maximal calcium concentration, and $g_{K\text{-ATP}}$ is the K-ATP conductance. The K-ATP conductance is independent of both calcium and voltage, but it is directly blocked by glucose metabolism.

The key point is that the slow potassium conductance increases with intracellular calcium. We reemphasize that it is not the channels themselves that are slow, but the total conductance, because it is dependent on slowly varying intracellular calcium. We will see that bursting is initiated and terminated by oscillations in Ca_i through its effect on the slow potassium conductance.

In the absence of glucose $g_{K\text{-ATP}}$ is high, so $g_{K\text{-slow}}$ is large, the membrane potential rests near V_K, and bursting cannot occur. As glucose is increased from 0 to 7 mM the membrane slowly depolarizes, presumably because $g_{K\text{-ATP}}$ is reduced. Figure 5 in Ashcroft et al.[20] indicates that the K-ATP channel is mostly blocked by 7 mM glucose, the threshold for bursting. Above 7 mM glucose, $g_{K\text{-ATP}}$ drops off very slowly and the influence of intracellular calcium on the K-Ca channel dominates. We therefore take $g_{K\text{-ATP}}$ to be independent of glucose in the superthreshold range. Later we will explore the consequences if $g_{K\text{-ATP}}$ depends on pH.

We now consider the regime in which glucose is sufficiently high that further increases in glucose have a negligible effect on $g_{K\text{-ATP}}$. The dominant effect of glucose on $g_{K\text{-slow}}$ is now indirect, through the calcium removal rate. In order for bursting to occur, glucose must be high enough to stimulate calcium removal sufficiently. If the removal rate is too low, Ca_i will be high, $g_{K\text{-}Ca}$ will be large, and the membrane potential will remain below the threshold for the voltage-gated channels. In very high glucose, the removal rate is large, Ca_i is low, $g_{K\text{-}slow}$ is small, and the membrane potential remains high enough to activate the voltage-gated calcium and potassium channels continuously. The result is continuous spike activity.

There is an intermediate range of glucose levels in which there is an alternation between silent and active phases. Figure 2A illustrates this regime where the relative proportion of time spent spiking increases gradually with glucose. At all levels of glucose in this intermediate range, the same periodic cycle of events occurs: During the silent phase calcium efflux exceeds influx causing Ca_i and $g_{K\text{-}slow}$ to decrease and the membrane to depolarize slowly. When Ca_i reaches a critical minimum value Ca_1, and $g_{K\text{-}slow}$ reaches a corresponding minimum value g_1, the membrane begins to depolarize rapidly as the threshold for activating the calcium channels is reached. The increased potential also turns on the delayed-rectifier potassium channels and establishes a repetitive oscillation of inward calcium and outward potassium currents. This is the spiking or active phase of a burst. The spikes bring in calcium faster than it can be removed, which gradually increases $g_{K\text{-}slow}$. When Ca_i reaches a critical maximum value Ca_2, and $g_{K\text{-}slow}$ reaches its maximum g_2, the combined outward delayed-rectifier and slow potassium currents drive the membrane potential below the threshold for the calcium channel. The delayed-rectifiers also switch off, but the slow potassium current repolarizes the membrane to near V_K. This restores the silent phase of a burst. The efflux of calcium once again dominates the influx, the membrane slowly depolarizes, and the cycle is repeated. This process is illustrated in the first five panels of Fig. 2B, where V and Ca_i are plotted as functions of time.

4. GLUCOSE DOSE RESPONSE

4.1. Membrane Potential

The glucose dose response is defined electrically by the threshold for depolarization which occurs at about 7 mM and the graded increase in the active phase (active phase percent) which occurs between 7 and 22 mM glucose. The normal threshold and range of response is illustrated graphically in Fig. 3A in which steady-state records such as those in Fig. 2A were analyzed and the active phase percent was plotted as a function of increasing glucose concentration. Figure 3A also compares the electrical dose response in the presence of physiologically buffered HCO_3/CO_2 solutions (filled circles) to that in Na-Hepes buffered solution (open circles), both with external pH 7.4. It can be seen that solutions buffered with Na-Hepes buffer substituting for HCO_3/CO_2 induced a shift in the threshold and the range of the graded glucose response. This will be discussed in detail in the following sections. The mathematical model also predicts a threshold and a graded response to glucose as shown in Fig. 2B. The dose response of insulin secretion as a function of glucose also exhibits a sigmoidal relationship; however, the onset of electrical activity occurs before appreciable increase in secretion.[6,31]

4.2. Modeling Dose Response as Calcium Removal Rate

To model the effect of changing glucose levels we follow Chay and Keizer[10] in interpreting an increase in glucose as an increase in the calcium removal rate, k_{Ca} [see equation (1)]. Figure 2B was produced by allowing k_{Ca} to vary while keeping g_{K-ATP} constant at 50 pS. Figure 3B summarizes the dose response for the model. Note that the active phase percent is plotted against removal rate, not glucose, and that equal increments in glucose do not have the same effect on percent active phase as equal increments in k_{Ca}. The active phase percent is an approximately linear function of k_{Ca}, with sudden jumps at the bursting threshold and at the onset of continuous spiking. We believe that such sudden jumps would not be seen in a real islet because they would be washed out by statistical variation in individual cell properties. Himmel and Chay[32] carried out similar simulations and obtained sigmoidal curves by postulating a Hill relationship between k_{Ca} and glucose. We do not assume any particular relationship between k_{Ca} and glucose, only that they increase together.

The critical levels of g_{K-slow}, g_1 and g_2, at which the cell switches between the silent and active phases are determined by the characteristics of the fast channels, which depend only on voltage, not calcium or glucose. *This implies that in the bursting regime the cell experiences the same range of slow potassium conductance regardless of the glucose level.* In our formulation, using equation (6) with g_{K-ATP} fixed, this means that the critical minimum and maximum values of intracellular calcium, Ca_1 and Ca_2, corresponding to g_1 and g_2, respectively, are also independent of glucose levels in the bursting regime. This was first pointed out by Rinzel.[29] Thus, the minimum and maximum calcium levels are the same in each of the first five panels of Fig. 2B; only the frequency of the calcium oscillation changes. A counterintuitive consequence of the model is that the levels of calcium are the same during the silent and active phases; only the direction of change differs. The existence of oscillations in intracellular calcium is a key theoretical prediction of the model. Although intracellular calcium has been measured in cell suspensions, it has not been measured in single islets, and thus oscillations in calcium have not yet been seen experimentally. There is, however, substantial indirect experimental evidence for calcium oscillations. Dawson et al.[33] showed that silent phase duration increases with extracellular calcium and decreases with glucose, suggesting that silent phase duration depends on the levels of intracellular calcium at the end of a burst and on a glucose sensitive removal rate. Perez-Armendariz and Atwater[34] found oscillations in extracellular potassium and calcium that are synchronized with the electrical bursts. They found that there is a net efflux of calcium from the cells during the silent phase and a net influx during the active phase, and vice versa for potassium. Chay and Keizer[35] incorporated the effects on membrane potential of oscillations in extracellular potassium into their mathematical model.

To understand how varying the removal rate controls the percent active phase, consider the following: When k_{Ca} is small, it takes a long time to bring Ca_i down to Ca_1, but only a short burst to bring it back up to Ca_2. Therefore, the active phase percent is small. When k_{Ca} is large, Ca_i comes down fast but many spikes are needed to overcome the large efflux and bring it up. Therefore, the active phase percent is large. If k_{Ca} is large enough, the amount of calcium brought in by a single spike balances the amount removed during the spike and there is continuous spiking (Fig. 2B, bottom panel).

5. EFFECTS OF pH

5.1. Membrane Potential

Insulin release from rat islets of Langerhans has been shown to be significantly inhibited in HCO_3/CO_2 free solutions.[36] We have observed that removal of HCO_3/CO_2 buffered solution and substitution with Na-Hepes at the same $[pH]_0$ (7.4) caused a hyperpolarization and abolished the characteristic burst pattern exhibited in 11 mM glucose.[37] We hypothesized that this might be due to a shift in the threshold and/or the graded glucose-evoked electrical response. This was indeed the case as demonstrated by the comparison of the electrical dose response curves in the two buffers shown in Fig. 3A.

We hypothesized that the shift in the threshold might be due to an increase in K-permeability, possibly mediated by an intracellular pH change. Although the external pH, $[pH]_0$, is the same in the two cases, Lindstrom and Sehlin[38] have shown that in mouse islets of Langerhans, $[pH]_i$ was increased in Na-Hepes buffered solutions when compared to HCO_3/CO_2 buffered solutions, and that changes in $[pH]_i$ followed changes in $[pH]_0$ to some extent.

Because both the K-ATP and K-Ca channels have been proposed to play a role in the threshold and graded response to glucose, we used the technique of cell-attached patch-clamp in isolated cells in culture to investigate the effect on the activity of these two channels of altering the buffer to Na-Hepes and then altering the $[pH]_0$ of the Na-Hepes buffer. K-ATP channel activity was higher in the Na-Hepes buffered solution pH 7.4 and significantly decreased by HCO_3/CO_2 solutions.[37] This effect to decrease K-ATP channel activity was mimicked by lowering the $[pH]_0$ of the Na-Hepes buffer to 7.0, as shown in Fig. 4. The K-Ca channel was not significantly altered by these changes.

Using the technique of intracellular voltage recordings in whole islets, we also investigated whether changing $[pH]_0$ would shift the threshold and graded response to glucose in the Na-Hepes buffered solutions. Figure 5A compares the electrical dose response curve to glucose at different $[pH]_0$ in the Na-Hepes buffer. Lowering $[pH]_0$ restored the threshold and increased the steepness of the response.

The patch-clamp and intracellular recording experiments together support the hypothesis that the shifts in the dose response curves in Figs. 3A and 5A are due to a reduction in K-ATP channel activity. Changing pH, however, could have manifold, complex effects in the cell. We therefore turn to the model where the effect of K-ATP channel block on the dose response can be isolated and studied.

5.2. Modeling and Theoretical Analysis

To simulate the effects of pH on glucose dose response, the value of $g_{K\text{-ATP}}$ was arbitrarily increased to 100 pS for alkalinization and reduced to 0 pS for acidification. As before, k_{Ca} was varied to model the effect of glucose. The theoretical curves in Fig. 5B show the same qualitative shifts as the experimental ones: the threshold is shifted to the left and the curve becomes steeper as pH is lowered.

For further insight into these processes, we have reproduced three sample threshold solutions in Fig. 6, representing the 50% active points on the dose response curves for the acid, normal, and alkaline cases. The voltage time courses are nearly identical, but

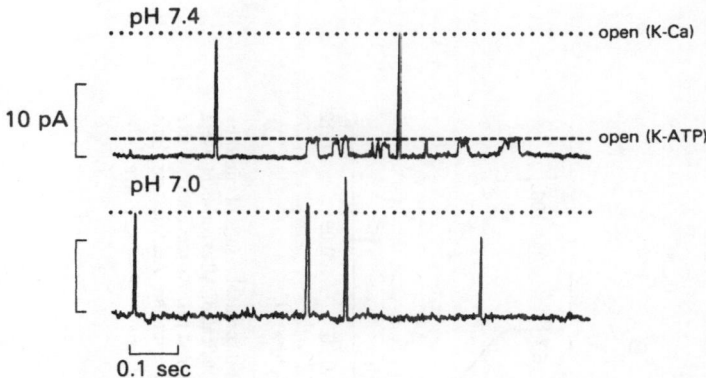

Figure 4. Effect of lowering $[pH]_0$ on K-channel activity in 0-HCO_3 HEPES buffered solutions. Results are portions of a continuous record from the same experiment. The experiment was obtained from cultured β-cells from normal rats using the cell-attached configuration as shown in the insert of Fig. 1. As before, K-current flowing from the cell to the pipet (outward current) is shown by upward deflections. K-ATP and K-Ca openings are as indicated by the lines. (Solutions (mM:) pipet: 135 NaCl, 5 KCl, 10 HEPES pH 7.4, bath 125 KCl, 15 NaCl, 3.6 $MgCl_2$, 1μM $CaCl_2$, 10 HEPES pH 7.4 (upper panel) or 7.0 (lower panel). Glucose was absent throughout. Pipet holding potential was − 50 mV. Lowering $[pH]_0$ depresses the activity of the K-ATP channel without significantly affecting the K-Ca channel.

the minimum and maximum calcium levels decrease as $g_{K\text{-}ATP}$ increases. This effect was previously described by Himmel and Chay,[32] although they interpreted the reduction of $g_{K\text{-}ATP}$ as an increase in glucose rather than as an increase in pH. Calcium levels drop as $g_{K\text{-}ATP}$ rises because the minimum and maximum values of the total slow potassium conductance, $g_{K\text{-}slow}$, g_1 and g_2, are independent of $g_{K\text{-}ATP}$. Therefore, as $g_{K\text{-}ATP}$ rises, the contribution of $g_{K\text{-}Ca}$ to $g_{K\text{-}slow}$ falls, and thus Ca_1 and Ca_2 fall as well. A geometric interpretation of this phenomenon was given in Rinzel et al.[39] Here we show that the magnitudes of the changes in Ca_1 and Ca_2 can be predicted algebraically by solving equation (6) for calcium. We can simplify the calculation by using the fact that $K_d \gg Ca_i$, so that $p_{open} \approx Ca_i/K_d$. Then

$$Ca_1 \approx \frac{g_1 - g_{K\text{-}ATP}}{\bar{g}_{K\text{-}Ca}} K_d \qquad (7)$$

and similarly for Ca_2. The reader may use Table I to check that the results of the numerical simulation agree with the predictions of equation (7).

From the change in calcium levels we can deduce the shift in the threshold as $g_{K\text{-}ATP}$ varies. Because the calcium efflux is $k_{Ca} \cdot Ca_i$, if Ca_i decreases, k_{Ca} will have to increase in order to balance a given calcium influx. To be more precise, observe that in each of the simulations in Figs. 2B and 6 there is a threshold value of V, call it V_1, at which the membrane potential suddenly shoots up. The value of V_1 is the same for all values of $g_{K\text{-}ATP}$ and k_{Ca}. The criterion for initiation of spiking is that net calcium flux be outward at V_1 (otherwise potassium conductance would be increasing and the membrane would repolarize toward rest). It is also evident from Figs. 2B and 6 that when V

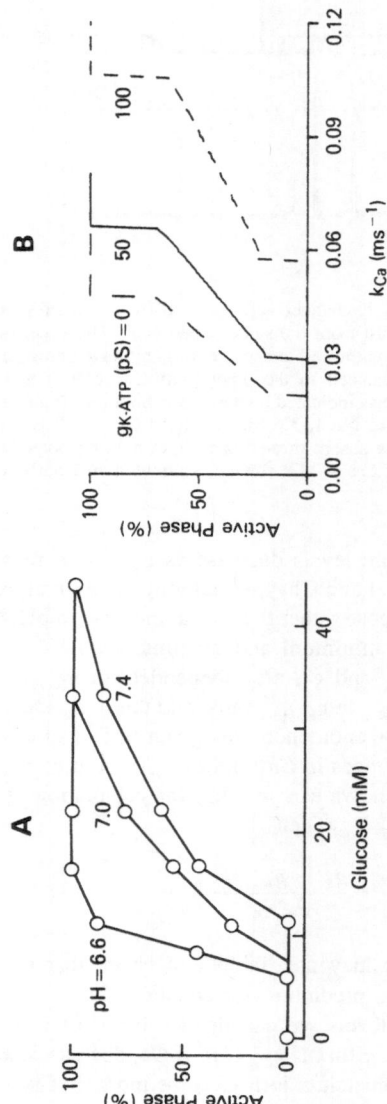

Figure 5. Effect of varying [pH]$_0$ on the measured graded glucose-evoked response compared with theoretical results obtained by altering g_{K-ATP} in the model. (A) The results obtained from several experiments performed at [pH]$_0$ as indicated. All experiments were performed in 0-HCO$_3$ (20 mM Na-HEPES) buffered solutions equilibrated with 100% O$_2$ to the indicated pH. As previously, active phase (%) was calculated from the last 3 min of steady-state recording obtained at each glucose concentration. Acidification shifts threshold to left and steepens curve. (B) Glucose dose response for the model at various values of g_{K-ATP}. Increasing g_{K-ATP} shifts curve to right while decreasing g_{K-ATP} shifts it to the left.

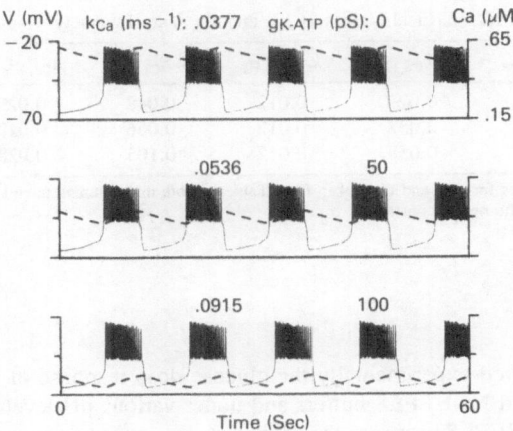

Figure 6. Voltage and calcium time courses for the mathematical model. Curves represent the 50% active cases at three values of $g_{K\text{-ATP}}$. The voltage records are almost identical, but calcium levels drop as $g_{K\text{-ATP}}$ increases.

$= V_1$, intracellular calcium is at its minimum value, Ca_1. Therefore, the minimum value of k_{Ca} at which bursting can occur, denoted $k_{Ca,1}$, is that removal rate at which influx just balances efflux:

$$k_{Ca,1} \, Ca_1 = -\alpha I_{Ca}(V_1) = \text{const} \qquad (8)$$

From equation (8) we conclude that $k_{Ca,1}$ will vary inversely with Ca_1 (i.e., increase with $g_{K\text{-ATP}}$), and the numerical simulations confirm this. See Table II. A similar result holds for the maximal value of k_{Ca} for bursting, $k_{Ca,2}$. In fact, because the percent active phase is very nearly linear in k_{Ca} between the maximal and minimal values, we could have obtained the dose response curves for the cases $g_{K\text{-ATP}} = 0$ and $g_{K\text{-ATP}} = 100$ from the baseline case to a good approximation by a simple algebraic transformation.

Table I. Predicted Dependence of Minimum and Maximum Ca_i on $g_{K\text{-ATP}}{}^a$

$g_{K\text{-ATP}}$	Ca_1	Ca_2
0	0.53	0.61
50	0.36	0.44
100	0.20	0.27

$^a g_1 = 159$ pS, $g_2 = 182$ pS, $K_d = 100$ μM, $\bar{g}_{K\text{-Ca}} = 30{,}000$ pS. Calcium values taken from numerical simulations.

Table II.   Critical Values of k_{Ca} and Ca_i Vary Inversely[a]

g_{K-ATP}	$k_{Ca,1}$	$k_{Ca,1} \cdot Ca_1$	$k_{Ca,2}$	$k_{Ca,2} \cdot Ca_2$
0	0.022	0.012	0.048	0.029
50	0.032	0.012	0.066	0.029
100	0.058	0.012	0.105	0.028

[a]Values for Ca_1 and Ca_2 taken from Table I. Note that columns three and five are nearly constant.

6. DISCUSSION

We have explored experimentally the glucose dose response of pancreatic β-cells with HCO_3/CO_2 and Na-HEPES buffers and under various pH levels. Replacement of HCO_3/CO_2 by Na-HEPES broadens the dose response curve (measured as percent active phase) and shifts its threshold to the right (higher glucose levels). The effect can be partially reversed by acidification. While acidification increases electrical activity and metabolism in general, insulin secretion is inhibited.[40,41] Changes in pH have complex effects intracellularly, altering enzymatic reactions, protein binding, pump rates, and presumably other processes that are important to the secretion process.

We have also demonstrated, in cell-attached patches in the absence of glucose, that lowering $[pH]_0$ blocks K-ATP channel activity, but has no effect on the K-Ca channel. The results are in agreement with recent observations showing that in cell-attached patches, the K-ATP channel is sensitive to manipulations that alter $[pH]_i$ (NH_4Cl, Na-proprionate).[42] Also, in agreement with the hypothesis that pH affects the K-ATP channel, but not the K-Ca channel, are experiments showing that addition of HCO_3/CO_2, which has been shown to acidify β-cells,[38] blocked the K-ATP channel, but did not affect the K-Ca channel.[37,43] HCO_3/CO_2 did not have a direct effect on the K-ATP channel in excised patches, whereas lowering $[pH]_0$ to 6.5 or below could directly block the channel in excised patches.[44] Our studies are different than a previous report indicating that the K-Ca channel is pH sensitive[45]; those studies, however, were performed on excised patches in the presence of EGTA. Our experimental protocol allows us to subject both channels to the same treatment in conditions where they are simultaneously active.

Based on our experiments we have proposed that the effects of lowering pH are due to alterations of K-ATP channel activity. In patch-clamp experiments the K-ATP channel is completely blocked by pH 7.0, while in intracellular potential recordings further lowering of pH continues to induce shifts in the glucose dose response curve. While the patch-clamp provides a powerful tool to investigate effects at the single-channel level directly, there are many differences between the techniques used in intracellular voltage recordings and single-channel current recordings that make correlations of observations in the two systems difficult. Nonetheless, by changing to Na-HEPES buffered solutions, we observe a shift in the threshold as measured by intracellular recordings that correlates with an increase in K-ATP channel activity measured by patch-clamp current recordings. The threshold can be shifted back by lowering the $[pH]_0$ of the Na-Hepes buffered solutions that correlates with a block of K-ATP channel activity in current recordings.

We cannot assess the relative participation of the two potassium channels in the graded glucose response in these studies; however, the data support a role for the K-ATP channel in the threshold response as has been suggested by others.[46] Another piece of evidence is that blocking this channel with sulfonylureas, which are currently believed to be selective blockers of the K-ATP channel, can elicit continuous electrical activity below the usual threshold of glucose-evoked electrical activity.[47–49]

Our hypothesis is supported by numerical simulations with a variant of the Chay–Keizer model in which acidification is identified with reducing the total conductance (g_{K-ATP}) of the K-ATP channels, and increase of glucose is identified with increased calcium removal rate (k_{Ca}). We cannot directly compare the shapes of the experimental and theoretical dose response curves because we have not attempted to relate pH to g_{K-ATP} or glucose to k_{Ca} quantitatively. The model does confirm, however, that under acidification the dose response curve steepens and its threshold shifts to the left. The model further predicts that there is no significant shift in burst frequency at equivalent percent active levels. Preliminary experimental results confirm this, and further analysis of the model (see below) suggests that this is an important feature to study systematically in order to choose between alternative theories of pH action in the β-cell.

We see that the model can be used to guide the interpretation and design of experiments. Simulations with the model show the correct qualitative behavior as "pH" is varied. Further, the quantitative dependence on g_{K-ATP} of minimum and maximum calcium levels and of the minimum and maximum k_{Ca} for bursting can be determined by analysis of the model. These results give us more confidence in, but do not prove, our hypothesis that the pH effects are due to a block of the K-ATP channel. Strictly speaking, the above results apply only to the idealized model of the β-cell, but the insights may shed light on the behavior of the actual islet. Indeed, it is only the simplifications of the model that make analysis possible at all.

The model also makes predictions about phenomena that are not yet directly accessible to experiment. One such prediction is that intracellular calcium levels should rise as pH is lowered. (There is no contradiction between this effect and the failure of lower pH to stimulate the K-Ca channel in single-cell, cell-attached patch-clamp experiments (Fig. 4). In the latter case calcium influx is clamped because voltage is clamped by high K^+ in the bath. In the whole islet case, the potential is not clamped and lowering pH results in longer bursts of action potentials which bring in more calcium.) The dependence of intracellular calcium on g_{K-ATP} was reported by Himmel and Chay[32] who carried out similar parameter studies, and was analyzed by Rinzel et al.,[39] but they attributed the reduction of g_{K-ATP} to increase of glucose. We prefer to consider g_{K-ATP} as essentially independent of glucose above threshold and to model the primary effect of glucose as an increase in calcium removal rate. It is possible that glucose affects both g_{K-ATP} and k_{Ca} simultaneously. In that case, one would view Fig. 5B as the result of pure variations in g_{K-ATP} and k_{Ca}. From that diagram one can predict the effect of any proposed joint variation of the two parameters. If g_{K-ATP} is reduced by lowering pH as well as by increasing glucose, the results about shifting the threshold and steepening the dose response curves would still hold. The dose response curves would merely become a little steeper, and increasing glucose would increase intracellular calcium levels. Because intracellular calcium has not yet been measured experimentally in an intact islet during bursting, we cannot choose between these possibilities.

6.1. Alternative Models

There remains a question of whether the shift in dose response curves seen in Fig. 5B could be achieved theoretically in other ways. One alternative is to assume that $g_{K\text{-ATP}}$ is independent of pH, but that the dependence of the theoretical removal rate on glucose is pH-dependent. For example, one could follow Himmel and Chay[32] and assume that k_{Ca} depends on glucose as follows:

$$k_{Ca} = k_{Ca}^* \frac{G^n}{K_G + G^n}$$

where G represents glucose and k_{Ca}^* the maximal removal rate. If k_{Ca}^* increases with acidification, then a qualitatively correct shift in the threshold and steepening of the dose response curves would be obtained. A difficulty with this model is that if lowering pH enhances calcium removal, then it should suppress K-Ca activity in Fig. 4, just as glucose does in Fig. 1. We cannot rule out the possibility that there is an effect, but that it is masked by compensating pH dependence of calcium binding and of the calcium affinity of the K-Ca channel.

In fact, it has been suggested[32] that the primary effect of acidification is reduction of the affinity of the K-Ca channel for calcium (i.e., K_d is increased). Simple scaling arguments, confirmed by simulations, show that this would have the desired effect on the dose response curves. A side effect, however, is that burst frequency would decrease as pH is lowered, and this is not observed experimentally. Also, the patch-clamp experiment of Fig. 4 shows that pH does not affect the K-Ca channel, at least in 0 glucose. The same qualitatively correct shift in dose responses curves, together with a reduction in frequency, would be found in a model in which the K-ATP channel is indirectly calcium sensitive[27] and its affinity for calcium is reduced by lowering pH. Chay and Kang[50] present a model with no slow potassium conductance in which bursting is obtained by slow calcium inactivation of the voltage-gated calcium channel. They model the effect of acidification as a suppression of the calcium inactivation. They report an increase in calcium as pH is lowered, just as we do. Similar scaling arguments account for the rise in calcium and also predict a decrease in burst frequency for their model, which we did not observe experimentally.

6.2. Conclusions

While pH has multiple effects in cells, and we cannot definitively rule out other explanations of the experimental data, our modeling does indicate that the shift in the threshold and the steepening of the glucose dose response curve could be explained by a direct block of the K-ATP channel by acidification.

ACKNOWLEDGMENTS.   We wish to thank John Rinzel for contributing to the theoretical discussions that led to this article and making helpful editorial suggestions. We thank Rosa Santos for producing Fig. 2A and MinXu Li for performing the patch clamp experiments represented in Figs. 1 and 4. Arthur Sherman was supported by a National Research Council–N. I. H. Research Associateship. Patricia Carroll was supported by an E.

Clarence Rice Fellowship of the American Diabetes Association, Washington, D.C. affiliate.

REFERENCES

1. Halter, J. B., Ward, W. K., Porte, D., Best, J. D., and Pfeiffer, M. A., 1985, Glucose regulation in non-insulin-dependent diabetes mellitus, *Am. J. Med.* **79**(2B): 6–12.
2. Meissner, H. P., and Schmidt, H., 1976, The electrical activity of pancreatic β-cells of diabetic mice, *FEBS Lett.* **67**: 371–374.
3. Rosario, L. M., Atwater, I., and Rojas, E., 1985, Membrane potential measurements in islets of Langerhans from ob/ob obese mice suggest an alteration in $[Ca^{2+}]$-activated K^+ permeability, *Q. J. Exp. Physiol.* **70**: 137–150.
4. National Diabetes Data Group, 1979, Classification and diagnosis of diabetes mellitus and other categories of glucose intolerance, *Diabetes* **28**: 1039–1057.
5. Dean, P. M., and Mathews, E. K., 1970, Glucose-induced electrical activity in pancreatic islet cells, *J. Physiol. (Lond.)* **210**: 255–264.
6. Scott, A. M., Atwater, I., and Rojas, E., 1981, A method for the simultaneous measurement of insulin release and β-cell membrane potential in single mouse islets of Langerhans, *Diabetologia* **21**: 470–475.
7. Henquin, J. C., 1978, D-glucose inhibits potassium efflux from pancreatic islet cells, *Nature* **271**: 271–273.
8. Atwater, I., Ribalet, B., and Rojas, E., 1978, Cyclic changes in potential and resistance of the β-cell membrane induced by glucose in islets of Langerhans from mouse, *J. Physiol. (Lond.)* **278**: 117–139.
9. Atwater, I., Dawson, C. M., Scott, A., Eddlestone, G., and Rojas, E., 1980, The nature of the oscillatory behavior in electrical activity for pancreatic β-cell, *Horm. Metab. Res. (Suppl.)* **10**: 100–107.
10. Chay, T. R., and Keizer, J., 1983, Minimal model for membrane oscillations in the pancreatic β-cell, *Biophys. J.* **42**: 181–190.
11. Rorsman, P., and Trube, G., 1986, Calcium and delayed potassium currents in mouse pancreatic β-cells under voltage clamp conditions, *J. Physiol. (Lond.)* **374**: 531–550.
12. Cook, D. L., and Hales, C. N., 1984, Intracellular ATP directly blocks K^+ channels in pancreatic β-cells, *Nature* **311**: 271–273.
13. Ashcroft, F. M., Harrison, D. E., and Ashcroft, S. J. H., 1984, Glucose induces closure of single potassium channels in isolated rat pancreatic β-cells, *Nature* **312**: 446–448.
14. Trube, G., Rorsman, P., and Ohno-Shosaku T., 1986, Opposite effects of tolbutamide and diazoxide on the ATP-dependent $K+$ channel in mouse pancreatic β-cells, *Pflügers Arch.* **407**: 493–499.
15. Misler, S., Falke, L. C., Gillis, K., McDaniel, M. L., 1986, A metabolite-regulated potassium channel in rat pancreatic β-cells, *Proc. Natl. Acad. Sci. USA* **83**: 7119–7123.
16. Ashcroft, F. M., 1987, Adenosine 5'-triphosphate-sensitive potassium channels, *Annu. Rev. Neurosci.* **11**: 97–118.
17. Findlay, I., Dunne, M. J., and Peterson, O. H., 1985, High-conductance K^+ channel in pancreatic islet cells can be activated and inactivated by internal calcium, *J. Membr. Biol.* **83**: 169–175.
18. Atwater, I., Li, M-X., Rojas, E., and Stutzin, A., 1988, Glucose reduces both ATP-blockable and Ca-activated K-channel activity in cell-attached patches from rat pancreatic β-cells in culture, *Biophys. J.* **53**(2), Pt. 2: 145a (abstr.).
19. Ribalet, B., Eddelstone, G. T., Ciani, S., 1988, Glucose modulation of two K-channels in an insulin-secreting cell line, *Biophys. J.* **53**(2), Pt. 2: 460a [abstr.].
20. Ashcroft, F. M., Harrison, D. E., and Ashcroft, S. J. H., 1986, A potassium channel modulated by glucose metabolism in rat pancreatic β-cells, *Adv. Exp. Med. Biol.* **211**: 53–62.

21. Plant, R. E., 1978, The effects of Ca^{2+} on bursting neurons: A modeling study, *Biophys. J.* **21:** 217–237.

22. Frankenhaeuser, B., and Hodgkin, A. L., 1956, The after-effects of impulses in the giant nerve fibres of *Loligo*, *J. Physiol. (Lond.)* **131:** 341–376.

23. Velasco, J. M., and Petersen, O. H., 1987, Voltage-activation of high-conductance K^+ channel in the insulin-secreting cell line RINm5F is dependent on local extracellular Ca^{2+} concentration, *Biochim. Biphys. Acta* **896:** 305–310.

24. Sherman, A., Rinzel, J., and Keizer, J., 1988, Emergence of organized bursting in clusters of pancreatic β-cells by channel sharing, *Biophys. J.* **54:** 411–425.

25. Chay, T. R., 1986, On the effect of the intracellular calcium-sensitive potassium channel in the bursting pancreatic β-cell, *Biophys. J.* **50:** 765–777.

26. Chay, T. R., 1987, The effect of inactivation of calcium channels by intracellular Ca^{2+} ions in the bursting pancreatic β-cell, *Cell Biophys.* **11:** 77–90.

27. Keizer, J. E., 1988, Electrical activity and insulin release in pancreatic β-cells, *Math. Biosci.* **90:** 127–138.

28. Atwater, I., and Rinzel, J., 1986, The β-cell bursting pattern and intracellular calcium, in: *Ionic Channels in Cells and Model Systems* (R. Latorre, ed.), Plenum, New York, pp. 353–362.

29. Rinzel, J., 1985, Bursting oscillations in an excitable membrane model. in: *Ordinary and Partial Differential Equations* (B. D. Sleeman and R. J. Jarvis, eds.), Springer-Verlag, New York, pp. 304–316.

30. Hodgkin, A. L., and Huxley, A. F., 1952, A quantitative description of membrane current and its application to conduction and excitation in nerve, *J. Physiol. (Lond.)* **117:** 205–249.

31. Atwater, I., Gonçalves, A., Herchuelz, A., Lebrun, P., Malaisse, W. J., Rojas, E., and Scott, A., 1984, Cooling dissociates glucose-induced insulin release from electrical activity and cation fluxes in rodent pancreatic islets, *J. Physiol. (Lond.)* **348:** 614–627.

32. Himmel, D., and Chay, T. R., 1987, Theoretical studies on the electrical activity of pancreatic β-cells as a function of glucose, *Biophys. J.* **51:** 89–107.

33. Dawson, C. M., Atwater, I., and Rojas, E., 1984, The response of pancreatic β-cell membrane potential to potassium-induced calcium influx in the presence of glucose, *Q. J. Exp. Physiol.* **69:** 819–830.

34. Perez-Armendariz, E., and Atwater, I., 1986, Glucose-evoked changes in $[K^+]$ and $[Ca^{2+}]$ in the intercellular spaces of the mouse islet of Langerhans, in: *Biophysics of the Pancreatic β-cell* (I. Atwater, E. Rojas, and B. Soria, eds.), Plenum, New York, pp. 31–51.

35. Chay, T. R., and Keizer, J., 1985, Theory of the effect of extracellular potassium on oscillations in the pancreatic β-cell, *Biophys. J.* **48:** 815–827.

36. Henquin, J. C., and Lambert, A. E., 1976, Bicarbonate modulation of glucose-induced biphasic insulin release by rat islets, *Am. J. Physiol.* **231:** 713–721.

37. Carroll, P., Li, M-X., Rojas, E., and Atwater, I., 1988, Physiological bicarbonate buffer inhibits the activity of the ATP-sensitive potassium channel in pancreatic β-cells, *FEBS Lett.* **234:** 208–212.

38. Lindstrom, P., and Sehlin, J., 1986, Effect of intracellular alkalinization on pancreatic islet calcium uptake and insulin secretion, *Biochem. J.* **239:** 199–204.

39. Rinzel, J., Chay, T. R., Himmel, D., and Atwater, I., 1986, Prediction of the glucose-induced changes in membrane ionic permeability and cytosolic Ca^{2+} by mathematical modeling, in: *Biophysics of the Pancreatic β-Cell* (I. Atwater, E. Rojas, and B. Soria, eds.), Plenum, New York, pp. 247–263.

40. Smith, J. S., and Pace, C. S., 1983, Modification of glucose-induced insulin release by alteration of pH, *Diabetes* **32:** 6106.

41. Hutton, J. C., Sener, A., Herchuelz, A., Valverde, L., Boschero, A. C., Malaisse, W. J., 1980, The stimulus-secretion coupling of glucose-induced insulin release: Effects of extracellular pH on insulin release: their dependency on nutrient concentration, *Horm. Metab. Res.* **12:** 294–299.

42. Gillis, K., Tabcharani, J., Hammoud, A., and Misler, S., 1988, Effects of ammonium chloride (NH_4Cl) and sodium proprionate (NaPr) on the activity of a metabolite-regulated K^+ channel in rat pancreatic islet and RIN insulinoma cells, *Biophys. J.* **53(2),** Pt. 2: 530a [abstr.].

43. Li, M-X., Carroll, P. B., Li, M-Y., Rojas, E., 1988, Patch-clamp measurements show removal of HCO_3/CO_2 and changes in pH alter a glucose-sensing mechanism of the β-cell: K-ATP channel activity, *Diabetes* **37**, Supp. 1, p. 194a, Abstract No. 750.
44. Atwater, I., Carroll, P. B., and Li, M-X., 1989, Electrophysiology of the pancreatic β-cell, in: *Insulin Secretion: Molecular and Cellular Biology of Diabetes Mellitus* (B. Draznin, S. Melmed, and D. Leroith, eds.), Alan R. Liss, Inc., New York, pp. 49–68.
45. Cook, D. L., Ikeuchi, M., and Fujimoto, W. Y., 1984, Lowering of pH inhibits calcium-activated potassium channels in isolated rat pancreatic islet cells, *Nature* **311:** 269–271.
46. Ribalet, B., and Ciani, S., 1987, Regulation by cell metabolism and adenine nucleotides of a K-channel in insulin-secreting β-cells (RINm5F), *Proc. Natl. Acad. Sci. USA* **84:** 1721–1725.
47. Meissner, H. P., and Atwater, I. J., 1976, The kinetics of electrical activity of β-cells in response to a "square wave" stimulation with glucose of glibenclamide, *Horm. Metab. Res.* **8;** 11–16.
48. Henquin, J. C., and Meissner, H. P., 1982, Opposite effects of tolbutamide and diazoxide on 86 Rb+ fluxes and membrane potential in pancreatic β-cells, *Biochem. Pharmacol.* **31:** 1407–1415.
49. Ferrer, R., Atwater, I., Omer, E. M., Goncalves, A. A., Croghan, P. C., and Rojas, E., 1982, Electrophysiological evidence for the inhibition of potassium permeability in pancreatic β-cells by glibenclamide, *Q. J. Exp. Physiol.* **69:** 831–839.
50. Chay, T. R., and Kang, H. S., 1987, Multiple oscillatory states and chaos in the endogenous activity of excitable cells: Pancreatic β-cell as an example, in: *Chaos in Biological Systems* (H. Degn, A. V. Holden, L. F. Olsen, eds.), Plenum, New York, pp. 173–181.

Chapter 9

Electrical and Secretory Response to Cholinergic Stimulation in Mouse and Human Adrenal Medullary Chromaffin Cells

Verónica Nassar-Gentina, Harvey B. Pollard, and Eduardo Rojas

1. INTRODUCTION

Chromaffin cells from the adrenal gland secrete catecholamines in response to acetylcholine (ACh).[1,2] ACh induces a transient depolarization of the adrenal chromaffin cell membrane, which in some animals is due to the opening of nicotinic-receptor channels.[3–7] It has been observed that the ACh-induced depolarization is often accompanied by the generation of action potentials or by a marked increase in the frequency of spontaneously occuring action potentials.[3,6–9] These action potentials are presumably due to the activation of both Na^+ and Ca^{2+} voltage-gated ionic channels. Blocking Na^+-channels with tetrodotoxin (TTX) leads to a partial inhibition of the stimulated release of catecholamines.[10,11] This result suggests that Ca^{2+} entry associated with the ACh-evoked depolarization is reduced in the presence of TTX.[10,11] The rapid depolarization resulting from the activation of Na^+-channels should enhance Ca^{2+} entry by recruitment of Ca^{2+}-channels with a more positive potential for activation. This is presumably the physiological pathway for Ca^{2+} entry at low concentrations of ACh (10 μM). At high concentrations of ACh (55 μM), however, additional Ca^{2+} entry occurs through the ACh nicotinic-receptor channel.[12] While Ca^{2+} entry through the ACh-channel is restricted to

VERÓNICA NASSAR-GENTINA • Laboratorio de Fisiología Celular, Facultad de Ciencias y Facultad de Medicina, Universidad de Chile, Viña del Mar, Chile. HARVEY B. POLLARD and EDUARDO ROJAS • Laboratory of Cell Biology and Genetics, National Institute of Diabetes and Digestive and Kidney Diseases, National Institutes of Health, Bethesda, Maryland 20892.

the small region of clustered nicotinic-receptor channels, voltage-dependent Ca^{2+}-channels are probably evenly distributed over the entire cell surface as patch-clamp[9] and other studies[13] appear to indicate.

Previous electrophysiological studies of adrenal medullary cells have been carried out predominantly on cultured cells. To learn more about the electrical behaviour of chromaffin cells in situ, we have used ultrafine-tipped microelectrodes, which permitted us to hold the electrode inside the cell for longer times, thus enabling us to run both the test and the control protocol in the same chromaffin cell in the gland. We studied the dependence of the resting membrane potential on $[K^+]_0$ and concluded that it is controlled by K^+-channels. Input resistance measurements of medullary cells in situ suggested that the chromaffin cells in the mouse were coupled electrically. Furthermore, ACh induced a substantial decrease in input resistance which may be due, at least in part, to a decrease in cell-to-cell junctional resistance. In the human adrenal gland we studied both the membrane electrical response and, in separate experiments, the secretory response to different cholinergic agonists. Our results indicate that both in the mouse and the human adrenal gland, ACh action on electrical activity is mediated predominantly, but not exclusively, by muscarinic receptors. We suggest that ACh-evoked catecholamine secretion in these preparations is also under muscarinic receptor control.[14]

2. METHODOLOGICAL CONSIDERATIONS

To inhibit liberation of catecholamines during and after removal of the adrenal glands from a female mouse, the dissection was performed in the presence of a modified high-Mg^{2+} Krebs solution[15] of the following composition (mM): 110 NaCl, 25 $NaHCO_3$, 5 KCl, 10 $MgCl_2$, 2.56 $CaCl_2$, 10 Na-HEPES. The glands were sliced in half and one part was placed in a perspex chamber. As described previously,[16] the chamber was perfused continuously with a modified Krebs solution of the following composition (mM): 120 NaCl, 25 $NaHCO_3$, 5 KCl, 1.1 $MgCl_2$, 2.56 $CaCl_2$, 5.6 glucose. The solution was equilibrated continuously at 37°C with 95% O_2/5% CO_2 to give a pH of 7.5. The high-K^+ Krebs solutions were made by equimolar substitution of Na^+ by K^+. The human adrenal gland was kept in a high K^+ saline at about 4°C during the transportation to the laboratory. Glands remained in this medium for up to 6 hr. Small pieces (5 to 25 mg) of the medullary tissue were microdissected and incubated in Krebs bicarbonate buffer at 37°C for electrophysiological recordings and secretion studies.

Electrophysiological Recordings

The methods used for intracellular recordings of membrane potential (V_m) and current injection have been described previously.[16] V_m was measured between two Ag-AgCl electrodes, one in the external solution and the other in the intracellular microelectrode.

To estimate the input resistance, three consecutive measurements were carried out: First, the voltage displacement induced by the injection of a current pulse (10^{-10} A) was measured with the tip of the microelectrode in the solution. This signal (used to measure the microelectrode resistance before impalement of the cell) was then cancelled using a summing amplifier. Second, the voltage deflection after penetration of the chromaffin membrane was recorded. This voltage deflection was used to calculate the input resistance. Third, the electrode was withdrawn, and the voltage displacement induced by the injection of 10^{-10}A pulses was recorded. As a rule, the initial cancellation of the voltage displacement recorded prior to the penetration of the cell remained acceptable. Otherwise, corrections of the signal obtained with the microelectrode tip inside the cell were applied.

Ionic currents during electrical activity were estimated from the time derivative of the intracellularly recorded action potentials. The intracellular volume was assumed to be isopotential and the membrane capacity C_m was taken as 6 pF. Selected electrical events were digitized using a digital storage oscilloscope (Nicolet, Model 4562). A mathematical software package provided smoothing and time derivative programs which calculated dV/dt. The total ionic current records were calculated as $-C_m \, dV/dt$.

3. PROPERTIES OF THE RESTING CHROMAFFIN CELL MEMBRANE

3.1. Potassium Dependence of the Resting Potential

Resting membrane potential V_m was recorded in situ from mouse and human adrenal chromaffin cells. Under physiological conditions, that is, adrenal tissue superfused with Krebs bicarbonate solution at 37°C, V_m values obtained immediately after penetration of the cell membrane remained constant for at least 60 min. The mean value of the resting membrane potential V_m in nonstimulated cells was -54.3 ± 8.8 mV ($n = 21$). This V_m value falls within the range reported for other chromaffin cell preparations.[3,6] The mean value of the resting potential from human chromaffin cells under the same experimental conditions was -43.5 ± 9.5 mV ($n = 3$).

The dependence of the resting membrane potential on $[K^+]_0$ was studied by measuring the response of the cell membrane to modified Krebs solutions of increasing concentration of $[K^+]_0$ (from 5 to 140 mM) keeping the sum $[Na^+]_0 + [K^+]_0$ constant. The time course of the response for a change in $[K^+]_0$ from 5 to 12.5 mM is illustrated in Fig. 1A. The membrane potential changed from -48 mV (measured just before the application of 12.5 mM K^+) to -35.7 mV measured towards the end of the record shown in Fig. 1A. Steady-state changes were always achieved in about 10 sec. These values, plotted in Fig. 1B as a function of the logarithm of $[K^+]_0$, were analyzed using the constant-field equation.[17,18] Assuming that P_{Cl} is negligible[19] and taking $[Na^+]_i$ as 10 mM,[14] the equation takes the familiar form

$$V_m = RT/F \, (\ln[K^+]_0 + \alpha \, [Na^+]_0)/([K^+]_i + \alpha \, [Na^+]_i)$$

where α represents the permeability ratio P_{Na}/P_K. The curve in Fig. 1B, drawn to fit the data points from all experiments, was calculated with the above equation. The average value of α obtained was 0.09. Thus, the resting membrane of the mouse chromaffin cell is about 11 times more permeable to K^+ than Na^+.

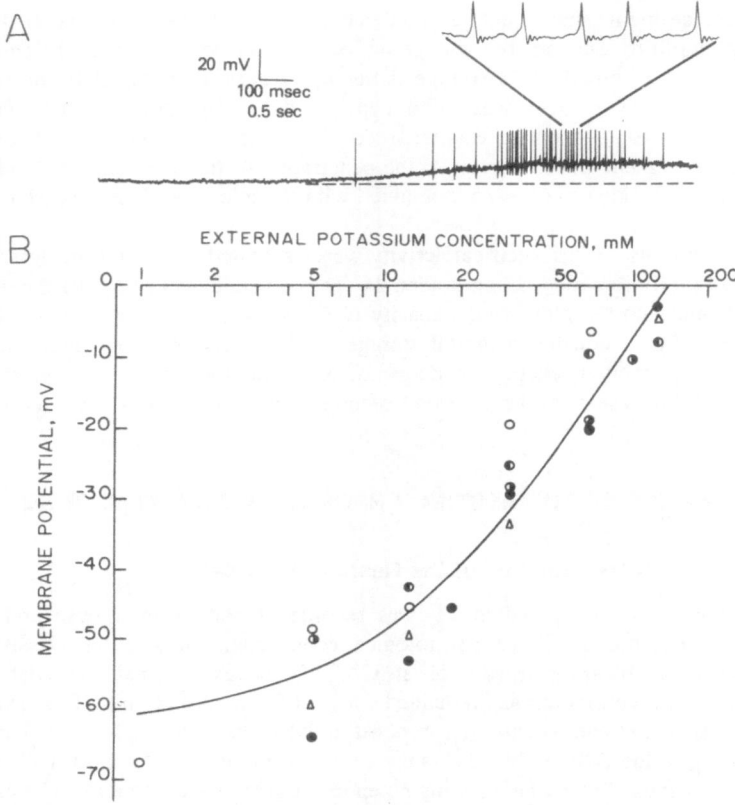

Figure 1. Dependence of resting membrane potential on the external potassium concentration. (A) Increasing $[K^+]_0$ from 5 to 12.5 mM depolarized the membrane from -48 to -35.7 mV. (B) Different symbols represent V_m measurements in different cells. Curve was drawn to fit the points and was calculated using the constant-field equation in the form given in the text. (From Ref. 14.)

3.2. The Chromaffin Cell in situ Exhibits Normal Rectification of the Intracellularly Injected Current

Current–voltage relationships obtained by intracellular injection of rectangular pulses of current (in the range from $\pm$ 0.1 to $\pm$ 0.4 nA) showed a clear rectification of the injected current (Fig. 2). The input resistance (R_A) values estimated from the slope of the linear portion of the current–voltage relationship were 102 $\pm$ 26 $M\Omega$ ($n = 12$) for the inward currents and 44.4 $\pm$ 26 $M\Omega$ ($n = 6$) for outward currents. The rectification ratio of 0.4 (= 41.4/102.8) was not significantly affected by external application of tetraethylammonium (TEA) and/or TTX. The membrane resistance of a chromaffin cell in isolation has been measured to be about 5 GΩ[6]. This value is greater than the range of values found here by more than one order of magnitude.

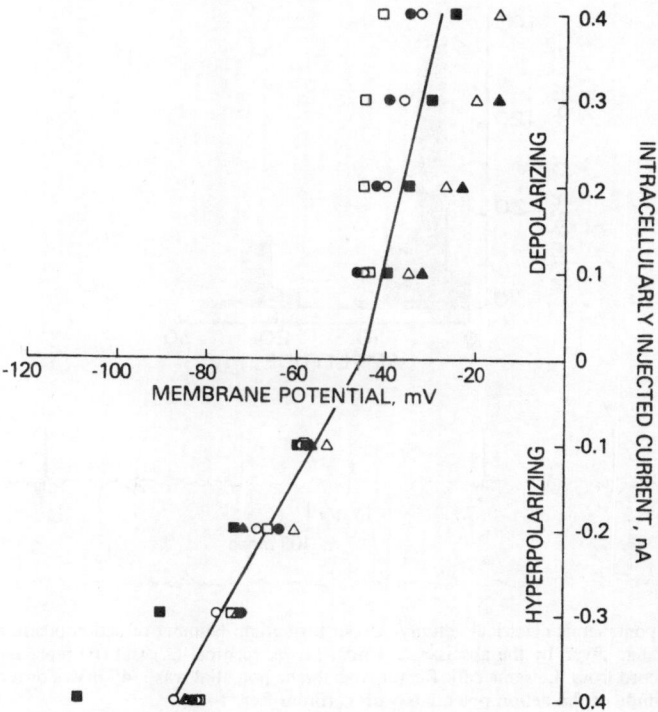

Figure 2. Rectification of the intracellularly injected current by the chromaffin cell system. Current pulses (1 sec) of variable amplitude (vertical axis) were applied. Membrane potential during the current pulse was measured toward the end of the membrane response (horizontal axis). Straight lines were calculated to fit the experimental points. Different symbols represent measurements from different cells. Input resistance values were 41 ± 26 $M\Omega$ ($n = 6$) for positive currents and 103 ± 15 $M\Omega$ for negative currents. (From Ref. 14.)

4. ELECTRICAL ACTIVITY IN MOUSE AND HUMAN ADRENAL MEDULLARY CELLS

4.1. Spike Activity Evoked by Current Injection

Up to 50% of the adrenal medullary cells in the mouse showed spontaneous activity in the form of action potentials occurring at a frequency of about 6 Hz. The amplitude distribution of these action potentials was often bimodal with mean values around 10 and 30 mV (see insert in Fig. 3). The simplest explanation for this bimodal distribution is that the larger spikes originate in the impaled cell and the smaller spikes originate in neighboring, electrically coupled cells.[20]

For all the cells examined in the present work (from mouse or human adrenal gland), electrical activity could be induced by intracellular injection of current pulses. For depolarizing current pulses spikes were elicited at the onset of the pulse and immedi-

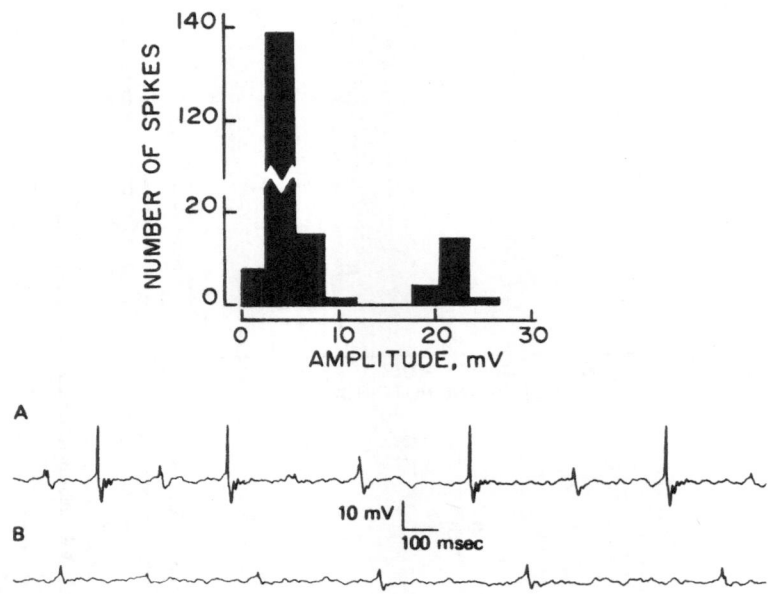

Figure 3. Spontaneous electrical activity. Upper histogram: Number of action potentials with amplitudes in the range given by the abscissa ± 1 mV. Lower records: (A) and (B) represent portions of a continuous record from the same cell. Resting membrane potential was −45 mV. For record (A) notice that the amplitude of the action potentials varies. (From Ref. 14.)

ately after the pulse for hyperpolarizing currents. Cells with spontaneous electrical activity consisting of small action potentials (about 10 mV) responded to depolarizing current pulses with a train of larger spikes (about 30 mV). Furthermore, in the presence of normal Krebs solution, induced or spontaneous action potentials did not reach positive membrane potentials (overshoot). As shown later, only in the presence of TEA or Ba^{2+} did the membrane potential reach positive values.

Figure 4 depicts typical responses of a human chromaffin cell to intracellular injection of depolarizing current pulses. The resting potential of this cell was −30 mV. Application of both depolarizing and hyperpolarizing current pulses elicited repetitive electrical activity. The mean amplitude of the spikes was 30 mV.

4.2. Na⁺ and Ca²⁺ Dependence of the Action Potentials

The role of Na^+ and Ca^{2+} in the genesis of the electrical activity was studied by measuring the effects of lowering $[Na^+]_0$ or $[Ca^{2+}]_0$ on both spike amplitude and membrane currents. The chromaffin cell resting membrane potential was unaffected by lowering $[Na^+]_0$ (choline in place of Na^+) or $[Ca^{2+}]_0$.

Furthermore, electrical activity, in those chromaffin cells with spontaneous activity, was unaffected by the absence of either Na^+ (Fig. 5B) or Ca^{2+} (Fig. 5C). Although the amplitude of the action potentials did not change significantly, membrane currents were

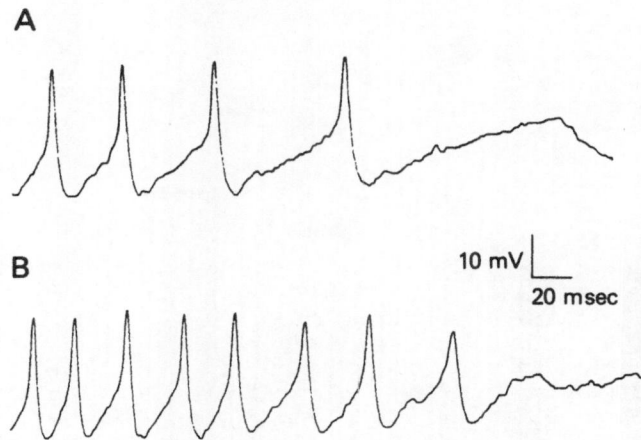

Figure 4. Electrical activity evoked by depolarizing current pulses in human chromaffin cells. Resting potential was −30mV. (A) Electrical activity during the injection of a depolarizing current pulse (0.1 nA). (B) Electrical activity from the same cell after the injection of a hyperpolarizing current pulse (−0.1 nA).

profoundly affected. Indeed, inward currents were reduced by the removal of Na⁺ from −40 to −7.7 pA (Fig. 5B) and reduced to a lesser extent in the absence of Ca^{2+} (Fig. 5C). It should be noted that the negative after-potential (often referred to as action potential undershoot) was reduced by both treatments (Figs. 5B and 5C). This result suggests that this component of the outward current not only depends on $[Ca^{2+}]_0$ but also on $[Na^+]_0$.

Since choline may act as a weak agonist at the nicotinic receptor, the contribution of Na⁺-channels to the genesis of the action potential was also examined using TTX to

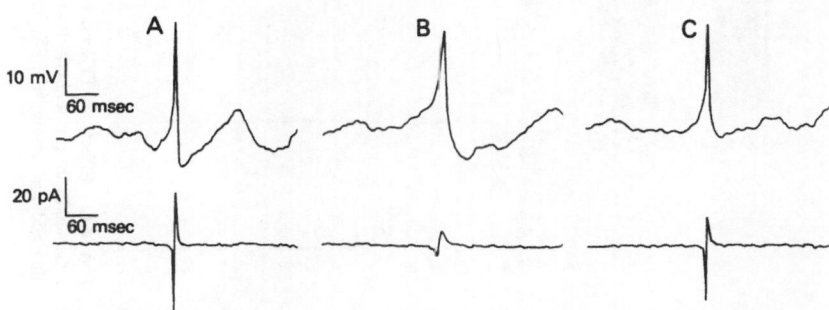

Figure 5. Effects of Na⁺ or Ca^{2+} removal on spontaneous electrical activity. Upper records: Individual action potentials recorded (A) in the presence of Na⁺ and Ca^{2+}, (B) in the absence of Na⁺ (choline⁺ in place of Na⁺), and (C) in the absence of Ca^{2+}. Lower records: Ionic currents calculated as $-C_m \, dV/dt$, where C_m was taken as 6 pF. (From Ref. 14.)

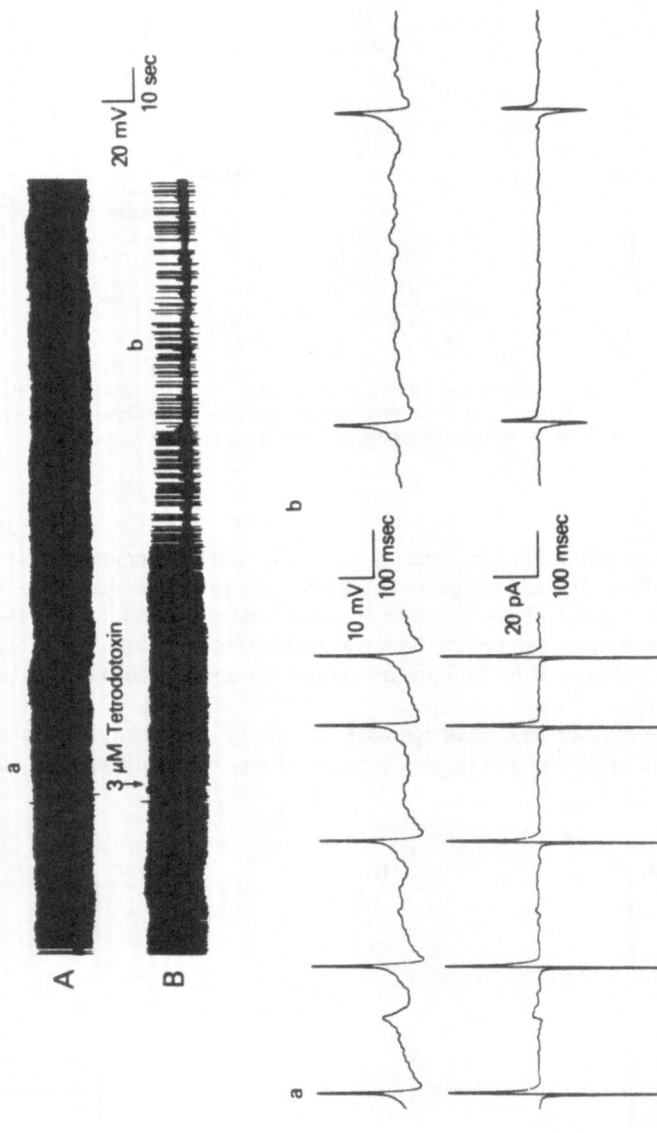

Figure 6. Effects of TTX on spontaneous electrical activity. Record (B) is the continuation of record (A). TTX was added as indicated by the arrow. In the lower part, (a) and (b) show details of the action potentials and calculated ionic currents (a) before and (b) after addition (arrow) of TTX. Resting membrane potential was −50 mV. (From Ref. 14.)

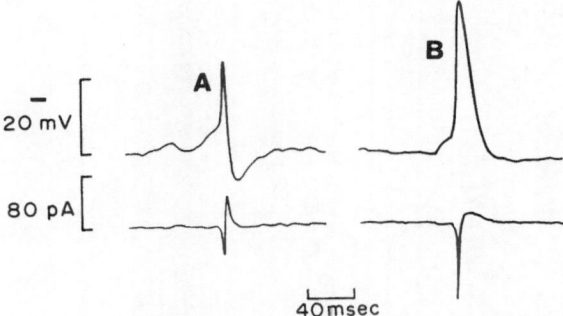

Figure 7. Effects of TEA on the shape of the action potential and on the size of the derived currents. (A) Control and (B) in the presence of 20 mM TEA. Resting membrane potential was -50 mV. (Fron Ref. 14.)

block voltage-dependent Na$^+$ channels. As shown in Fig. 6, inward currents were reduced from -58 to -20 pA, a 66% reduction in the presence of TTX (3 μM).

However, in addition to the inhibition of the inward currents, outward currents were also diminished. Since TTX is a highly specific Na$^+$-channel blocker, the inhibition of the negative after-potential evoked by TTX together with the inhibition obtained in the absence of external Na$^+$ indicates the presence of a component of the outward current that depends on Na$^+$ entry.

Figure 7 shows that the amplitude of spontaneously occurring action potentials (A) was greatly increased by the external application of TEA (B) presumably due to the

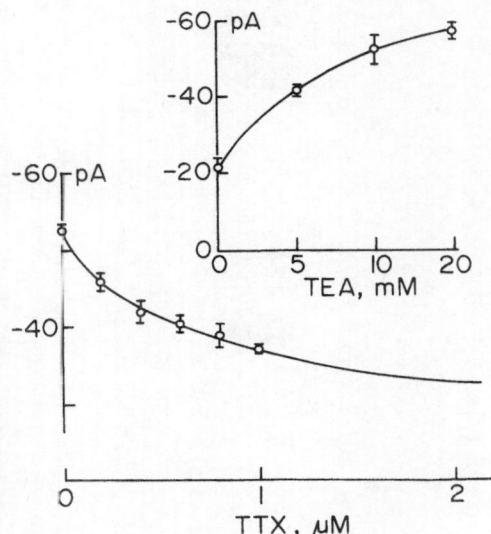

Figure 8. Blockade of the inward currents by TTX in the presence of TEA. Upper part: Peak value of the calculated inward current as a function of [TEA]$_0$. Vertical bars represent ± S.D. Lower part: Peak value of the calculated inward current as a function of [TTX]$_0$ in the presence of 10 mM TEA. Vertical bars represent ± S.C. (From Ref. 14.)

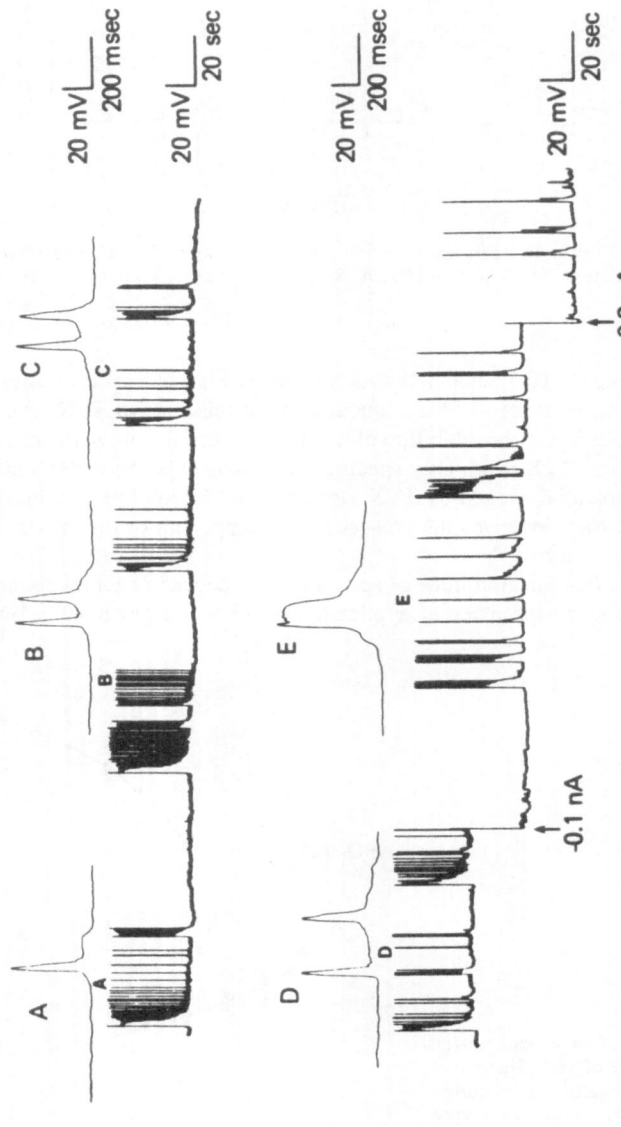

Figure 9. Bursting electrical activity induced by TEA. Resting membrane potential was −51 mV. (A)–(E) show details of action potentials on an expanded time scale. Hyperpolarizing current was injected as indicated by the arrows in th lower record. (From Ref. 14.)

blockade of outward currents through K^+-channels. Figure 7 also shows that TEA (20 mM) blocked the negative after-potential.

The comparison of the inward and outward membrane currents shown in Fig. 6 revealed that an important fraction of the inward current remained in the presence of TTX (3 μM). The effects of TTX on the size of the inward current in the presence of TEA are summarized in Fig. 8 (lower graph). Also shown in Fig. 8 is a TEA dose–effect curve (upper graph). In the absence of TTX, increasing $[TEA]_0$ increased the size of the inward current from about -20 to -55 pA (see upper graph). Increasing $[TEA]_0$ increases the fraction of blocked K^+-channels. This blockade in turn reduces the rate of depolarization of the individual spikes and, thus, the time spent in the depolarized state of the action potentials increases. At constant $[TEA]_0$ (10 mM), increasing $[TTX]_0$ decreased the size of the maximum inward current from about -52 to -32 pA. Presumably this TTX-resistant component of the current is carried by Ca^{2+}.

Thus, the mouse adrenal chromaffin cell action potential results from the activation of both Na^+- and Ca^{2+}-channels, confirming similar conclusions based on patch-clamp measurements of the ionic currents in cultured bovine chromaffin cells.[3,9]

TEA (20 mM) always induced spike activity in cells without spontaneous electrical activity, as depicted in Fig. 9. At regular intervals the membrane depolarized spontaneously, causing bursts of electrical activity (upper records). Although intracellular injection of hyperpolarizing current reduced the duration of the bursts, bursting electrical activity persisted.

4.3. Electrical Activity in the Presence of Ba^{2+}

Ba^{2+} is a potent blocker of outward K^+ currents[21] and carries inward current through the Ca^{2+}-channel better than Ca^{2+}. In the presence of Ba^{2+} the cell membrane depolarized slowly (11 mV), and electrical activity was generated. This activity was often in bursts of action potentials (Fig. 10) and lasted as long as Ba^{2+} was present. The

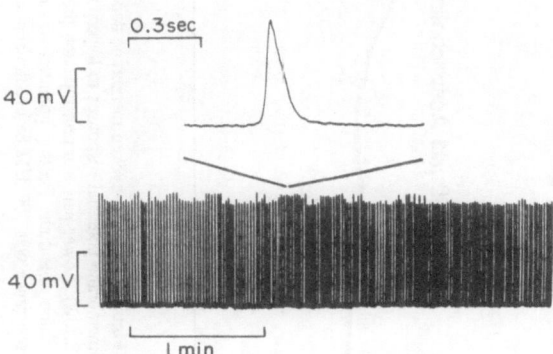

Figure 10. Electrical activity evoked by Ba^{2+}. Upon the addition of Ba^{2+} (6 mM) to the perfusion solution, the resting membrane potential decreased from -62.5 to -51.5 mV. Spontaneous action potentials were generated while Ba^{2+} was present. Upper part: Single action potential on an expanded time base. (From Ref. 14.)

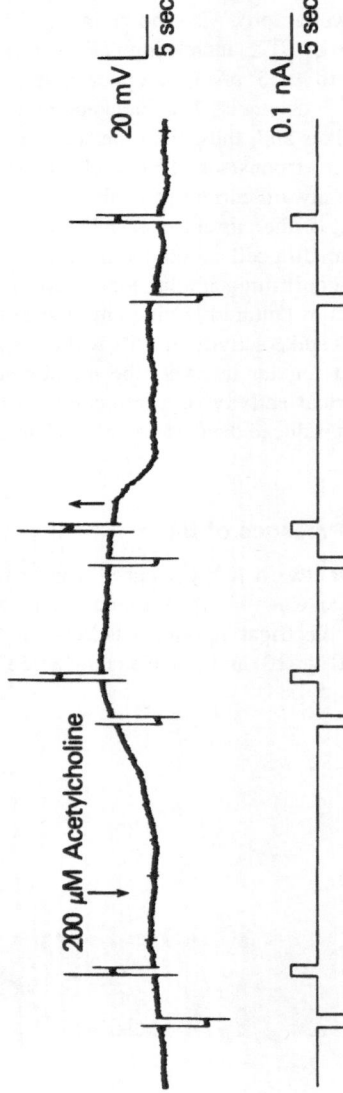

Figure 11. ACh-evoked membrane depolarization in the human chromaffin cell. Upper part: Segment of the chart record of the membrane potential responses to hyperpolarizing and depolarizing current pulses (± 0.1 nA) and to the application of ACh (first arrow). Lower part: Current injection protocol. The cell responded to the injection of depolarizing current pulses by firing a few action potentials. Resting potential was held at −50 mV by injecting a constant hyperpolarizing current of 0.05 nA. The input resistance was 190 $M\Omega$ before the application of ACh and decreased to 175 $M\Omega$ toward the end of the exposure to ACh.

spikes were long-lasting (see record in Fig. 10), and at the peak the membrane potential reached positive values (10 to 20 mV). It should be mentioned here that electrical activity induced by Ba^{2+} could not be blocked by application of hyperpolarizing current. It is well established that Ba^{2+} causes secretion of catecholamines from adrenal medullary chromaffin cells.[4] Thus, the spike activity evoked by Ba^{2+} is indicative of the close relationship between electrical activity and secretion.

4.4. ACh-Induced Depolarization

As a rule, application of ACh to both mouse and human chromaffin cells depolarized the membrane in a dose-dependent manner. Maximal depolarization of about 20 mV was obtained at ACh concentration greater than 100 μM (see Figs. 11 and 12). Three different types of ACh responses were distinguished. In some cells without spontaneous activity, ACh either depolarized without causing spike activity (Fig. 11) or it induced repititive action potentials (Fig. 12A). In both instance the input resistance decreased, from about 98 to about 72 $M\Omega$. In cells with spontaneous activity, ACh transiently increased the spike frequency from 6 (measured in the absence of externally applied ACh) to 13 spikes/sec^{-1} (measured 30 sec after the introduction of ACh).[14] In these cells, after the initial transient increase in spike frequency, the frequency decreased from 10 to 3 sec^{-1} during the period of exposure to ACh. This gradual decrease in the response could be due to inactivation of Na^+- and Ca^{2+}-channels.

ACh-induced depolarization and generation of spikes were not blocked by D-tubocurarine (50 to 100 μM). However, atropine (10 to 100 μM) blocked completely both depolarization and spike activity (Fig. 12). Nicotine (20 to 200 μM) had no noticeable effect on the membrane potential, while muscarine depolarized the membrane in a concentration-dependent manner (data not shown). These results indicate that both

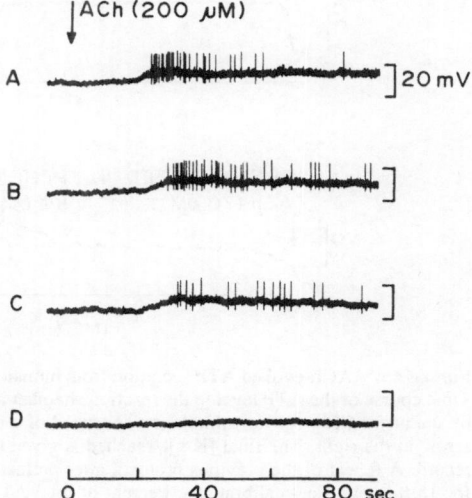

Figure 12. Effects of D-tubocurarine and atropine on ACh-induced depolarization. (A)–(D) are segments from the same chart record. Resting membrane potential at the start of each segment was constant at −46 mV. ACh was applied as indicated by the arrow either (A) alone (200 μM); (B) in the presence of D-tubocurarine (10 μM); (C) in the presence of D-tubocurarine (100 μM); or (D) atropine (100 μM). (From Ref. 14.)

mouse and human medullary chromaffin cells possess muscarinic receptors that control
critical membrane channels.

5. REAL-TIME MEASUREMENTS OF ACh-INDUCED RELEASE
OF ATP FROM HUMAN MEDULLARY CHROMAFFIN CELLS

In many types of secretory cells the secretory granules contain ATP in addition to
the specific hormones. We have recently shown[22] that it is possible to obtain quantitative measurements of the kinetics of secretion of granule contents by monitoring ATP
release using the luciferase-catalyzed luminescent oxidation of luciferin. Using this
technique we studied the secretory competence of the chromaffin cells present in pieces
of human adrenal medullary tissue from the same gland used for the electrophysiological
measurements.

5.1. ACh-Induced ATP Release from Human Adrenal Chromaffin Cells
Mediated by Both Nicotinic and Muscarinic Receptors

In mouse chromaffin cells, the secretion of catecholamines (CA) is evoked by
stimulation of cholinergic receptors and depends on $[Ca^{2+}]_0$.[21] As shown in the upper
part of Fig. 13, this is also the case for stimulation of ATP release from human
chromaffin cells. It may be seen that immediately after the application of ACh ($70\mu M$),
the signal increased rapidly and reached a steady state after 60 sec. Because the ATP
released by the cells is confined to the tube, the light emitted represents the time integral

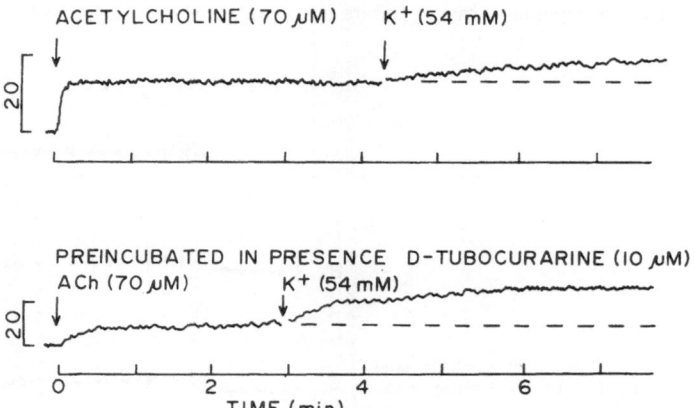

Figure 13. ACh-evoked ATP secretion from human adrenal medullary chromaffin cells. Upper record:
Time course of the ATP level in the reaction chamber as a function of time. ACh was applied as indicated
by the arrow. The second stimulation consisted of a sudden elevation of the $[K^+]_0$ as indicated by the
arrow in the right. The final $[K^+]_0$ reached is given (in mM) by the number in the parentheses. Lower
record: A repeat of the previous protocol after preincubation in the presence of D-tubocurarine (10 μM)
for 15 min. Vertical calibrations: Percent of total ATP in the tissue (about 22 mg w.w.).

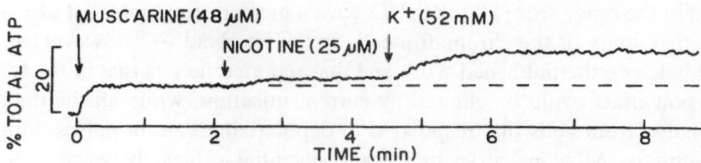

Figure 14. ATP release evoked by sequential cholinergic stimulation of the human adrenal medullary tissue (about 27 mg w.w.). Time course of the ATP level in the reaction chamber.

of the ATP released. Thus, the rate of ATP secretion (the time derivative of the records shown in Fig. 13) decays rapidly, consistent with data for the rate of secretion of CA.[23]

ACh-evoked ATP secretion could be partially inhibited by pretreatment of the cells with D-tubocurarine (10 μM), a nicotinic cholinergic receptor blocker (Fig. 13, lower record). It is interesting to note that the initial rate of ATP release was greatly reduced in the presence of D-tubocurarine.

Figure 14 shows that direct activation of muscarinic receptors by the cholinergic agonist muscarine (25 μM) induced ATP secretion (6–7% of total ATP in the cells) from pieces (ca. 1 mm³) of the human adrenal medulla. Niconitic stimulation (25 μM) after muscarinic stimulation evoked further ATP release. These properties of the ATP secretion indicate that both cholinergic receptors control CA secretion in the human adrenal medulla.

5.2. ATP Release Evoked by Membrane Depolarization in Human Chromaffin Cells

A sudden elevation of the $[K^+]_0$, keeping the sum of $[K^+]_0$ plus $[Na^+]_0$ constant, caused prompt release of ATP (Figs. 13 and 14). The extent of this response depended on both $[K^+]_0$ and $[Ca^{2+}]_0$.

ACh-stimulated ATP release thus shares many important properties with the process of CA secretion in chromaffin cells from other species, and we therefore presume that release of ATP from human cells also marks catecholamine secretion. For example, in cultured bovine chromaffin cells, both processes depend on extracellular $[Ca^{2+}]_0$ and are inhibited by Ca^{2+}-channel blockers such as Cd^{2+}, and both secretory processes are blocked by low concentrations (μM) of trifluoroperazine and promethazine[24].

6. CONCLUDING REMARKS

6.1. The Dependence of Membrane Potential on $[K^+]_0$

The mean value of the resting membrane potential of -54.3 mV obtained in this work is more negative than other published values (-10 to -50 mV)[3,25] possibly due to the use of microelectrodes with higher resistances (250–300 $M\Omega$) and thus smaller tip diameters. The mean resting membrane potential recorded with intracellular microelectrodes of chromaffin cells from the rat was -49 ± 6 mV[3]. Microelectrodes with resistances in the range from 50 to 60 $M\Omega$ gave a mean resting potential of -49.6 ± 1.9 mV in another study of the chromaffin cells in the rat gland.[26] However, an important difference between the published work and that reported here is that in the former study no action potentials could be elicited by current injection, while all the data presented here originated from cells that responded to depolarizing current pulses with repetitive action potentials. More negative membrane potential values, between -50 and -80 mV, have been obtained using patch-clamp micropipettes in the whole cell configuration.[6] Indeed, Friedman et al.[27] used a potentiometric-dye method and estimated a value of -55 mV for isolated bovine adrenal chromaffin cells.

The resting membrane potential in chromaffin cells from mouse adrenal gland depends on the external concentration of potassium, and the experimental data could be fitted with the Goldman–Hodgkin–Katz equation neglecting Cl^- permeability. This assumption was justified by the finding that replacement of Cl^- by the impermeable anion SO_4^{2-} has little effect on the membrane potentials of various secretory cells.[19,28] A permeability ratio P_{Na}/P_K of 0.09 was calculated from these data, a similar ratio being reported in rat adrenal chromaffin cells in situ[26] and in pancreatic β-cells.[28] TEA, a rather selective blocker of the membrane potential-dependent K^+ permeability in nerve membranes[27] did not affect the dependence of the membrane potential on extracellular K^+. This means that either the K^+ channel is refractory to external TEA, or another K^+ permeability mechanism may operate in the chromaffin cell membrane. In support of this concept was our observation that the membrane potential of the non stimulated chromaffin cell was not affected by TEA. Since there are several pharmacologically distinguishable K^+ permeabilities present in the cell membrane of other secretory cells,[28–30] another TEA-insensitive K^+ permeability must indeed be operating in this membrane.

6.2. Modulation of the Cell Input Resistance by Secretagogues

The use of high-resistance microelectrodes to measure the membrane potential of adrenal medullary chromaffin cells in the intact gland has enabled us to study the membrane responses to different secretagogues such as acetylcholine, K^+, and Ba^{2+}. The current–voltage relationship obtained in this work exhibited normal rectification in agreement with previous reports.[3,6] The mean value of the input resistance was estimated to be 102.8 ± 15 $M\Omega$. However, measurements of input resistances in cultured cells with intracellular microelectrodes or with patch-clamp micropipettes give values in the G range.[3,6] The input resistances of rat chromaffin cells in situ are about 32 $M\Omega$.[26] Even though low input resistance values could result from slightly damaged cells,[13] our controls (i.e., stability of the records and preservation of the ability of the cells to respond to depolarizing current pulses with electrical activity throughout the entire duration of the experiment) suggest that the input resistance is genuinely low. The implication of this result can be fully realized if one considers the following expression for the input resistance of an aggregate of coupled cells:

$$A = (R_m + R_j)/[(1 + k) + R_j/R_m] \tag{2}$$

where R_m represents the membrane resistance of each cell in the aggregate, and R_j represents the junctional resistance between adjacent cells. Both R_m and R_j are assumed to be the same for each of the k cells in the aggregate. From this equation it is clear that it is possible to account for the low R value measured by assuming a junctional resistance R_j in G range and a finite number K of coupled cells.[16,25] The higher input resistance values measured in cultured cells may be indicative of electrical uncoupling.[31] The substantial decrease of the input resistance (26 $M\Omega$) produced by ACh can be explained either by a decrease in R_m due to activation of ACh-channels, a decrease in R_j, or a decrease in both parameters. Taking 44 pS as the single ACh-channel conductance in the bovine chromaffin cell in culture,[6] about 1000 channels would have to be activated by ACh to account for the measured decrease in R_A. This value is an order of magnitude higher than estimated for the chromaffin cell.[6] Although this calculation is based on measurements made in two different preparations—that is, chromaffin cells from mouse that respond to muscarinic receptor activation, and bovine chromaffin cells that respond to nicotinic receptor activation—it is unlikely that the decreased input resistance is solely the result of ACh-channel activity. A muscarinic receptor mediated ACh action on R_j may modulate the electrical coupling among the chromaffin cells in the medulla.

6.3. Ionic Components of the Spike Activity

We have seen that the amplitudes of the action potential and the inward ionic current depend on both $[Na^+]_0$ and $[Ca^2]_0$. We have also seen that blockers of Na^+-channels and removal of Ca^{2+} also reduce the spike height and associated ionic currents. These results strongly suggest that chromaffin cells have both Na^+ and Ca^{2+} components, confirming results obtained in cultured medullary chromaffin cells.[3] Since $[K^+]_0$-induced depolarization of the cell membrane evokes both catecholamine release[26] and repetitive electrical activity (Fig. 1A), it is likely that activation of the voltage-gated Ca^{2+}-channel is involved.[31,32] Activation of the fast Na^+-channel should accelerate the depolarization of the membrane and thus accelerate the activation of slower Ca^{2+}-channel. This idea is supported by the observation that TTX ($6\mu M$) partially inhibits adrenaline release from rat adrenal gland.[10] Thus, the Na component of the action potential may have an important role in stimulus–catecholamine secretion coupling.

6.4. ACh-Evoked Depolarization is Mediated by Muscarinic Receptor Stimulation

ACh-evoked depolarization was unaffected by D-tubocurarine, and blocked by atropine. These results suggest that ACh stimulation of mouse adrenal medullary cells is mediated by ACh muscarinic receptor activation as in the rat[33] and hamster.[34] Our data does not rule out the possibility that stimulation of mouse adrenal medullary cells is also mediated by an additional weak nicotinic component as in the cat.[35] In cells with spontaneous electrical activity, application of ACh increased the frequency of the spike activity.[14] Similar results have been reported for adrenal medullary cells in culture.[3,7] The increase in spike frequency was probably the result of the depolarizing action of ACh on the chromaffin cell membrane, activating in this way Na^+ and Ca^{2+} voltage-dependent channels. The mechanism by which muscarinic receptor stimulation evokes catecholamine release remains unknown, although recent evidence suggest a different pathway for elevation of cytosolic calcium than that used by the nicotonic receptors.[22]

6.5. Cholinergic Receptor Operated ATP Secretion in Human Adrenal Chromaffin Cells

Human chromaffin cells secrete ATP in response to ACh. Because the cholinergic agonists nicotine and muscarine are secretagogues in this preparation, it is clear that nicotinic as well as muscarinic receptors control the secretory response in the human adrenal medulla.

We have shown that muscarine, but not nicotine, can induce electrical activity in adrenal chromaffin cells from the mouse. Therefore, the mouse is different from the human chromaffin cell in this respect. In contrast, in bovine chromaffin cells in culture, nicotinic but not muscarinic stimulation can evoke secretion of CA and ATP.[22] Nonetheless, prestimulation with a subthreshold dose of nicotine can allow muscarinic stimulation to be observed. Thus, human, cow, and mouse chromaffin cells possess both muscarinic and nicotinic receptors, the cells differing with respect to whether muscarine alone (mouse, human) or nicotine alone (cow) is sufficient to evoke secretion. To our knowledge, this is the first instance of the measurement of stimulus–secretion coupling in human chromaffin cells, or of the characterization of the different cholinergic receptors in these cells able to mediate this secretion event.

ACKNOWLEDGMENT.   The authors are pleased to thank Dr. M. R. Alijani, Department of Transplant Surgery, Georgetown University School of Medicine, Washington, D.C.

REFERENCES

1. Edwards, A. V., Furness, P. N., and Helle, K. B., 1980, Adrenal medullary responses to stimulation of the splanchnic nerve in the conscious calf, *J. Physiol.* **308:** 15–27.
2. Feldberg, W., Minz, B., and Tsudzimura, H., 1934, The mechanism of the nervous discharge of adrenaline, *J. Physiol.* **81:** 286–304.
3. Brandt, B. L., Hagiwara, S., Kidokoro, Y., and Miyasaki, S., 1976, Action potentials in the rat chromaffin cell and effects of acetylcholine, *J. Physiol.* **263:** 417–439.
4. Douglas, W. W., and Rubin, R. P., 1964, Stimulant action of varium on the adrenal medulla, *Nature (London)*, **203:** 305–307.
5. Douglas, W. W., Kanno, T., and Sampson, S. R., 1967, Effects of acetylcholine and other medullary secretagogues and antagonists on the membrane potential of adrenal chromaffin cells: An analysis employing techniques of tissue culture, *J. Physiol.* **188:** 107–120.
6. Fenwick, E. M., Marty, A., and Neher, E., 1982a, A patch-clamp study of bovine chromaffin cells and of their sensitivity to acetylcholine, *J. Physiol.* **331:** 577–597.
7. Kidokoro, Y., Miyazaki, S., and Ozawa, S., 1982, Acetylcholine-induced membrane depolarization and potential fluctuations in the rat chromaffin cell, *J. Physiol.* **324:** 203–220.
8. Biales, B., Dichter, M., and Tischler, A., 1976, Electrical excitability of cultured adrenal chromaffin cells, *J. Physiol.* **262:** 743–753.
9. Fenwick, E. M., Marty, A., and Neher, E., 1982b, Sodium and calcium channels in bovine chromaffin cells, *J. Physiol.* **331:** 599–635.
10. Kidokoro, Y., Ritchie, A. K., and Hagiwara, S., 1979, Effect of tetrodotoxin on adrenal secretion in perfused rat adrenal medulla, *Nature* **278:** 63–65.

11. Kidokoro, Y., and Ritchie, A. K., 1980, Chromaffin cell action potentials and their possible role in adrenaline secretion from rat adrenal medulla, *J. Physiol.* **307**: 199–216.

12. Blaschke, E., and Uvnas, B., 1981, No effect of tetrodotoxin on catecholamine release from the perfused cat adrenal gland, *Acta Phy. Scandinavica* **113**: 267–269.

13. Kidokoro, Y., 1985, Electrophysiology of adrenal chromaffin cell, in: *The Electrophysiology of the Secretory Cell* (A. M. Poisner and J. M. Trifaro, eds.), Elsevier, Amsterdam, pp. 195–218.

14. Nassar-Gentina, V., Pollard, H. B., and Rojas, E., 1988, Electrical activity in chromaffin cells of intact mouse adrenal gland, *Am. J. Physiol.* **254**: 675–683.

15. Douglas, W. W., and Rubin, R. P., 1963, The mechanism of catecholamine release from the adrenal medulla and the role of calcium in stimulus–secretion coupling, *J. Physiol.* **167**: 288–310.

16. Atwater, I., Ribalet, B., and Rojas, E., 1978, Cyclic changes in potential and resistance of the β-cell membrane induced by glucose in islets of Langerhans from mouse, *J. Physiol.* **278**: 117–139.

17. Hodgkin, A. L., and Katz, B., 1949, The effect of sodium ions on the electrical activity of the giant axon of the squid, *J. Physiol.* **108**: 37–77.

18. Goldman, D. E., 1943, Potential, impedance, and rectification in membranes, *J. Gen. Physiol.* **27**: 37–60.

19. Williams, J. A., 1970, Origin of transmembrane potential in nonexcitable cells, *J. Theor. Biol.* **28**: 287–296.

20. Ferrer, R., Soria, B., Dawson, C. M., Atwater, I., and Rojas, E., 1984, Effects of Zn on glucose-induced electrical activity and insulin release from mouse pancreatic islets, *Am. J. Physiol.* **246**: C520–C527.

21. Hagiwara, S., Fukuda, J., and Eaton, D. C., 1974, Membrane currents carried by Ca, Sr, and Ba in barnacle muscle fiber during voltage clamp, *J. Gen. Physiol.* **63**: 564–578.

22. Rojas, E., Pollard, H. B., and Heldman, E., 1985, Real-time measurements of acetylcholine-induced release of ATP from bovine medullary chromaffin cells, *FEBS Lett.* **185**: 323–327.

23. Oka, M., Isosaki, M., and Watanabe, J., 1980, Calcium flux and catecholamine release in isolated bovine adrenal medullary cells: Effects of nicotinic and muscarinic stimulation, in: *Synthesis, Storage and Secretion of Adrenal Catecholamines,* Advances in the Biosciences, Vol. 36, pp. 29–36.

24. Sussman, K. E., Pollard, H. B., Leitner, J. W., Nesher, R., Adler, J., and Cerasi, E., 1983, Differential control of insulin secretion and somatostatin receptor recruitment in isolated islets, *Biochem. J.* **214**: 225–230.

25. Eddlestone, G. T., Goncalves, A., Bangham, J., and Rojas, E., 1984, Electrical coupling between cells in islets of Langerhans from mouse, *J. Memb. Biol.* **77**: 1–14.

26. Ishikawa, K., and Kanno, T., 1978, Influences of extracellular calcium and potassium concentrations on adrenaline release and membrane potential in the perfused adrenal medulla of the rat, *Jpn. J. Physiol.* **28**: 275–289.

27. Friedman, J. E., Lelkes, P. I., Lavie, E., Rosenheck, K., Schneeweiss, F., and Schneider, A. S., 1985, Membrane potential and catecholamine secretion by bovine adrenal chromaffin cells: Use of tetraphenylphosphorium distribution and carbocyanine dye fluorescence, *J. Neurochem.* **44**: 1391–1402.

28. Atwater, I., Ribalet, B., and Rojas, E., 1979, Mouse pacreatic β-cells: Tetraethylammonium blockage of the potassium permeability increase induced by depolarization, *J. Physiol.* **288**: 561–574.

29. Cook, D. L., and Hales, C. N., 1984, Intracellular ATP directly blocks K^+ channels in pancreatic β-cells, *Nature* **311**: 271–273.

30. Findlay, I., Dunne, M. J., Ullrich, S., Wollheim, C. B., and Petersen, O. H., 1985, Quinine inhibits Ca-dependent K^+ channels, whereas tetraethylammonium inhibits Ca-activated K-channels in insulin-secreting cells, *FEBS Lett.* **185**: 4–8.

31. Grynszpan-Wynograd, O., and Nicolas, G., 1980, Intercellular junction in the adrenal medulla: A comparative freeze-fracture study, *Tissue Cell* **12**: 661–672.

32. Kilpatrick, D. L., Slepetis, R., and Kirshner, N., 1981, Ion channels and membrane potential in stimulus–secretion coupling in adrenal medulla cells, *J. Neurochem.* **36**: 1245–1255.

33. Wakade, A. R., and Wakade, T. D., 1979, Contribution of nicotonic and muscarinic receptors in the secretion of catecholamines evoked by endogenous and exogenous acetylcholine, *Neuroscience* **10:** 973–981.

34. Wilson, S. P., and Kirshner, N., 1977, The acetylcholine receptor of the adrenal medulla, *J. Neurochem.* **28:** 687–692.

35. Lee, F. L., and Trendelenburg, U., 1967, Muscarinic transmission of preganglionic impulses to the adrenal medulla of the cat, *J. Pharmacol. Exp. Therap.* **158:** 73–79.

Chapter 10

Calcium Currents and Hormone Secretion in the Rat Pituitary Gonadotroph

Andres Stutzin, Stanko S. Stojilkovic, Kevin J. Catt, and Eduardo Rojas

1. INTRODUCTION

The release of pituitary gonadotrophin hormones, luteinizing hormone (LH), and follicle-stimulating hormone (FSH) is specifically stimulated by the decapeptide gonadotrophin-releasing hormone (GnRH), which is secreted from the hypothalamus and is borne by blood to the pituitary gland.

It is well established that calcium is required for GnRH-stimulated LH secretion[1-3] by the pituitary gonadotroph, but the mechanism of GnRH action and the calcium dependence of LH secretion remains to be elucidated. It has been shown that calcium ionophores (ionomycin and A23187) stimulate LH secretion,[4] and that EGTA-buffered media and Ca^{2+}-channel blockers decrease GnRH-induced LH release.[5-7] Direct measurements of intracellular Ca^{2+} levels by fluorescence techniques have demonstrated that the calcium signal increases significantly during GnRH stimulation.[8,9] This increase can be modified by Ca^{2+}-channel blockers, suggesting that at least an important fraction of GnRH effect depends on the entry of external calcium.[9]

ANDRES STUTZIN and EDUARDO ROJAS • Laboratory of Cell Biology and Genetics, National Institute of Diabetes and Digestive and Kidney Diseases, National Institutes of Health, Bethesda, Maryland 20892. STANKO S. STOJILKOVIC and KEVIN J. CATT • Endocrinology and Reproduction Research Branch, National Institute of Child and Human Development, National Institutes of Health, Bethesda, Maryland, 20892. *Present address for A. S.:* Laboratorio de Fisiopatología Molecular, Departamento de Medicina Experimental, Facultad de Medicina, Universidad de Chile, Santiago, Chile.

In view of experimental evidence that suggests the involvement of dihydropyridine-sensitive pathways for Ca^{2+} entry, we decided to study these pathways directly by electrophysiological techniques in order to clarify their role in hormone secretion.

2. METHODOLOGICAL CONSIDERATIONS

2.1. Cell Preparation

Anterior pituitaries were obtained from adult female Sprague–Dawley rats 2 weeks after ovariectomy and were then dispersed into single cells by controlled trypsinization.[10] The cells were kept in primary culture and were assayed for LH secretion on the third day. Electrophysiological measurements were usually made at the sixth day. The large size of the gonadotrophs (15–20 μm in diameter) and their polymorphism allowed ready identification of the cells.

2.2. LH Release

The cells were washed thoroughly and incubated for 3 hr at 37° C in the experimental medium containing the appropriate secretagogue and/or drugs.[11] At the end of the incubation period, media were removed, and the LH content was measured by radioimmunoassay.[12] In addition, cells were solubilized for determination of cellular LH. Data were expressed in terms of the RP-2 Rat LH Reference Preparation as nanograms of gonadotropin content and release per 10^6 cells.

2.3. Fura-2 AM Assays for Cytosolic Calcium

Measurements of intracellular Ca^{2+} concentration were performed in dispersed pituitary cells. The cells were incubated for 1 hr at 37° C in a medium containing 2 μM Fura-2 AM.[13] After the incubation, the cells were washed and kept on ice before use. Fluorescence output was determined in a dual excitation mode (340 and 390 nm), and the intracellular calcium concentration was calculated after corrections of emission data measured at 500 nm for dye leakage and autofluorescence.[14]

2.4. Electrophysiological Measurements

The whole-cell variant of the patch-clamp technique was used.[15] Pipettes were prepared from micro-hematocrit capillary tubes and had an open-tip resistance in physiological saline in the range of 2.5 to 5 MΩ. All experiments were carried out at room temperature. After sealing the surface of the cell (resistance > 20 GΩ), the membrane patch was broken by suction, and the current transients were recorded under voltage-clamp conditions using a List EPC-7 amplifier (List Electronics, Darmstadt-Eberstadt, F.R.G.). The current records were filtered by means of an 8-pole 902LPF Bessel filter (Frequency Devices, Haverhill, MA) set at 8 KHz (-3dB 4KHz) and stored on-line on a computer (System 90, Computer Instrumentation Limited, Southampton, U.K.), for later analysis.[16]

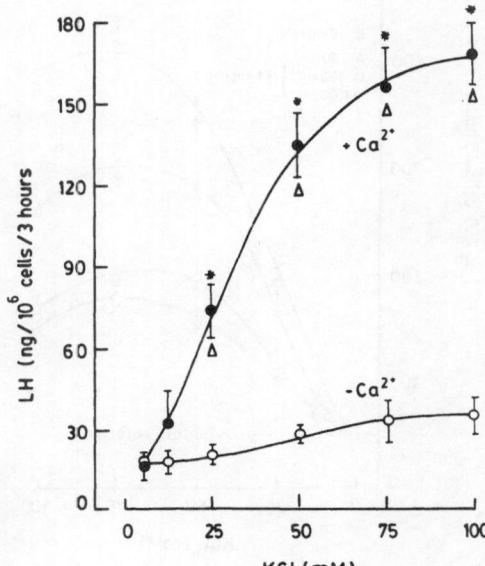

Figure 1. Extracellular Ca^{2+}- and K^+-induced LH release. Points represent the mean ± SE of data from five experiments. Comparison between normal (1.25 mM) calcium and Ca^{2+}-deficient media (< 20 μM).

3. LH SECRETION: EFFECTS OF GnRH, K⁺, AND DIHYDROPYRIDINES

LH secretion in static cultures stimulated by K^+ was both dose and calcium dependent, as depicted in Fig. 1. The secretory response reached a half-maximum at 30 mM K^+ concentration and tended to saturate at values close to 100 mM. In calcium-deficient media, LH secretion was significantly inhibited and the highest K^+ concentration (100 mM) elicited only about 20% of the secretory response obtained at physiological extracellular calcium concentrations.

Addition of increasing doses (10 nM to 1 μM) of the Ca^{2+}-channel antagonist, nifedipine, caused a significant dose-dependent decrease in K^+-stimulated LH release, as shown in Fig. 2. In contrast, BK 8644 not only enhanced basal LH release, but also increased the secretory response to 25 mM K^+ and 10 nM GnRH. Conversely, Ca^{2+}-channel blockers inhibited LH secretion (Fig. 3).

GnRH-induced LH secretion also depends on extracellular calcium, as shown in Fig. 4. However, the GnRH-induced response was not completely abolished even at subnanomolar Ca^{2+} concentrations, as compared to the K^+-induced LH secretion.

4. INTRACELLULAR CALCIUM SIGNALS

The changes induced by GnRH in intracellular Ca^{2+} concentration can be described as biphasic.[9] A rapid peak increase is reached within 20 sec, followed by a sustained increase at a lower level for about 10 min, as depicted in Fig. 5A. The first phase is not affected by incubation of the cells in a EGTA-buffered extracellular medium

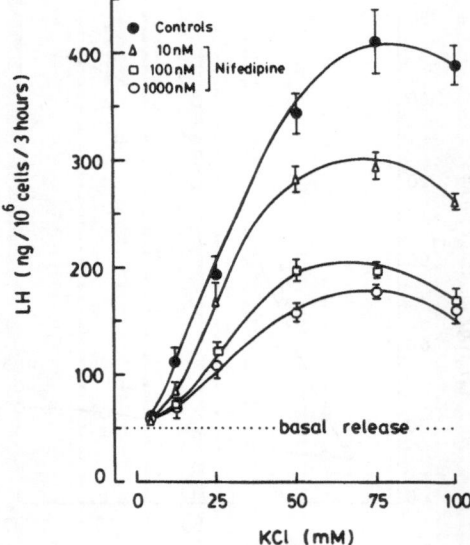

Figure 2. K⁺-induced LH secretion and inhibition by nifedipine. Effect of increasing doses of nifedipine on LH secretion induced by high K⁺ in a medium containing 5.4 mM Ca^{2+}.

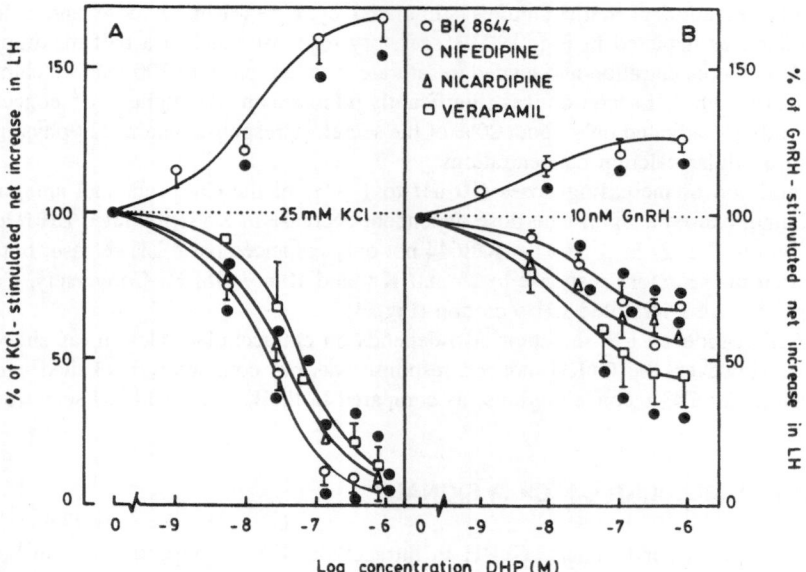

Figure 3. Stimulatory and inhibitory actions of dihydropyridines on (A) K⁺, 25 mM and (B) GnRH, 10 nM (induced LH secretion). Points (mean ± SE of three or five experiments, ●, $P < 0.05$).

(Fig. 5B). In contrast, the second phase is completely abolished. This rapid phase was essentially not modified by the Ca^{2+}-channel blocker nifedipine (Fig. 5C). However, the sustained phase of the calcium increase upon GnRH addition was reduced by nifedipine ($53 \pm 4\%$). Thus, the time course of GnRH-induced increase in intracellular Ca^{2+} differed both quantitatively and qualitatively, and especially in its sensitivity to external calcium and blockers of the L-type calcium channels.

5. CALCIUM CURRENTS IN THE RAT PITUITARY GONADOTROPH

A family of ionic currents from a gonadotroph is depicted in Fig. 6. The cell was depolarized from a holding potential of -80 mV to the values shown in the figure. We were mainly interested in the inward currents carried by Ca^{2+} ions, and have therefore eliminated potassium and sodium currents by replacing virtually all the potassium in the pipette by caesium and external sodium by choline. In addition, to prevent passage of Ca^{2+} ions through Na^+ channels, tetrodotoxin (20 μM) was added to the bath.

Under these experimental conditions it can be assumed that the inward current shown in Fig. 6 is carried by Ca^{2+} ions. The simplest explanation for the complex temporal course of the current is that it represents the flow of Ca^{2+} ions through more than one pathway. The current activates at relatively low depolarizing potentials (-60 mV) and increases up to membrane potential values around 10 mV. The temporal course of this

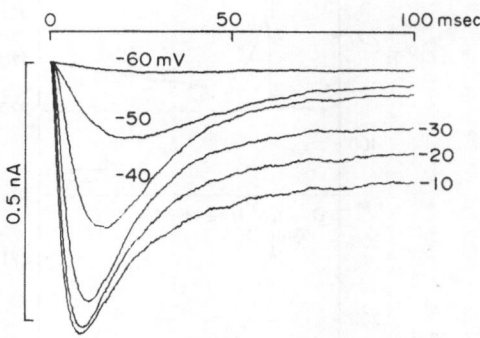

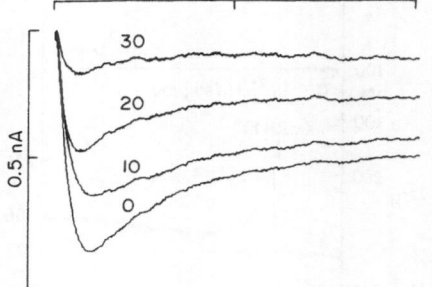

Figure 6. Family of calcium currents in the gonadotroph. The medium contained (in mM) 140 choline chloride, 10 HEPES-K, 2.6 CaCl$_2$, 1 MgCl$_2$, 20 μM tetrodotoxin, pH 7.4. The intracellular dialyzing solution contained 140 CsCl, 10 HEPES-K, 2.5 MgATP, 1 MgCl$_2$, 0.1 Na-EGTA, pH 7.2. The cell was kept at a holding potential of -80 mV and depolarized to the potentials shown in the figure with a pulse lasting 100 msec.

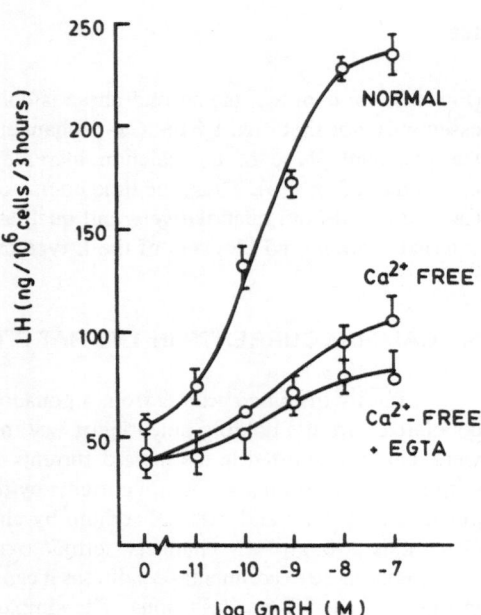

Figure 4. Extracellular Ca^{2+} and GnRH-induced LH secretion. Points (mean $\pm$ SE of six experiments).

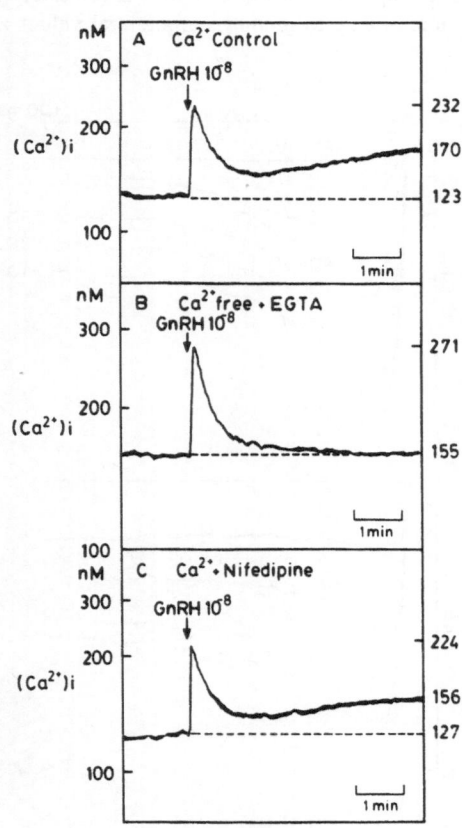

Figure 5. Intracellular calcium changes in pituitary cells stimulated by 10 nM GnRH. (A) Medium containing 1.25 mM calcium; (B) medium with no calcium added plus 100 μM EGTA; (C) medium containing 1.25 mM calcium plus 1 μM nifedipine.

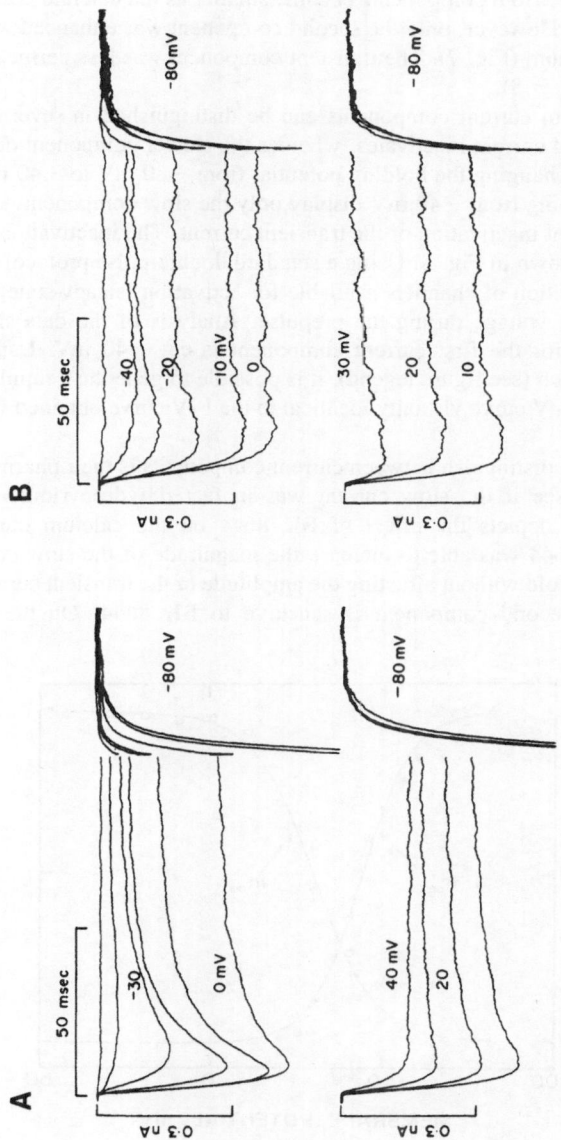

Figure 7. Current records obtained with (A) 2.6 CaCl$_2$ in the bath and (B) 2.6 BaCl$_2$. The experimental conditions are the same as described for Fig. 6.

inward current reveals two distinct components, a fast one, with activation and inactivation kinetics triggered at low depolarizing pulses, and a delayed one, with slower activation. The first component reaches its maximum amplitude at -20 mV and the second one at 10 mV. Both components become smaller as the calcium concentration is lowered in the bath. However, only the second component was enhanced when calcium was replaced by barium (Fig. 7). The transient component was less permeable to Ba^{2+} than Ca^{2+} $(P_{Ca}/P_{Ba} = 8)$.

The two calcium current components can be distinguished in several ways. The low-voltage activated current inactivates, whereas the slower component does not. This can be explored by changing the holding potential from -80 mV to -40 mV. Currents obtained when stepping from -40 mV display only the slow component demonstrating the voltage-dependent inactivation of the transient current. The inactivation curve of the fast component is shown in Fig. 8. Using a standard double-pulse protocol it is possible to determine the fraction of channels available for activation (steady-state inactivation) as a function of the voltage during the prepulse. Analysis of the data show that the transition potential for the first current component is ca. -40 mV. Using the same experimental approach (see figure legend), it is possible to generate a family of currents with a steady-state I–V curve virtually identical to the I–V curve obtained with a single-pulse protocol.

Another way to distinguish between current components is their pharmacology. We were interested to see if the slow current was in fact dihydropyridine-sensitive, as expected. Figure 9 depicts the effect of BK 8644 on the calcium currents of the gonadotroph. BK 8644 was able to increase the magnitude of the slow component (at steady state) by fivefold without affecting the amplitude of the transient current, suggesting that only the second component is sensitive to BK 8644. On the other hand,

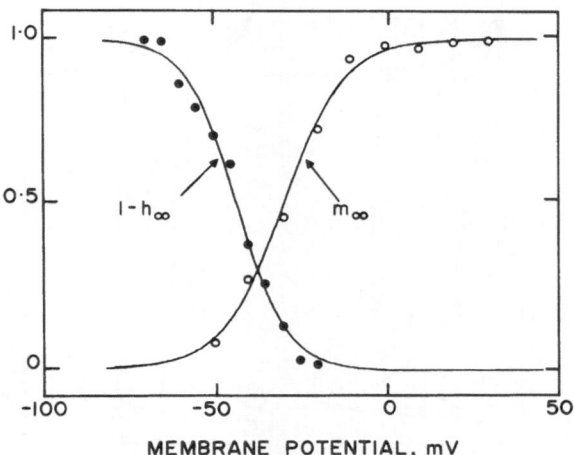

MEMBRANE POTENTIAL, mV

Figure 8. Steady-state inactivation curve for the treatment calcium current. The double-pulse protocol consisted in a prepulse with a duration of 150 msec and increasing amplitudes from -80 mV to 10 mV in 5 mV steps. After a delay of 10 msec at -80 mV, a second pulse stepped the membrane potential to 0 mV. The experimental points were fitted with a Boltzmann distribution function. $V_{0.5}$ is -40.1 mV.

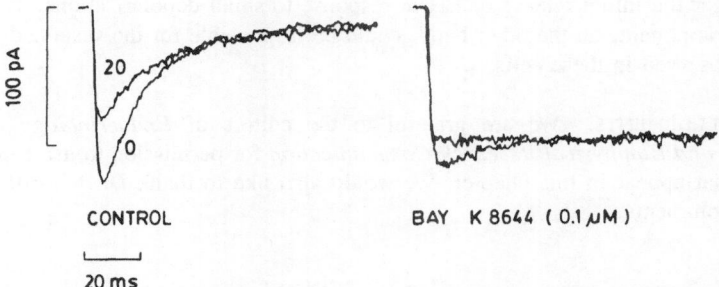

Figure 9.   Effect of BK 8644 on the calcium currents of the gonadotroph. Left-hand panel: Control, two depolarizing pulses from a holding potential of −80 mV to the membrane potential shown. Right-hand panel: BK 8644, 0.1 μM. BK 8644 was added (same cell as control) and the cell was kept depolarized at −40 mV for 5 min. After incubation with the drug, the holding potential was returned to −80 mV and the voltage was stepped to the same values shown for the control.

nifedipine (0.1 μM) significantly reduced the current amplitude of the second component, without modifying the current flowing through the transient channels (not shown).

6. CONCLUSION

As discussed earlier in this chapter, rat pituitary gonadotrophs kept in primary culture are able to secrete LH upon stimulation with various secretagogues. In a number of secretory systems, the activation of the release process is related to changes in the concentration of intracellular calcium.[17,18] In the rat pituitary gonadotroph, secretion is supported by an elevation of the cytosolic calcium level, and the source for Ca^{2+} is both extracellular and intracellular.[11,19] However, extracellular calcium is mandatory to achieve a full and sustained secretory response. The purpose of this work was to examine the calcium entry pathways present on the gonadotroph surface membrane and to correlate their kinetics and pharmacological sensitivity to the secretory events.

Our studies reveal two kinetically different calcium channels available for Ca^{2+} influx in these cells.[20] The kinetic properties, the pharmacological sensitivity, and the Ba^{2+} permeability of the current allow us to identify these channels as being from the T-type and L-type.[21] These Ca^{2+} channels have been found in a variety of tissues.[22,23] Our results suggest that an important fraction of the Ca^{2+} required for secretion enters the cell from the extracellular space via these Ca^{2+}-channels. Moreover, the effects of the dihydropyridine derivatives on LH release evoked by high potassium or GnRH is only partially blocked by dihydropyridines, suggesting that another, dihydropyridine-insensitive pathway is also present. The fast, transient calcium current component is in fact insensitive to these drugs,[24] thus providing another pathway for calcium entry upon membrane depolarization.

More data will need to be collected to define the physiological role of the two calcium conductances present in the cell membrane of the gonadotroph and their relationship to secretion. However, it is reasonable to speculate that the fast conductance may trigger the initial release of LH in response to small depolarizations. The slower current component, on the other hand, could be responsible for the sustained hormone release observed in these cells.

ACKNOWLEDGMENTS. We are grateful to the editors of *Endocrinology* and *Biochemical and Biophysical Research Communication* for permission to use some of the figures that appear in this chapter. We would also like to thank Dr. Ricardo Bull for helpful comments.

REFERENCES

1. Conn, P. M., Marian, J., McMillian, M., Stern, J., Rogers, D., Hamby, M., Penna, A., and Grant, E., 1981, Gonadotropin-releasing hormone action in he pituitary: a three-step mechanism, *Endocr. Rev.* **2:** 174–185.
2. Catt, K. J., Loumaye, E., Wynn, P. C., Iwashita, M., Hirota, K., Morgan, R. O., and Chang, J. P., 1985, GnRH receptors and actions in the contro of reproductive function, *J. Steroid Biochem.* **23:** 677–689.
3. Conn, P. M., 1986, The molecular basis of gonadotropin-releasing hormone action, *Endocr. Rev.* **7:** 3–10.
4. Kiesel, L., and Catt, K. J., 1984, Phosphatidic acid and the calcium-dependent actions of gonadotropin-releasing hormone in pituitary gonadotrophs, *Arch. Biochem. Biophys.* **231:** 202–210.
5. Conn, P. M., Rogers, D. C., and Sheffield, T., 1981, Inhibition of gonadotropin-releasing hormone and stimulated luteinizing hormone release by pimozide: evidence for a site of action after calcium mobilization, *Endocrinology* **109:** 1122–1126.
6. Borgeat, P., Garneau, P., and Labrie, F., 1975, Calcium requirement for stimulation of cyclic AMP accumulation in anterior pituitary gland by LH–RH, *Mol. Cell. Endocr.* **2:** 117–124.
7. Naor, Z., Leifer, A. M., and Catt, K. J., 1980, Calcium-dependent actions of gonadotropin-releasing hormone on pituitary guanosine $3',5'$-monophosphate production and gonadotropin release, *Endocrinology* **107:** 1438–1445.
8. Leong, D. A., Beshoar, D. F., Sullivan, J. A., Mandell, G. L., and Thorner, M. O., 1986, Changes in intracellular free $[Ca^{2+}]$ measured directly in individual LH secretory cells stimulated with LHRH, *68th Annual Meeting of The Endocrine Society*, Anaheim, CA, Abstract 27, p. 40.
9. Tasaka, K., Stojilkovic, S. S., Izumi, S. I., and Catt K. J., 1988, Biphasic activation of cytosolic free calcium and LH responses by gonadotropin-releasing hormone, *Biochem. Biophys. Res. Comm.* **145:** 398–405.
10. Hyde, C. L., Childs (Moriarty) G., Wahl, L. M., Naor, Z., and Catt, K. J., 1982, Preparation of gonadotroph-enriched cell populations from adult rat anterior pituitary cells by centrifugal elutriation, *Endocrinology* **111:** 1421–1426.
11. Chang, J. P., Stojilkovic, S. S., Graeter, J. S., and Catt, K. J., 1988, Gonadotropin-releasing hormone stimulates luteinizing hormone secretion by extracellular calcium-dependent and -independent mechanisms, *Endocrinology* **122:** 87–97.
12. Loumaye, E., Naor, Z., and Catt, K. J., 1982, Binding affinity and biological activity of gonadotropin-releasing agonists in isolated pituitary cells, *Endocrinology* **111:** 730–736.
13. Grynkiewicz, G., Poenie, M., and Tsien, R. Y., 1985, A new generation of Ca^{2+} indicators with greatly improved fluorescence properties, *J. Biol. Chem.* **260:** 3340–3350.
14. Rink, T. J., and Pozzan, T., 1985, Using Quin-2 in cell suspensions, *Cell Calcium* **6:** 133–144.
15. Hamill, O. P., Marty, A., Neher, E., Sakmann, B., and Sigworth, F. J., 1981, Improved patch-clamp techniques for high-resolution current recording from cells and cell-free membrane patches, *Pfluegers Arch.* **391:** 85–100.

16. Quinta-Ferreira, E. M., Arispe, N., and Rojas, E., 1982, Sodium currents in the giant axon of the crab *Carcinus maenas*, *J. Membrane Biol.* **66:** 159–169.
17. Feinman, R. D., and Detwiler, T. C., 1974, Platelet secretion induced by divalent cation ionophores, *Nature (London)* **249:** 172–173.
18. Rubin, R. P., Sink, L. E., and Freer, R. J., 1981, Activation of (arachidonyl)phosphatidylinositol turnover in rabbit neutrophils by the calcium ionophore A23187, *Biochem. J.* **194:** 497–505.
19. Berridge, M. J., and Irvine, R. F., 1984, Inositol trisphosphate, a novel second messenger in cellular signal transduction, *Nature (London)* **312:** 315–321.
20. Stutzin, A., Stojilkovic, S. S., Catt, K. J., and Rojas, E., 1989, Two types of calcium channels in the membrane of the pituitary gonadotroph, *Am. J. Physiol.* (in press).
21. Nowycky, M. C., Fox, A. P., and Tsien, R. Y., 1985, Three types of neuronal calcium channels with different calcium agonist sensitivity, *Nature (London)* **316:** 440–443.
22. Bean, B. P., 1985, Two kinds of calcium channels in canine atrial cells. Differences in kinetics, selectivity, and pharmacology, *J. Gen. Physiol.* **86:** 1–30.
23. Armstrong, C. M., and Matteson, D. R., 1985, Two distinct populations of calcium channels in a clonal line of pituitary cells, *Science* **4682:** 65–67.
24. Tsien, R. Y., Hess, P., and Nilius, B., 1987, Cardiac calcium currents at the level of single channels, *Experientia* **43:** 1169–1172.

Chapter 11

Synexin-Driven Membrane Fusion: Molecular Basis for Exocytosis

Eduardo Rojas, A. Lee Burns, and Harvey B. Pollard

1. INTRODUCTION

Secretion by exocytosis involves fusion of a secretory vesicle membrane with the plasma membrane of the secreting cell. In many endocrine cells, including chromaffin cells and pancreatic β-cells, such simple exocytosis is followed by contact and fusion of more deeply situated secretory vesicles with the initially fused secretory vesicle membranes. The latter process is called compound exocytosis, and presumably allows for additional secretion without moving secretory granules long distances through the cytoskeleton to reach the plasma membrane.

Both simple and compound exocytosis have usually been closely associated with a requirement for extracellular calcium. With the introduction of fluorescent Ca^{2+} indicators to monitor intracellular calcium, the concept that exocytotic secretion in many types of cells requires a substantial increase in the intracellular $[Ca^{2+}]$ has been firmly established. For this reason, most model studies on the mechanism of membrane contact and fusion focus on how calcium might promote the process.

Because several results with model systems indicated that calcium alone did not promote fusion, we began a search for cytosolic proteins that could mediate calcium-dependent contact and fusion of chromaffin granules. The initial result of this search was

EDUARDO ROJAS, A. LEE BURNS, and HARVEY B. POLLARD • Laboratory of Cell Biology and Genetics, National Institute of Diabetes and Digestive and Kidney Diseases, National Institutes of Health, Bethesda, Maryland 20892.

the discovery of the protein synexin.[1] This review is a summary of what ensued over the following 10 years as we pursued a better understanding of the mechanism of synexin action.

2. BIOCHEMICAL PROFILE OF SYNEXIN

2.1. Purification of Synexin

The initial isolation of synexin from bovine adrenal gland included precipitation in ammonium sulfate, gel filtration, and hydroxylapatite chromatography.[1] For all the experimental results presented here, however, highly purified synexin was prepared from bovine liver using a protocol involving the usual ammonium sulfate fractionation and chromatography on ultragel steps, followed by a step of purification to near homogeneity by chromatofocussing.[2]

2.2. Synexin-Driven Aggregation of Secretory Vesicles from Bovine Adrenal Medullary Cells

Synexin was initially characterized as a cytosolic protein that caused chromaffin granules to aggregate in a Ca^{2+}-dependent manner. Synexin activity was measured quantitatively by monitoring the increase in turbidity associated with granule aggregation. Other divalent cations including Mg^{2+}, Sr^{2+}, and Ba^{2+} were tested, but only Ca^{2+} proved effective. The concentration of free calcium ($[Ca^{2+}]$) required to induce 50% aggregation of the chromaffin-secretory granules ($K_{\frac{1}{2}}$) was found to be 200 μM. We later learned that synexin self-associated in the presence of calcium with the same $K_{\frac{1}{2}}$.[2] Thus, the calcium effect seemed to be on synexin, and not on the membranes of the chromaffin-secretory vesicles. Electron micrographs showed that aggregated granules were connected by quite close "pentalaminar" membrane contacts. Because these contacts were similar to those observed in secreting cells by Palade[3] and others, we were somewhat encouraged in our expectation that synexin action might actually have something to do with exocytosis.

However, at the time of the discovery of synexin, cytosolic Ca^{2+} concentration in resting chromaffin cells was believed, and later shown to be, in the range of 50 to 150 nM. Stimulation caused the calcium concentration to rise perhaps two- or three-fold, a value well below the $K_{\frac{1}{2}}$ for synexin, and substantially below the threshold calcium ion concentration needed to activate synexin (ca. 5 μM). For years, we and others[4] were puzzled over the physiological meaning of the rather high $K_{\frac{1}{2}}$ value for calcium dependence of granule aggregation. However, with the help of better intracellular Ca^{2+} indicators, it was clearly established that, following stimulation of chromaffin cells, the concentration of free calcium increased transiently in the range of 10 to 100 μM in the volume immediately beneath the plasma membrane.[5,6] Although the calcium dependence for binding of synexin to granular membranes can be shifted to the low micromolar range by raising the pH of the assay from 6 to 7,[7] the magnitude of granule aggregation under these conditions is less than at pH 6, possibly indicating that the processes mediating synexin binding to membranes and granule aggregation could be distinct in some meaningful ways.

2.3. Synexin-Driven Fusion of Chromaffin Cell Granule and Granule Ghosts

In studies with intact chromaffin granules, fusion per se could only be observed if relatively low concentrations (ca. 5 μM) of arachidonic acid were added to the aggregated granules.[8,9] However, other fatty acids possessing at least one *cis*-unsaturated double bond could replace arachidonic acid in supporting fusion. In addition, while calcium was required for the synexin-driven granule aggregation step, subsequent removal of calcium by addition of EGTA left the granules still aggregated and still susceptible to fusion by added fatty acids. Thus, the fusion step itself was calcium independent.

The possible physiological relevance of this process is indicated by the fact that chromaffin cells synthesize substantial amounts of free arachidonic acid when stimulated by endogenous secretagogues.[10,11] A more compelling argument for the relevance of this mechanism is the nature of the fusion structures formed by arachidonic acid treatment of synexin-aggregated granules. The granules fuse into vacuolar structures nearly identical in character to compound exocytotic structures observed in stimulated chromaffin cells.[12,13]

However, when synexin was used to promote fusion of liposomes[14-16] or of chromaffin granule ghosts[17-19] arachidonic acid was not required. Other studies[14-16,20] revealed that liposomes prepared from the acidic phospholipids phosphatidylserine (PS), phosphatidic acid (PA), or phosphatidylethanolamine (PE), or mixtures of these, would fuse directly if calcium and synexin were added to them in solution. Synexin was completely ineffective when the liposomes were prepared from phosphatidylcholine (PC), while liposomes prepared from phosphatidylinositol (PI) aggregated.[15] Thus, it seemed clear that liposome aggregation and eventual fusion was quite dependent on the specific lipid used, and that acidic phospholipid seemed to be preferred over neutral phospholipids. Consistent with these ideas[21], Hong et al.[16] showed in a kinetic study that the rate-limiting step for synexin-induced fusion was the liposome aggregation step.

Chromaffin granule ghosts, prepared by a freeze-thaw technique, were also found to fuse spontaneously upon the addition of synexin. As illustrated in Fig. 1, this process can be followed by a volume-mixing assay in which some ghosts are loaded with self-quenching concentrations of FITC-dextran (fluorescein-isothiocyanate dextran) and allowed to fuse with empty ghosts. This process is, quantitatively different, however, from liposome fusion in several regards. By contrast with the liposome experiments, the freeze-thaw ghosts fused in a manner only partially dependent on Ca^{2+} concentration. Furthermore, membrane carboxyl groups, possibly on protein, seemed important for the process. However, Nir et al.[19] showed that the rate-limiting step was the aggregation event itself, rather that the membrane fusion event, as shown for liposomes.

These data, therefore, indicate that synexin can indeed cause contact and fusion of membranes that are relevant to the exocytotic process. However, the fact that acidic phospholipid liposomes are also substrates indicates that synexin may not only be directing its attention toward specific membranes in the secreting cell. Nonetheless, it is a fact that acidic phospholipids such as PS, PE, PA and PI are primarily localized to the inner leaflets of plasma membranes and to the outer leaflets of granule membranes. Choline-containing phospholipids, such as PC and sphingomyelin, by contrast, are localized to the outer leaflets of plasma membranes and the inner leaflets of granule

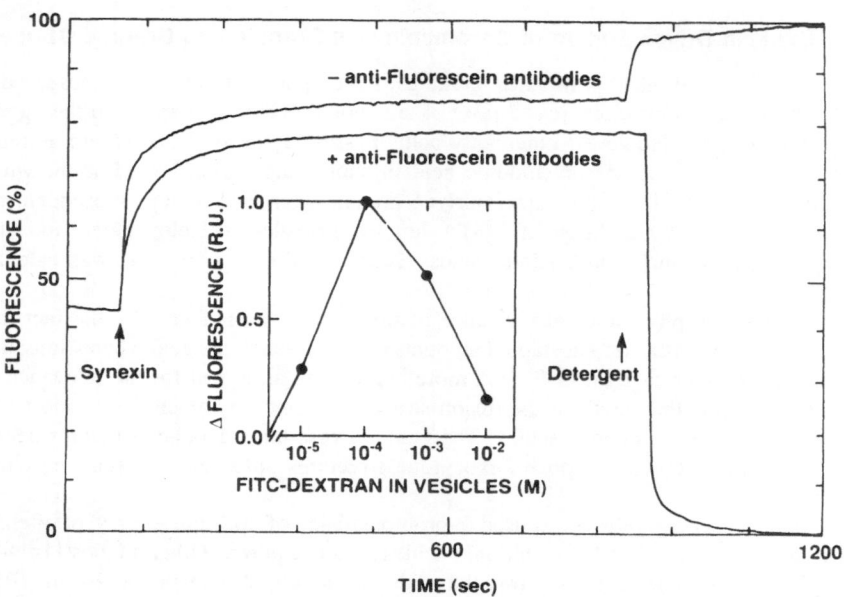

Figure 1. Synexin induces fusion of chromaffin granule ghosts. The increase of fluorescence due to addition of synexin (34.08 μg, first arrow) is shown in the absence and presence of antifluorescein antibody. Reaction medium (mM): 140 KCl, 10 K-HEPES, pH 6, pCA 7, at 37°C. Detergent (NP-40) was added at the second arrow in the presence or absence of the antibody to calibrate the signal. An increase in fluorescence of the system indicates fusion of the vesicle containing self-quenching concentrations of FITC-dextran with an empty vesicle. Inset: The self-quenching curve for FITC-dextran. Granules were loaded with different concentrations of FITC-dextran by freezing and thawing in liquid nitrogen. Granule ghosts were loaded in 30 mM FITC-dextran. Data are a composite from Ref. 17.

membranes. Thus, calcium-activated synexin may aggregate and fuse granules involved in exocytosis merely because of proximity of the appropriate phospholipids at the sites where calcium concentration is highest.

This interpretation of how synexin might be directed to act during exocytosis is also consistent with our experiments on inside-out plasma membranes from chromaffin cells, attached to poly-L-lysine coated beads.[22] In these experiments we found that synexin would bind in a calcium-dependent manner to the inner leaflets of chromaffin cell plasma membranes, but not to the PC-rich outer leaflets of intact chromaffin cells, otherwise also attached to the beads.

2.4. Synexin-Driven Exchange of Membrane Lipids from Chromaffin Granule Ghosts

Substantial evidence exists that membrane lipid mixing occurs either immediately before or simultaneously with synexin-driven membrane fusion processes. For example, self-quenching concentrations of octadecylrhodamine (R-18) have been used to label the outer leaflets of chromaffin granule ghosts. When synexin is added to such labelled

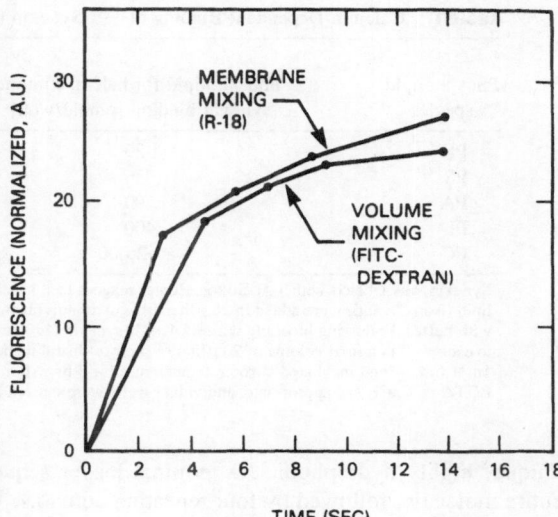

Figure 2. Comparison of rates of synexin-driven chromaffin granule ghosts fusion measured by membrane mixing (R-18) and volume mixing (FITC-dextran).

ghosts in the presence of unlabelled ghosts, lipid mixing occurs, as measured by acquisition of a fluorescence signal.[17] Separate evidence using the FITC-volume mixing assay verifies that the same ghosts are undergoing true fusion. True fusion, as defined by volume mixing, occurs slightly slower than membrane mixing as defined by the R-18 assay (Fig. 2).

3. CLONING AND SEQUENCING OF HUMAN SYNEXIN

To establish the nature of synexin–membrane interactions at a molecular level, we needed information on the complete primary structure of synexin. For this reason, during the last 3 years we initiated a successful effort to purify synexin from human liver, and then purify and sequence tryptic peptides. From these peptides we prepared oligonucleotide probes, and successfully searched human lung, liver, and retina libraries for synexin-specific clones.[23]

3.1. General Structure of the Human Synexin Clone

Human synexin proved to be virtually identical to bovine synexin in terms of calcium-dependent chromaffin granule aggregation properties. Furthermore, derived amino acid sequences from a human synexin clone proved to have substantial homology with our known bovine sequences obtained by sequence analysis of tryptic fragments. Examples of these homologies are shown in Table I, and we will discuss these relationships further after describing the structure of our human synexin clone.

The general structure of our initial human synexin clone is shown in Fig. 3, and included characteristic 5′ and 3′ noncoding regions. The coding region consisted of a

Table I.   Calcium-Dependent Binding of ^{125}I-Synexin to Membrane Phospholipids[a]

Phospholipid species	Phospholipid for half-maximum synexin binding [pmole/well]	Synexin bound at half-maximum lipid [pmole/well]
PE	29	4.8
PS	36	6.0
PA	90	3.0
PI	400	13.5
PC	>>2,000	None

[a]Synexin was labeled with ^{125}I-Bolton-Hunter reagent to a specific activity of 526 cpm/pmole. The lipid (from Avanti) were added in 20 μliter ethanol to individual wells, vacuum-dried, and rehydrated with buffer. Following blocking with BSA, ^{125}I-synexin (110 pmole or about 60,000 cpm) was added to each well in a final volume of 20 μliter of sucrose–histidine buffer, pH 6, in either 1 mM EGTA or 1mM CaCl$_2$, and incubated at room temperature for 3 hr. After a rapid washing procedure, in either EGTA or CaCl$_2$, as appropriate, individual wells were cut out and counted. (From Ref. 45).

unique, highly hydrophobic N- terminal leader sequence comprising ca. 30% of the entire molecule, followed by four repeating domains. We have labelled these domains I, II, III, and IV, and have noted that while they are homologous with one another they are by no means identical. However, within each of these repeating domains are characteristic core sequences with a much higher homology to one another. We have labelled these α, β, σ, and δ. These repeats also have substantial regions of strong hydrophobicity, as well as regions of hydrophilicity.

We conclude from these data that the N-terminal segment might satisfy, at least in principle, the requirement for a highly hydrophobic structure able to interact in a profound manner with the fusing membranes. Of course, we can not, at present, conclude that this hydrophobic N-terminal segment is or is not the key domain to explain the interactions between synexin and membranes. In fact, the C-terminal segment containing the fourfold repeat could also be involved in these interactions.[(23)]

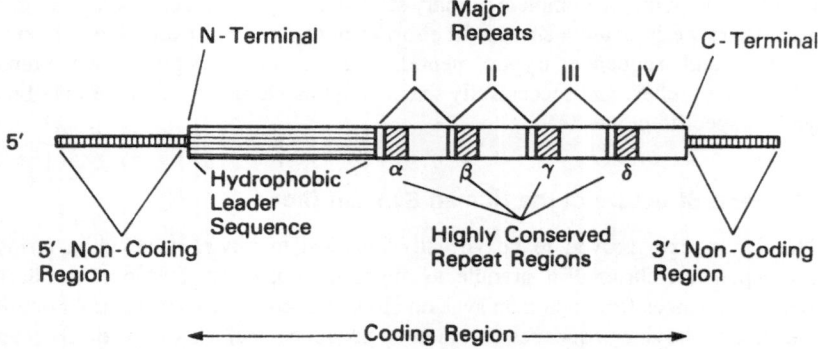

Figure 3.   Structure of the human synexin cDNA clone, Hu-Syn. Details are given in the text and are taken from Ref. 50. The N-terminal represents a unique, hydrophobic sequence comprising about 30% of the molecule. The remainder is homologous with a fourfold repeat observed in other members of the synexin gene family.

3.2. Synexin-Related Proteins

A search of the protein data banks revealed significant homology between human and bovine synexin and a unique set of calcium-dependent membrane binding proteins. These proteins include lipocortin I,[4] endonexin II,[25,26] calpactin heavy chain/ P36,[27-30] Protein II,[31] and Calelectrin 67K.[32] Data in Table II summarize the homologies in the four 16-amino-acid core repeats, also labelled α, β, γ, and δ in Fig. 3. The key to the similarity is a characteristic GXGTDE sequence, found with varying degrees of fidelity in the different core repeats. Clearly evident from these comparisons is that bovine and human synexin share many common sequences. For these specific examples, the homology was α (15 of 15), δ (14 of 14), and σ (9 of 11) in the three of four sequences available for comparison.

The similarities between human and bovine synexin indicated to us that synexin varies less *across* species lines that it varies with regard to the other members of the synexin gene family *within* species lines. In fact, within the four highly conserved core repeats the other proteins did not give homologies of more than ca. 50%, except in the case of the β segment. The β segment, however, is substantially different from the others in the C-terminal half of the sequence, where the characteristic LIEIL sequence is found. All the proteins share part or all of this sequence to varying degrees.

Table II. Highly Conserved Core Segments in Repeats I, II, III, and IV of Human Synexin: Comparison with Other Members of the Synexin Gene Family

Protein[a]	Core segment/repeat	Amino acic sequence	Sequence similarity to human synexin
hu-synexin	α-I	G F G T D E Q A I V D V V A N R	
bov-synexin	α-I	G F G T D E Q A I I D V V A N R	15/16
hu-endonexin II	α-I	G L G T D E E S I L T L L T S R	6/16
hu-calpactin I	α-I	T K G V D E V T I V N I L T N R	7/16
hu-lipocortin I	α-I	V K G V D E A T I I D I L T K R	7/16
bov-calelectrin (67 kDa)	α-I	G F G S D K E A I L D I I T S R	8/16
hu-synexin	α-I	G A G T Q E R V L I E I L C T R	
hu-endonexin II	β-II	G A G T N E K V L T E I I A S R	10/16
hu-calpactin II	β-II	G L G T D E D S L I E I I C S R	10/16
hu-lipocortin I	β-II	G L G T D E D T L I E I L A S R	10/16
bov-calelectrin	α-I	G I G T D E K C L I E I L A S R	10/16
hu-synexin	γ-III	R L G T D E S C F N M I L A T R	
hu-endonexin II	γ-III	K W G T D E E K F I T I F G T R	9/16
hu-calpactin I	γ-III	R K G T D V P K W I S I M T E R	7/16
hu-lipocortin I	γ-III	R K G T D V N V F N T I L T T R	10/16
bov-celectrin	γ-III	K W G T D E A Q F I Y I L G N R	8/16
hu-synexin	δ-IV	G A G T D D S T L V R I V V T R	
hu-endonexin II	δ-IV	G A G T D D H T L I R V M V S R	8/16
hu-calpactin I	δ-IV	G H G T R D K V L I R I M V S R	8/16
hu-lipocortin I	δ-IV	G V G T R H K A L I R I M V S R	6/16
bov-calelectrin	δ-iv	G L G T R D N T L I R I M V S R	8/16

[a]Hu = human; bov = bovine.

The most common characteristic of the core repeats in the synexin gene family is the fact that peripheral portions of the 15 amino acid sequence are quite hydrophobic, while carboxylic amino acids occur at internal sites. In synexin, three negatively charged amino acids occur in the α core and two occur in the β, δ, and σ cores. Most characteristic is the DE or DD sequence in positions 5 and 6 from the N-terminal end of the core repeats.

This characteristic location of negative charges in a hydrophobic nest has been noted previously by Taylor and Geisow[33] in an analysis of some previously known members of the synexin gene family. The repeats were considered to represent the Ca^{2+} binding sites, possibly analogous to the E–F hands in the parvalbumin–calmodulin gene family. In the specific case of synexin, calcium interaction sites cannot have too great an affinity for Ca^{2+}, or else the protein would be a Ca^{2+} binding protein rather than a Ca^{2+}-channel protein (see Section 4.3).

However, there exists one possible interpretation of this concept with interesting structural consequences. If these core sequences did indeed bind or transmit calcium in or through the synexin molecule, then these regions would have to span the membrane, possibly protected from the low dielectric medium by hydrophobic domains in the neighboring parts of the synexin molecule. A further consequence of this structural interpretation is that far more of the synexin molecule than the N-terminal leader sequence might be available to provide hydrophobic surfaces to drive membrane fusion processes.

4. THE NATURE OF SYNEXIN–MEMBRANE INTERACTIONS

During the brief time that synexin has been available for analysis, two distinct views have been pursued as to how the molecule might interact with target membranes. We had previously been of the opinion that Ca^{2+} interacted with individual synexin molecules, thereby modifying the conformation, and thus rendering synexin able to interact with the membrane in some way (summarized in Pollard et al.[21]). Alternatively, others suggested that synexin acted by causing close approach of membrane pairs, and that calcium per se induced fusion through a dehydration step.[15].

4.1. Ca²⁺-Dependent Binding of Synexin to Acidic Phospholipids

To distinguish between these possibilities we studied further the association of synexin with membrane lipids—in particular with acidic phospholipids known to be present in the cytosolic aspect of cell membranes.

Two different methods were used here. The first method involved placing different phospholipids onto 96-well plates, constructed from a quite hydrophobic plastic. After drying away the methanol solvent, we added buffer, presumably allowing the phospholipids to orient on the plate with hydrophobic tails toward the plastic surface. We found that ^{125}I-labelled synexin bound to acidic phospholipids such as PS in a Ca^{2+}-dependent manner. However, labelled synexin did not bind to PC or to cholesterol. Furthermore, synexin binds to different acidic lipids in a specific order as summarized in Table II.

These data thus indicated that synexin could bind to specific acidic phospholipids in a Ca^{2+}-dependent manner, and that the mechanism of fusion induced by synexin could involve such binding. However, the details of this interaction remained to be elucidated. For example, did synexin bind to Ca^{2+}, and thus became competent to interact with specific phospholipids? Alternatively, could Ca^{2+} itself modify the membrane and make molecules on it competent to bind synexin? As a final alternative, both processes could occur simultaneously. To answer these questions we turned to a method that could allow us to measure the association of $^{45}Ca^{2+}$ and synexin to acidic phospholipid mono-molecular films at the air–water interface.[34,35]

As shown in Fig. 4, a soft β-radiation detector was placed above a dish containing sucrose–histidine buffer, pH 6, $CaCl_2$ (200 μM, $^{45}Ca^{2+}$), and synexin, where appropriate. Uptake of Ca^{2+} by the monolayer was measured by counting the number of β-particles emanating from the hypophase containing $^{45}Ca^{2+}$ before and after spreading

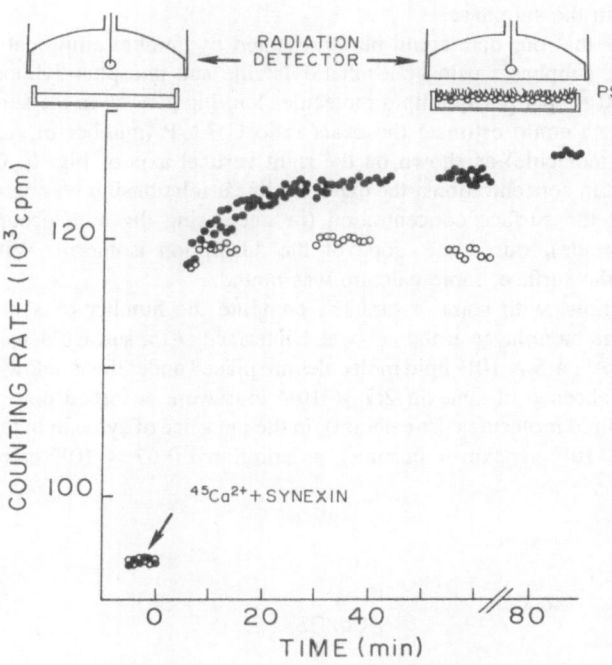

Figure 4. Association of synexin molecules with PS monolayers. A detector of the β-radiations was placed over a tissue culture dish (50 mm in diameter) containing $^{45}Ca^{2+}$-labelled solution (drawings shown in the upper part). Open circles (O) represent $^{45}Ca^{2+}$ adsorption in the absence of synexin. Closed circles (●) were obtained in the presence of synexin (18 μg/5 ml). Counts emanating from a 660 μm thick surface layer. All surfaces were coated with bovine serum albumin (1 mg/ml) dissolved in buffer prior to the experiment, and we ascertained that ^{14}C-BSA was inert to all experimental manipulations described here.

of the monolayer (open circles). From the data in Fig. 4 it is apparent that the increase in the number of β-particles emanating from the surface (from ca. 93,000 to 119,000 cpm) does not change over the 90-min course of the experiment. Spreading the monolayer in the presence of synexin in the hypophase (closed circles) induced an identical initial increase in cpm. However, with time the counting rate augmented. Thus, if synexin is present in the hypophase, more radioactivity is acquired at the surface, and it takes more than 1 hr to approach equilibrium. Taken at face value, these data support the concept that synexin binds $^{45}Ca^{2+}$ and brings it to the phospholipid monolayer at the air–water interface. Consistent with this interpretation is the observation that in the absence of phospholipid synexin does not induce measurable changes in the radioactivity emanating from the surface.

Knowing the specific activity of the $^{45}Ca^{2+}$ in the subphase, after verification that $^{45}Ca^{2+}$ adsorption to the monolayer followed a predictable isotherm, it was possible to estimate the number of synexin molecules interacting with the phospholipid monolayer from the excess β-radiation acquired at equilibrium. We used the Temkin empirical equation,[36] which states that for charged substances the number of absorbed particles per unit area is proportional to the logarithm of the concentration of charged species (i.e., Ca^{2+}) in the subphase.

To verify that our data could be represented by Temkin empirical law we varied $[Ca^{2+}]$ in the subphase, using phosphatidylserine and phosphatidylinositol packed at either 40 or 50 $Å^2$ per phospholipid molecule. Varying $[Ca^{2+}]$ in the subphase from 10 to 1000 μM we could estimate the exact ratio Ca^{2+}/P (number of calcium ions per phospholipid molecule) as shown on the right vertical axis of Fig. 5. Over the entire range of calcium concentrations, the expected linear relationship was observed. Furthermore, raising the surface concentration (or decreasing the area occupied per phospholipid molecule), raised the slope of the adsorption isotherm. With more phospholipids on the surface, more calcium was bound.

We can now, with some assurance, compute the number of synexin molecules adsorbed to the monolayer at the air–water interface. At a surface density of one lipid molecular/40 $Å^2$, 4.5×10^{15} lipid molecules are placed under the window of the detector. While in the absence of synexin 2.7×10^{15} ions were adsorbed onto the monolayer (roughly two lipid molecules per calcium), in the presence of synexin in the subphase (18 μg or 0.23×10^{15} synexin molecules), an additional 0.67×10^{15} calcium ions were

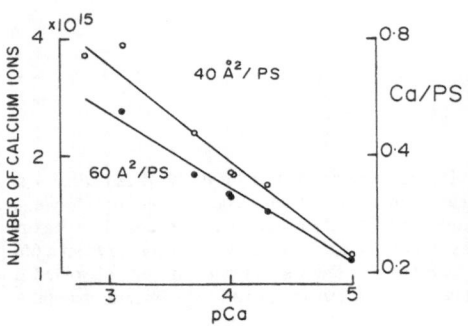

Figure 5. Temkin isotherms of $^{45}Ca^{2+}$ adsorption to PS monolayers as a function of $[Ca^{2+}]$ in the subphase. The straight line with higher slope was obtained with the PS molecules packed at 40 $Å^2$/PS molecule. The other line at 60$Å^2$.

adsorbed onto the monolayer. Since we know that calcium binds synexin ($2\,Ca^{2+}$/synexin[37]), the calculated amount of $^{45}Ca^{2+}$-labelled synexin adsorbed to the surface would be half of the measured number of additional $^{45}Ca^{2+}$ adsorbed, that is, 0.3×10^{15} synexin molecules. We can conclude that nearly all the synexin molecules added to the subphase first interacted with Ca^{2+} and then diffused toward the monolayer where they interacted with the phospholipid molecules.

Furthermore, though still indirect, evidence that the $^{45}Ca^{2+}$ activity reaching the interface is borne by synexin can be obtained by measuring the apparent diffusion constant τ_{app} of the $^{45}Ca^{2+}$-labelled synexin molecules. From the time course of the $^{45}Ca^{2+}$-synexin complex adsorption isotherm (Fig. 4) we estimated the diffusional time constant τ as 3.73×10^3 sec. Taking the thickness of the layer of solution from which β radiations were detected (δ) as 0.065 cm,[36] from the definition of apparent diffusion coefficient, that is,

$$\tau_{app} = \delta^2/_\tau$$

we calculated τ_{app} for the $^{45}Ca^{2+}$-synexin complex as 11.2×10^{-7} cm²/sec. Although we do not know the diffusion constant of Ca^{2+}-activated synexin as yet, the value obtained should correspond to the diffusion constant of a ribonuclease molecule (mol. wt. $= 13.7 \times 10^3$; $\tau_{H_2O} = 11.9 \times 10^{-7}$ cm²/sec). The molecular weight of synexin is 47×10^3.

4.2. Ca^{2+}-Activated Synexin Insertion into PS Bilayers

The results presented in the preceeding section clearly indicate that Ca^{2+}-activated synexin molecules were preferentially adsorbed onto monolayers of acidic phospholipids. Furthermore, modelling studies based on the sequence indicate that the cytosolic molecule synexin could, once activated by Ca^{2+}, adopt a conformation that would permit the penetration of substantial portions of the synexin molecule into the core of bilayer membranes. One way to distinguish between synexin adsorption to a membrane and synexin inserting into the membrane is to study the effect of Ca^{2+}-activated synexin on membrane capacitance, a method that we have recently exploited with some success.[38]

To this end we prepared PS bilayers at the tip of a patch pipette using the now classical double-dip methods. The results showed that synexin could profoundly change the capacity of the membrane by nearly tenfold if calcium were also present in the *cis* side.[38] An example of such data is shown in Fig. 6, where C is measured using a time-based method. Upon application of a voltage pulse, current flows across the membrane and, due to the separation of charges across the membrane dielectric, charges Q and $-Q$ are placed on either side of the bilayer. In the example, two voltage pulses of identical size but opposite polarity have been applied to the membrane and the capacitative currents of the synexin-supplemented membrane subtracted from those of the membrane before synexin addition. The records shown in the lower part of Fig. 6 are therefore *synexin-specific* capacitative currents. The advantage of this method is that it allows all frequencies below 10^4 Hz to be sampled and the charge displacement in the bilayer to be quantitatively determined. Since Q can be easily calculated from the time integral of the

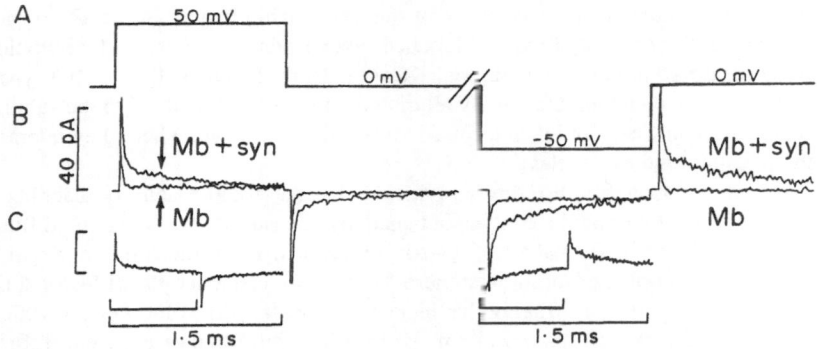

Figure 6. Displacement currents across PS bilayers: protocol used to measure membrane capacity. Voltage pulses of identical size but opposite polarity were applied to a PS bilayer made at the tip of a patch pipette. The solution in the pipette (*trans* side) was (mM): 105 NaCl, 30 CaCl$_2$, 10 Na-HEPES at pH 7.4. The *cis* side contained a solution of similar composition but only 1 mM CaCl$_2$. The records made prior to the application of synexin are labeled as Mb and those made after the addition of synexin (5 μg/ml) are labeled Mb + Syn. Upper records represent the voltage levels applied to the solution in the pipette. The two pairs of superimposed records shown in the middle represent the capacity current transients in response to the pulses before (Mb) and 2 min after (Mb + Syn) addition of Ca^{2+} (1 mM) in the presence of synexin. The difference between the current record made in the presence of synexin (Mb + Syn) minus the current record made before the addition of synexin (MB) is plotted beneath each pair of current transients and represents the synexin-specific displacement currents at different *trans*-membrane potentials. The data are from Ref. 38 with minor modifications.

current that flows across the membrane over the time course of the voltage pulse V, the capacity of the membrane is just the ratio

$$C = Q/V$$

Figure 7A demonstrates that with the bilayer membrane alone Q is proportional to V with a slope c equal to 0.25 pF. To a first approximation, C is proportional to the area A and inversely proportional to the thickness δ of the low dielectric region, that is,

$$C = (\epsilon/\epsilon_0)(A/\delta)$$

where ϵ_0 is the permitivity of the free space and $\epsilon \cdot \epsilon_0$ represents the relative dielectric constant. In the case of a bilayer formed by the double-dip method, the geometrical parameters are relatively fixed, and thus C should be constant. This was verified (Fig. 7B) at all potentials in the range from -100 to 100 mV. However, after exposure of the bilayer to Ca^{2+}-activated synexin molecules the Q–V relationship obtained was sigmoidal (Fig. 6C). This result indicates that mobile charges have been incorporated into the bilayer which is equivalent to an increase in the ratio ϵ/ϵ_0. Indeed, the slope of the Q–V relationship representing C is voltage dependent with a maximum of about 0.7 pF at ca. 0 mV. The dramatic increase in the capacity of the bilayer indicates that a large number of mobile charges had been incorporated into the bilayer.

The dielectric constant of a phospholipid bilayer is ca. 2, and if protein dipoles were to be inserted into the membrane the value of the dielectric constant would rise. It is possible to estimate the increase in capacity caused by the insertion of a dipole molecule with an apparent dipole moment p using the Clausius–Mosotti relationship,

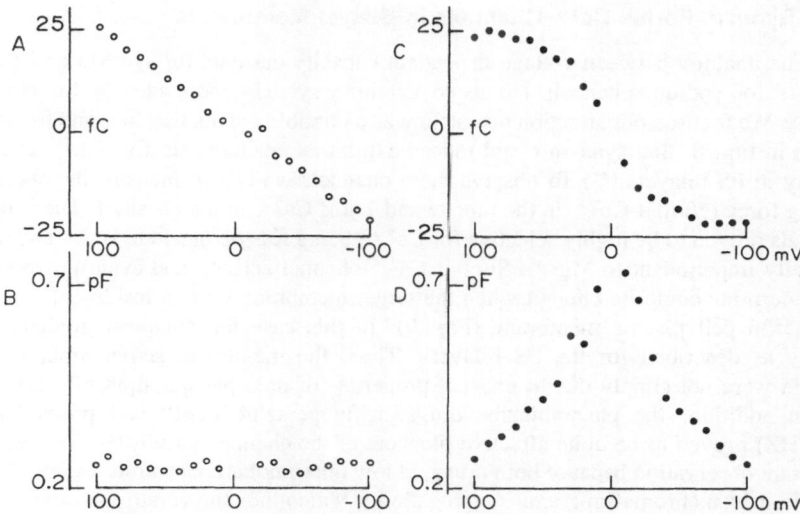

Figure 7. Effects of Ca^{2+}-activated synexin on charge movement (Q) and bilayer capacity (C) as a function of pipette potential (V). Charge displacement across the bilayer in the absence (A) and in the presence of Ca^{2+}-activated synexin (C). Bilayer capacity (B) derived from the slope of the $Q-V$ curve shown in part (A). Bilayer capacity from the data in part (C). (○): In the absence of synexin. (●): In the presence of synexin (3 μg/ml). Membrane resistance: 12 G for (A) and (B), 8 G for (C) and (D). The pipette, which was coated with sylgard, had an open-tip resistance of 20 M.

$$(\epsilon - 1)/(\epsilon + 2) = \tfrac{4}{3}\pi N(p)^2/3KT$$

where ϵ represents the dielectric constant, N is the number of dipoles inserted in the bilayer, and KT equals 25.3 meV at 20°C.[39] From the area of the bilayer facing the *cis* side (1 μm^2) we estimate that 2.6×10^6 PS molecules were available to bind synexin. We also know that, in the presence of Ca^{2+}, four PS molecules in the monolayer can accommodate single Ca^{2+} synexin complexes. Therefore, the total number of synexin molecules N which could be inserted into the bilayer could be as large as ca. 6×10^5. Assuming a dipole moment p of 250 D (Debye) per segment of the synexin molecule inserted in the bilayer, we might expect to see substantial increases in capacity C (tenfold or more).[39] Thus, synexin is clearly able to enter the bilayer, and does not merely adhere to it.

In addition, we also found that the synexin-dependent charge movement and bilayer capacity were strongly voltage dependent (Fig. 7C,D). This voltage dependence could be fit to a Boltzmann's distribution.[38] The possible meaning of this result is that the synexin dipole(s) can be moved within a reaction coordinate which is defined in some way by the membrane, but that the dipole(s) cannot be moved out of the confines of the membrane. This structural limitation is our interpretation of the saturation of change in capacitance at negative and positive extremes of the applied voltage. In the case of sodium channels, this voltage-dependent capacity current has been called "gating current."[41] The conventional interpretation of gating current has been that part or all of the channel protein moves within the membrane to allow the channel to open.[41]

4.3. Synexin Forms Ca²⁺-Channels in Bilayer Membranes

The analogy between voltage-dependent capacity currents for synexin and gating currents for sodium channels led us to examine synexin more closely for channel activity. We focused our attention on calcium as a possible conducting ion, and found, as shown in Fig. 8, that synexin could indeed exhibit exquisitely selective Ca^{2+}-channel activity in PS bilayers.[41] To observe these channels we had to increase the chemical driving force (50 mM Ca^{2+} in the pipette and 1 µM Ca^{2+} in the *cis* side). The synexin channels proved to be highly selective for Ca^{2+}, being less permeant to Ba^{2+} and Sr^{2+} or totally impermeant to Mg^{2+}. Similar Ca^{2+}-channel activity and even macroscopic Ca^{2+} currents could be elicited when the target membrane was an inside-out patch of chromaffin cell plasma membrane (Fig. 9). In this case the chemical gradient was exactly as described for the PS bilayers. Thus, the membrane active properties of synexin were not simply due to unusual properties of pure phospholipid bilayers.

In addition, the phenothiazine drugs, trifluoperazine (TFP) and promethazine (PMTHZ) proved to be quite effective blockers of the channel activity.[41] This was an important observation because both drugs, at low micromolar concentrations, also block synexin-driven chromaffin granule aggregation,[43] nicotine- and veratridine-driven exocytosis from chromaffin cells,[43] and glucose-driven insulin secretion from rat or *Psommys obesus* islets of Langherhans.[51] By contrast, TPF has no reported effect on con-

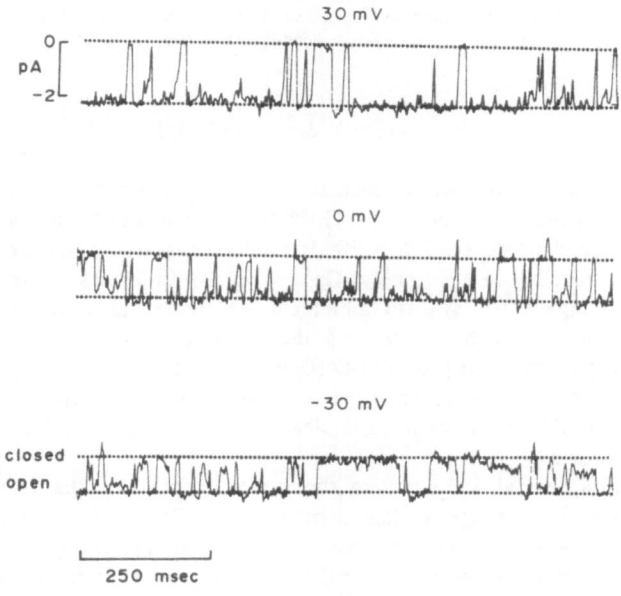

Figure 8. Synexin calcium-channel activity. Synexin channel currents from a multichannel bilayer. A Pl bilayer was exposed to bovine synexin in the presence of 1 mM $CaCl_2$. Synexin channel activity was recorded in the presence of 50 mM $MgCl_2$ in the chamber. Single-channel current varied with the potential as follows: Top: 2.29 ± 0.21 pA; middle: −1.10 ± 0.18 pA; bottom: −1.21 ± 0.19 pA. Chord conductance 18.3 ± 3.5 pS.

Figure 9. Ca^{2+}-activated synexin inserts itself into inside-out patches of membrane from chromaffin cells and forms Ca^{2+}-channels. Chromaffin cells were cultured for 3 days, and electrically silent cell-attached patches were obtained. For patch-clamp experiments the cells were incubated first in a modified buffer containing (mM): 250 TEA-Cl, 1 $CaCl_2$, 10 TEA-HEPES, pH 6.8. After formation of a >5 Gohm seal the patch of membrane was excised and, prior to the application of synexin, voltage-clamp pulses of increasing size (−100 through 100 mV in 10 mV increments) were applied and the current transients recorded. If no current events were recorded during the application of depolarizing pulses (holding pipette potential was 50 mV), a stream of solution containing synexin was puffed onto the cytosolic aspect of the patch of membrane. $[Ca^{2+}]$ was then lowered to less than 1 μM. Another series of pulses were applied. Synexin-specific capacity and ionic current transients were then obtained by subtraction of the control current transients obtained in the absence of synexin from the current

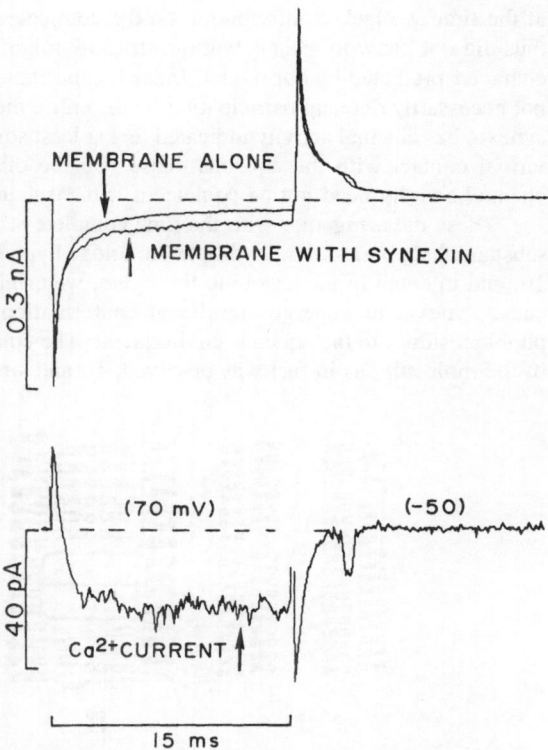

transients recorded after the exposure of the patch to Ca^{2+}-activated synexin. Records below the zero current dashed line represent Ca^{2+} current flowing from the external to the internal side of the membrane.

ventional calcium channels (L, T, or N). Finally, neither Cd^{2+} nor nifedipine were able to block synexin channels at pharmacologically relevant doses. These properties taken together thus effectively exclude synexin channel activity from being due to inadvertent contamination of the preparation by conventional membrane calcium channels (see Miller[44] for pharmacology of conventional calcium channels).

Although the physiological significance of synexin channels remains to be determined,[41] the *operational* meaning of the synexin channel observation is clearly that calcium allows all or part of the synexin molecule not only to *enter* the bilayer but also to *span* the bilayer. This conclusion has had important significance for our thinking about the mechanism by which synexin might cause fusion of two target membranes.

5. THE "HYDROPHOBIC BRIDGE HYPOTHESIS"

Several years ago we proposed that synexin might mediate membrane fusion by forming a hydrophobic bridge between the fusing membranes partners.[45,46] However, at the time we lacked information on the complete primary sequence of synexin, and thus did not know to what extent the structure might express the profound hydrophobic character predicted by our model. Indeed, capacitance changes and channel activity did not necessarily demand participation by the entire molecule. In fact the observation that synexin has channel activity indicated that at least some parts of the molecule must be in virtual contact with the aqueous phase. On the other hand, structures involved with channel activity need not be coincident with structures involved in fusion activity.

These data, together with the now complete sequence of human synexin,[23] lend substantial support to the "hydrophobic bridge hypothesis.[45,46] As summarized in Fig. 10, and in detail in the legend to the figure, we originally proposed that calcium would cause synexin to undergo significant conformational changes, thus exposing hydrophobic residues to the aqueous environment. The consequence would be polymerization of the molecule, as in fact was observed,[37] and simultaneous insertion of the synexin

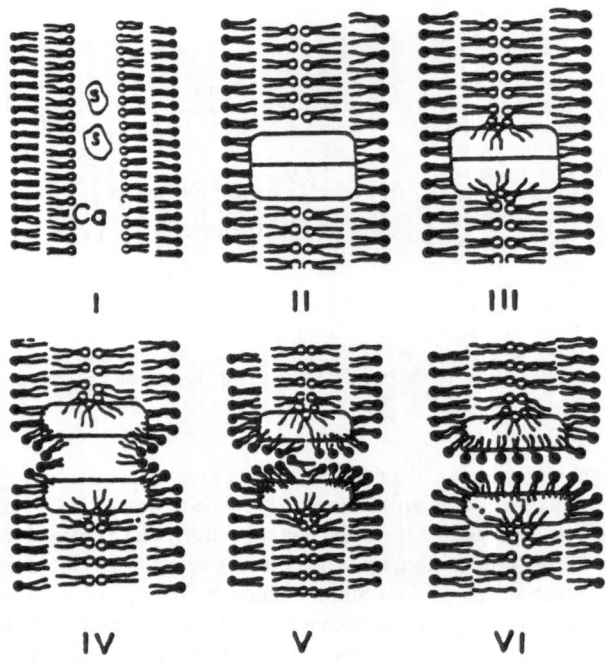

Figure 10.   Hydrophobic bridge hypothesis for synexin-driven membrane fusion. (I) Two membranes are poised to fuse. Open circles represent acidic phospholipids on the cytoplasmic (cis) leaflet of a membrane. Closed circles represent phospholipids (e.g., PC or sphingomyelin) on the opposing (*trans*) leaflet. Calcium causes synexin monomers (labeled "S") to polymerize. (II) Calcium-activated synexin polymer binds to the acidic phospholipids on both *cis* surfaces, and penetrates both bilayers and crosslinks the membranes. (III) Phospholipids cross the "hydrophobic bridge" created by synexin, thereby allowing the *cis* leaflets of the fusing membranes to mix. (IV) Synexin polymer dissociates in the low dielectric environment, thereby providing a hydrophobic pathway for the *trans* leaflets to approach. (V) Trans leaflets complete their reorientation, leaving (VI) synexin in the bilayer of the newly fused membranes. Some of all of the synexin may leave the membrane, but we have little information as yet on this process. (From Refs. 45 and 46.)

polymer into two adjacent target membranes. The capacitance and channel data support the concept that synexin can do this to a single membrane. The only additional suggestion here is that synexin can also do this to two membranes at once. Finally, fusion should ensue when the hydrophobic bridge destabilizes both bilayers and provides a pathway for lipids on facing leaflets of both membranes to cross and mix.

The concept of hydrophobic defect being the driving force for membrane fusion is a common proposal in many hypothetical mechanisms for this process.[47] Once this defect is in place there have also been a variety of possible pathways proposed to achieve final fusion. The contribution of synexin is to provide at least one concrete biological example of a bona fide driving force and direction for the hydrophobic defect.

The sequence in Fig. 10 (stages IV, V, and VI) merely outlines one possible pathway that is consistent with our knowledge of synexin. We do not consider this pathway to be exclusive, since there is little compelling evidence for the proposed details. However, the concept that synexin might depolymerize or otherwise move about within the bilayer is certainly consistent with our capacitance data, as well as with time constants (on the order of milliseconds) determined for opening and closing of synexin channels.

6. CONCLUSIONS

Synexin initially attracted our attention because it provided a mechanism for Ca^{2+}-dependent membrane contact and fusion processes which we know to occur during exocytosis. For some years after its discovery, synexin was the only protein with the appropriate properties, and for this reason held our unwavering attention. We have now been rewarded by a quite detailed biophysical picture of how synexin interacts with membranes, and have had the opportunity to apply this knowledge to formulate the hydrophobic bridge hypothesis.

More recently, other proteins have been discovered that have shared with synexin a calcium-dependent affinity for acidic phospholipids and, in the cases of calpacting polymer[48] and calectrin 67K,[49] the ability to aggregate chromaffin granules. We might therefore have reasonably anticipated that some or all of these proteins might share other properties of synexin. But with the successful cloning and sequencing of human synexin we now appreciate that all of these proteins are members of a common gene family, and that some properties in common might be expected.

It follows that if any of these proteins, or other yet to be discovered members of the synexin gene family, also proves to have membrane fusion properties, we might reasonably expect it (them) to do so by mechanisms similar to those described above for synexin itself. In previous discussions we have only considered the synexin molecule for possible involvement in exocytosis. However, we may eventually have to consider some or all of the synexin gene family members (annexins) as a class in this process. Time will tell.

REFERENCES

1. Creutz, C. E., Pazoles, C. J., and Pollard, H. B., 1978, Identification and purification of an adrenal medullary protein (synexin) that causes calcium-dependent aggregation of isolated chromaffin granules, *Biol. Chem.* **253:** 2858–2866.

2. Scott, J. H., Kelner, K. L., and Pollard, H. B., 1985, Purification of synexin by pH step elution from chromatofocusing media in the absence of amphols, *Anal. Biochem.* **149**: 163–165.
3. Palade, G., 1975, Intracellular aspects of the process of protein synthesis, *Science* **189**: 347–358.
4. Morris, S. J., Hughes, J. M. X., and Whittaker, V. P., 1982, Purification and mode of action of synexin: A protein-enhancing calcium-induced membrane aggregation, *J. Neurochem.* **39**: 529–536.
5. Simon, S. M., and Llinas, R. R., 1985, Compartmentalization of the submembrane calcium activity during calcium influx and its significance in transmitter release, *Biophys. J.* **48**: 485–489.
6. Tsien, R. W., Hess, P., McCleskey, E. W., and Rosenberg, R. L., 1987, Mechanisms of selectivity, permeation, and block, *Ann. Rev. Biophys. Biophys. Chem.* **16**: 265–290.
7. Creutz, C. E., and Sterner, D. C., 1983, Calcium dependence of the binding of synexin to isolated chromaffin granules, *Bioch. Biophys. Res. Commun.* **114**: 355–364.
8. Creutz, C. E., 1981, *Cis*-unsaturated fatty acids induce the fusion of chromaffin granules aggregated by synexin, *J. Cell Biol.* **91**: 247–256.
9. Creutz, C. E., and Pollard, H. B., 1982, Development of a cell-free model for compound exocytosis using components of the chromaffin cell, *J. Auton. Nerv. Syst.* **7**: 13–18.
10. Hotchkiss, A., Pollard, H. B., Scott, J., and Axelrod, J., 1981, Release of arachidonic acid from adrenal chromaffin cell cultures during secretion of epinephrine, *Fed. Proc.* **40**: 256.
11. Frye, R. A., and Holz, R. W., 1984, The relationship between arachidonic acid release and catecholamine secretion from culture bovine adrenal chromaffin cells, *J. Neurochem.* **43**: 146–150.
12. Pollard, H. B., Creutz, C. E., Fowler, V. M., Scott, J. H., and Pazoles, C. J., 1982, Calcium-dependent regulation of chromaffin granule movement, membrane contact, and fusion during exo-cytosis, *Cold Spring Harbor Symp. Quant. Biol.* **46**: 819–834.
13. Ornberg, R. L., Duong, L. T., and Pollard, H. B., 1986, Intergranular vesicles: New organelles in the secretory granules of adrenal chromaffin cells, *Cell and Tissues Res.* **245**: 547–553.
14. Hong, K., Duzgunes, N., and Papahadjopoulos, D., 1981, Role of synexin in membrane fusion: Enhancement of calcium-dependent fusion of phospholipid vesicles, *J. Biol. Chem.* **256**: 3641–3644.
15. Hong, K., Duzgunes, N., Ekert, R., and Papahadjopoulos, D., 1982, Synexin facilitates fusion of specific phospholipid vesicles at divalent cation concentrations found intracellularly, *Prod. Nat. Acad. Sci. USA* **79**: 4642–4644.
16. Hong, K., Ekert, R., Bentz, J., Nir, S., and Papahadjopoulos, D., 1983, Kinetics of synexin-facilitated membrane fusion, *Biophys. J.* **41**: 31a.
17. Stutzin, A., 1986, A fluorescence assay for monitoring and analyzing fusion of biological mem-branes vesicles in vitro, *FEBS Lett.* **197**: 274–280.
18. Stutzin, A., Cabantchik, I., Lelkes, P. I., and Pollard, H. B., 1987, Synexin-mediated fusion of bovine chromaffin granule ghosts: Mechanism of pH dependence, *Biophys. Biochem. Acta* **905**: 205–212.
19. Nir, S., Stutzin, A., and Pollard, H. B., 1987, Effect of synexin on aggregation and fusion of chromaffin granule ghosts at pH 6, *Biochemistry, Biophys. Biochem. Acta* **903**: 309–318.
20. Hong, K., Duzgunes, N., and Papahadjopoulos, D., 1982, Modulation membrane fusion by cal-cium-binding proteins, *Biophys. J.* **37**: 297–306.
21. Pollard, H. B., Ornberg, R., Levine, M., Heldman, E., Morita, K., Kelner, K., Lelkes, P., Brocklehurst, K., Forsberg, E., Duong, L., Levine, R., and Youdim, M. B. H., 1985, Hormone packaging and secretion by exocytosis: A view from the chromaffin cell, *Vitamins and Hormones* (G. Aurbach, ed.), **42**: 109–196.
22. Scott, H. H., Creutz, C. E., Pollard, H. B., and Ornberg, R. O., 1985, Synexin binds in a calcium-dependent fashion to oriented chromaffin cell plasma membranes, *FEBS Lett.* **180**: 17–23.
23. Burns, A. L., Magendzo, K., Shirvan, A., Srivastava, J., Rojas, E., Alijani, M. R., and Pollard, H. B., 1989, Calcium channel activity of purified human synexin and structure of the human synexin gene, *Proc. Nat. Acad. Sci. USA* **86**: 3798–3802.
24. Wallner, B. P., Mattaliano, R. J., Hession, C., Cate, R. L., Tizard, R., Sinclair, L. K., Foeller, C., Chow, E. P., Browning, J. L., Ramachandrau, K. L., and Pepinsky, R. B., 1986, Cloning and

expression of human lipocortin, a phospholipase A_2 inhibitor with potential anti-inflammatory activity, *Nature* **320:** 77–81.

25. Schlaepfer, D. D., Mehlman, T., Burgess, W. H., and Haigler, H. T., 1987, Structural and functional characterization of endonexin II, a calcium- and phospholipid-binding protein, *Biophys. J.* **48:** 485–492.
26. Kaplan, R., Jaye, M., Burgess, W. H., Schlaepfer, D. D., and Haigler, H. T., 1988, Cloning and expression of cDNA for endonexin II, a Ca^{2+} and phospholipid-binding protein, *J. Biol. Chem.* (in press).
27. Glenney, J. R., 1986, Two related but distinct forms of the M_r 36,000 tyrosine kinase substrate (calpactin) that interact with phospholipid and acting in a Ca^{2+}-dependent manner, *Proc. Nat. Acad. Sci. USA* **83:** 4258–4262.
28. Huang, K-S., Wallner, B. P., Mattaliano, R. J., Tizard, R., Burne, C., Frey, A., Hession, C., McGray, P., Sinclair, L. K., Chow, E. P., Browning, J. L., Ramachandran, K. L., Tang, J., Smart, J. E., and Pepinsky, R. B., 1986, Two human 35-kD inhibitors of phospholipase are related to substrate of pp60^{V-src} and of the epidermal growth factor receptor kinase, *Cell* **46:** 191–199.
29. Kirstensen, T., Saris, C. J. M., Hunter, T., Hicks, L. J., Noonan, D. J., Glenney, J. R., Jr., and Tack, B. F., 1986, Primary structure of bovine calpactin I heavy chain (p36), a major cellular substrate for retroviral protein–tirosine kinases: Homology with the human phospholipase A^2 inhibitor lipocortin, *Biochem.* **25:** 4497–4503.
30. Saris, C. J. M., Tack, B. F., Kristensen, T., Glenney, J. R., Jr., and Hunter, T., 1986, The cDNA sequence for the protein–tyrosine kinase substrate p36 (calpactin I heavy chain) reveals a multidomain protein with internal repeats, *Cell* **46:** 201–212.
31. Weber, K., Johnsson, N., Plessmann, U., Van, P. N., Soling, H-D., Ampe, C., and Vandekerckhove, J., 1987, The amino acid sequence of protein II and its phosphorylation site for protein kinase C: The domain structure (of) Ca^{2+}-modulated lipid-binding proteins, *EMBO J.* **6:** 1599–1604.
32. Sudhof, T. C., Slaughter, C. A., Leznicki, I., Barjon, P., and Reynolds, G. A., 1988, Human 67-kDa calelectrin contains a duplication of four repeats found in 35-kDa lipocortins, *Proc. Nat. Acad. Sci. USA* **85:** 664–668.
33. Taylor, W. R., and Geisow, M. J., 1987, Predicted structure for the calcium-dependent membrane-binding proteins p35, p36, and p32, *Prot. Engin.* **1**(3): 183–187.
34. Rojas, E., and Tobías, J. M., 1965, Membrane model: Association of inorganic cations with phospholipid monolayers, *Biochem. Biophys. Acta* **94:** 394–404.
35. Santis, M., and Rojas, E., 1969, On the chemistry of ion exchange in monomolecular layers of lipids, *Biochim. Biophys. Acta* **193:** 319–332.
36. Davies, J. T., and Rideal, E. K., 1961, *Interfacial phenomena*, Academic, New York.
37. Creutz, C. E., Pazoles, C. J., and Pollard, H. B., 1979, Self-association of synexin in the presence of calcium: Correlation with synexin-induced membrane fusion and examination of the structure of synexin aggregates, *J. Biol. Chem.* **254:** 553–558.
38. Rojas, E., and Pollard, H. B., 1987, Membrane capacity measurements suggest a calcium-dependent insertion of synexin into phosphatidylserine bilayers, *FEBS Lett.* **217:** 25–31.
39. Debye, P., 1929, *Polar Molecules*. Dover, New York.
40. Rojas, E., 1976, Gating mechanism for activation of the sodium conductance in nerve membranes, *Cold Spring Harbor Symp. Quant. Biol.* **XL:** 305–320.
41. Pollard, H. B., and Rojas, E., 1988, Calcium-activated synexin forms highly selective, voltage-gated calcium channels in phosphatidylserine bilayer membranes, *Proc. Nat. Acad. Sci. USA* **85:** 2974–2978.
42. Pollard, H. B., Creutz, C. E., and Pazoles, C. J., 1981, Mechanisms of calcium action and hormone release during exocytosis, *Rec. Prog. in Horm. Res.* R. O. Greep, ed. Academic, New York **37:** 299–322.
43. Pollard, H. B., Scott, J. H., and Creutz, C. E., 1983, Inhibition of synexin activity and exocytosis from chromaffin cells by phenothiazine drugs, *Biochem. Biophys. Res. Comm.* **113:** 908–915.
44. Miller, R. J., 1987, Multiple calcium channels and normal function. *Science* **235:** 46–52.

45. Pollard, H. B., Rojas, E., and Burns, A. L., 1987, Synexin and chromaffin granule membrane fusion: A novel "hydrophobic bridge" hypothesis for driving and directing the fusion process, *Ann. New York Acad. Sci.* **493:** 524–551.

46. Pollard, H. B., Rojas, E., Burns, A. L., and Parra, C., 1988, Synexin calcium and the hydrophobic bridge hypothesis for membrane fusion, in: *Molecular Mechanisms of Membrane Fusion* (S. Ohki, D. Doyle, T. Flanagan, S. W. Hui, and E. Mayhew, Eds.) Plenum, New York, pp. 341–355.

47. Blumenthal, R., 1987, Membrane fusion, *Currents Topics in Membrane and Transport* **253:** 2858–2866.

48. Drust, D. S., and Creutz, C. E., 1988, Aggregation of chromaffin granules by calpactin at micromolar levels of calcium, *Nature* **331:** 88–91.

49. Sudhof, T. C., Ebbecke, M., Walker, J. H., Fritsche, U., and Boustead, C., 1984, Isolation of mammalian calelectrins: A new class of ubiquitous Ca^{2+}-regulated proteins, *Biochemistry* **23:** 1103–1109.

50. Pollard, H. B., Burns, A. L., and Rojas, E., 1988, A molecular basis for synexin-driven calcium-dependent membrane fusion, *J. Exptl. Biology* **139:** 267–286.

51. Sussman, K. E., Pollard, H. B., Leitner, J. W., Nesher, R., Adler, J., and Cerasi, E., 1983, Differential Control of insulin secretion and somatostation receptor recruitment in isolated islets, *Biochem. J.*, **214:**225–230.

Chapter 12

Signal Transduction and Ion Permeability in Adrenal Glomerulosa Cells

Elisa T. Marusic and Maria V. Lobo

1. INTRODUCTION

In recent years it has become apparent that many types of endocrine cells exhibit electrical activity, an important property with regard to stimulus–secretion coupling. This is also true for adrenal glomerulosa cells; from the early work of Matthews and Saffran,[1] it has been known that adrenal cells have a negative resting potential that is influenced by the binding of ACTH to specific receptors. Adrenacorticotropin (ACTH) stimulates aldosterone secretion from adrenal glomerulosa cells. The secretion of this hormone is under the control of at least three stimuli: angiotensin II, external potassium ions, and ACTH. Different transduction mechanisms have been proposed in relation to the effects of these factors and include cAMP formation, regulation of cytosolic free calcium through the phosphatidylinositol system, and ion channel modification, protein kinase C activation, and modification of the membrane potential. This situation is summarized in Fig. 1.

The stimulatory action of angiotensin II is initiated through the binding to specific, high-affinity receptors in the plasma membrane. Recent evidence suggests that the angiotensin II receptors may be coupled to more than one GTP-binding protein, one that stimulates phospholipase-C and another cAMP generation.[2]

ACTH stimulation involves the activation of adenylate cyclase with an increase in

ELISA T. MARUSIC and MARIA V. LOBO • Departamento de Fisiología y Biofísica, Facultad de Medicina, Universidad de Chile, Santiago, Chile.

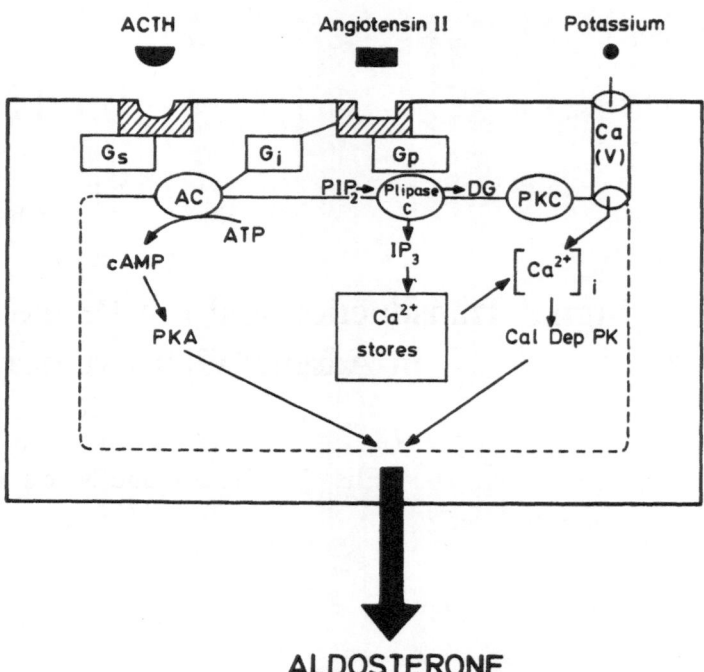

ALDOSTERONE

Figure 1. Comparison of the action of three agonists that regulate aldosterone production from the adrenal glomerulosa cells. ACTH activates the G_s-cAMP system by hormone–receptor interaction. Angiotensin II activates phospholipase-C and probably inhibits cAMP generation. External potassium, by modifying plasma membrane potential, opens voltage-dependent Ca-channels. In the case of angiotensin II, hormone receptor activation of phospholipase-C leads to two changes: hydrolysis of PIP, which gives rises to two messengers, DG and IP_3, and an increase in plasma membrane Ca influx. As a consequence of the rise of IP_3 there is a transient rise in intracellular free Ca^{2+} and an increase in Ca efflux from the nonmitochondrial Ca pool. This rise leads to the activation of the Cal Dep PK. In addition, the increase in DG leads to the activation of PKC. In the case of an increase in extracellular K^+ concentration, a decrease in membrane potential leads both to an increase in plasma membrane Ca influx and in cAMP content. In the case of ACTH, the possibility of interaction of the hormone with two receptors has been postulated: One leads to an increase in cAMP and the other increases plasma membrane Ca influx. These activate, respectively, PKA and Cal Dep PK. In the case of potassium stimulus, the increase in Ca influx is relatively large and the increase in cAMP is small, whereas the opposite is true in the case of ACTH. (KEY: AC, adenylate cyclase; PKA, protein kinase-A; G_s, stimulatory G protein; G_i, inhibitory G protein; G_p, unidentified protein G; Plipase C, phospholipase-C; PIP_2, phosphatidylinositol 4,5-bisphosphate; DG, diacylglycerol; IP_3, inositol 1,4,5-trisphosphate; PKC, protein kinase-C; Ca(V), voltage-dependent Ca-channel; Cal Dep PK, calmodulin-dependent protein kinases.

intracellular cyclic AMP. The possibility of interaction of ACTH with a second receptor has been postulated; this receptor could be a calcium-channel.[3] Some studies suggest that the response of the glomerulosa cells to ACTH at concentration lower than $10^{-10}\,M$ is mediated by phospholipase-C activation.[3] On the other hand, external potassium acts at the plasma membrane by modifying the membrane potential.[4] The membrane depolarization due an increment in potassium concentration activates voltage-dependent calcium channels in the plasma membrane as the initial transduction step (see Fig. 1 and legend). It is now generally accepted that of the three stimuli of aldosterone production from the adrenal glomerulosa cells, mainly angiotensin II and potassium trigger steroidogenesis by a mechanism involving changes in cytosolic Ca^{2+} concentration. In fact, when the extracellular concentration of calcium is reduced, the aldosterone response to the three secretagogues is impaired and is completely abolished in the absence of extracellular calcium.[5]

The mechanism involved in the action of angiotensin II is of special interest because this polypeptide is the main physiological factor controlling aldosterone secretion. In this chapter, we shall concentrate on the role of angiotensin II on ion permeability. Angiotensin II, by stimulating the hydrolysis of phosphatidylinositol 4,5-biphosphate initiates a bifurcating signal pathway that functions to control the secretory process. Inositol 1,4,5-trisphosphate regulates calcium release from internal nonmitochondrial stores but the role of diacylglycerol-C kinase limb on the ionic permeability of adrenal glomerulosa cells is not known. Therefore, the potential regulatory role of these two signals pathways on the potassium permeability of the glomerulosa cells was studied.

2. THE CALCIUM MESSENGER SYSTEM IN THE ADRENAL GLOMERULOSA CELLS

Over the past quarter-century, it has become increasingly clear that the calcium ion messenger system is a nearly universal means by which extracellular messengers regulate cell function. There are two branches in this system, a calmodulin branch and a C-kinase branch, which are mainly regulated in the adrenal glomerulosa cells by potassium ions and angiotensin II, respectively. Also, the cAMP messenger system regulated by ACTH in the adrenal cells is intimately related to and modifies the functional expression of the calcium messenger system.[5]

Angiotensin II plays the major role in regulating aldosterone secretion under physiological circumstances. This peptide hormone, as with the other aldosterone secretagogues, is effective only if extracellular calcium is present. Also, the sustained aldosterone stimulation by angiotensin II is blocked by calcium channel blocking agents such as verapamil, D-600, La^{3+}, and dihydropyridines.[6] The role of calcium entry in the action of angiotensin II has been demonstrated by direct measurements of Ca influx, even though contradictory results have been reported.[7,8] This polypeptide hormone has been shown to depolarize glomerulosa cells in a dose-dependent manner,[9] but the ionic mechanisms underlying angiotensin II-mediated depolarization are not known. Therefore, we decided to investigate the potassium permeability under the stimulation by this hormone, because changes in the conductance to this ion can regulate membrane potential and hence modulate the activity of voltage-dependent calcium channels.

3. ACTIVATION OF POTASSIUM CHANNELS BY ALDOSTERONE SECRETAGOGUES

In many endocrine cells, the potassium channel acts as a link between the cytosolic Ca^{2+} and the membrane potential. In adrenal glomerulosa cells, we have described a close link between ^{45}Ca and ^{86}Rb fluxes immediately after angiotensin II stimulus.[10] In

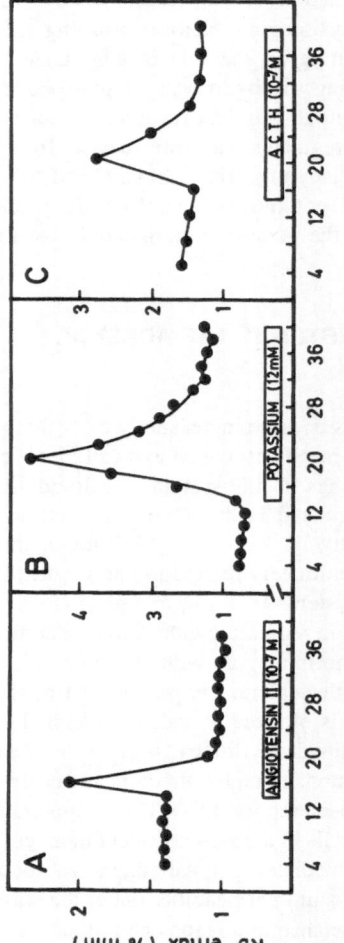

Figure 2. Comparative effect of angiotensin II (A), potassium ions (B), and ACTH (C) on the rate coefficient efflux of ^{86}Rb from perfused glomerulosa cells. On this and subsequent experiments, collagenase dispersed bovine adrenal glomerulosa cells were preequilibrated with ^{86}Rb for 60 min in a Krebs–Ringer–bicarbonate–glucose medium (KRBG). Cells were transferred into perfusion chambers and equilibrated with a KRBG medium containing 2 mg/ml albumin during 30 min before starting samples collections; the potassium concentration was 4 mM (control value). The agonists were added at 14 min and were present throughout the experiment: 100 nM of each peptide hormone and 12 mM K. Each point is a mean from two separate experiments with similar results.

fact, angiotensin II stimulus on perfused bovine adrenal glomerulosa cells elicited an initial increase in ^{86}Rb efflux from cells previously equilibrated with the radioisotope. When ^{45}Ca fluxes were measured under similar conditions, it was observed that Ca and Rb effluxes occurred within the first 30 sec of the addition of the hormone and were independent of the presence of external Ca. The ^{86}Rb efflux due to angiotensin II was inhibited by quinine and apamin. The hypothesis that the initial increase in the K permeability of the glomerulosa cell membrane is triggered by an increase in cytosolic Ca is supported by the finding that the divalent cation ionophore A23187 also initiated ^{86}Rb or K loss (as measured by an external K electrode). This increased K conductance was also seen in the presence of external ATP, which is known to move calcium ions.[10] Quinine and apamin greatly reduced the effect of ATP or A23187 on ^{86}Rb or K release in adrenal glomerulosa cells.[11] The increased potassium efflux due to angiotensin II, however, was transient and was followed by a permanent reduction on K permeability.

Figure 2 includes a comparative study of the effects of angiotensin II, potassium ions, and ACTH on the constant-rate coefficient of ^{86}Rb efflux in bovine adrenal glomerulosa cells. The three aldosterone secretagogues elicited a significant increase on ^{86}Rb efflux. The increment was transient with each stimulus; nevertheless, a different pattern was observed in each case, as shown in the figure:

(A) ACTH produces only the initial transient phase of increased ^{86}Rb efflux with the rate returning to base line within 6 to 8 min.

(B) In the presence of high external potassium the peak response is larger than that seen with ACTH and the baseline rate of efflux remains slightly higher than controls after the initial increment on ^{86}Rb efflux.

(C) Angiotensin II causes a significant and prolonged reduction in the ^{86}Rb release

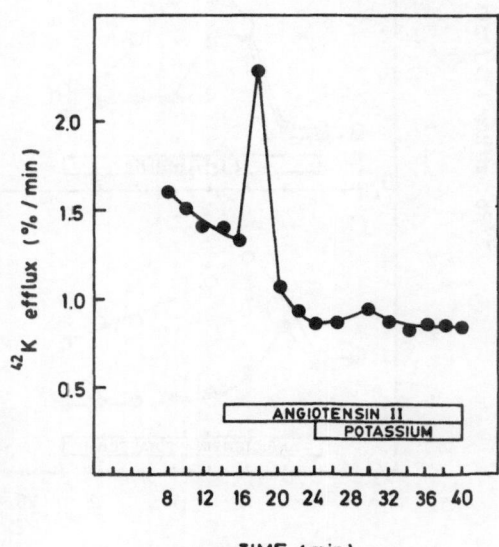

Figure 3. Effect of angiotensin II on the rate coefficient efflux of ^{42}K from perfused glomerulosa cells. From 14 to 40 min 100 nM angiotensin II was present. Potassium (12 mM) was added 24 to 40 min. The experimental conditions were similar to those described in Fig. 2.

immediately after the transient rise in the efflux. As shown by us,[12] this inhibition was dose related. Apparent first-order rate coefficients (percentage per minute) revealed a significant inhibitory effect of 10^{-8} and $10^{-7}M$ angiotensin II on ^{86}Rb efflux. We have recently repeated the same protocol in the presence of ^{42}K instead of ^{86}Rb, confirming the dual effect of angiotensin II on potassium permeability (Fig. 3). The patterns of angiotensin II effect is qualitatively similar to those previously described: Both rates increased significantly within the first 2 min of stimulation, followed by a second phase in which a sustained inhibition of the efflux is observed. Also shown in the same figure is the lack of effect of a depolarizing concentration of potassium added during the inhibitory period induced by angiotensin II. The longer radionuclide half-life of ^{86}Rb compared to that of ^{42}K makes it technically more convenient to use. Therefore, most experiments were done with trace amounts of ^{86}Rb as an indicator of the movement of K ions across the plasma membrane.

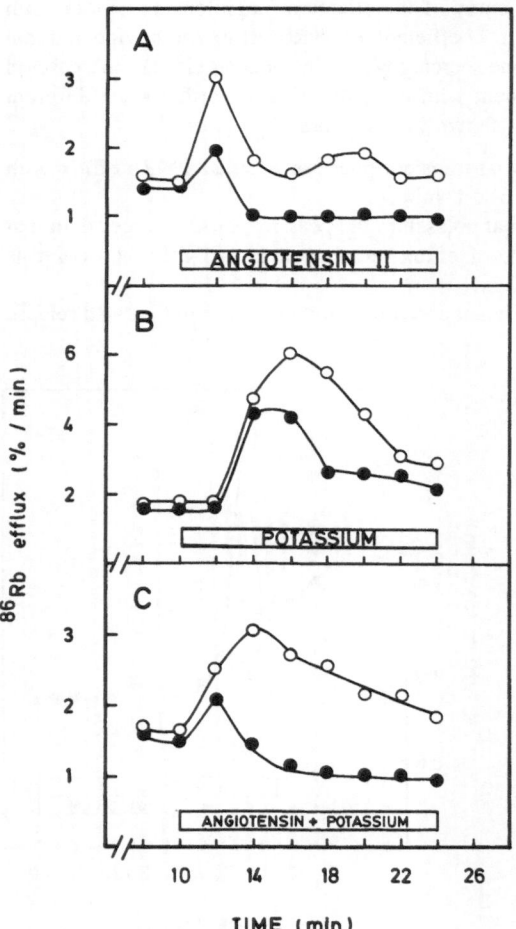

Figure 4. Comparison of the effect of potassium ions and angiotensin II on ^{86}Rb efflux from perfused adrenal glomerulosa cells in the presence of Ca and when Ca was replaced by Sr. (●), cells perfused with a standard medium (0.6 mM Ca); (○) cells perfused with a medium containing 0.6 mM Sr in the absence of Ca. Cells were perfused with the respective medium during 30 min before starting samples collection. Angiotensin II (100 nM) and/or potassium (12 mM) were present 10–24 min. The ^{86}Rb efflux under different experimental conditions was measured simultaneously in parallel chambers.

4. ANGIOTENSIN II MODULATION OF POTASSIUM PERMEABILITY

The dual effect of angiotensin II on the rate of ^{86}Rb efflux correlates with the brief hyperpolarization observed in the glomerulosa cells immediately after the introduction of this polypeptide hormone,[13] whereas the long-lasting depolarization may correspond to the sustained decrease in ^{86}Rb efflux.

Figure 4 contrasts the results obtained with the standard Krebs–Ringer–bicarbonate–glucose medium (KRBG) containing 0.6 mM Ca, to the results obtained when cells are perfused with KRBG that contains no added Ca and 0.6 mM Sr. In the absence of external Ca, angiotensin II produces the initial transient phase of increased efflux and a second small peak at 8 to 10 min, with the rate returning to baseline within 12 to 14 min. These results would suggest that external Ca could be involved in the inhibitory effect of angiotensin II on K channels. Also, in the absence of external Ca the peak response is significantly higher as compared to that observed when extracellular Ca is present. The same was true in the case of stimulation with high external potassium concentration, as shown Fig. 4B. The simultaneous presence of both aldosterone secretagogues, high potassium (12 mM), and angiotensin II (10^{-7} M) completely blocked the potassium response in the presence of Ca, and the typical angiotensin II pattern was observed (Fig. 4C).

5. CHARACTERISTICS OF THE ANGIOTENSIN II-MODULATED POTASSIUM PERMEABILITY

Channels-selective toxins and drugs have proved of great value in the analysis of membrane ionic permeability. Table I summarizes the results obtained with classical K-channel blockers on the angiotensin II-mediated ^{86}Rb efflux from isolated bovine adrenal glomerulosa cells. As shown in the table, apamin, quinine, and TEA effectively block the initial transient peak, but the three compounds are ineffective on the late phase of the hormone action. The opposite is true when 1 mM Ba is added to the perfusion medium: Barium ions enhance the initial rise in ^{86}Rb efflux, whereas a partial reversion of the inhibitory phase is observed.

Table I. Effect of Different Inhibitors of K-Channels on the Angiotensin II-Mediated [86]Rb Efflux from Perfused Bovine Adrenal Glomerulosa Cells[a]

	Angiotensin II-mediated ^{86}Rb efflux	
K-channel blocker	Initial transient increase	Sustained decrease
Apamin (10^{-7} M)[b]	Blocked	Unaltered
Quinine (10^{-5} M)[c]	Blocked	Unaltered
TEA (10^{-2} M)[b]	Blocked	Unaltered
Ba^{2+} (10^{-3} M)[d]	Increased	Partial-reverse
Charybdotoxin (10^{-8} M)[d]	No effect	—

[a]Cells were preequilibrated in the presence of ^{86}Rb for 60 min, as indicated in Fig. 2.
[b]From Refs. 10.
[c]From Ref. 12.
[d]Lobo, unpublished.

Table II. Effect of Induced Changes in Cellular-Ca Homeostasis on the Angiotensin II-Mediated [86]Rb Efflux in Isolated Bovine Adrenal Glomerulosa Cells

	Angiotensin II-mediated ^{86}Rb efflux	
Incubation conditions	Initial transient rise	Sustained decrease
No external Ca^{2+}[a]	Maintained	Reversed
TMB-8 $(10^{-4}\ M)$[b]	No effect	No effect
Dantrolene $(10^{-5}\ M)$[b]	Blocked	No effect
K-depolarization (12 mM)[a]	No effect	No effect
Veratridine $(10^{-6}\ M)$[b]	No effect	No effect

[a]From Ref. 12.
[b]Lobo, unpublished.

In the excitable cells like nerve and muscle, apamin, a peptide from bee venom selectively blocks Ca-activated K channels (low conductance), whereas TEA blocks a large voltage- and Ca-dependent K conductance. The contribution made by Ca-activated K-channels is not only important in the regulation of neuronal or muscle activity, but also in some endocrine tissues, as first shown in pancreatic islet β-cells.[14] In the glomerulosa cells, we have modified cellular calcium homeostasis by different maneuvers, in order to study the modulatory effect of angiotensin II on potassium permeability. A summary of the results are included in Table II.

The main result of these studies was the suppression of the inhibitory phase of angiotensin II-mediated ^{86}Rb efflux from bovine adrenal glomerulosa cells. As shown in Table II, the omission of calcium in the perfusion medium is the only condition in which the inhibitory phase of ^{86}Rb efflux is not observed. The transient initial peak is due to the release of intracellular calcium from nonmitochondrial stores mediated by inositol

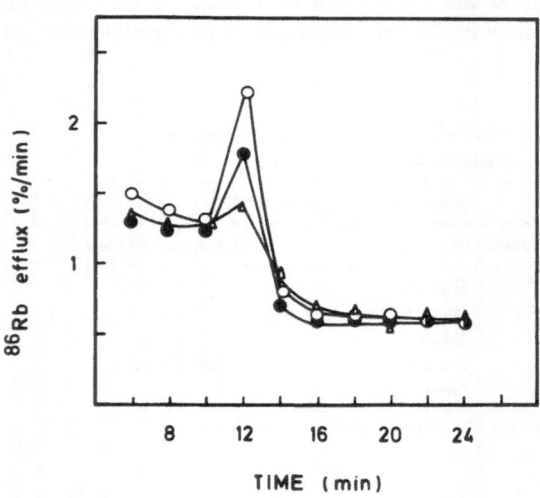

Figure 5. Effect of atrial natriuretic peptide (ANP) on angiotensin II-mediated ^{86}Rb efflux from bovine adrenal glomerulosa cells. The cells were perfused with standard medium during 30 min of equilibration, as indicated in Fig. 1. From 8 to 24 min ANP was present; angiotensin II (100 nM) was added from 10 to 24 min. One experiment representative of four is shown. (●), 100 nM angiotensin II alone; (○), angiotensin II plus 10 nM ANP; (△), angiotensin II plus 100 nM ANP.

1,4,5-triphosphate.[10] Nevertheless, in the absence of external Ca, angiotensin II-mediated aldosterone secretion is rapidly turned off.

Further studies were directed to determine whether atrial natriuretic peptide (ANP) could modify potassium permeability in angiotensin II-stimulated cells. ANP consists of a family of peptides with natriuretic and vasorelaxant properties that have a direct inhibitory effect on the adrenal glomerulosa cell steroidogenesis. It is believed that ANP inhibits agonist-stimulated aldosterone secretion by activating a guanylate cyclase system, but the mechanism is not yet understood. Recently, it has been postulated that its effect does not involve alterations in calcium homeostasis.[15] Figure 5 includes the results in the rate coefficient of ^{86}Rb efflux observed in the presence of ANP. The experiments were done with angiotensin II alone or angiotensin plus ANP at two different concentrations. As shown in Fig. 5, in the presence of ANP, angiotensin II induces a rise in ^{86}Rb efflux that is smaller than the one observed with the hormone alone. However, the inhibitory phase is not modified by the presence of ANP, suggesting that potassium permeability is not the locus of ANP's inhibitory action.

6. PERSPECTIVES

Angiotensin II, increased serum potassium concentration, and ACTH trigger the secretion of aldosterone. Elevation in the concentration of each of these agents is associated with membrane depolarization in adrenal glomerulosa cells, suggesting that ionic membrane permeability may play a regulatory role in the secretion processes of these cells. Potassium permeability, as measured by ^{86}Rb efflux from preloaded cells, was increased by these three secretagogues. An interesting feature in these studies is the dual effect of angiotensin II on potassium permeability, namely, an initial increase, followed by a sustained decrease in ^{42}K or ^{86}Rb efflux. This dual effect was not observed with the other two aldosterone secretagogues.

The modulatory role of angiotensin II on potassium permeability supports the evidence for membrane potential changes previously described by us and others. Figure 6 includes a schematic diagram of a possible mechanism involved in the potassium permeability change mediated by angiotensin II: During the initial phase the rapid hydrolisis of phosphatidylinositol 4,5-biphosphate leads to the production of inositol trisphosphate, which induces the release of calcium from a nonmitochondrial pool. This transient rise in the cytosolic Ca^{2+} opens Ca-activated K-channels that can be inhibited by apamin. The hallmark of this inositol lipid signaling system is that there is a bifuration in the flow of information through the formation of diacylglycerol, which is an activator of protein kinase-C, or it can be hydrolyzed by a diacylglycerol lipase to form monoacylglycerol, and from this compound it can release arachidonic acid.

Because the latter is the precursor of the eicosanoids (prostaglandins, thromboxanes, and leukotrienes) the primary second messenger diacylglycerol can give rise to additional messengers. It is possible that some of these compounds or protein kinase-C may play a role in the sustained inhibitory phase of the potassium permeability mediated by the angiotensin II (see Fig. 6). In the nonactivated cell the protein kinase-C is largely in its Ca^{2+}-insensitive form in the cytosol; the rise in the diacylglycerol content of the plasma membrane, due to angiotensin II and the rise in the Ca^{2+} concentration in the

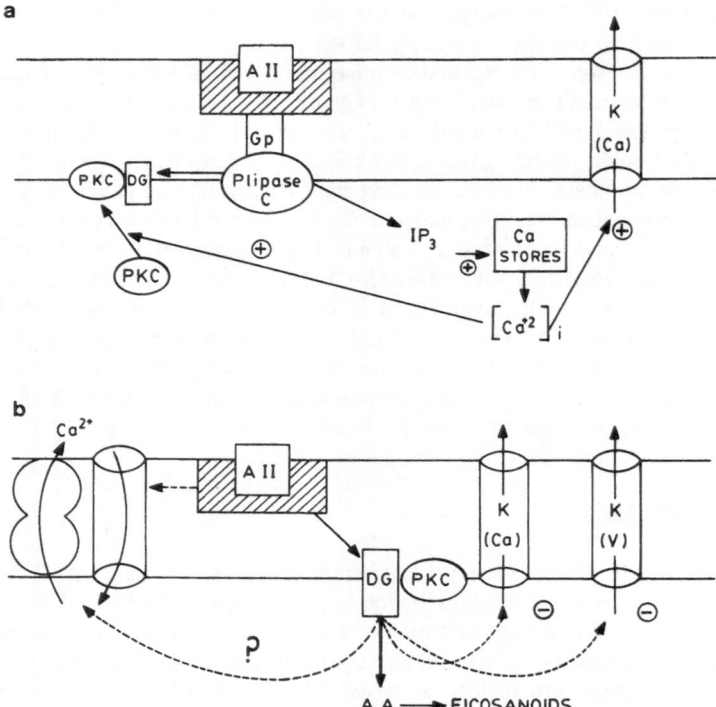

Figure 6. Schematic model of the two stages in angiotensin II activation of bovine glomerulosa cells. (a) During the initial phase of cell activation, phospholipase-C action (Plipase C) involves two signals. The rise in inositol 1,4,5-triphosphate (IP$_3$) mediated-Ca^{2+} concentration in the cell cytosol ([Ca^{2+}]$_i$) leads to the opening of Ca-dependent potassium channels. This rise in cytosolic Ca acts synergistically with the rise in the diacylglycerol (DG) content of the plasma membrane to cause translocation of the protein kinase-C (PKC) to the plasma membrane. The activity of this Ca^{2+}-sensitive form of the kinases is controlled by the rates of Ca^{2+} influx and efflux across the plasma membrane; therefore, external Ca is necessary to maintain the Ca^{2+} concentration in the subdomain of the plasma membrane. (b) Associated with this effect on kinase-C, the primary second messenger DG can give rise to additional messengers—arachidonic acid (AA) and eicosanoids, which could modify potassium and/or calcium channel activity.

cell cytosol, act synergistically to cause the translocation of the enzyme to the plasma membrane in its active state.

It is known that the active enzyme requires cycling of Ca across the plasma membrane;[16] therefore, its activation is not supported in the absence of external calcium. Protein kinase-C has been shown to control a variety of physiological processes in different tissues; for example, kinase-C activation prolongs Ca-dependent inactivation of K-currents in *Hermissenda*,[17] and has an inhibitory effect on voltage-dependent K conductance in a neuroblastoma–glioma hybrid cell line.[18] In this cell line, bradykinin—an activator of phospholipase-C—leads to the appearance of two different sequential membrane potassium conductance changes that are quite similar to those described here. Also, it is feasible that some eicosanoids or arachidonic acid itself, released by the continuous presence of angiotensin II, may play a role. Yet, the possibility of other intracellular inhibitors of potassium permeability, such as increased intracellular ATP[19] or angiotensin II-stimulated G proteins[20] should be considered in future studies.

ACKNOWLEDGMENTS. The authors' work described in this paper was supported by Fondecyt and DTI, Universidad de Chile research grants.

REFERENCES

1. Mathews, E. K., and Saffran, M., 1973, Ionic dependence of adrenal steroidogenesis and ACTH-induced changes in the membrane potential of adrenocortical cells, *J. Physiol.* **234:** 43–64.
2. Hausdorff, W., Sakura, R., Aguilera, G., and Catt, K., 1987, Control of aldosterone production by angiotensin II is mediated by two guanine nucleotide regulatory proteins, *Endocrinology* **120:** 1608–1627.
3. Quinn, S., and Williams, G. H., 1988, Regulation of aldosterone secretion, *Ann. Rev. Physiol.* **50:** 409–426.
4. Foster, R., Lobo M. V., Rasmussen, H., and Marusic, E. T., 1982, The effect of calcium on potassium-induced depolarization of adrenal glomerulosa cells, *FEBS Lett.* **149:** 253–256.
5. Rasmussen, H., 1986, The calcium messenger system, *N. Engl. J. Med.* **314:** 1164–1170.
6. Kojima, K., Kojima, I., and Rasmussen, H., 1984, Dihydropyridine calcium antagonist and agonist: Effect on aldosterone secretion, *Am. J. Physiol.* **247:** E645–E650.
7. Foster, R., Lobo, M. V., Rasmussen, H., and Marusic, E. T., 1981, Calcium: Its role in the mechanism of action of angiotensin II and potassium in aldosterone production, *Endocrinology* **109:** 2196–2201.
8. Elliott, M. E., Siegel, F. L., Hadjokas, N., and Goodfriend, T. L., 1985, Angiotensin effects on calcium and steroidogenesis in adrenal glomerulosa cells, *Endocrinology* **116:** 1051–1059.
9. Natke, E., and Kabela, E., 1979, Electrical responses in adrenal cortex: Possible relation to aldosterone secretion, *Am. J. Physiol.* **237:** E158–E162.
10. Lobo, M. V., and Marusic, E. T., 1986, Effect of angiotensin II, ATP, and ionophore A23187 on potassium efflux in adrenal glomerulosa cells, *Am. J. Physiol.* **250:** E125–E130.
11. Marusic, E. T., and Lobo, M. V., 1988, Steroidogenesis and ionic permeability in adrenal glomerulosa cells, *Arch. Biol. Med. Exp.* **21:** 171–176.
12. Lobo, M. V., and Marusic, E. T., 1988, Angiotensin II causes a dual effect on potassium permeability in adrenal glomerulosa cells, *Am. J. Physiol.* **254:** E144–E149.
13. Quinn, S. J., Cornwall, M. C., Williams, G. H., 1987, Electrophysiological responses to angiotensin II of isolated rat adrenal glomerulosa cells, *Endocrinology* **120:** 1581–1589.
14. Mattews, E. K., 1986, Calcium and membrane permeability, *Br. Med. Bull.* **42:** 391–398.
15. Apfeldorf, W., Isales, C., and Barrett, P., 1988, Atrial natriuretic peptide inhibits the stimulation of aldosterone secretion but not the transient increase in intracellular free calcium concentration induced by angiotensin II addition, *Endocrinology* **122:** 1460–1465.
16. Kojima, I., Kojima, K., and Rasmussen, H., 1985, Role of calcium fluxes in the sustained phase of angiotensin II-mediated aldosterone secretion for adrenal glomerulosa cells, *J. Biol. Chem.* **260:** 9177–9184.
17. Alkon, D. L., Kusata, M., Neary, J., Naito, S., Coulter, D., and Rasmussen, H., 1986, Kinase-C activation prolongs Ca-dependent inactivation of K currents, *Biochem. Biophys. Res. Commun.* **134:** 1245–1253.
18. Higashida, H., and Brown, J. A., 1986, Two polyphosphatidylinositide metabolites control two K^+ currents in neural cells, *Nature* **225:** 323–333.
19. Cook, D. C., and Hales, N., 1984, Intracellular ATP directly blocks K-channels in pancreatic β-cells, *Nature* **311:** 271–273.
20. Litosh, I., 1987, Regulatory GTP-binding proteins: Emerging concepts on their role in cell function, *Life Sciences* **41:** 251–258.

C. Other Transduction Mechanisms

The Target of Inositol 1,4,5-Trisphosphate in Nonmuscle Cells: Calciosome or Endoplasmic Reticulum?

Pompeo Volpe, Mariangela Bravin, Barbara H. Alderson, Daniel P. Lew, Jacopo Meldolesi, and Tullio Pozzan

> . . . *la natura non multiplica le cose senza necessità, e che ella si serve de' mezi più facili e semplici nel produrre i suoi effetti, e che ella non fa niente indarno.* . . . (". . . nature never multiplies things without need, and it employs the easiest and simplest means to produce its effects, and it does nothing in vain. . .")
>
> G. Galilei, *Dialogo sopra i due massimi sistemi del mondo Tolemaico e Copernicano*

1. INTRODUCTION

The total calcium (Ca^{2+}) content of mammalian cells is of the order of 1 to 3 mmoles/liter of cell water, that is, it is similar to that of the extracellular milieu. On the other

POMPEO VOLPE and BARBARA H. ALDERSON • Department of Physiology and Biophysics, The University of Texas Medical Branch, Galveston, Texas 77550. MARIANGELA BRAVIN and TULLIO POZZAN • Centro di Studio per la Fisiologia dei Mitocondri del CNR, Istituto di Patologia Generale dell'Università di Padova, 35131 Padua, Italy. DANIEL P. LEW • Division des Maladies Infectieuses, Hôpital Cantonal Universitaire, 1211 Geneva 4, Switzerland. JACOPO MELDOLESI • Dipartimento di Farmacologia dell' Università di Milano, Centro di Studio, di Cito Farmacologia del CNR, Istituto Scientifico San Raffaele, 20132 Milan, Italy.

hand, the extracellular free Ca^{2+} concentration is about 1 mM and four orders of magnitude higher than that of the cytoplasm, which is around 100 nM. Although little information is available on the distribution of total cytoplasmic Ca^{2+}, it is reasonable to assume that total cytoplasmic Ca^{2+} represents no more than 10 to 20% of cellular Ca^{2+} content.[1] Thus, the vast majority of the cell Ca^{2+} must be sequestered within intracellular, membrane-bound compartments (Ca^{2+} pools).

As indicated by ^{45}Ca autoradiography,[2] electron probe x-ray microanalysis,[3] and, more recently, computerized Ca^{2+} imaging,[4] there are several intracellular Ca^{2+} pools. Some of them can be identified morphologically [e.g., secretory granules, mitochondria, lysosomes, sarcoplasmic reticulum (SR) of striated muscles, and, perhaps, some elements of the endoplasmic reticulum (ER)]. These Ca^{2+} pools also differ functionally in that there are slowly exchangeable Ca^{2+} pools (e.g., mitochondria, secretory granules, lysosomes) and rapidly exchangeable Ca^{2+} pools (e.g., sarcoplasmic reticulum).

Intracellular Ca^{2+} storage organelles capable of both high-affinity Ca^{2+} uptake and rapid Ca^{2+} release are known to be ubiquitous, but their identity has been established with certainty only in striated muscle fibers. In striated muscle fibers, the fluctuations of intracellular free Ca^{2+}, $[Ca^{2+}]_i$, that underly the contraction–relaxation cycle are due to a well-developed membrane network structure, the SR,[5] composed of two continuous yet distinct portions—a longitudinal network of cisternae that surround the contractile elements, and the terminal cisternae (TC), strategically located in close apposition to the transverse tubule (T-tubule) infoldings of the sarcolemma. SR is exquisitely suited for $[Ca^{2+}]_i$ control: Uptake is due to a high-affinity Ca^{2+}-ATPase,[6] which accounts for over 80% of the SR membrane proteins; storage is due to a high-capacity, medium-affinity Ca^{2+}-binding protein, calsequestrin (CS; Ref. 7), localized exclusively within the lumen of TC[8]; and release is due to a Ca^{2+} release channel, also localized in the TC, which is activated upon T-tubule depolarization. Organelles that might resemble the SR have been described, but not yet characterized in detail, in smooth muscle[9] and platelets.[10] In all other cells, the possible existence of an organelle(s) homologous to the SR was not even considered until recently. The structure responsible for $[Ca^{2+}]_i$ control was proposed to be the ER,[3,11,12] or at least a part of this endomembrane system.[13]

In this chapter we will review the evidence indicating that the rapidly exchangeable, IP$_3$-sensitive Ca^{2+} pool of nonmuscle cells is distinct from ER, and resides, instead, in a population of vesicles and small vacuoles distributed throughout the cytoplasm and collectively termed "calciosome."[14] The calciosome appears to be the nonmuscle counterpart of SR.

2. INOSITOL 1,4,5-TRISPHOSPHATE AND Ca^{2+} RELEASE

Eukaryotic cells share the ability to accumulate Ca^{2+} within a discrete, intracellular membrane-bound store, and to release it rapidly into the cytosol in response to adequate stimuli.[15] Transient increases of $[Ca^{2+}]_i$ trigger key cellular events such as contraction, secretion, growth, and fertilization. Several neurotransmitters, hormones, agonists, and growth factors act at plasma membrane receptors to stimulate the break-down of phosphatidylinositol 4,5-bisphosphate into diacylglycerol and inositol 1,4,5-trisphosphate (IP_3). The link between receptor activation and Ca^{2+} release from "intracellular stores" has been shown to be IP_3.[15] The role of IP_3 as a water-soluble intracellular messenger has been described in a very large number of different cells.[16]

Some progress has been made over the past 5 years concerning the characterization of the "intracellular (nonmitochondrial) Ca^{2+} store." The store appears to consist of a vesicular pool and is endowed with an ATP-dependent Ca^{2+} pump and an IP_3-sensitive Ca^{2+}-channel (or efflux pathway). IP_3 releases Ca^{2+} from an "intracellular store" that has been tentatively identified as ER, both rough and smooth ER, based on distribution of enzyme markers.[11,17,18] For instance, the excellent correlation between ER markers, ribonucleic acid content, and IP_3 sensitivity has led to the conclusion that the IP_3-sensitive Ca^{2+} stores coincide with the rough ER in exocrine pancreas microsomal fractions.[11,17] The possibility of copurification of ER with other specialized structures was never considered. During the past few years, several reports have instead indicated that ER per se cannot be the IP_3-sensitive Ca^{2+} store.

1. The amount of Ca^{2+} being released by IP_3 in a variety of cell types was very similar[19-22] and did not correlate with the relative development of either rough ER or smooth ER in each cell type (i.e., it was scanty in neutrophils and impressive in liver and pancreas).

2. Upon extensive fractionation, there was found to be no correlation between known ER markers and putative markers of the IP_3-sensitive Ca^{2+} store (i.e., IP_3 binding and IP_3-induced Ca^{2+} release). This has been shown in neutrophils[23] and in liver.[24,25]

3. Microinjection of IP_3 into the large *Limulus* photoreceptor cell preloaded with aequorin revealed intracellular release of Ca^{2+} only in a cell region containing no (or very few) ER elements, and not in the region containing typical and multiple ER stacks.[26] Moreover, analysis of sensory neurons preloaded with fura-2 and then treated with either bradykinin (an agent that causes receptor-triggered generation of IP_3) or caffeine, revealed two spatially distinct patterns of $[Ca^{2+}]_i$ increase.[27] Thus, the organelles sensitive to either IP_3 or caffeine are likely to be distinct.[27]

4. IP_3 was able to release only part (around 30–40%) of the Ca^{2+} accumulated by the high-affinity (nonmitochondrial) Ca^{2+} store[11,20,28,29] and this suggested that IP_3-sensitive and -insensitive Ca^{2+} pools are, at least, functionally distinct.

3. THE "CALCIOSOME"

Starting from these observations, we hypothesized that the IP_3-sensitive Ca^{2+} store of nonmuscle cells was a unique organelle, more similar to SR than to ER. Under this assumption, it seemed plausible to predict that the "nonmitochondrial Ca^{2+} store" would contain an intraluminal protein with characteristics similar to those of striated muscle CS to serve as a Ca^{2+} storage site between uptake–release cycles. Thus, in the fall of 1986, we started to investigate whether a CS-like protein also existed in nonmuscle cells and whether the subcellular distribution of this protein correlated with the Ca^{2+} uptake activity and IP_3-sensitive Ca^{2+} release. Three lines of experimental evidence are provided.

1. Proteins that bear immunological or biochemical resemblance to striated muscle CS are expressed not only in smooth muscles[30] but also in a variety of nonmuscle tissues, cells, and clonal cell lines (Table I). The CS-like protein(s) share a number of properties with skeletal muscle CS[14]: Similar apparent molecular weight after SDS-polyacrylamide gel electrophoresis, metachromatic blue staining with stains all, Ca^{2+}-binding capacity, alkali and detergent extractability, and pH dependence of electrophoretic mobility.

2. The availability of anti (skeletal muscle CS) antibodies, which cross-react with related proteins of nonmuscle tissues (see Table I and Ref. 14), allowed the subcellular localization of this putative marker in situ. Immunogold labeling of ultrathin cryosections has been carried out in rat liver and exocrine pancreas, in two cell lines, PC12 and HL60, and are currently being carried out in a variety of different tissues and cells. In ultrathin cryosections, anti-CS antibodies decorate specifically a population of small (50–250 nm in diameter) organelles, in no apparent continuity with typical elements of the ER and Golgi complex, and morphologically distinguishable from multivesicular bodies, lysosomes, mitochondria, nuclei, and secretory granules.[14]

These membrane-bound organelles are discrete and are distributed throughout the cytoplasm with some preference for the cytoplasmic rim beneath the plasma membrane. Within the hepatocytes, which have been more thoroughly investigated,[35,36] cal-

Table I. Immunological Detection of Calsequestrin-like Proteins (Apparent Mol. Wt. $\cong$ 65,000 da)

	Immunoblot	Immunocytochemistry	References
A. *Tissues/cells*			
Rat liver	a	a,c	Volpe et al. (14)
	a		Damiani et al. (31)
Rat exocrine pancreas	a	a,c	Volpe et al. (14)
Bovine brain	b		Benson and Fine (32)
Rat brain		b,d	Meldolesi et al. (33)
Bull spermatozoa	a		Volpe and Deana (unpublished)
Sea urchin eggs	b		Oberdorf et al. (34)
B. *Cell lines*			
Rat PC12	a	a,c	Volpe et al. (14)
Human HL60	a	a,c	Volpe et al. (14)
Rat RIN 2A37	a		Volpe, Ullrich and Wollheim (unpublished)
Rat RIN m5F		a,c	Hashimoto et al. (35)

KEY: a: anti (rabbit fast-twitch skeletal muscle calsequestrin) antibody; b: anti (cardiac muscle calsequestrin) antibody; c: immunogold labeling of ultrathin cryosections; d: immunofluorescence

TC   3
                    ◄─1st   1st─►

2nd↓   ↓2nd

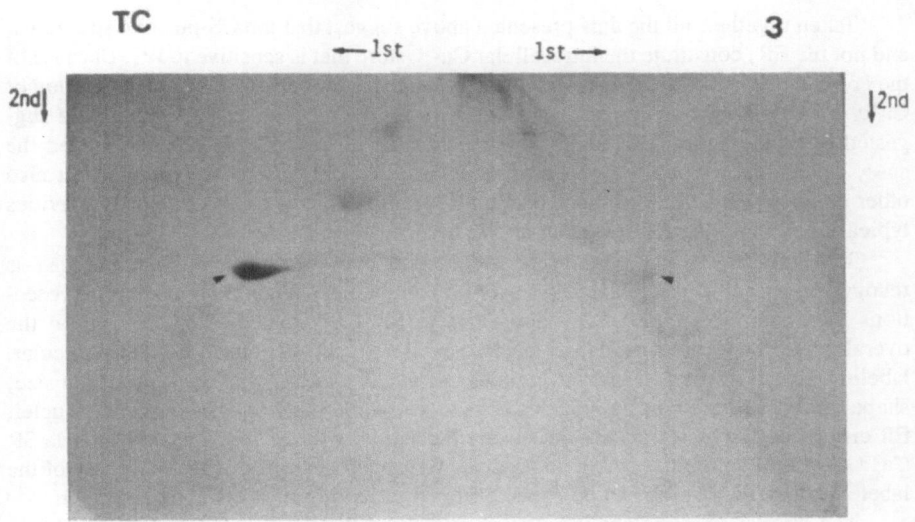

Figure 1. Two-dimensional polyacrylamide gel electrophoresis (PAGE) of rabbit skeletal muscle terminal cisternae (TC) and HL60 fraction 3. Two-dimensional PAGE was carried out as described by Michalak et al. [48] using 10% gels. Two disc gels, run as in Ref. 49, were placed head-to-head on top of the slab gel, that was run as in Ref. 46. Stains-All staining was carried out as described by Campbell et al. [51] TC, on the left, 20 μg of protein; HL60 fraction 3, [3] on the right, 70 μg of protein. Blue-staining calsequestrin (CS) and CS-like protein are indicated by arrowheads.

ciosomes are distributed throughout the cytoplasm interspersed with, and often adjacent to, other organelles. In the regions occupied by ER stacks, calciosomes are seen in between adjacent cisternae, often closely apposed to the cytosolic surface of one such cisterna. In no case, however, is obvious luminal continuity of CS-positive structures with ER cisternae observed in the cryosections. At other cell sites, calciosomes often appear apposed to other organelles, for example, mitochondria, or sandwiched in between one mitochondrion and an ER cisterna. In the cytoplasmic rim beneath the plasma membrane, calciosomes are seen isolated or adjacent to tubular structures (smooth ER ?). Apposition of calciosomes to the plasma membrane, although not common, was repeatedly seen at the basolateral portion of the cytoplasm. Quite a number of calciosomes were present in the Golgi area, often located at the *trans* Golgi or intermingled with other, CS-negative vesicles, lateral to the cisternal stacks.

3. Subcellular fractions of the human promelocytic leukemia cell line HL60 which contain an ATP-dependent Ca^{2+} pump and respond to IP_3 are separated from the markers of known subcellular organelles, such as granules, ER, mitochondria, Golgi complex, and endosomes, and are enriched in a Ca^{2+}-binding protein that has several properties similar to CS of rabbit skeletal muscle SR(14). Figure 1 shows that after two-dimensional polyacrylamide gel electrophoresis, the 63,000 blue-staining, cross-reactive CS-like protein of HL60 fractions (14) falls off the diagonal line (panel B), as in the case of rabbit skeletal muscle CS (panel A).

Taken together, all the data presented above suggest that the CS-positive structures, and not the ER, constitute the intracellular Ca^{2+} store that is sensitive to IP_3. Because of the content of a CS-like protein, and their hypothesized involvement in the regulation of Ca^{2+} homeostasis, the new structures were given the name of calciosome(s) and suggested to be the simplified, nonmuscle counterpart of the SR.[14] If this is indeed the case, calciosomes would be expected to express not only a CS-like protein, but also other components similar to those of the SR, and to exclude components and activities typical of different structures, such as the ER.

Table II shows that nonmuscle tissues and cells display a polypeptide that is recognized by anti (muscle SR Ca^{2+}-ATPase) antibodies. When liver ultrathin cryosections were immunodecorated by anti (muscle SR Ca^{2+}-ATPase) antibodies,[36] the overall pattern appeared similar to that observed with anti-CS antibodies. In particular, labeling was most often seen over vesicles and small vacuoles that were similar in size, shape, and distribution to CS-positive organelles, while other structures such as nuclei, ER cisternae, and plasma membranes were labeled very little. With the anti (muscle SR Ca^{2+}-ATPase) antibodies, the gold particles were mostly aligned at the periphery of the labeled organelles, as was to be expected for the labeling of a membrane antigen.

The similar immunolabeling pattern suggested that both anti (muscle SR Ca^{2+}-ATPase) antibodies and anti-CS antibodies recognize the same organelle—the calciosome. Double-labeling experiments showed that most CS-positive structures localized in all regions of the cell were in fact markedly labeled by anti (muscle SR Ca^{2+}-ATPase) antibodies as well.[36] In these specimens, the two labels often exhibited the distribution previously described in the samples labeled by either one of the antibodies separately: small gold particles addressed to CS decorated the content preferentially, with large gold particles addressed to the Ca^{2+}-ATPase often arranged in short rows and preferentially located at the organelle rim. In favorable sections, a partial topological segregation of the two labels was observed, with the CS immunodecoration concentrated toward one edge of the calciosome content, and the Ca^{2+}-ATPase distributed more evenly all around the organelle. In addition

Table II. Immunological Detection of Ca^{2+}-ATPase Proteins (Mol. Wt. $\cong$ 110,000 da)

	Immunoblot	Immunocytochemistry	References
A. *Tissues/cells*			
Rat liver	a		Damiani et al. (31)
		a,c	Hashimoto et al. (36)
Human platelets	a		Dean (37)
	d		Sarkadi et al. (38)
Rat brain	d		Sarkadi et al. (38)
(synaptosomes)			
B. *Cell lines*			
Rat PC12		a,c	Hashimoto et al. (35)
Human HL60	b		Krause and Campbell (39)

KEY: a: anti (rabbit fast-twitch skeletal muscle Ca^{2+}-ATPase) antibody; b: monoclonal anti (cardiac muscle Ca^{2+}-ATPase) antibody; c: immunogold labeling of ultrathin cryosections; d: anti (rat gastrocnemius muscle Ca^{2+}-ATPase) antibody.

to the calciosomes immunodecorated by both antibodies, other organelles positive for only one antigen were also seen in the double-labeled sections.

Preliminary cytochemical evidence indicating that calciosomes are distinct from ER[14] is now corroborated by double-immunolabeling of liver ultrathin cryosections.[36] In these experiments, the immunolocalization of CS was compared to that of other antigens which are bona fide markers of ER. In the liver, ER membranes contain cytochrome P_{450},[40] which is a predominant protein, and NADH-cytochrome b_5 reductase.[41] An additional, recently identified ER marker is endoplasmin,[42] a protein segregated within the ER lumen and expressed in a variety of cell types. Calciosomes were shown not to express any ER markers, that is, cytochrome P_{450}, NADH-cytochrome b_5 reductase, and endoplasmin.[36] Conversely, neither CS-like protein(s) nor Ca^{2+}-ATPase were observed in identifiable ER cisternae.[35,36] The latter observation casts serious doubt on the role of the ER in Ca^{2+} homeostasis inasmuch as this membrane system was found to be devoid of the Ca^{2+}-ATPase.

4. RELATIONSHIP BETWEEN CALCIOSOME(S) AND ER

An important question to be addressed is whether calciosomes are indeed discrete organelles, as suggested by their appearance in thin sections, or are connected to one another in a network. At the present time no evidence is available in favor of the calciosome network. This possibility cannot be ruled out, however, until three-dimensional reconstructions are made based on serial sections. These studies are not yet feasible because so far calciosomes are recognized only in ultrathin frozen sections which cannot be routinely prepared in series. Other immunocytochemical procedures, which allow serial sectioning, imply the resin embedding of the samples before immunodecoration. This has been found to cause the loss of anti-CS immunoreactivity. This problem might be circumvented when antibodies directed to each individual CS-like protein will be available.

This same technical limitation applies to the study of the relationships between calciosomes and other organelles, in particular the ER. In both the pancreas and liver, many calciosomes were indeed found to lie in close apposition to ER cisternae,[14,36] and this could be taken as a suggestion of luminal continuity between the two structures. We do not think this to be the case: First, calciosomes were seen not only in the ER-rich regions of the cytoplasm, but also in the Golgi and other areas. This argument is particularly cogent for the apical area of pancreatic acinar cells, which is almost exclusively occupied by zymogen granules. Second, close apposition of calciosomes was seen not only with the ER, but also with other structures such as mitochondria, zymogen granules, Golgi cisternae, and in no case was clear luminal continuity with any of these structures observed. Third, the continuity of calciosomes with the ER would be difficult to envisage in view of the demonstrated segregation for both membrane (Ca^{2+}-ATPase versus cytochrome P_{450} and NADH-cytochrome b_5 reductase) and content (CS versus endoplasmin) markers.

To our knowledge, the existence of highly heterogeneous subcompartments has never been demonstrated for the ER, whose two portions, rough- and smooth-surfaced, possess a number of common membrane components and the same intraluminal proteins. The relationship between ER and calciosomes, in the liver at least, might, on the contrary, resemble that between ER and SR in striated muscles. Although derived by vesicle budding from the rough ER,[43] the SR of skeletal muscle represents a separate cytological entity, different from the ER in structure, function, and composition. ER markers are absent from the SR of fast-twitch skeletal muscle.[44] Calciosomes resemble smooth ER elements in conventional thin sections, but differ from the ER because of their specific expression of proteins similar to those of the SR. Thus, we tend to believe that calciosomes are distinct from ER and constitute the nonmuscle version of the SR.

The assignment of membrane-bound organelles to the ER, particularly to its smooth-surfaced portion, has been made essentially based on the general architecture of the organelles themselves in conventional thin-section electron microscopy (e.g., tubules, cisternae, small vacuoles) rather than on the knowledge of their molecular composition. The so-called "smooth ER" may be, instead, composed by elements that, in spite of their overall similarity, can be heterogeneous in terms of composition and of function, and should therefore be considered as separate cytological entities. The calciosome(s) might be one of them.

The proposed heterogeneity of smooth ER might help to reconcile results which appear to be in conflict. There might be different, rapidly exchangeable Ca^{2+} pools (e.g., IP_3- and GTP-sensitive).[45] There might also be a distinct, genuine Ca^{2+} pool in rough ER cisternae as proposed by Somlyo et al.[3] The Ca^{2+} content of liver ER cisternae, as revealed by electron probe x-ray microanalysis, represents less than 25% of the total cellular Ca^{2+}.[3] This allows ample room for additional Ca^{2+} stores in the cell. Since "calciosomes" in hepatocytes are often localized near the ER cisternae,[14,36] they would be counted as part of the rough ER given the size of the electron probe used by Somlyo et al.[3] This latter argument also applies to the claim that vasopressin would mobilize Ca^{2+} from liver rough ER.[12]

5. ON THE FUNCTION OF THE CALCIOSOME(S)

Calciosomes have been observed in all cell types so far investigated (over 10). They might therefore be ubiquitous and, in view of their content of a Ca^{2+}-binding protein and their endowment, at least in liver,[36] with a high-affinity Ca^{2+}-ATPase, they would be expected to play a general role in Ca^{2+} homeostasis. The colocalization of CS with a high-affinity Ca^{2+}-ATPase in the calciosome is strongly suggestive for the identification of the latter with the IP_3-sensitive Ca^{2+} store. In fact, in nonmuscle cells the physiological low resting $[Ca^{2+}]_i$ is due to the high-affinity Ca^{2+} uptake into the same organelle that releases Ca^{2+} in response to IP_3. We lack, however, the formal and definitive proof that the calciosomes, as presently identified in situ, are endowed with the IP_3 receptor or IP_3-sensitive Ca^{2+}-channel. The IP_3 receptor protein (of the brain) has been recently identified and purified.[46] The study of its intracellular distribution in relation to the distribution of calciosome components, such as the CS-like protein, is expected to become possible in the near future.

An important question is whether the size of the calciosome compartment and its content of CS-like protein(s) are sufficient to sustain the $[Ca^{2+}]_i$ transients elicited by

IP_3. The extent of Ca^{2+} required to sustain the $[Ca^{2+}]_i$ transients due to release from intracellular stores is in the order of 0.2–0.4 mmoles/liter of cell volume. Calciosomes were found to account for ~1 and 2% of the cytoplasmic volume in acinar pancreatic and liver cells,[36] respectively. Thus, the calciosome internal Ca^{2+} content should lie in the 20–40 mmoles/liter range. Assuming that Ca^{2+} is bound to the CS-like proteins within the calciosome lumen and assuming that their Ca^{2+}-binding properties are similar to those of muscle CS, the Ca^{2+} buffering capacity of the calciosome has been evaluated and found to be compatible with the role of the organelle as the IP_3-sensitive Ca^{2+} store in liver and exocrine pancreas (see also Ref. 14 for HL60 and PC12 cells).

6. FUTURE DIRECTIONS

The biochemical, functional, and morphological characterization of the calciosome is being actively pursued in our laboratories. Short-term goals are as follows:

1. Purification of the calciosome. The identification of putative calciosome markers (Ca^{2+}-pump activity, IP_3-induced Ca^{2+} release, CS-like protein) will facilitate this task. Different fractionation procedures for each individual cell type will be implemented.

One such approach is depicted in Fig. 2. Rat exocrine pancreas microsomes were

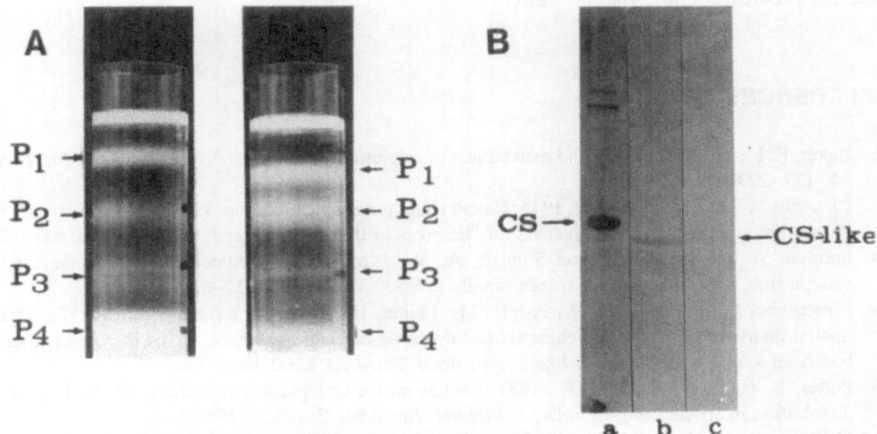

Figure 2. Fractionation of rat exocrine pancreas on a discontinuous sucrose gradient and distribution of calsequestrin-like protein. Effect of puromycin: (A) Fractionation of a high-speed pellet from rat pancreas homogenate was carried out according to Preissler and Williams [52] with slight modifications. *Left:* Pellet material was layered onto a gradient consisting of 27%, 32%, 38%, and 45% sucrose and centrifuged at 70,000 × g_{max} overnight. Fractions P1 through P4 were separated. *Right:* Pellet material was pretreated with 1 mM puromycin for 30 min as described by Adelman et al. [47] and then centrifuged as described above. (B) Skeletal muscle terminal cisternae proteins (lane a), pancreas puromycin P4 fraction (see panel A, right tube) proteins (lane b) and pancreas P4 fraction (see panel A, left tube) proteins (lane c) separated by SDS-PAGE [50] and then subjected to immunoblot with anti (skeletal muscle CS) antibodies as described previously. [14] Fifteen μg of protein was applied in lane a, 150 μg in lanes b and c.

fractionated on a discontinuous sucrose gradient. One of the samples (right-hand side in Fig. 2A) was incubated prior to fractionation with puromycin, an agent known to detach ribosomes from rough ER.[47] After such a treatment, the P_4 fraction (a rough-ER-enriched fraction in the absence of puromycin) lost the bulk of its protein content due to the changed density properties of the stripped rough ER membranes.[47] On the other hand, the content of CS-like protein was increased after puromycin treatment (Fig. 2B, lane b). These preliminary experiments indicate that a putative calciosome marker, that is, the CS-like protein, can be dissociated from rough ER markers.

2. Isolation and Ca^{2+}-binding properties of the CS-like protein(s). Attempts along those lines have been reported.[32,34,39] Data on both capacity and K_d for Ca^{2+} are crucial for the assessment of the role of CS-like protein(s) as the Ca^{2+} storage site within the calciosome lumen.

3. Biogenesis of the calciosome. These studies should clarify the relationship between the calciosome, on one hand, and ER and Golgi complex, on the other.

ACKNOWLEDGMENTS.   This work was supported in part by grants from the National Institutes of Health (GM 40068) to P.V., from the Italian CNR "Progetto Finalizzato Oncologia" (grant n.87.01410.44) and from the Italian Ministry of Public Education to T.P. and J.M., and from the Swiss National Research Foundation (3.990-0.84) to D.P.L. M.B. was supported by a grant from the ASSNE (Associazione per lo Sviluppo delle Scienze Neurologiche), Padua, Italy.

REFERENCES

1. Baker, P. F., 1972, Transport and metabolism of calcium ions in nerve, *Progr. Biophys. Mol. Biol.* **24:** 177–223.
2. Clemente, F., and Meldolesi, J., 1975, Calcium and pancreatic secretion. I. Subcellular distribution of calcium and magnesium in the exocrine pancreas of the guinea pigs, *J. Cell. Biol.* **65:** 88–102.
3. Somlyo, A. P., Bond, M., and Somlyo, A. V., 1985, Calcium content of mitochondria and endoplasmic reticulum in liver frozen rapidly *in vivo*, *Nature* **314:** 622–625.
4. Lipscombe, D., Madison, D. V., Poenie, M., Reuter, H., Tsien, R. Y., and Tsien, R. W., 1988, Spatial distribution of calcium channels and cytosolic calcium transients in growth cones and cell bodies of sympathetic neurons, *Proc. Natl. Acad. Sci. USA* **85:** 2398–2402.
5. Porter, K. R., and Palade, G. E., 1957, Studies on the endoplasmic reticulum. III. Its form and distribution in striated muscle cells, *J. Biophys. Biochem. Cytol.* **3:** 269–298.
6. MacLennan, D. H., 1970, Purification and properties of the adenosine triphosphatase from sarcoplasmic reticulum, *J. Biol. Chem.* **245:** 4508–4518.
7. MacLennan, D. H., and Wong, P. T. S., 1971, Isolation of a calcium-sequestering protein from sarcoplasmic reticulum, *Proc. Natl. Acad. Sci. USA* **68:** 1231–1235.
8. Jorgensen, A. O., Kalnins, V., and MacLennan, D. H., 1979, Localization of sarcoplasmic reticulum proteins in rat skeletal muscle by immunofluorescence, *J. Cell. Biol.* **80:** 372–384.
9. Somlyo, A. V., and Franzini-Armstrong, C., 1985, New views of smooth muscle structure using freezing, deep-etching, and rotatory shadowing, *Experimentia* **41:** 841–856.
10. Kaser-Glanzmann, R., Jakabova, M., George, J. N., and Luscher, E. F., 1977, Stimulation of calcium uptake in platelet membrane vesicles by adenosine 3'-5' cyclic monophosphate and protein kinase, *Biochim. Biophys. Acta* **466:** 429–440.

11. Streb, H., Bayerdoffer, E., Hasse, W., Irvine, R. F., and Schultz, I., 1984, Effect of inositol 1,4,5-trisphosphate on isolated subcellular fractions of rat pancreas, *J. Membrane. Biol.* **81:** 241–253.

12. Bond, M., Vadasz, G., Somlyo, A. V., and Somlyo, A. P., 1987, Subcellular calcium and magnesium mobilization in rat liver stimulated *in vivo* with vasopressin and glucagon, *J. Biol. Chem.* **262:** 15630–15636.

13. Leslie, B. A., Burgess, G. M., and Putney, J. W., 1988, Persistent inhibition by inositol 1,4,5-trisphosphate of oxalate-dependent as calcium accumulation in permeable guinea-pig hepatocytes, *Cell Calcium* **9:** 9–16.

14. Volpe, P., Krause, K.-H., Hashimoto, S., Zorzato, F., Pozzan, T., Meldolesi, J., and Lew, D. P., 1988, "Calciosome," a cytoplasmic organelle: The inositol 1,4,5-trisphosphate-sensitive Ca²⁺ store of nonmuscle cells? *Proc. Natl. Acad. Sci.* **85:** 1091–1095.

15. Berridge, M. J., 1984, Inositol trisphosphate and diacylglycerol as novel second messengers, *Biochem. J.* **220:** 345–360.

16. Berridge, M. J., 1987, Inositol trisphosphate and diacylglycerol: Two interacting second messengers, *Ann. Rev. Biochem.* **56:** 159–193.

17. Streb, H., Irvine, R. F., Berridge, M. J., and Schultz, I., 1983, Release of Ca²⁺ from a nonmitochondrial intracellular store in pancreatic acinar cells by inositol 1,4,5-trisphosphate, *Nature* **306:** 67–69.

18. Prentki, M., Biden, T. J., Janijc, D., Irvine, R. F., Berridge, M. J., and Wollheim, C. B., 1984, Rapid mobilization of Ca²⁺ from rat insulinoma microsomes by inositol 1,4,5-trisphosphate, *Nature* **309:** 562–564.

19. Joseph, J. K., and Williamson, J. R., 1986, Characterization of inositol trisphosphate-mediated Ca²⁺ release from permeabilized hepatocytes, *J. Biol. Chem.* **261:** 14658–14664.

20. Biden, T. J., Wollheim, C. B., and Schlegel, W., 1986, Inositol 1,4,5-trisphosphate and intracellular Ca²⁺ homeostasis in clonal pituitary (GH₃) cells, *J. Biol. Chem.* **261:** 3184–3192.

21. Lew, D. P., Wollheim, C. B., Waldvogel, F. A., and Pozzan, T., 1984, Modulation of cytosolic free Ca²⁺ transients by changes in intracellular calcium-buffering capacity: Correlation with exocytosis and O₂ production in human neutrophils, *J. Cell. Biol.* **99:** 1212–1220.

22. Bruzzone, R., Pozzan, T., and Wollheim, C. B., 1986, Caerulein and carbamoylcholine stimulate pancreatic amylase release at resting cytosolic free Ca²⁺, *Biochem. J.* **235:** 139–143.

23. Krause, K.-H., and Lew, D. P., 1987, Subcellular distribution of Ca²⁺ pumping sites in human neutrophils, *J. Clin. Inv.* **80:** 107–116.

24. Guillemette, G., Balla, T., Baukal, A. J., Spat, A., and Catt, K. J., 1987, Intracellular receptors for inositol 1,4,5-trisphosphate in angiotensin II target tissues, *J. Biol. Chem.* **262:** 1010–1015.

25. Guillemette, G., Balla, T., Baukal, A. J., and Catt, K. J., 1988, Characterization of inositol 1,4,5-trisphosphate receptors and calcium mobilization in a hepatic plasma membrane fractions, *J. Biol. Chem.* **263:** 4541–4548.

26. Payne, R., and Fein, A., 1987, Inositol 1,4,5-trisphosphate releases calcium from specialized sites within *Limulus* photoreceptors, *J. Cell. Biol.* **104:** 933–937.

27. Thayer, S. A., Perney, T. M., and Miller, R. J., 1988, Regulation of calcium homeostasis in sensory neurons by bradykinin, *J. Neurosci.* **8:** 4089–4097.

28. Burgess, G. M., Irvine, R. F., Berridge, M. J., McKinney, J. S., and Putney, J. W., 1984, Actions of inositol phosphates on Ca²⁺ pools in guinea pig hepatocytes, *Biochem. J.* **224:** 741–746.

29. Taylor, C. W., and Putney, J. W., 1985, Size of the inositol 1,4,5-trisphosphate-sensitive calcium pool in guinea pig hepatocytes, *Biochem. J.* **232:** 435–438.

30. Wuitack, F., Raeymakers, L., Verbist, J., Jones, L. R., and Casteels, R., 1987, Smooth-muscle endoplasmic reticulum contains a cardiaclike form of calsequestrin, *Biochim. Biophys. Acta* **899:** 151–158.

31. Damiani, E., Spamer, C., Heilmann, C., Salvatori, S., and Margreth, A., 1988, Endoplasmic reticulum of rat liver contains two proteins closely related to skeletal muscle sarcoplasmic reticulum Ca-ATPase and calsequestrin, *J. Biol. Chem.* **263:** 340–343.

32. Benson, R. J. J., and Fine, R. E., 1987, Purification of a calsequestrinlike protein from bovine brain, *Neuroscience* (Abstr.) 473.6.

33. Meldolesi, J., Volpe, P., and Pozzan, T., 1988, The intracellular distribution of calcium, *Trends Neurosci.* **11**: 449–452.

34. Oberdorf, J. A., Lebeche, D., Head, J. F., and Kaminer, B., 1987, Identification of a calsequestrinlike protein in sea urchin eggs, *J. Cell Biol.* **105**: 338a.

35. Hashimoto, S., Bruno, B., Volpe, P., Zorzato, F., Pozzan, T., Krause, K.-H., Lew, P. D., and Meldolesi, J., 1988, On the nature and function of the calciosome, a cytoplasmic organelle containing calsequestrinlike protein(s) which is expressed in nonmuscle cells, in: *Hormones and Cell Regulation*, Vol. 165 (J. Nunez, J. Ormont, and E. Carafoli, eds.), John Libbey Eurotext, London, pp. 167–180.

36. Hashimoto, S., Bruno, B., Lew, D. P., Pozzan, P. Volpe, P., and Meldolesi, J., 1988, Immunocytochemistry of calciosomes in liver and pancreas, *J. Cell Biol.* **107**: 2523–2531.

37. Dean, W. L., 1984, Purification and reconstitution of a Ca^{2+} pump from human platelets, *J. Biol. Chem.* **259**: 7343–7348.

38. Sarkadi, B., Enyedi, A., Penniston, J. T., Verma, A. K., Dux, L., Molner, E., and Gardos, G., 1988, Characterization of membrane calcium pumps by simultaneous immunoblotting and ^{32}P radiography, *Biochem. Biophys. Acta* **939**: 40–46.

39. Krause, K.-H., and Campbell, K. P., 1988, Sarcoplasmic reticulum proteins in phagocytes: Immunological and functional properties and subcellular distribution, *FASEB J.* **2**: 542a.

40. Garfinkel, J., 1963, A comparative study of electron transport in microsomes, *Comp. Biochem. Physiol.* **8**: 367–379.

41. De Pierre, J. W., and Dallner, G., 1975, Structural aspects of the membrane of the endoplasmic reticulum, *Biochim. Biophys. Acta* **415**: 411–472.

42. Koch, G., Smith, M., Macer, D., Webster, D., and Mortara, R., 1986, Endoplasmic reticulum contains a common, abundant calcium-binding glycoprotein, endoplasmin, *J. Cell. Sci.* **86**: 217–232.

43. Ezerman, E. B., and Ishikawa, H., 1967, Differentation of the sarcoplasmic reticulum and T system in developing chick skeletal muscle *in vitro*, *J. Cell. Biol.* **35**: 405–420.

44. Salviati, G., Betto, R., Salvatori, S., and Margreth, A., 1979, Evidence for the presence of the stearyl-CoA desaturase system in the sarcoplasmic reticulum of slow muscle, *Biochim. Biophys. Acta* **574**: 280–289.

45. Henne, V., Piiper, A., and Soling, H.-D., 1987, Inositol 1,4,5-trisphosphate and 5'-GTP induce calcium release from different intracellular pools, *FEBS Lett.* **218**: 153–158.

46. Supattapone, S., Worley, P. F., Baraban, J. M., and Snyder, S. H., 1988, Solubilization, purification, and characterization of an inositol trisphosphate receptor, **263**: 1530–1534.

47. Adelman, M. R., Sabatini, D. D., and Blobel, G., 1973, Ribosome–membrane interaction: Nondestructive disassembly of rat liver rough microsomes into ribosomal and membraneous components, *J. Cell. Biol.* **56**: 206–229.

48. Michalak, M., Campbell, K. P., and MacLennan, D. H., 1980, Localization of the high-affinity calcium-binding protein and an intrinsic glycoprotein in sarcoplasmic reticulum membranes, *J. Biol. Chem.* **255**: 1317–1326.

49. Weber, K., and Osborn, M., 1968, The reliability of molecular weight determinations by dodecyl sulphate polyacrylamide gel electrophoresis, *J. Biol. Chem.* **244**: 4406–4410.

50. Laemmli, U. K., 1970, Cleavage of structural proteins during the assembly of the head of bacteriophage-T4, *Nature* **227**: 680–685.

51. Campbell, K. P., MacLennan, D. H., and Jorgensen, A. D., 1983, Staining of the Ca^{2+}-binding proteins, calsequestrin, calmodulin, troponin-C, and S-100 with the cationic dye "Stains All," *J. Biol. Chem.* **258**: 11267–11273.

52. Preissler, M., and Williams, J. A., 1983, Localization of ATP-dependent calcium transport activity in mouse pancreatic microsomes, *J. Membr. Biol.* **73**: 137–143.

Chapter 14

Stimulus–Response Coupling in Mammalian Ciliated Cells: The Role of Ca^{2+} in Prostaglandin Stimulation

Manuel Villalon and Pedro Verdugo

1. OVERVIEW

Calcium has been thought to play an important role in the control of ciliary movement in a broad variety of ciliated cells. In mammalian ciliated cells Ca^{2+} depletion can cause reversible ciliary arrest. Also, ciliostimulation produced by prostaglandins in ciliated cells of the oviduct is thought to be coupled by release of intracellular Ca^{2+}. However, direct measurements of fluctuations of intracellular [Ca^{2+}] associated with stimulation of ciliary activity have not been reported.

The experiments presented here were designed to measure changes in intracellular [Ca^{2+}] following stimulation of ciliary activity with prostaglandin F$_{2\alpha}$ (PGF$_{2\alpha}$). Experiments were conducted in cultured ciliary cells of the rabbit oviduct. Changes in intracellular [Ca^{2+}] were measured by the fluorescent probe Fura-2.

Monolayers of ciliated cells were loaded with Fura-2 AM; equilibrated in Hanks' solution, pH 7.2, 37°C for 15 min; and mounted at a 45° angle in the cuvette of a double channel spectrofluorometer. Changes in fluorescence emission were detected at 500 nm before and after stimulation with PGF$_{2\alpha}$, using excitation wave lengths of 347 and 380 nm. Preliminary results indicate that a 2- to 3-fold transient increase of intracellular [Ca^{2+}] occurs within the first 10 sec, and lasts approximately 1 min after the infusion of 1 mM PGF$_{2\alpha}$ in the cuvette.

MANUEL VILLALON and PEDRO VERDUGO • Center for Bioengineering, University of Washington, Seattle, Washington 98195.

The present results provide the first direct measurement of stimulus-induced changes of intracellular $[Ca^{2+}]$ in mammalian ciliated cells. These results confirm previous observations that $PGF_{2\alpha}$ can strongly stimulate ciliary movement in oviductal ciliated cells in culture, and also agree with indirect evidence which suggests that ciliostimulation of oviductal ciliated cells by $PGF_{2\alpha}$ might be coupled by fluctuations of intracellular $[Ca^{2+}]$.

2. INTRODUCTION

Chemical messages and mechanical stimulation are known to modify the frequency of ciliary beat in epitheliary ciliated cells[1–5]; however, the cellular mechanisms that couple ciliostimulation are still poorly understood. In ciliated cells, variations in intracellular $[Ca^{2+}]$ may be an important step in the relay of chemical signals that control ciliary movement.[6–11] However, direct measurements of intracellular $[Ca^{2+}]$ associated with ciliostimulation have not been investigated.

Epitheliary ciliated cells of the trachea and the oviduct of the rabbit have been shown to respond to several hormones and chemical transmitters, including prostaglandins, β-adrenergic agonists, and ATP.[1,2,4,9,12–14] In the present experiments, prostaglandin $F_{2\alpha}$ was used to investigate the relationship between hormonal ciliostimulation and fluctuations of intracellular $[Ca^{2+}]$.

3. MATERIAL AND METHODS

Experiments were performed in cultured ciliated cells of the rabbit oviduct (Fig. 1). Mucus-free monolayers of ciliated cells were grown for 5–7 days in Rose chambers filled with Eagle's medium. Details of the culturing procedure have been published elsewhere.[1,2] The frequency of ciliary beat was monitored and recorded by dynamic laser scattering spectroscopy according to the method we developed previously.[15]

The Rose chambers containing the cultured cells were positioned in a Reichert inverted microscope modified for laser-scattering spectroscopy (Fig. 2). The attenuated beam of a 2 mW He-Ne laser was collimated to illuminate an area of approximately 10^4 μm^2 containing about 6×10^3 cilia. The scattered light was collected by the objective lens aperture of the microscope at an angle of 30°, and detected by a photomultiplier tube (PMT).

In the heterodyne detection scheme used here, two fields of scattered light are mixed on the cathode of the photomultiplier tube [see equation (1)]: (1) the local oscillator field, E_{Lo}, which is produced by the scattering at the air–glass interface and has a frequency equal to the frequency of the laser light, ω_0, and (2) the scattered field produced by the moving cilia, E_s. The photocurrent of the photomultiplier resulting from this optical mixing may be expressed as

$$i(t) = \beta[E_{Lo}(t) + E_s(t)] \cdot [E_{Lo}(t) + E_s(t)] \tag{1}$$

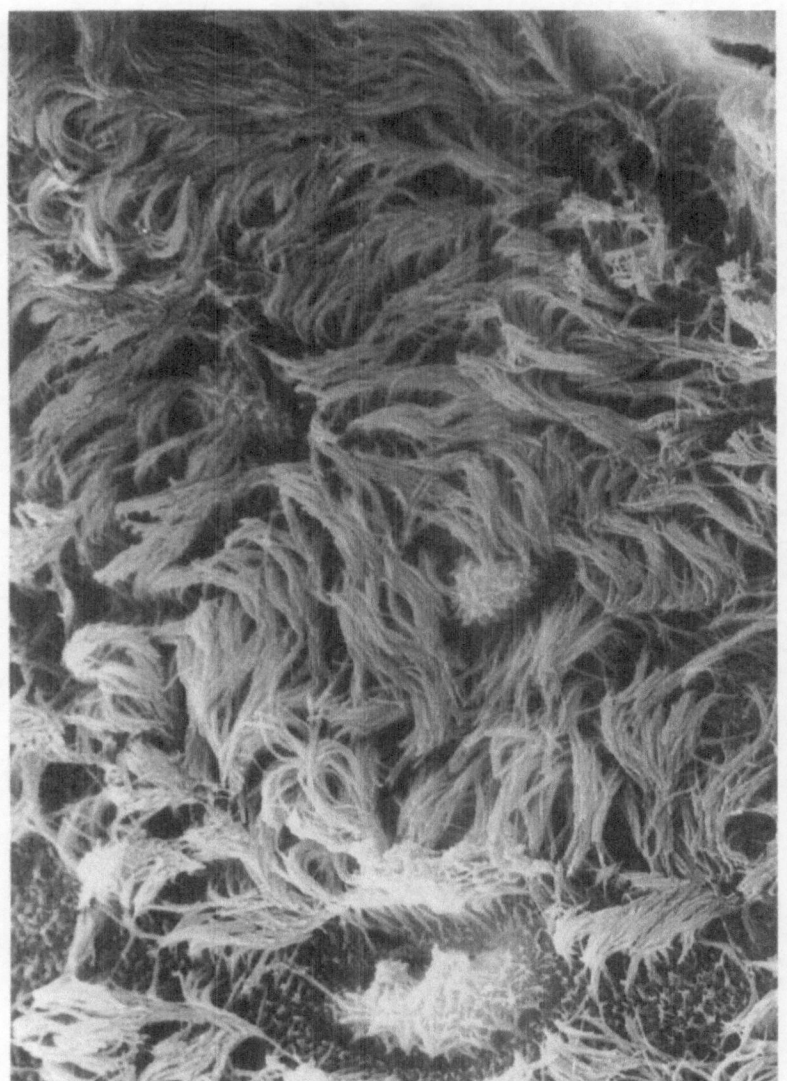

Figure 1. Scanning electron micrograph of a 7-day-old tissue cultured monolayer of ciliated mucosa of the rabbit oviduct. Notice that epithelial cells form a confluent monolayer, and the absence of secretory cells. (Magnification 2800×.)

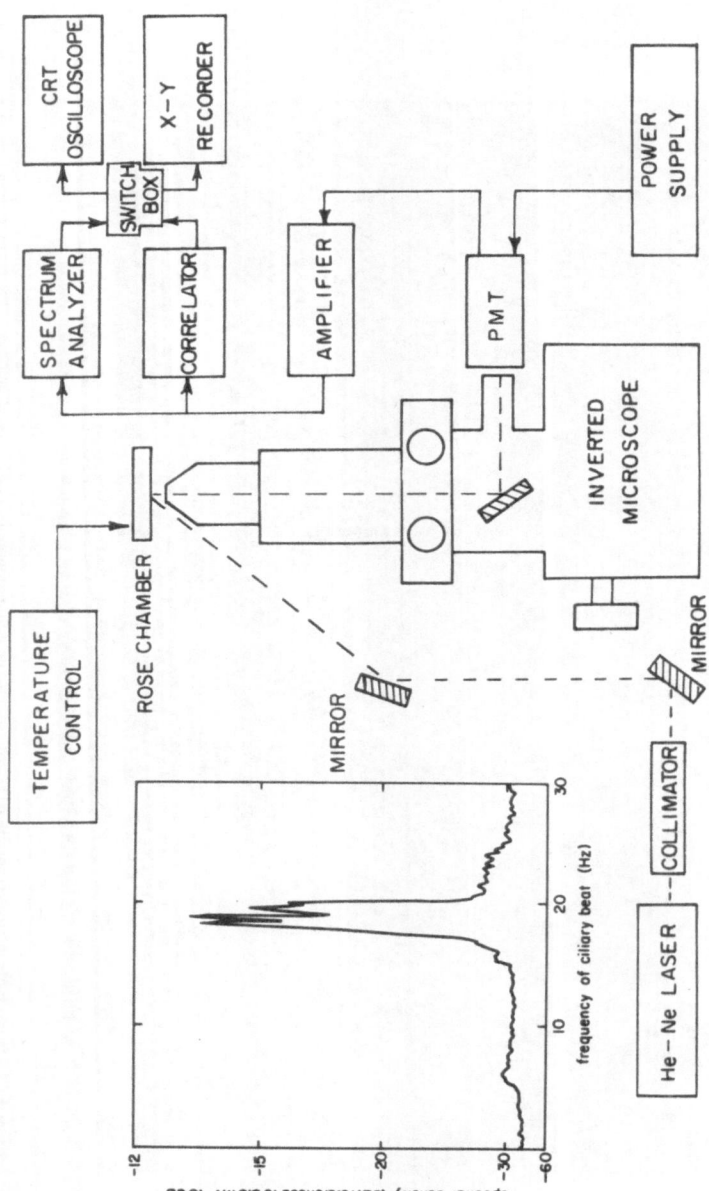

Figure 2. Schematic diagram of the laser spectrometer. Inset shows a typical spectral distribution of the frequency of ciliary beat.

where β is an instrumental constant that includes both the quantum and the collection efficiency of the photomultiplier tube. The heterodyne mixing mode also makes the results independent of the nongaussian nature of the light scattered by the ciliated cell. The spectral structure of the scattered light intensity can be expressed by the time correlation of the photocurrent, which can be shown as[15]

$$C(\tau) = ei_{Lo}\delta(\tau) + i^2_{Lo} \left[1 + \frac{2i_s}{i_{Lo}} \epsilon^2 \exp \left(-\frac{\tau}{\tau_R} \cos \omega_c \right) \right] \quad (2)$$

where $\delta(\tau)$ is the Dirac delta function, defined as $\delta(\tau) = 0$ if $\tau \neq 0$ and $\int \delta(\tau)d\tau = 1$. The currents i_s and i_{Lo} are the average photocurrents of the scattered light and the local oscillator, respectively, and $i_{Lo} \gg i_s$ is assumed. The heterodyne mixing efficiency, ϵ, represents the degree to which the local oscillator and the scattered wavefront have matched phase fronts over the detector's surface. The decay constant τ_R is an indirect index of the coherency of ciliary movements, and ω_c is the frequency of ciliary beat.

Therefore, the photocurrent autocorrelation function consists of a shot noise term, a d.c. term, and a third term that is a damped cosine function with damping coefficient, τ_R, and whose frequency equals the frequency of ciliary beat, ω_c. A more convenient way to record the statistical distribution of the frequency of ciliary beat in a population of moving cilia is to perform the spectral analysis of the output of the mixer (see inset in Fig. 2), in which case the frequency modulations of the Doppler signal equals the frequency of motion of the cilia. Spectral analysis renders, in the frequency domain, the same information that the autocorrelation function renders in the time domain. Either one can be determined from the other by Fourier transformation.

The spectral structure of the scattering fluctuations was processed on-line by a fast Fourier transform digital spectrum analyzer designed in our own laboratory. The averaged spectrum was recorded in an x–y plotter.

The dose–response relationship of $PGF_{2\alpha}$ was investigated in 16 tissue cultures. Ciliated cells were first equilibrated in Hanks' solution at 37°C for a control period of 15 min, while ciliary activity was continuously monitored. The concentration of $PGF_{2\alpha}$ in the Rose chamber was increased in random sequence between 10^{-8} to 10^{-5} M. The frequency of ciliary beat (FCB) was monitored for 15 min following each increase in agonist concentration. The effect of 14 μM $PGF_{2\alpha}$ on intracellular $[Ca^{2+}]$ and FCB was investigated in 8 tissue cultures. The same protocol was repeated in ciliated cells preincubated for 5 min in Hanks' solution containing 0.6 mM $LaCl_3$. Previous observations have indicated that $LaCl_3$ (0.6 mM) can block both the increase in cytosolic $[Ca^{2+}]$, and the ciliostimulation produced by ATP, which depend on transmembrane influx of Ca^{2+}.[14]

Measurements of intracellular $[Ca^{2+}]$ were performed using the method introduced by Tsien et al.[16,17] The cultures of ciliated cells were incubated for one hour at 37°C in Hanks' solution containing 1 μM Fura2-AM. Monolayers of ciliated cells were mounted at a 45° angle in the cuvette of a double-channel spectrofluorometer (SLM 4200). The preparation was excited with 347 and 380 nm wavelengths, while changes in fluorescence emission were detected at 500 nm before and after stimulation with 14 μM $PGF_{2\alpha}$.

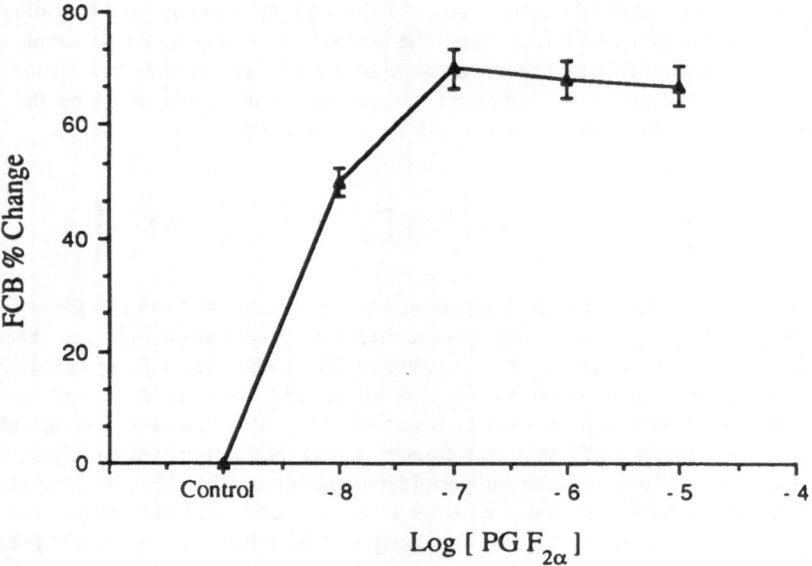

Figure 3. Dose–response relationship of the stimulating effect of $PGF_{2\alpha}$ on ciliary activity in cultured ciliated cells of the rabbit oviduct.

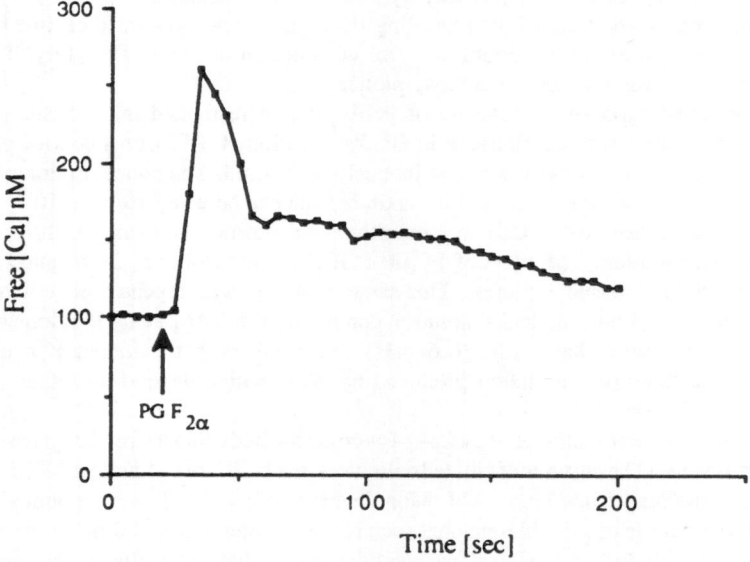

Figure 4. Time course of the effect of 14 μM $PGF_{2\alpha}$ on cytosolic $[Ca^{2+}]$ in cultured ciliated cells of the rabbit oviduct, as detected by Fura-2.

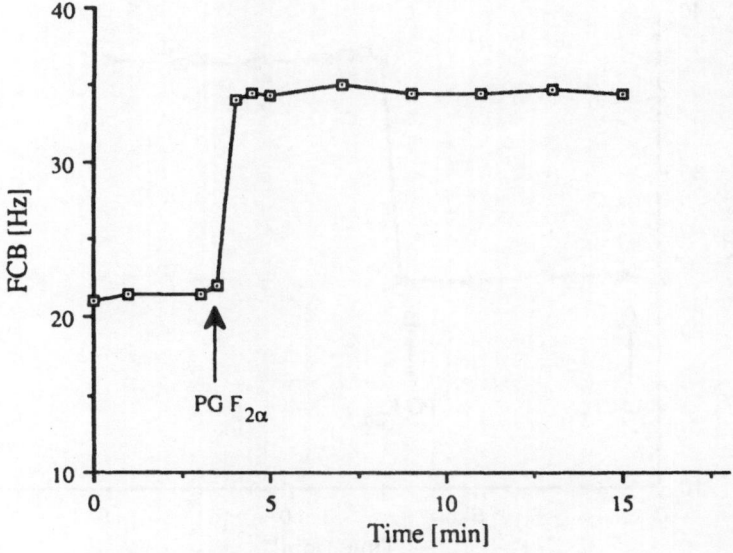

Figure 5.   Time course of the effect of 14 μM PGF$_{2\alpha}$ on the frequency of ciliary beat, in cultured cells of the rabbit oviduct.

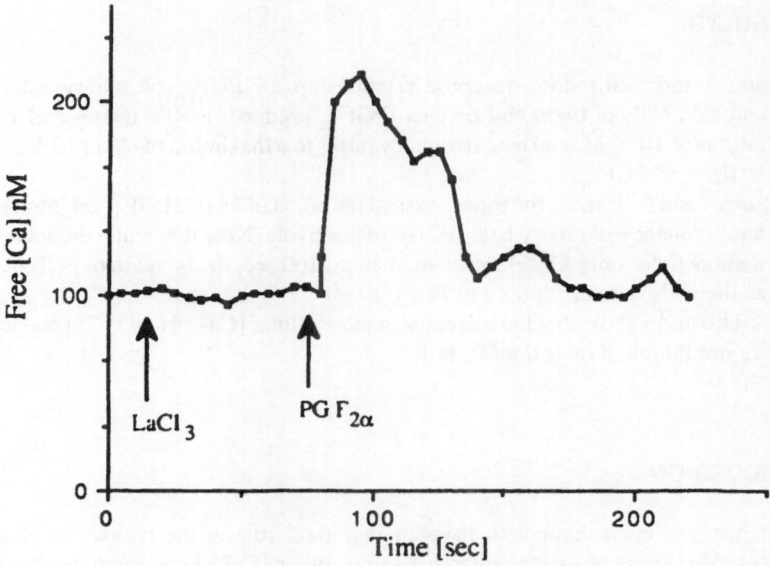

Figure 6.   Effect of 14 μM PGF$_{2\alpha}$ on cytosolic $[Ca^{2-}]$ in cultured ciliated cells, pretreated with 600 μM of the Ca-channel blocker LaCl$_3$.

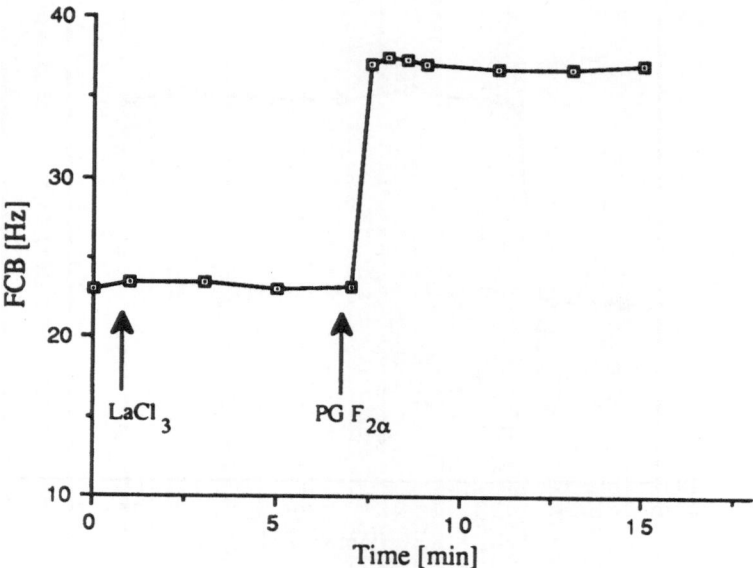

Figure 7. Ciliostumulating effect of 14 μM PGF$_{2\alpha}$ on cultured cells of the rabbit oviduct, pretreated with 600 μM of the Ca-channel blocker LaCl$_3$.

4. RESULTS

Figure 3 shows the dose–response relationship for PGF$_{2\alpha}$ on ciliary activity in cultured ciliated cells of the rabbit oviduct. PGF$_{2\alpha}$ produces a 50% increase of FCB at concentrations of $10^{-8}\,M$, and beat frequency raises to a maximum of about 70% over the control at $10^{-7}\,M$ PGF$_{2\alpha}$.

Figures 4 and 5 illustrate the time course of the effect of 14 μM PGF$_{2\alpha}$ on intracellular $[Ca^{2+}]$ and frequency of ciliary beat (FCB), respectively. Note that while the increase in $[Ca^{2+}]$ demonstrates only a brief transient of about 100 sec, the increase of FCB lasts for as long as the cells remain exposed to PGF$_{2\alpha}$.

Figures 6 and 7 show that the increases in intracellular $[Ca^{2+}]$ and FCB produced by PGF$_{2\alpha}$ are not inhibited by 0.6 mM LaCl$_3$.

5. DISCUSSION

Calcium has been thought to have an important role in the regulation of ciliary movement. Variations of extracellular and intracellular $[Ca^{2+}]$ can produce changes in ciliary activity in a variety of ciliated cells.[6–11,18,19] In the presence of $2 \times 10^{-8}\,M$

calmodulin, demembranated axonemal models of epitheliary ciliated cells can also exhibit Ca-dependent variations of beat frequency on reactivation with ATP.[20]

Fluctuations of intracellular $[Ca^{2+}]$ have been thought to relay a broad range of cellular responses including excitation–contraction coupling, excitation–secretion coupling, and also stimulus–response coupling in epitheliary ciliated cells.[9,21–23]

In oviductal ciliated cells PGs stimulate ciliary movement.[1] Previous observations have indicated that prostaglandins' stimulation of oviductal ciliated cells is independent of extracellular $[Ca^{2+}]$. Also, ciliostimulation switches from a long-lasting to a transient response when extracellular $[Ca^{2+}]$ is below 10^{-8} M, suggesting that prostaglandins stimulation might be coupled by release of intracellular calcium rather than by inflow of Ca^{2+} across the cell membrane.[9] However, direct measurements of PG-induced changes of cytosolic $[Ca^{2+}]$ had not been investigated.

The results presented here are the first direct indication that ciliostimulation by $PGF_{2\alpha}$ is associated with increased intracellular $[Ca^{2+}]$. The observations that the blockage of Ca-channels by $LaCl_2$ does not prevent the increase of intracellular $[Ca^{2+}]$, nor the ciliostimulation produced by $PGF_{2\alpha}$, is in agreement with our previous findings.[9] It suggests that $PGF_{2\alpha}$ must raise cytosolic $[Ca^{2+}]$ by inducing the release of this cation from intracellular sources.

The rise of cytosolic Ca^{2+} concentration induced by $PGF_{2\alpha}$ is equivalent to the $[Ca^{2+}]$ previously found to produce calmodulin-dependent stimulation of ciliary movement in ATP-reactivated axonemes of epitheliary ciliated cells.[20] Cytosolic $[Ca^{2+}]$ increases only transiently following exposure of the cells to $PGF_{2\alpha}$; however, ciliostimulation takes place for as long as the agonist remains in the medium. This findings suggest that the Ca^{2+} transient may act only as an enabling signal, in a way analogous to an "and" gate in a logic network. Thus, both the transient elevation of intracellular $[Ca^{2+}]$ and the continued occupancy of the receptor might be necessary, yet not sufficient, to produce ciliostimulation. A similar pattern of response has been observed following the stimulation of vascular smooth muscle with angiotensin II. It has been suggested that the observed Ca^{2+} transient might result in the activation of protein kinase-C that would then transduce the message for contraction.[23] Although this idea proposes a role for the observed Ca^{2+} transients, it does not explain why the agonist needs to remain in the medium to produce ciliostimulation. An alternative explanation is that in the presence of the agonist, $[Ca^{2+}]$ remains at levels that are sufficient to implement signal transduction but yet are beyond the sensitivity of the fluorescent probe.

In conclusion, the present results indicate that $[Ca^{2+}]$ increases transiently following $PGF_{2\alpha}$ stimulation of ciliated cells. However, a rise in Ca^{2+} concentration might not be sufficient to implement the instruction received by the membrane receptor. The significance of a second component in signal transduction—that is, the receptor occupancy in the coupling of chemical messages in epitheliary ciliated cells—is still uncertain and needs to be further investigated.

ACKNOWLEDGMENTS.   Supported by grants HL 38494 from NIH and R 010-7-01 from the Cystic Fibrosis Foundation. We are grateful to Mrs. Lynn Langley for her technical assistance, and to the famous Dr. Thomas Hinds for his help with the fluorescence measurements.

REFERENCES

1. Verdugo, P., Rumery, R. E., and Tam, P. Y., 1980, Hormonal control of oviductal ciliary activity: Effect of prostaglandins, *Fertil. Steril.* **33:** 193–196.
2. Verdugo, P., Johnson, N. T., and Tam, P. Y., 1980, β-adrenergic stimulation of respiratory ciliary activity, *J. Appl. Physiol.* **48:** 868–871.
3. Murakami, A., and Machemer, H., 1982, Mechanoreception and signal transmission in the lateral ciliated cells on the gill of *Mytilus*, *J. Comp. Physiol.* **145:** 351–362.
4. Villalón, M., and Verdugo, P., 1982, Hormonal regulation of ciliary function in the oviduct: The effect of β-adrenergic agonists, *Cell Motility*, Suppl. 1, 59–65.
5. Sanderson, M. J., and Dirksen, E. R., 1986, Mechanosensitivity of cultured ciliated cells from the mammalian respiratory tract: Implications for the regulation of mucociliary transport, *Proc. Natl. Acad. Sci. USA* **83:** 7302–7306.
6. Naitoh, Y., Eckert, R., and Friedman, K., 1972, A regenerative calcium response in *Paramecium*, *J. Exp. Biol.* **59:** 667–681.
7. Brehm, P., Dunlap, K., and Eckert, R., 1978, Calcium-dependent repolarization in *Paramecium*, *J. Physiol.* **274:** 639–654.
8. Satir, P., Reed, W., and Wolf, D. I., 1976, Ca^{2+}-dependent arrest of cilia without uncoupling epithelial cells, *Nature* **263:** 520–521.
9. Verdugo, P., 1980, Ca^{2+}-dependent hormonal stimulation of ciliary activity, *Nature* **283:** 764–765.
10. Stommel, E. W., 1984, Calcium regenerative potentials in *Mytilus edulis* gill abfrontal ciliated epithelial cells, *J. Comp. Physiol.* **155A:** 445–456.
11. Stommel, E. W., 1984, Calcium activation of mussel gill abfrontal cilia, *J. Comp. Physiol.* **155A:** 457–469.
12. Vorhaus, E. F., and Deyrup, I. J., 1953, The effect of adenosinetriphodphate on the cilia of the faringeal mucosa of the frog, *Science* **118:** 553–554.
13. Murakami, A., Machemer, H., and Eckert, R., 1978, Stimulation of ciliary activity by low levels of extracellular adenine nucleotides in amphibian oviduct, *Exp. Cell Res.* **85:** 154–158.
14. Villalón, M., Hinds, T., and Verdugo, P., 1988, Stimulus–response coupling in mammalian ciliated cells: Intracellular $[Ca^{2+}]$ transients detected by Fura-2, *Biophys. J.* **53:** 602a.
15. Lee, W. I., and Verdugo, P., 1976, Laser light-scattering spectroscopy: A new application in the study of ciliary activity, *Biophys. J.* **16:** 1115–1119.
16. Tsien, R. Y., Rink, T. J., and Poenie, M., 1985, Measurement of cytosolic free Ca^{2+} in individual small cells using fluorescence microscopy with dual excitation wavelength, *Cell Calcium* **6:** 145–157.
17. Grynkiewicz, G., Poenie, M., and Tsien, R. Y., 1985, A new generation of Ca^{2+} indicators with greatly improved fluorescent properties, *J. Biol. Chem.* **260:** 3440–3450.
18. Naitoh, Y., and Kaneko, H., 1973, Control of ciliary activity by adenosine triphsophate and divalent cations in triton extracted models of *Paramecium caudatum*, *J. Exp. Biol.* **58:** 657–676.
19. Brehm, P., and Eckert, R., 1978, Calcium entry leads to inactivation of calcium channels in *Paramecium*, *Science* **202:** 1203–1206.
20. Verdugo, P., Raess, B. V., and Villalón, M., 1983, The role of calmodulin in the regulation of ciliary movement in epithelial cells, *J. Submicrosc. Cytol.* **15:** 95–97.
21. Rubin, R. P., 1974, *Calcium and the Secretory Process*, Plenum, New York.
22. Szent-Gyorgyi, A. G., 1976, Comparative survey of the regulatory role of calcium in muscle, in: *Calcium in Biological Systems*, SEB Symp. (C. J. Duncan, ed.), Cambridge University Press, New York, **30:** 335–348.
23. Rasmussen, H., Kojima, I., Kojima, K., Zawalich, W., and Apfeldorf, W., 1984, Calcium as intracellular messenger: Sensitivity modulation, C-kinase pathway, and sustained cellular response, in: *Advances in Cyclic Nucleotide and Protein Phosphorylation Research, Vol. 18* (P. Greengard, G. A. Robison, Eds.), Raven Press, New York.

Chapter 15

Hepatocyte Gap Junctions: Metabolic Regulation and Possible Role in Liver Metabolism

Juan C. Sáez, Michael V. L. Bennett, and David C. Spray

1. INTRODUCTION

Over the past three decades, it has become clear that cells of most tissues communicate through specialized intercellular structures called gap junctions,[1] which have also been termed nexus or maculae communicantes. One of the most exhaustively studied examples of this type of intercellular communication is that between hepatocytes, which is emphasized in this chapter. In electron micrographs of thin sections, gap junctions are seen as specialized regions of contact where apposed plasma membranes of adjacent cells are separated by a gap of 2–3 nm (Fig. 1A). In electron micrographs of freeze fracture replicas, hepatocyte gap junctions show arrays or plaques of 8.5–9.5 nm intramembrane particles cleaving with the P face (Fig. 1B); complementary pits appear on the E face. In the center of the fractured particles a dimple is commonly discernible that presumably represents the central aqueous lumen of the gap junction channel. Channels of isolated gap junctions can form a regular hexagonal array, and application of Fourier transform techniques reveal substructure in which each hemichannel or connexon is made up of six subunits.[2] The hemichannels or connexons crossing each plasma membrane protrude into the extracellular gap, where they connect to those in the other half of the junction to form the complete aqueous channel[3] (Fig. 1C).

JUAN C. SÁEZ, MICHAEL V. L. BENNETT, and DAVID C. SPRAY • Department of Neuroscience, Albert Einstein College of Medicine, Bronx, New York 10461.

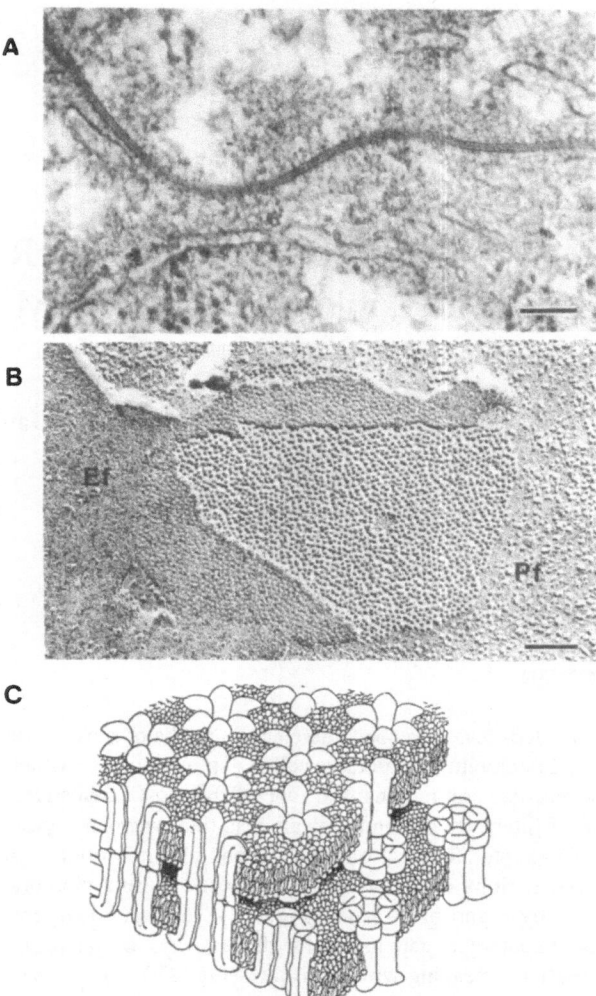

Figure 1.   Hepatocyte gap junctions. (A) Gap junctions from adjacent hepatocytes in thin section show parallel membrane appositions. (B) Freeze fracture replica of hepatocyte plasma membrane. Gap junctions are commonly seen as plaquelike arrays of *P*-face particles 8.5–9.5 nm in diameter; complementary pits are on the *E*-face. (C) Models of gap junction structure have the following as common features: six subunits cross the plasma membrane of one cell, forming a connexon that protrudes into the extracellular gap, where it connects with a connexon of an adjacent cell, thereby forming an intercellular channel. A cluster of these channels forms a gap junction plaque (from Makowski et al., 1984). Gap junctions in (A) and (B) were closely located to bile canaliculii which served as a marker of hepatocyte plasma membrane (calibration bar: 0.1 μm).

In most excitable cells, ionic flow through gap junction channels serves to synchronize the electrical activity. Furthermore, these channels are permeable to ions and molecules with molecular weights below about 1,000 Da. Molecules with a range of signaling functions can diffuse from one cell to another, driven by the electrochemical gradient across the junction. Therefore, the functions of the gap junctions in various cell types depend on the roles of whichever diffusible molecules are present.

Gap junctions from liver and various other tissues have been isolated by similar procedures that include solubilization of the nonjunctional membrane with detergent or alkali, and isolation of the junctional membranes in sucrose density gradients.[4] Fractions enriched in gap junctions have been identified by electron microscopy and their protein components analyzed by polyacrylamide gel electrophoresis (PAGE). These gap-junction-enriched fractions have been utilized to raise antibodies that recognize the gap junction proteins and also to obtain sequences of the proteins. The availability of antibodies and partial primary structure made it possible to obtain the complementary DNA encoding the 27 kDa protein and low stringency hybridization made it possible to find several other cDNAs. Through these studies, it has become clear that gap junctions are formed by a family of homologous but clearly distinct proteins.[5-7] Presumably, each cell type expresses the gap junctional protein(s) that provide a channel structure appropriate for the function and regulation required. In hepatocytes two major proteins are expressed, 27 kDa and 21 kDa, which are colocalized in the same junction,[8] but it remains to be determined whether the proteins form homomeric or heteromeric channels.

Communication between cells mediated by gap junctions depends on gating, which is a rapid process that determines the fraction of channels that are open, and on expression and removal, which are slower processes that determine the total number of channels between the cells. Our main interest has been to study the effect of signals transduced at the cytoplasmic membrane on both short- and long-term regulation of hepatocyte gap junctions.

2. SHORT-TERM REGULATION

2.1. Glucagon Effect on Junctional Conductance

In hepatocytes the transduction of glucagon signals to second messenger systems probably represents one of the best-characterized hormone–hepatocyte interactions. Glucagon causes rapid increase (max. effect at 5 min) of adenylate cyclase activity and cAMP levels in hepatocytes.[9] cAMP binds to the regulatory subunit of the cAMP-dependent protein kinase (cAMP-dPk), releasing the catalytic subunit that phosphorylates various protein substrates, including several metabolic regulatory enzymes.

We have reported that in primary cultures of rat hepatocytes the addition of membrane-permeant cAMP derivatives increases junctional conductance (g_j) over a period of 3 to 5 min.[10] Similarly, the addition of glucagon increases g_j (Fig. 2), supporting the view that elevation of intracellular cAMP levels increases g_j. Because the majority of cAMP effects are mediated by cAMP-dPk, we microinjected through the recording pipettes in each cell a protein inhibitor of cAMP-dPk (Walsh inhibitor) to prevent the

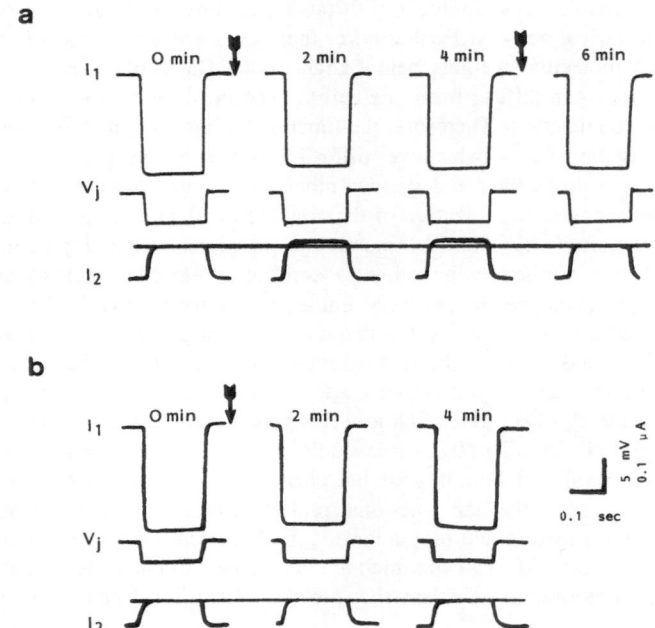

Figure 2. Glucagon increases junctional conductance (g_j), and its effect is prevented by the Walsh inhibitor. Representative regions of chart records from single experiments of glucagon (20 μg/ml) application with (b) or without (a) prior injection of Walsh inhibitor (W.I.) into both cells of a pair. Time is indicated in minutes. I_1, current in cell 1; V_j, voltage in cell 1, the transjunctional voltage. I_2 with reference line, current in cell 2 equal to junctional current since potential in cell 2 was held constant. g_j increased after glucagon addition (first arrow), and was reversed when excess glucagon was washed out (second arrow). When W.I. was injected through the recording pipettes, the response to glucagon was blocked (lower record). (Modified from Ref. 10.)

glucagon effect. The effect of glucagon addition was completely prevented and the g_j was slightly reduced below control after about 5 min of recording (Fig. 2). These results suggest that the glucagon effect on g_j was mediated by phosphorylation of proteins forming gap junction channels and that it was not due to a direct action of cAMP on the channels.

2.2. Phosphorylation of the 27 kDa Gap Junction Protein

The major protein component of isolated gap junctions from rat or mouse liver has an electrophoretic mobility in PAGE of 27 kDa.[4] We have shown previously that the 27 kDa protein contained in purified junctions is phosphorylated by the catalytic subunit of cAMP-dPk.[10] To prove that this covalent modification occurs in intact cells, freshly isolated rat hepatocytes were radiolabeled with [^{32}P]PO$_4$, and the 27 kDa protein was

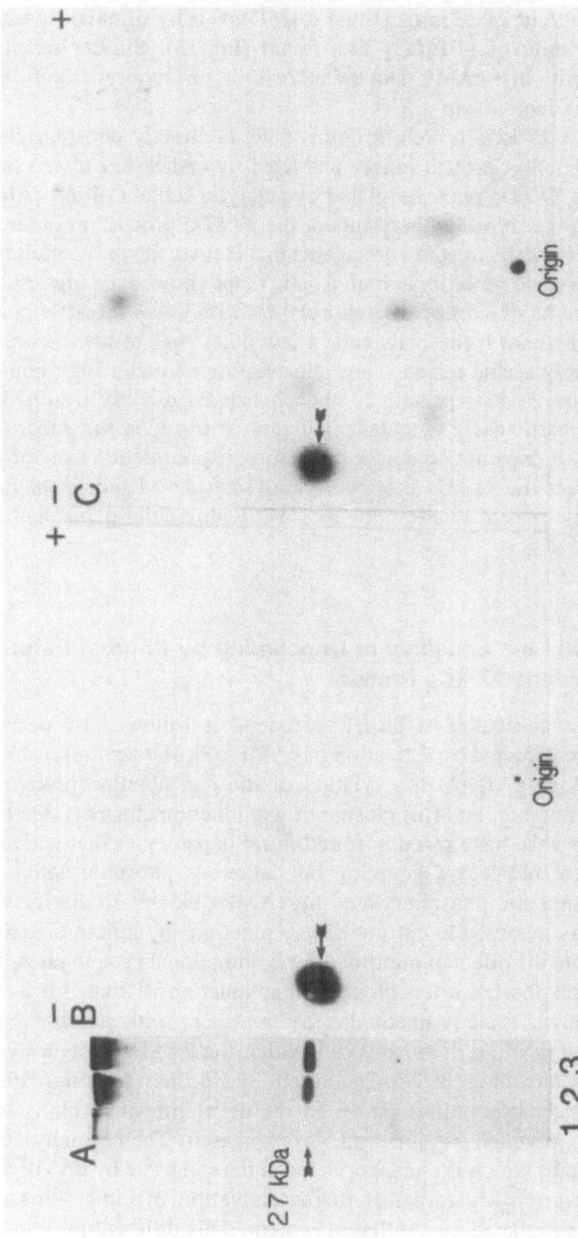

Figure 3. Forskolin enhances phosphorylation of the 27 kDa protein primarily within the same tryptic peptide phosphorylated by cAMP-dPk in isolated junctions. Cells labeled with [^{32}P]PO$_4$ were treated with forskolin (25 μM) and disrupted after 5 min; the 27 kDa protein was immunoprecipitated. (A) Autoradiograph of the immunoprecipitated 27 kDa protein separated in 12.5% PAGE. Precipitation with preimmune serum from untreated cells (lane 1), and with anti 27 kDa protein antibody from untreated cells (lane 2) and forskolin-treated cells (lane 3). The 27 kDa is a phosphoprotein (lane 2), and its phosphorylation state is enhanced by a treatment that elevates cAMP concentration (lane 3). Two-dimensional tryptic fingerprints of (B) the 27 kDa protein phosphorylated in isolated gap junctions by the catalytic subunit of cAMP-dPk and of (C) the 27 kDa protein immunoprecipitated from [^{32}P]PO$_4$-labeled cells after 5 min treatment with forskolin (25 μM). Arrows show the main tryptic phosphopeptide, whose mobility is identical in the two maps.

immunopreciptated and analyzed for radioactivity content. The 27 kDa protein was labeled indicating that it is a phosphoprotein (Fig. 3). In cells prelabeled with [32P]PO$_4$ and treated with forskolin to increase intracellular cAMP levels by stimulating adenylate cyclase, higher incorporation of [32P]PO$_4$ was found (Fig. 3). Similar results were obtained in cells treated with 8Br-cAMP, and the increase in phosphorylation followed a time course similar to the increase in g_j.[10]

To study whether the 27 kDa protein in intact cells is directly phosphorylated by cAMP-dPk (and not by another protein kinase activated by cAMP-dPk), we compared the tryptic fingerprints of 27 kDa phosphorylated by catalytic subunit of the cAMP-dPk in isolated junctions with the tryptic fingerprints of the 27 kDa protein immunoprecipitated from [32P]PO$_4$-labeled cells treated with agents that elevate the intracellular cAMP levels. A peptide with the same mobility in both tryptic maps showed the highest content of [32P]PO$_4$, and the patterns of sites of phosphorylation with lower incorporation were similar, although in the material from intact cells a few other sites of low incorporation were present (Fig. 3). Only serine residues are phosphorylated under both conditions. These results indicate that 27 kDa protein is phosphorylated directly by cAMP-dPk.

A 21 kDa protein, particularly abundant in livers of mice, is the second major junctional protein,[8] and it does not contain a putative phosphorylation site for cAMP-dPk.[7] In isolated junctions the 21 kDa protein was not phosphorylated by the catalytic subunit of cAMP-dPk, by protein kinase-C or by Ca^{2+}/calmodulin-dependent protein kinase.[11]

2.3. A Hepatocyte Cell Line Sensitive to Uncoupling by Phorbol Esters Does Not Express the 27 kDa Protein

In several cell types activation of protein kinase-C is followed by decrease in coupling,[12] which is counteracted by activation of cAMP-dPk.[13] Several mechanisms have been proposed including (i) phosphorylation of the gap junction protein, (ii) a decrease in gap junction number, and (iii) closure of gap junction channels due to a rise of intracellular free Ca^{2+}. We have recently found in rat hepatocytes that activation of protein kinase-C does not block dye coupling but increases phosphorylation of the 27 kDa protein at the same site phosphorylated by cAMP-dPk.[14] To clarify whether protein phosphorylation is involved in closure of gap junctions by kinase-C activation, we attempted to investigate the effect of phorbol ester on junctional protein phosphorylation in a cell type in which phorbol esters block the gap junctions. Clone 9 is a cell line derived from rat hepatocytes that is uncoupled by treatment with phorbol esters or diacylglycerol.[15] We did not find in clone 9 cells either the 27 kDa protein by immunofluorescence or mRNA encoding the 27 kDa protein by Northern blotting with a full-length cDNA (unpublished observation). Thus, in the development of clone 9, there must have been a change in which gap junction was expressed. The normally abundant 27 kDa gap junction protein, which is phosphorylated at the same site by cAMP-dPk and protein kinase-C, and presumably responds to the activation of either kinases with increased conductance, is replaced by a different protein. This different protein may be phosphorylated differentially by the two kinases or regulated by a different mechanism.

3. LONG-TERM REGULATION

3.1. cAMP Prolongs the Lifetime of Gap Junctions

Freshly isolated rat hepatocytes are well coupled, but coupling decreases abruptly beginning 5 to 8 hr after plating cells on plastic culture dishes in physiological saline containing insulin and fetal calf serum (Fig. 4). Loss of coupling is associated with the

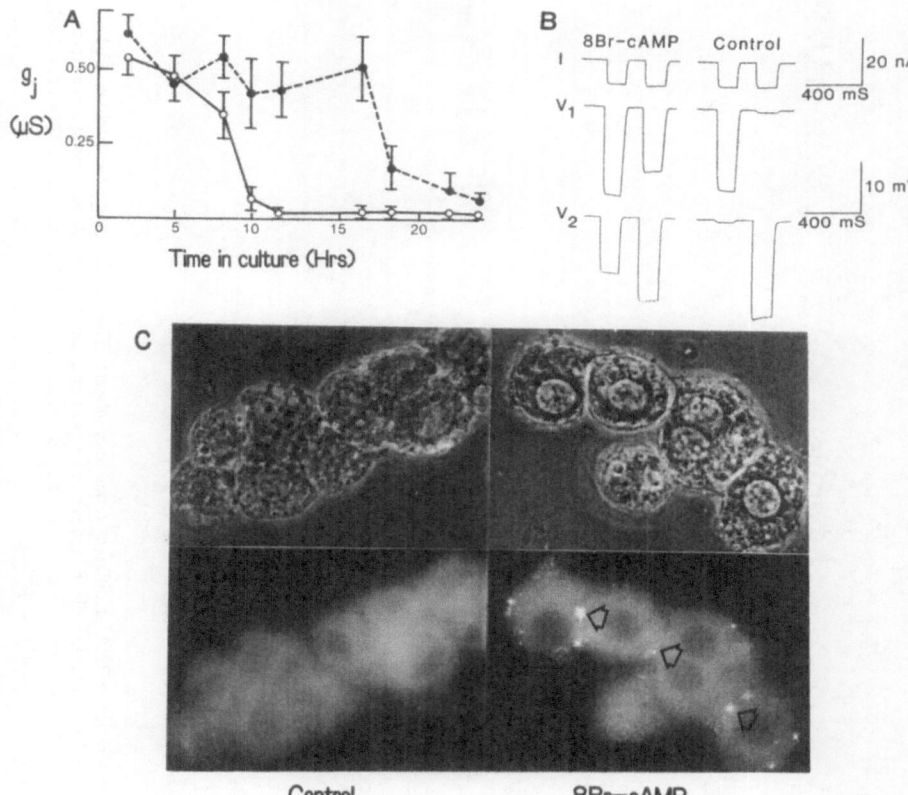

Figure 4. 8Br-cAMP delayed the disappearance of gap junctions in primary cultures of adult rat hepatocytes. (A) Time course of reduction in junctional conductance (g_j). Open circles and continuous lines: cell pairs cultured in control medium (Waymouth's medium plus 10% fetal calf serum and 25 μg/l insulin). Filled circles and dashed lines: cell pairs from cultures maintained in control medium supplemented with 1 mM 8Br-cAMP 45–60 min after plating. Each point represents the mean g_j value for 35 cell pairs in each of 5 sister cultures; bars represent standard deviations. The time axis represents time in culture after the dissociation procedure had been completed. (B) Records of electrical coupling 16 hr after plating. The traces represent current (1st pulse in cell 1, 2nd pulse in cell 2) and voltage in the first (V_1) and second (V_2) cell, respectively. In control medium the cells were weakly coupled (g_j values for these records: control, 0.03 μS; 8Br-cAMP, 0.7 μS). (C) Fluorescence micrographs of representative untreated and 8Br-cAMP (1 mM)-treated hepatocytes, respectively. Virtually all abutting cAMP-treated cells showed punctate fluorescence at appositions (arrows), while very few spots were seen at the interfaces of untreated cells.

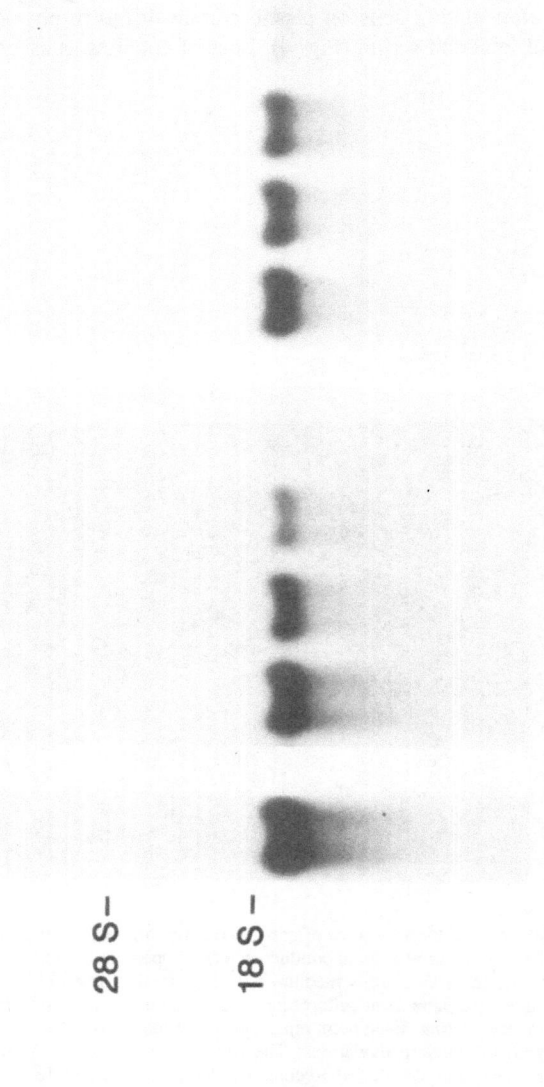

Figure 5. Levels of gap junction mRNA at different times after plating. The mRNA disappeared within 16 hr in primary hepatocytes cultured in control medium. In cells treated with (1 mM) 8Br-cAMP the mRNA was lost at a slower rate, which is especially evident at 9–16 hr. Densitometric measurements showed gap junction mRNA level in 8Br-cAMP-treated cells to be at least twice that of controls at 9 and 16 hr after plating.

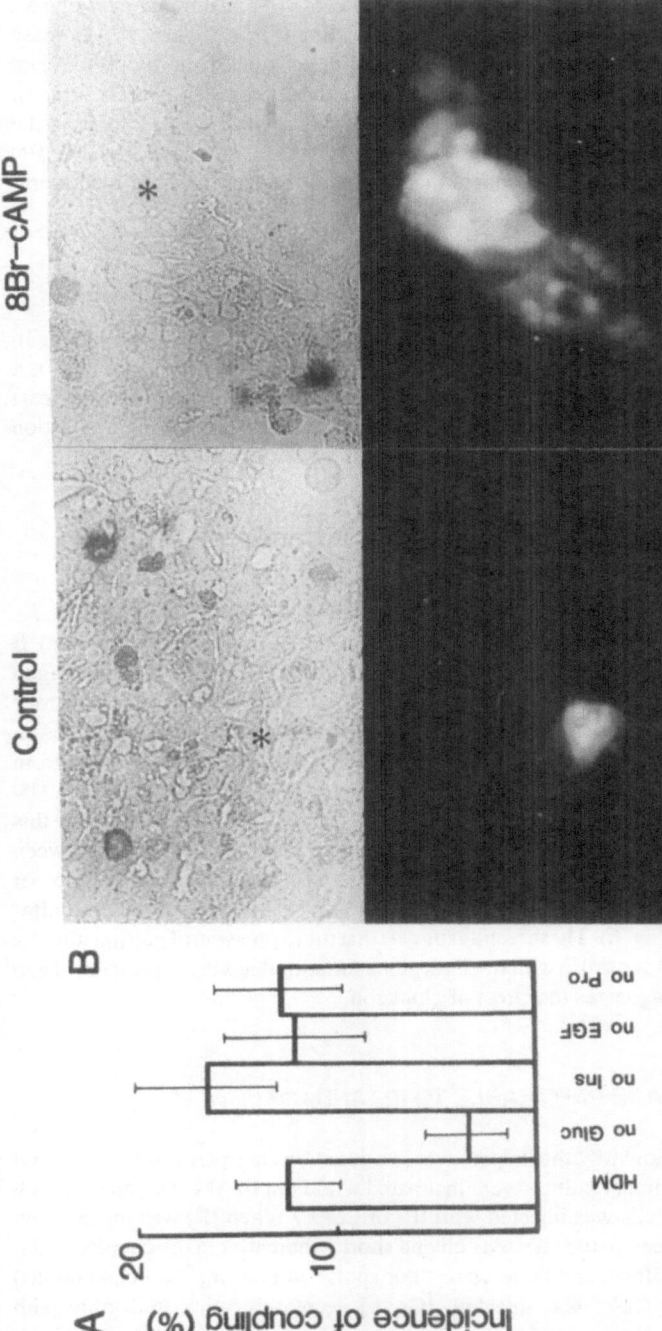

Figure 6. Incidence of dye coupling between hepatocytes in primary culture subjected to various hormonal treatments and treated with 8Br-cAMP. (A) Incidence of dye coupling (Lucifer yellow) 5 days after plating in hormonally defined medium (HDM), or HDM minus glucagon (no Gluc), minus insulin (no Ins), minus EGF (no EGF) or minus prolactin (no Pro). The exclusion of glucagon significantly reduced the incidence of coupling ($p < 0.005$), the exclusion of insulin slightly increased the incidence of coupling, and the exclusion of EGF or prolactin did not affect the number of coupled cells. Each point represents the mean of incidence of coupling for 30 cell pairs in each of 3 sister cultures. (B) In cells cultured in serum-free medium (RPMI, Gibco) for 3 days, 8Br-cAMP (1 mM) added for 10 hr leads to extensive dye coupling compared to untreated (control) cells. Vertical bars indicate standard deviation. The asterisk (*) indicates the site of injection in the field.

disappearance of gap junctions seen electron-microscopically[16] or immunocytological-
ly, and of the 27 kDa protein determined by Western blotting (Fig. 4). Thus, the decrease
in g_j after plating is due to removal and degradation of the channel-forming protein and
not due either to closure of the channels or cellular redistribution of the 27 kDa protein.
The disappearance of functional channels is delayed approximately 8 hr by treating the
cells with membrane-permeant derivatives of cAMP (Fig. 4) but not of cGMP.[16]
Levels of gap junction protein[16] and anatomically identified junctions are also maintained by 8Br-cAMP (Fig. 4).

3.2. cAMP Increases Stability of the mRNA Encoding the 27 kDa Protein

The level of mRNA encoding the gap junction protein was maintained longer in
cells treated with 8Br-cAMP than in untreated cells (Fig. 5), but 8Br-cAMP did not
detectably increase transcription rate.[16] Apparently, prolongation of coupling is at least
partially due to extension of the lifetime of the gap junction mRNA, allowing translation
of message and assembly of channels for a longer period after plating.

3.3. Glucagon and Insulin Affect Gap Junction Expression
in Opposite Ways

The expression of the 27 kDa protein in primary cultures of adult rat hepatocytes is
turned off within 8 hr after plating in medium containing serum and insulin[16] (see above). If
the medium is replaced by hormonally defined medium [HDM, which contains glucagon
(10 μg/ml), insulin (265 mU/ml), prolactin (2 mU/ml), EGF (50 μg/ml), and trace
elements] the 27 kDa protein is reexpressed more rapidly and more abundantly than in serum
containing medium.[17] Recently, it was reported that exclusion of glucagon or insulin from
the HDM decreases or enhances stability of mRNA encoding the 27 kDa protein.[18]
Omission of other components of HDM did not significantly affect the stability of this
mRNA. Exclusion of single components of HDM tends to affect dye coupling between
hepatocytes (Fig. 6) and mRNA stability[18] in the same direction. Moreover, addition of
8Br-cAMP to serum-free medium causes reappearance of extensive dye coupling after
about 10 hr of treatment (Fig. 6). These results suggest that the expression of gap junctions is
under a delicate hormonal control in which at least glucagon stimulates the expression of gap
junctions and insulin antagonizes the effect of glucagon.

4. GAP JUNCTIONS ARE PERMEABLE TO IP₃ AND Ca²⁺

We have recently shown[19] that hepatocyte gap junctions are permeable to IP_3 and
Ca^{2+}. Hepatocytes in primary culture were incubated with Fura II/AM, and one cell of a
small group of coupled cells was injected with IP_3 or Ca^{2+}. When IP_3 was injected, an
increase in $[Ca^{2+}]$ was seen in the injected cell and shortly thereafter in the coupled cells.
The increase was not uniform and there were "hot spots" in both injected and coupled
cells. In contrast, when Ca^{2+} was injected, $[Ca^{2+}]$ decreased rather uniformly with
distance as it diffused to the coupled cells. These data suggest that both IP_3 and Ca^{2+}

diffuse through gap junctions and that IP_3 causes local release of Ca^{2+} from intracellular stores. In all experiments $[Ca^{2+}]$ was increased by no more than a few hundred nanomolar, which is the range over which $[Ca^{2+}]$ is changed in hepatocytes by hormonal stimulation.

5. DISCUSSION

There are approximately 20 hepatocytes between the vascular axis and the terminal hepatic venule (termed an hemiacinus). Moreover, electrical coupling has been recorded between cells separated by at least 500 μm.[20] Because hepatocytes have a diameter of approximately 20 μm, it is probable that hepatocytes along an hemiacinus are connected by gap junctions.

The effects of cAMP on g_j imply that cAMP and agents that elevate it, including various hormones and neurotransmitters, may quickly increase the strength of coupling between communicating cells. The early increase, presumably involving gating, is reversible and is followed by a later increase that may be due in part to stabilization of the mRNA encoding the gap junction protein, thereby permitting a prolonged synthesis of the channel-forming protein. One would expect increased sharing of intracellular ions and small molecules following cAMP increases. Conversely, agents that decrease cAMP levels, such as insulin,[21] or that decrease ATP concentration, would favor dephosphorylation of the gap junction, which would presumably reduce g_j. A reduction in intercellular coupling might be useful during acute metabolic intoxication,[22] irreversible cell injury, or cell division.

The exocrine function of liver is another physiological target for hormone modulation. Many hepatic enzyme systems are activated via cAMP or Ca^{2+}, causing, for example, increased glycogenolysis and gluconeogenesis.[23,24] Cell communication via gap junctions would tend to equalize the distribution of second messengers including cAMP, IP_3 and Ca^{2+}, causing, for example, increased glycogenolysis and gluconeogenesis. Furthermore, cAMP and presumably diacylglycerol will increase the spread of this response by increasing g_j. Thus, the unit containing the metabolic machinery activated by the hormonal stimulus (e.g. glucagon) is the sum of the contents of the coupled cells in spite of any heterogeneity between cells in terms of receptor or effector content. The net effect of this averaging could be increased glucose release. This interpretation is supported by the fact that glucagon, which stimulates glucose release, also stimulates the expression of gap junctions, whereas insulin antagonizes both its metabolic effect and its effect on g_j.

In addition to its action through the adenylate cyclase system, glucagon activates the IP_3 signal-transduction system in hepatocytes.[25] IP_3 induces release of Ca^{2+} from intracellular stores and causes canalicular contractions which, by their pumping effect, facilitate bile flow in the canalicular system.[26] Recently, it has been shown that glucagon mediates the choleresis induced by cholecystokinin.[27] Therefore, open gap junctions, by equalizing the distribution of Ca^{2+} and IP_3,[19] would lead to a more synchronous and effective pumping of bile.

ACKNOWLEDGMENTS.   The experiments described here have been carried out over the last 4 years with the capable technical assistance of Christine Roy and Laura Cipriani. We would like to thank Drs. Elliot L. Hertzberg, Angus C. Nairn, Paul Greengard, Rolf Dermietzel, Lola M. Reid, and John A. Connor for continued collaborative investigation. This work was supported by N.I.H. grants HL38449, NS16524, NS07512 and a grant-in-aid from the New York chapter of the American Heart Association.

REFERENCES

1. Bennett, M. V. L., and Spray, D. C., (Eds.), 1985, *Gap Junctions,* Cold Spring Harbor Laboratory, Cold Spring Harbor, N.Y.
2. Unwin, P. N. T., and Ennis, P. D., 1984, Two configurations of the channel-forming membrane protein, *Nature (London)* **307:** 609–613.
3. Makowski, L., Caspar, D. L. D., Phillips, W. C., and Goodenough, D. A., 1984, Gap junction structure. V. Structural chemistry inferred from x-ray diffraction measurements on sucrose accessibility and trypsin susceptibility, *J. Mol. Biol.* **174:** 449–481.
4. Hertzberg, E. L., 1984, A detergent-independent procedure for the isolation from rat liver, *J. Biol. Chem.* **259:** 9936–9943.
5. Paul, D., 1986, Molecular cloning of cDNA for rat liver gap junction protein, *J. Cell Biol.* **103:** 123–134.
6. Beyer, E. C., Paul, D., and Goodenough, D. A., 1987, Connexin 43: A protein from rat heart homologous to gap junction protein from liver, *J. Cell Biol.* **105:** 2621–2629.
7. Nicholson, B. J., and Zhang, J-T., 1988, Multiple protein components in a single gap junction: Cloning of a second hepatic gap junction protein (M_r 21,000), in: *Modern Cell Biology,* Vol. 7 (E. L. Hertzenberg and R. G. Johnson, Eds.), Alan R. Liss, New York, pp. 207–218.
8. Nicholson, B. J., Dermietzel, R., Teplow, D. B., Traub, O., Willecke, K., and Revel, J.-P., 1987, Two homologous protein components of hepatic gap junctions, *Nature* **329:** 732–734.
9. Exton, J. H., Cherington, A. D., Blackmore, P. F., Dehaye, J.-P., Strickland, W. G., Jordan, J. E., and Chisman, T. D., 1986, Hormonal regulation of liver glycogen metabolism, in: *Protein Phosphorylation,* Cold Spring Harbor Conferences on Cell Proliferation, Vol. 8. (O. M. Rosen and E. G. Krebs, Eds.), Cold Spring Harbor Laboratory, Cold Spring Harbor, N.Y., pp. 503–528.
10. Sáez, J. C., Spray, D. C., Nairn, A. C., Hertzberg, E. L., Greengard, P., and Bennett, M. V. L., 1986, cAMP increases junctional conductance and stimulates phosphorylation of the 27 kDa principal gap junction polypeptide, *Proc. Natl. Acad. Sci. USA* **83:** 2473–2477.
11. Sáez, J. C., Nairn, A. C., Spray, D. C., Hertzberg, E. L., Greengard, P., and Bennett, M. V. L., 1987, The major 27 kD gap junction protein is phosphorylated by cAMP dependent and Ca^{2+}-dependent protein kinases, *Soc. Neurosc.* **13:** 1133.
12. Yamasaki, H., and Mesnil, M., 1987, Cellular communication in cell transformation, in: *Biochemical Mechanisms and Regulation of Intercellular Communication* (M. A. Mehlman, Ed.), Princeton Scientific Publication Co., Inc., Princeton, N.J., pp. 181–207.
13. Enamoto, T., Martel, N., Kanno, Y., and Yamasaki, H., 1984, Inhibition of cell communication between Balb/c 3T3 cells by tumor promoters and protection by cAMP, *J. Cell. Physiol.* **121:** 323–333.
14. Sáez, J. C., Nairn, A. C., Czernick, A. J., Spray, D. C., Hertzberg, E. L., Greengard, P., and Bennett, M. V. L., 1990, Phosphorylation of connexin 32, the main hepatocyte gap junction protein, by cAMP-dependent protein kinase, protein kinase-C and Ca^{2+}/calmodulin-dependent protein kinase. (Submitted for publication.)
15. Yada, T., Rose, B., and Loewenstein, W. R., 1985, Diacylglycerol down regulates membrane permeability: TMB-8 blocks this effect, *J. Membr. Biol.* **88:** 217–232.
16. Sáez, J. C., Gregory, W. A., Dermietzel, R., Hertzberg, E. L., Watanabe, T., Reid, L. M., Bennett, M. V. L., and Spray, D. C., 1989, cAMP extends the functional lifespan of gap junctions in cultured rat hepatocytes, *Am. J. Physiol.* **257:** C1–C11.

17. Spray, D. C., Fujita, Y., Sáez, J. C., Choi, H., Rosenberg, L. C., and Reid, L. M., 1987, Glycosaminoglycans and proteoglycans induce gap junction synthesis and function in primary liver cultures, *J. Cell Biol.* **105:** 541–551.

18. Watanabe, T., Sáez, J. C., Spray, D. C., and Reid, L. M., 1987, Heparin potentiates the regulation by hormones and growth factors of liver-specific mRNA expression in cultured hepatocytes, *J. Cell Biol.* **105:** 356.

19. Sáez, J. C., Connor, J. A., Spray, D. C., and Bennett, M. V. L., 1989, Hepatocyte gap junctions are permeable to the second messenger, inositol 1,4,5-triphosphate, and to calcium ions, *Proc. Natl. Acad. Sci. USA* **86:** 2708–2712.

20. Graf, J., and Petersen, O. H., 1978, Cell membrane potential and resistance in liver, *J. Physiol.* **284:** 105–126.

21. Marchmount, R. J., and Houlay, M. D., 1980, Insulin triggers cAMP-dependent activation and phosphorylation of a plasma membrane cAMP phosphodiesterase, *Nature (London)* **286:** 904–906.

22. Sáez, J. C., Bennett, M. V. L., and Spray, D. C., 1987, Carbon tetrachloride at hepatotoxic levels blocks reversibly gap junctions between rat hepatocytes, *Science* **236:** 967–969.

23. Exton, J. H., 1980, Mechanisms involved in α-adrenergic phenomena: Role of calcium ions in actions of catecholamines in liver and other tissues, *Am. J. Physiol.* **238:** E3–E12.

24. Garrison, J. C., and Borland, M. K., 1979, Regulation of mitochondrial pyruvate carboxylation and gluconeogenesis in rat hepatocytes via an α-adrenergic adenosine 3′,5′-monophosphate-independent mechanisms, *J. Biol. Chem.* **254:** 1129–1133.

25. Wakelam, M. J. O., Murphy, G. J., Hruby, V. J., and Houslay, M. D., 1986, Activation of the two signal-transduction systems in hepatocytes by glucagon, *Nature (London)* **323:** 68–71.

26. Phillips, M. J., Oshio, C., Miyairi, M., Watanabe, S., and Smith, C. R., 1983, What is actin doing in the liver? *Hepatology* **3:** 433–436.

27. Kaminski, D. L., Deshpande, Y. G., and Beinfeld, M. C., 1988, Role of glucagon in cholecystokinin-stimulated bile flow in dogs, *Am. J. Physiol.* **254:** G864–G869.

Chapter 16

Mechanisms of Frequency Tuning in the Internal Ear

Luis Robles

1. INTRODUCTION

Much of the research work in the internal ear during the last 20 years has been concerned with the mechanisms involved in signal transduction and frequency tuning. It has long been known that the peripheral auditory system operates separating complex stimuli into its component frequencies, and that in this process it not only analyzes the stimulus into components but it also amplifies those frequency components, thus improving its capacity to transduce low-level stimuli. There is recent evidence showing that, in the internal ear, frequency tuning and signal transduction are intimately related processes.

In vertebrates the internal ear is constituted by several auditory and vestibular organs, each of which possesses various specialized structures adapted to their specific functions. In all of these organs, however, the process of mechano-electrical transduction takes place in basically identical receptor cells, named the hair cells. These are cylindrical cells characterized by the presence of a tuft of pseudocilia, known as the stereocilia, at their apical surface. In mammals the auditory organ is the cochlea, a fluid-filled double chamber about 18 mm in length (in the chinchilla), coiled in the shape of a snail. At the basal end of the cochlea the two chambers, the scala vestibuli and the scala tympani, communicate with the middle ear through membrane-covered openings, called

LUIS ROBLES • Departamento de Fisiología y Biofísica, Facultad de Medicina, Universidad de Chile, Santiago, Chile.

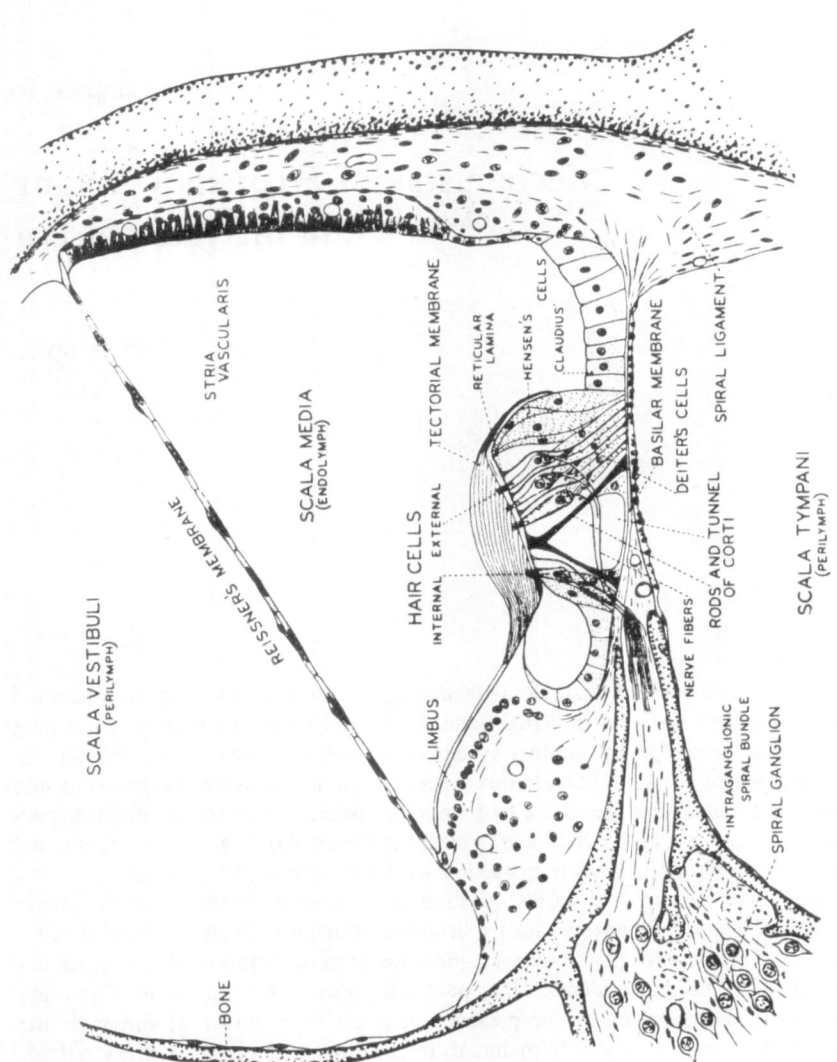

Figure 1. Camera lucida drawing of a cross section of the cochlear partition in the guinea pig cochlea, showing the inner and outer hair cells in the organ of Corti (from Ref. 1, Fig. 2).

the oval and round windows. In the scala media, a third compartment located between the two main ones, is the organ of Corti resting on the BM*. As seen in Fig. 1, the organ of Corti is formed by the receptor cells—distributed in one row of inner and three rows of outer hair cells—the tectorial membrane, and supporting cells and structures.

The acoustic pressure wave, the auditory stimulus, reaches the tympanic membrane at the external ear, and is transmitted through the middle ear ossicles to the cochlea. One of the ossicles, the stapes, in a pistonlike action at the oval window, produces pressure differences between the two cochlear scalae that displace the organ of Corti. From the pioneering work of von Békésy[2] it is known that the vibration of the organ of Corti propagates along the cochlea in the form of a travelling wave, which reaches its maximum amplitude of vibration at different positions along the cochlea depending on the frequency of stimulation. This tonotopic organization of the mechanical response, with the high frequencies localized at the base and the low ones at the apex, is mainly produced by a hundredfold increase of elasticity of the BM from base to apex.[2] The transverse vibration of the organ of Corti translates into a radial relative motion between the apical surface of the hair cells and the tectorial membrane. This radial motion displaces the hair bundle of the hair cells, producing the opening and closing of transduction channels, probably located at the stereocilia, thus generating the hair cells' receptor potential.[3]

The broadly tuned BM responses obtained in the classical measurements of von Békésy contrasted with the sharp tuning observed in recordings from isolated fibers of the auditory nerve.[4] This discrepancy between mechanical and neural frequency tuning remained unresolved for several years and gave origin to theories postulating the existence of a "second" cochlear filter that could sharpen the poorly tuned mechanical responses. Recently, new studies in mammals with minimal surgical damage have obtained mechanical tuning curves as sharply tuned as those recorded from afferents from the same regions of the cochlea,[5-7] thus obviating the need for a "second" filter.

In other vertebrates, like reptiles, amphibia, and birds, auditory nerve fibers are known to have response properties comparable to those of mammals. Cochlear fibers are also sharply tuned and arranged in a tonotopic order. Yet the auditory organs of these animals either have short BMs that do not display location-dependent mechanical tuning,[8] or lack any structure that could function as the BM in mammals.[9] Therefore, mechanisms of cochlear tuning different from those in mammals should be present in these animals.

In this chapter we review some of the new data showing that mechanical tuning can account for the neural frequency selectivity observed in mammals, and recent evidence on the ionic mechanisms determining electrical resonance of individual hair cells in amphibia and reptiles.

2. MECHANICAL TUNING IN MAMMALS

The most complete and reliable data on the mechanical response of the mammalian cochlea *in vivo* have been obtained by applying the Mössbauer technique to measure the

(*)BM: basilar membrane.

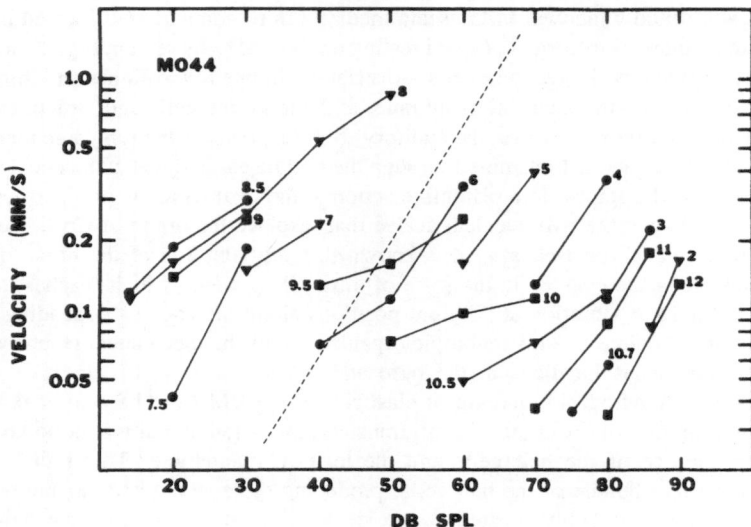

Figure 2. Basilar membrane intensity functions obtained in a single animal for various stimulus frequencies (indicated in kHz). A linear increase of basilar membrane velocity with sound pressure level (SPL) is indicated by the dashed line (from Ref. 10, Fig. 1).

velocity of vibration of the BM at the base of the cochlea.[5,10] A small radioactive source (a metal piece about 80×100 μm, cut from a 6-μm-thick rhodium foil doped with Co-57) is gently placed on the BM through a small hole drilled at the base of the scala tympani. When the source is at rest, most of its γ-radiation has the exact energy level required for absorption by the absorber, a Fe-57-enriched palladium foil mounted between the source and the γ-ray detector. Due to the very narrow spectral lines of emission and absorption, even low-velocity movements of the source cause a Doppler shift in energy that reduces the probability of absorption and increases the rate of detection of γ-rays. The Mössbauer γ-radiation detected and binned into period histograms locked to the stimulus makes it possible to determine the magnitude and phase of the BM velocity.

Using the Mössbauer technique, we measured[10] the mechanical response of the chinchilla BM at a location about 3.5 mm from its basal end. The intensity functions in Fig. 2 display, for one animal at several stimulus frequencies, the BM velocity response as a function of stimulus intensity, in dB SPL.[*] The BM location studied in this animal is most sensitive at 8.5 kHz. This frequency, which requires the lowest intensity for a given response, is known as CF.[**] Figure 2 shows a frequency-dependent nonlinearity similar to that discovered some years ago in the squirrel monkey.[11,12] Intensity functions at frequencies around CF (7–10.5 kHz) are clearly nonlinear, as they have slopes

---

[*]SPL: sound pressure level.
[**]CF: characteristic frequency.

less than unity (dashed line). In contrast, intensity functions at frequencies below 7 kHz and above 10.5 kHz are linear.

The frequency tuning curve corresponding to the BM location under study can be derived from the intensity functions by plotting the intensity required to obtain a given response at the various frequencies. The 0.1 mm/sec isovelocity tuning curve corresponding to the data in Fig. 2 is shown in closed circles in Fig. 3. This tuning curve, obtained early in the experiment, displays sharp tuning, with a minimum at 13 dB SPL and an abrupt intensity increase for frequencies higher than and lower than CF (8.5 kHz). At frequencies above 11 kHz and below 3 kHz the intensity required to obtain the same 0.1 mm/sec velocity response is more than 70 dB higher than at CF.

We compared the tuning of the BM responses with the tuning of frequency threshold curves obtained from chinchilla auditory nerve fibers with appropriate CFs. Figure 4 shows a mean BM isovelocity tuning curve (open circles) computed from the data of five experiments and the corresponding isodisplacement tuning curve (closed circles) for an amplitude of 1.9 nm. For the neural data, the figure includes a neural tuning curve (triangles) computed as the mean of frequency threshold curves from several auditory nerve fibers with CFs between 7 and 10 kHz, after shifting them to correspond to the mean mechanical CF. The comparison shows that, at the region near CF, the mechanical and neural tuning curves are remarkably similar. Comparisons have also been made between mechanical and neural tuning curves in the cat,[6] and between the tuning of BM motion and of inner hair cell receptor potentials in the guinea pig.[13] All of these comparisons support the conclusion that, in mammals, the sharp frequency tuning observed in auditory afferent fibers originates at the mechanical response.

The sharp tuning of the mechanical response, however, is quite labile. Even slight injury to the preparation produces broadly tuned curves with low sensitivity at CF; in addition, initially good preparations always show a progressive loss of sharpness of tuning and a decrease in sensitivity at frequencies around CF. An example of the

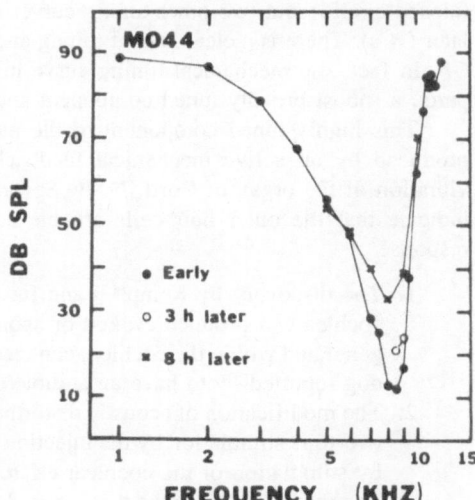

Figure 3. Basilar membrane isovelocity (0.1 mm/sec) tuning curves obtained at different times during an experiment. (•) Tuning curve drawn from the data in Fig. 2, collected within the first hour after placement of the source. (○, ×) Tuning curves obtained 3 hr and 8 hr later, respectively.

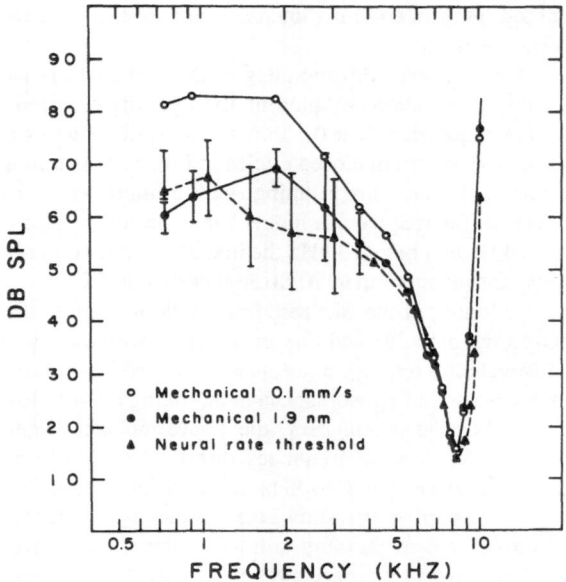

Figure 4. Comparison between tuning at the basilar membrane and in auditory nerve fibers. (○) Average isovelocity tuning curve computed from data of five experiments. (●) Isodisplacement tuning curve for 1.9 nm, computed from the average isovelocity tuning curve. (▲) Average single-unit rate threshold curve based on data from several fibers with CFs between 7 and 10 kHz (adapted from Ref. 10, Fig. 11).

progressive loss of tuning that we routinely observed is shown in Fig. 3. The figure displays the tuning curve obtained within the first hour after placement of the source (closed circles) and two other tuning curves recorded 3 hr later (open circles) and 8 hr later (×'s). There is a clear loss of tuning and sensitivity with time at frequencies about CF. In fact, the mechanical tuning curve in mammals seems to be comprised of two parts, a robust broadly tuned component and a very labile, sharply tuned component.

This highly tuned component of the mechanical response is now believed to be produced by an active mechanical feedback that would reduce the damping in the vibration of the organ of Corti.[14-16] Several lines of experimental evidence seem to indicate that the outer hair cells are the active elements modifying the mechanical response:

1. The discovery by Kemp[17] and further confirmation by others,[18,19] that the cochlea can produce evoked or spontaneous acoustic emissions. These tones, generated within the cochlea, can reach fairly high intensities, as in the case of a dog reported[20] to have an audible emission of 52 dB SPL.
2. The modification of acoustic distortion products, recorded at the ear canal under two-tone stimulation by the injection of electrical current into the scala media or by stimulation of the cochlear efferent system.[21,22]
3. The recent observations that isolated *in vitro* outer hair cells undergo changes of length in response to electrical and chemical stimulation.[23-25]

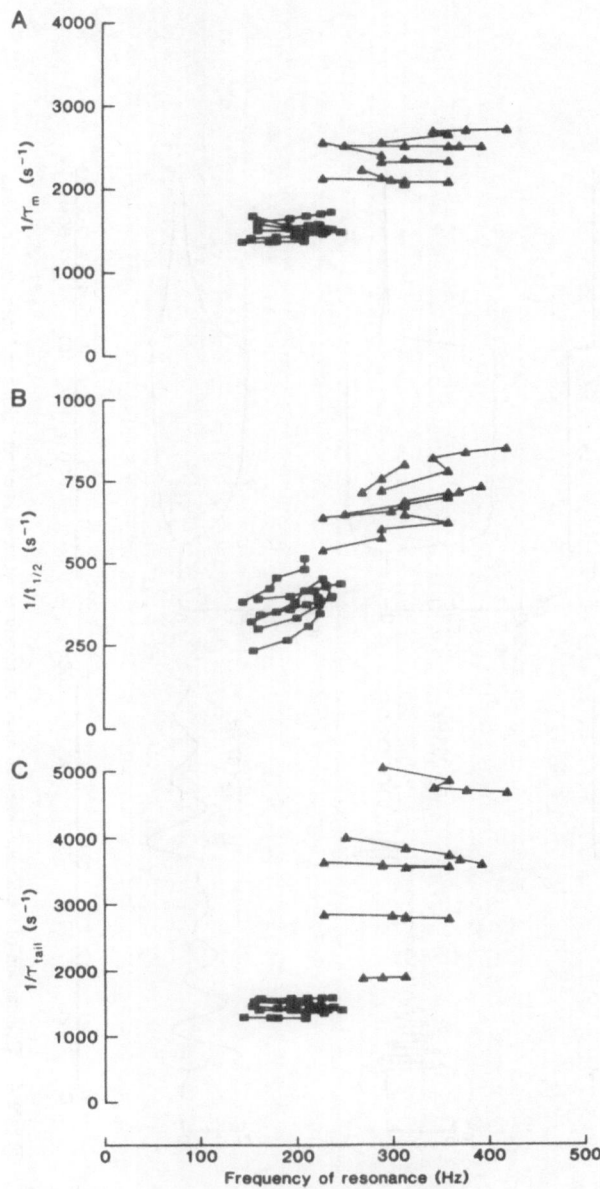

Figure 5. Correlation between kinetic parameters and resonant frequencies for hair cells from the sacculus (■) and amphibian papilla (▲). (A) Reciprocals of the time constants of the onset of the Ca^{2+} current (τ_m). (B) Reciprocals of the times to reach one-half of the steady-state value of the K^+ currents ($t_{1/2}$). (C) Reciprocals of the decay time constants of the K^+ tail currents (τ_{tail}). (From Ref. 32, Fig. 3).

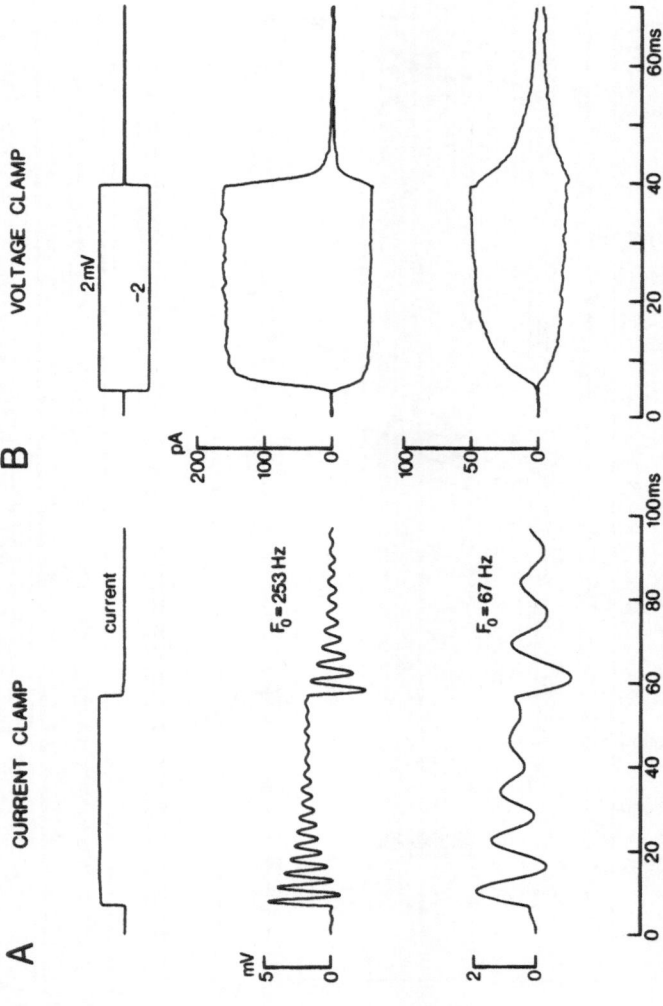

Figure 6. Comparison of membrane-potential resonant responses recorded in current clamp and membrane currents recorded in voltage clamp for two isolated hair cells of different resonance frequency. (A) Average membrane-potential responses to injection of small current steps. Resonant frequencies (F_0) estimated from terminal oscillations. (B) Average membrane currents for 2 mV depolarizing and hyperpolarizing pulses from a holding potential equal to the resting potential. Outward K^+ currents plotted upwards correspond to depolarization pulses. Notice the large difference in kinetics between the K^+ currents of the two cells with different resonance frequency. (From Ref. 33, Fig. 3).

3. ELECTRICAL TUNING IN HAIR CELLS

In 1981 Crawford and Fettiplace,[26] recording intracellularly in the turtle cochlea, reported that hair cells stimulated by a current pulse responded with a damped sinusoidal oscillation in membrane voltage. They found that individual hair cells had different resonant frequencies, which varied with the position of hair cells according to the known tonotopic organization of the turtle cochlea.[27] Their data therefore suggested that in the turtle cochlea frequency tuning is achieved by electrical resonance of the hair cells.

A similar electrical resonance was observed by Lewis and Hudspeth[28] in isolated hair cells from a vestibular organ, the sacculus of the bullfrog. Using whole-cell tight-seal recording techniques,[29] they found that the oscillatory response of the membrane potential could be explained by the interaction of two conductances, a voltage-dependent Ca^{2+} conductance and a Ca^{2+}-activated K^+ conductance.[30] They proposed a model[31] for the interaction between the two ionic currents I_{Ca} and $I_{K(Ca)}$, in which resonance is produced mainly by the delay existing between membrane depolarization and the activation of $I_{K(Ca)}$. The model's prediction that the frequency of oscillation of individual hair cells depends on the kinetic properties of its ionic channels was tested[32] in a study of hair cells from both the sacculus and the amphibian papilla, an auditory organ that responds to a wider range of stimulus frequencies.

In each isolated hair cell we measured both membrane-potential resonant frequencies elicited by current pulses and the kinetics of ionic currents produced by voltage-clamp pulses. The three kinetic parameters investigated (Fig. 5) were (i) the time constant of the onset of the Ca^{2+} current (τ_m), (ii) the risetime to one-half of the steady-state amplitude of the K^+ current ($t_{\frac{1}{2}}$) and (iii) the decay time constant of the K^+ tail current (τ_{tail}). Clear differences were found for these kinetic parameters, in both Ca^{2+} and K^+ currents, between cells from the sacculus and the amphibian papilla (Fig. 5). These results suggest that the resonant frequency of individual hair cells is determined by the kinetics of their ionic channels. A similar study in isolated hair cells from the turtle cochlea,[33] including cells with a wider range of resonant frequencies, also concluded that one of the major determinants of the hair-cell resonant frequency is the kinetics of the K^+ current (Fig. 6).

4. CONCLUSIONS

It is now becoming clear that sharp frequency tuning and tonotopic organization are achieved in the internal ear of different animal species by quite different mechanisms, in spite of the great similarities observed in the tuning curves recorded at their auditory nerves. In mammals,[5,6,10] frequency tuning and tonotopic organization are accomplished by mechanical resonance of the BM at different frequencies along the cochlea. In some reptiles,[33] amphibia,[32,34] and birds[35] they are obtained by electrical resonance of the membrane potential of individual hair cells. Finally, in the alligator lizard, a reptile in which the lengths of free-standing stereociliary bundles vary monotonically with longitudinal position in the basal region of the cochlea,[36] still a third mechanism of tuning has been reported. In this animal neural frequency tuning and tonotopic organization appear to be determined by the mechanical tuning of stereociliary bundles in the individual hair cells[37,38]

The evidence discussed here show that hair cells are not only involved in transduction, as previously believed, but that they also play an important role in determining cochlear tuning. As we have seen, in some vertebrates hair cells directly determine cochlear tuning by electrical resonance of their membrane potentials or by mechanical resonance of their stereociliary bundles. In mammals, the outer hair cells seem to be involved in an active mechanical feedback that produces sharply tuned basilar membrane responses.

ACKNOWLEDGMENTS.   The author thanks M. A. Ruggero and J. Bacigalupo for their valuable comments on the manuscript. Parts of the work reported here were supported by CONICYT, Chile, and NINCDS and NSF, U.S.A.

REFERENCES

1. Davis, H., 1961, Peripheral coding of auditory information, in: *Sensory Communication* (W. A. Rosenblith, Ed.), M.I.T. Press, Cambridge, MA, pp. 119–141.
2. Békésy, G. von, 1960, *Experiments in Hearing* (E. G. Wever, Ed.), McGraw-Hill, New York.
3. Holton, T., and Hudspeth, A. J., 1986, The transduction channel of hair cells from the bullfrog characterized by noise analysis, *J. Physiol.* 375: 195–227.
4. Kiang, N. Y. S., Watanabe, T., Thomas, E. C., and Clark, L., 1965, *Discharge Patterns of Single Fibers in the Cat's Auditory Nerve*, MIT Press, Cambridge, MA.
5. Sellick, P. M., Patuzzi, R., and Johnstone, B. M., 1982, Measurement of basilar membrane motion in the guinea pig using the Mössbauer technique, *J. Acoust. Soc. Am.* 72: 131–141.
6. Khanna, S. M., and Leonard, D. G. B., 1982, Basilar membrane tuning in the cat cochlea, *Science* 215: 305–306.
7. Robles, L., Ruggero, M. A., and Rich, N. C., 1984, Mössbauer measurements of the basilar membrane tuning curves in the chinchilla, *J. Acoust. Soc. Am.* (Suppl. 1) 76: 835.
8. Peake, W. T., and Ling, A., 1980, Basilar membrane motion in the alligator lizard: Its relation to tonotopic organization and frequency selectivity, *J. Acoust. Soc. Am.* 67: 1736–1745.
9. Lewis, E. R., Leverenz, E. L., and Kojama, H., 1982, The tonotopic organization of the bullfrog amphibian papilla, an auditory organ lacking a basilar membrane, *J. Comp. Physiol.* 145: 437–445.
10. Robles, L., Ruggero, M. A., and Rich, N. C., 1986, Basilar membrane mechanics at the base of the chinchilla cochlea. I. Input–output functions, tuning curves, and response phases, *J. Acoust. Soc. Am.* 80: 1364–1374.
11. Rhode, W. S., 1971, Observations of the vibration of the basilar membrane in squirrel monkeys using the Mössbauer technique, *J. Acoust. Soc. Am.* 49: 1218–1231.
12. Rhode, W. S., and Robles, L., 1974, Evidence from Mössbauer experiments for nonlinear vibration in the cochlea, *J. Acoust. Soc. Am.* 55: 588–596.
13. Sellick, P. M., Patuzzi, R., and Johnstone, B. M., 1983, Comparison between the tuning properties of inner hair cells and basilar membrane motion, *Hearing Res.* 10: 93–100.
14. Kim, D. O., Neely, S. T., Molnar, C. E., and Matthews, J. W., 1980, An active cochlear model with negative damping in the partition: Comparison with Rhode's ante- and postmortem observations, in: *Psychophysical, Physiological, and Behavioural Studies in Hearing* (G. V. D. Brink and F. A. Bilsen, Eds.), Delft U.P., Delft, pp. 7–14.
15. Neely, S. T., and Kim, D. O., 1983, An active cochlear model showing sharp tuning and high sensitivity, *Hearing Res.* 9: 123–130.
16. de Boer, E., 1983, Power amplification in an active model of the cochlea: Short-wave case, *J. Acoust. Soc. Am.* 73: 577–579.
17. Kemp D. T., 1978, Stimulated acoustic emissions from within the human auditory system, *J. Acoust. Soc. Am.* 64: 1386–1391.

18. Zurek, P. M., 1981, Spontaneous narrowband acoustic signals emitted by human ears, *J. Acoust. Soc. Am.* **69:** 514,523.
19. Wilson, J. P., and Sutton, G. J., 1981, Acoustic correlates of tonaltinnitus, in: *Tinnitus* (D. Evered and G. Lawrenson, Eds.), Pitman, London.
20. Ruggero, M. A., Kramek, B., and Rich, N. C., 1982, Otoacustic emissions in man and dog: Association with cochlear pathology, *Soc. Neurosci. Abstr.* **8:** 43.
21. Mountain, D. C., 1980, Changes in endolymphatic potential and crossed olivocochlear bundle stimulation alter cochlear mechanics, *Science* **210:** 71–72.
22. Siegel, J. H., and Kim, D. O., 1982, Efferent neural control of cochlear mechanics? Olivocochlear bundle stimulation affects cochlear biomechanical nonlinearity, *Hearing Res.* **6:** 171–182.
23. Brownell, W. E., Bader, C. R., Bertrand, D., and de Ribaupierre, Y., 1985, Evoked mechanical responses of isolated cochlear outer hair cells, *Science* **227:** 194–196.
24. Zenner, H. P., Zimmerman, U., and Schmitt, U., 1985, Reversible contraction of isolated mammalian cochlear hair cells, *Hearing Res.* **18:** 127–133.
25. Ashmore, J. F., 1987, A fast motile response in guinea pig outer hair cells: The cellular basis of the cochlear amplifier, *J. Physiol.* **388:** 323–347.
26. Crawford, A. C., and Fettiplace, R., 1981, An electrical tuning mechanism in turtle cochlear hair cells, *J. Physiol.* **312:** 377–412.
27. Crawford, A. C., and Fettiplace, R., 1980, The frequency selectivity of auditory nerve fibres and hair cells in the cochlea of the turtle, *J. Physiol.* **306:** 79–125.
28. Lewis, R. S., and Hudspeth, A. J., 1983, Voltage- and ion-dependent conductances in solitary vertebrate hair cells, *Nature* **304:** 538–541.
29. Hamill, O. P., Marty, A., Neher, E., Sakmann, B., and Sigworth, F. J., 1981, Improved patch-clamp techniques for high-resolution current recording from cells and cell-free membrane patches, *Pflügers. Arch.* **391:** 85–100.
30. Hudspeth, A. J., and Lewis, R. S., 1988, Kinetic analysis of voltage- and ion-dependent conductances in saccular hair cells of the bullfrog, *Rana catesbeiana, J. Physiol.* **400:** 237–274.
31. Hudspeth, A. J., and Lewis, R. S., 1988, A model for electrical resonance and frequency tuning in saccular hair cells of the bullfrog, *Rana catesbeiana, J. Physiol.* **400:** 275–297.
32. Roberts, W. M., Robles, L., and Hudspeth, A. J., 1986, Correlation between the kinetic properties of ionic channels and the frequency of membrane-potential resonance in hair cells of the bullfrog, in: *Auditory Frequency Selectivity* (B. C. J. Moore and R. D. Patterson, Eds.), Plenum, New York, pp. 89–95.
33. Art, J. J., Crawford, A. C., and Fettiplace, R., 1986, Electrical resonance and membrane currents in turtle cochlear hair cells, *Hearing Res.* **22:** 31–36.
34. Pitchford, S., and Ashmore, J. F., 1987, An electrical resonance in hair cells of the amphibian papilla of the frog *Rana temporaria, Hearing Res.* **27:** 75–83.
35. Fuchs, P. A., and Mann, A. C., 1986, Voltage oscillations and ionic currents in hair cells isolated from the apex of the chick's cochlea, *J. Physiol.* **371:** 31P.
36. Mulroy, M. J., 1974, Cochlear anatomy of the alligator lizard, *Brain Behav. Evol.* **10:** 69–87.
37. Holton, T., and Hudspeth, A. J., 1983, A micromechanical contribution to cochlear tuning and tonotopic organization, *Science* **222:** 508–510.
38. Frishkopf, L. S., and DeRosier, D. J., 1983, Mechanical tuning of free-standing stereociliary bundles and frequency analysis in the alligator lizard cochlea, *Hearing Res.* **12:** 393–404.

II. EXCITATION–CONTRACTION COUPLING IN STRIATED MUSCLE

A. Sodium Channels and Sodium Pump

Chapter 17

Coexistence of Different Types of Sodium Channels in Striated Muscle and Nerve

Richard E. Weiss

1. INTRODUCTION

The multiplicity of sodium channels found in excitable tissues has been represented most commonly as differences in binding affinities of the channels for certain highly specific neurotoxins, notably tetrodotoxin (TTX), saxitoxin (STX), and their respective derivatives. The discovery of functional differences between channel types is more recent and there are still relatively few well-described cases. Examples of functional differences between sodium channels may be categorized as follows: (i) differences between sodium channels in different tissues, for example, innervated muscle and nerve versus heart muscle; (ii) differences between distinct populations of sodium channels in the same cell membrane, and (iii) different open states of a single sodium channel type that interconverts.

Neurotoxins have been the most convenient way to distinguish between sodium channel types, although toxin binding affinity by itself is not necessarily a property of great physiological significance. TTX and STX bind with high affinity to the typical TTX-sensitive sodium channel of nerve and adult innervated skeletal muscle. It has also long been known that action potentials of cardiac muscle and denervated and developing mammalian skeletal muscle are resistant to blockage by TTX and STX [1–5] and that the sodium currents underlying these action potentials have equilibrium dissociation

RICHARD E. WEISS • Department of Pediatrics, Division of Cardiology, University of California–Los Angeles, Los Angeles, California 90024-1743.

constants for TTX and STX 10^2–10^3 times greater than TTX-sensitive sodium currents.[1]

More recently it has been shown that the macroscopic and single-channel properties of the TTX-resistant and TTX-sensitive currents are different.[6–10] These studies provide a well-balanced view of two different types of sodium currents because the data now spans action potential, macroscopic current, and single-channel measurements. What is still lacking is a clear indication of the physiological role of the TTX-resistant channels in skeletal muscle, and if these channels represent the same protein in cardiac and skeletal muscle. There have been different sodium channel types reported in neural tissue and, in some cases, they have been related to cell function. These questions and other examples of sodium channel diversity will be discussed.

2. NERVE

In nerve tissues there are now several reports of diverse sodium channel behavior coexisting in the same cell membrane. In some cases the differences have been interpreted as arising from distinct sodium channel proteins, in other cases, as interconversions of a single type of protein between different open states. As an example of the former case, it is thought that a separate class of channel, called threshold channels, coexists with typical sodium channels in squid giant axons,[11] in cockroach ganglia,[12] and in mammalian brain slices.[13,14]

A suggestion that sodium conductance might participate in the initiation of action potentials came from measurements of membrane potentials of cerebellar Purkinje cell dendrites in mammalian brain slices.[13] Long-lasting sodium-dependent plateau potentials were distinguished from spikes, suggesting two distinct sodium conductances with functions of initiation and driving the rapid upstroke of the action potential, respectively. Both spikes and plateau potentials were sensitive to TTX. In squid giant axons a subpopulation of sodium channels has been described which inactivates very slowly and activates at more negative voltages than typical sodium channels.[11] Although they comprise only a few percent of the total sodium permeability, the kinetic properties of these channels suggest that they predominate in the voltage range where action potentials are initiated, and hence they are called threshold channels. Threshold channels were reported to be sensitive to TTX. Channels with similar behavior have also been reported in rat hippocampal brain slices[14] and in cockroach ganglia.[12]

In an earlier study of *Myxicola* axons, it was found that the sodium current that recovered most rapidly from TTX block had more rapid activation and slower inactivation kinetics.[15] No estimate was given of the fraction of total sodium conductance this subpopulation might represent, though it appeared to be small (less than 10%). Because of the similarity of inactivation kinetics, one might speculate that the slowly inactivating channels in *Myxicola* are a form of the threshold channels found in squid. However, the slowly inactivating current of *Myxicola* giant axons were observed only during recovery from blockage by TTX and not while entering the blocked state. This suggests that the slowly inactivating channels are not normally present and that TTX in some way modified channel behavior. The study did not attempt to see if the altered behavior disappeared with time.

Another sodium channel behavior different from that of typical or threshold channels has been reported in studies of single-channel recordings. The occurrence of small amplitude single-channel openings in recordings of typical sodium channel activity was first reported in nerve by Nagy et al.[16] The small-amplitude openings had different kinetic behavior, and the discussion centered on whether the data represented two distinct channels or two open states of one channel. Because the ratio of small to large events never reversed in patches with few channels, the authors suggested that interconversion between two open states might underlie the observations. Small-amplitude channel openings have also been seen in other tissues and, in the case of TTX-resistant sodium channels discussed later in this chapter, have been interpreted as representing a different channel protein.[9,10]

3 INNERVATED SKELETAL MUSCLE

In normal innervated adult skeletal muscle, sodium channels in the surface membrane appear to be homogeneous with regard to functional properties. The sodium currents of frog skeletal muscle and myelinated nerve have been compared and found nearly identical in vaseline gap voltage clamp measurements.[17] Mammalian and frog skeletal muscle sodium currents have also been compared and found to be similar; mammalian sodium currents activate at more negative voltages and inactivate more slowly.[6,18] Using a double sucrose gap voltage clamp, Caille et al.[19] claimed to measure differences in kinetics between sodium currents arising from the surface membrane and the transverse tubular membranes. This conclusion may not be warranted, however, because the most detailed analyses of the sucrose and vaseline gap voltage clamp techniques concur that transverse tubules cannot be held under voltage control, rendering kinetic measurements of tubular sodium currents grossly inaccurate.[20,21] In addition, Furman et al.[22] have found that the single-channel properties of purified tubular sodium channels from rabbit are very similar to those seen with native sodium channels from rat sarcolemma under similar experimental conditions. On the basis of electrophysiological studies, it must therefore be concluded that there is no compelling evidence for more than one type of sodium channel in each type of adult innervated skeletal muscle.

In contrast, toxin-binding studies indicate that there may be diversity between transverse tubular and surface membrane sodium channels. While most of the naturally occurring forms of toxins, for example, TTX, STX, and *Leiurus* scorpion toxin, bind receptors with only one equilibrium dissociation constant[23–25], some modified forms of TTX, and the γ-toxin from the *Tityus* scorpion, are reported to have different affinities for sodium channels originating in the tubular versus surface membranes.[25,26] Jaimovich et al.[25] described differences in binding affinities of tubular and surface membrane sodium channels in frog skeletal muscle for ethylendiamine derivatives of TTX. Although the binding differences may be real, their suggestion that a correlation exists between the binding data and electrophysiological differences in kinetics obtained by the double sucrose gap voltage clamp technique in their own and a previous study[19] may be wrong. As discussed above, it is questionable whether current measurements of transverse tubular origin can be made accurately.

4. CARDIAC MUSCLE

There are two questions dealing with multiplicity of sodium channels in cardiac muscle that are of interest. First, do different types of sodium channels coexist in the membrane of individual cardiac cells? Second, are TTX-resistant sodium channels in cardiac muscle the same entity found in developing and denervated mammalian skeletal muscle?

The first question addresses a long-standing problem in cardiac electrophysiology. The plateau of the cardiac action potential is believed to be supported in part by a TTX-blockable sodium current. The characteristics of such a current must include an overlap of the steady-state activation and inactivation curves. This overlap has led to the assignation "window" current and has been described in sheep Purkinje fibers.[27] This current is controversial because it has not been found in all preparations, for example, in rabbit Purkinje fibers.[28] Macroscopically it appears to be a different current from the sodium current underlying the cardiac action potential upstroke, which shows little overlap of activation and inactivation.

Single-channel studies on dissociated rat ventricular cells by Patlak and Ortiz[29] detected a small component of mean channel current in which channels were slow to inactivate and opened repeatedly during depolarizing pulses. The authors proposed that the slow component was due either to the interconversion of single-channel type or to a separate class of channel. The former possibility was favored since this could be explained by modifying a single rate constant in an existing kinetic model[30] so that it varied between two values, spending the majority of time at the value corresponding to fast inactivating behavior. The subject of channel interconversion versus distinct types will be discussed again in a later section.

Another form of diverse channel activity in heart cells is the occurrence of channel openings of small amplitude in records of typical sodium channel activity.[31–33] Both Cachelin et al.[31] and Kunze et al.[32] reported small event amplitudes that were 60% as large as the typical channel events. Scanley and Fozzard[33] found that small events were approximately one-third the amplitude of typical channel events and that the relative difference in amplitudes was constant over a wide range of voltages. The kinetic properties of the two channels were not significantly different. The authors reasoned that the small- and large-amplitude events likely represented two distinct sodium channels based on two conjectures and the complimentary results. If the two events were different open states of one channel, then the probability of opening into one or the other open state should have been constant. The data showed that the ratio of small to large events was not constant from patch to patch. If two open states are accessible from the same closed state, then the probability of either open state will depend on the presence of the other open state. Analysis of the data indicated, however, that the probability of small-amplitude events was not affected by the presence of large events. There are two criticisms of these conclusions: (i) The analysis was based on only four patches, and (ii) the authors reported observing two instances of transitions from small to large amplitude openings with no apparent closing in between. The latter observation would seem to suggest that interconversion between different open states was occurring.

The most studied form of diversity among sodium channels has probably been sensitivity to TTX and STX. The TTX-resistance of sodium channels in cardiac and developing and denervated skeletal muscle raises the possibility that they represent a single gene product. There are similarities besides toxin sensitivity. For example, the same single-channel slope conductances were measured at 10°C and the amplitudes of open events were similar in rat heart cells[29] and primary rat myotubes.[9] Differences between the skeletal and cardiac sodium channels also exist, at least in the environment of the native membrane. For example, the voltage range over which channels open appears to be more negative in the case of myotubes than for heart cells. Experimental factors could account for this difference since the cardiac channels were observed in cell-attached patches[29,32] and myotube channels in outside-out patches.[9] It is not known how much the membrane environment of one cell type versus another affects channel properties. In the controlled environment of artificial bilayers, Moczydlowski et al.[34] have examined batrachatoxin-treated, TTX-resistant sodium channels from dog heart and denervated rat muscle to find identical single-channel conductances and characteristics of block by both TTX and μ-conotoxin. They concluded that the channels are likely coded for by the same gene. Whether all mammalian TTX-resistant channels share all properties and a common genotype will require more extensive investigations than those reviewed here. This subject is currently being pursued by many laboratories including my own.

5. CHANNEL SUBSTATES VERSUS DISTINCT CHANNELS

One of the most difficult questions to resolve is whether diverse channel behavior represents distinct channel types or interconversion between different open states. Interconversion may sometimes be demonstrated by standard electrophysiological techniques. Matteson and Armstrong[35] described sleepy channels as a different open state of the typical sodium channel in squid axons. Sleepy channels were distinguished by their slow activation and virtual lack of inactivation. The proportions of the total membrane current exhibiting the two behaviors shifted in a complementary fashion as temperature was varied or test pulses were preceded by large positive prepulses. The complementary shift in proportion is a persuasive statistical argument that sleepy and typical channels are different conformations of the same channel protein.

To answer definitively the question of channel states versus distinct proteins, the techniques of molecular biology would appear to hold the most promise. Using the methods of mRNA expression in *Xenopus* oocytes, or other expression systems, it is possible to identify channel electrical behavior with individual gene products. The cDNA for the channel protein must first be cloned, and this has been done for eel (*Electrophorus*) electroplax sodium channels[36] and for rat brain sodium channels.[37−39]

The primary structures of the eel sodium channel[36] and three rat brain sodium channel large polypeptides (α-subunit) are different.[37−39] The functional characteristics of the rat brain type II channel expressed in oocytes have been measured by Stuhmer et al.,[40] who identified it as the typical sodium channel found in mammalian peripheral nerve and innervated skeletal muscle. Only one open state behavior was reported. The molecular biology of the type II α-subunit was also investigated by Auld et al.[39] and found to contain 36 nucleotide sequence differences compared to the data of Noda et al.,[37] and hence was called type IIA. When expressed in oocytes alone, the inactivation kinetics of the type IIA α-subunit were slower than for typical channels. However, when low molecular weight mRNA was coinjected with type IIA mRNA, the inactivation kinetics were fast. The authors suggested that the low molecular weight mRNA codes for subunits of the sodium channel or for other factors which might modify the behavior of the α-subunit. The discrepancy with the results of Stuhmer et al.[40] is not resolved, however.

Molecular biological techniques have thus far not produced definitive examples of channel interconversion between states, but have shown that behavior can be modified by intrinsic cellular factors. We also know now that there are at least several distinct genes that code for sodium channels. From rat brain alone there are three identified cDNA's for the α-subunit of sodium channels. It remains to be demonstrated if type I or III sodium channels represent threshold channels.

6. DENERVATED AND DEVELOPING SKELETAL MUSCLE

The great majority of work on diversity of sodium channels in skeletal muscle has focused on TTX sensitivity. However, there have also been observations of small-amplitude sodium channels in skeletal muscle similar to those reported in nerve. DeCino and Kidokoro[41] reported channel openings of two magnitudes in single-channel studies of *Xenopus* myotubes in culture. The amplitude of the small events was approximately 70% of the typical open-channel current. A similar observation has been made in excised patches from mammalian rat myotubes in primary culture (Weiss, unpublished observation). No selectivity analysis or kinetics of the small events was performed in either case. However, the excised patches were exposed to CsF on the cytoplasmic side and the activity did not seem to wash out, which suggests that the currents were not carried by sodium ions through potassium channels or calcium channels. The functional significance of small channels, as well as their identity as unique proteins, remains unknown.

In developing mammalian and avian skeletal muscle, the TTX sensitivity of action potentials and the sodium currents that underlie them is initially quite low, but increases as the cells mature.[5,8,42-47] Innervation positively influences the development of TTX sensitivity[48] and, in adult fibers, denervation causes the reemergence of TTX-resistant sodium channels.[3,4,6] Pappone[6] measured sodium currents in a vaseline gap voltage clamp[20] from innervated and 5–7 day denervated rats. TTX-resistant current accounted for 25–30% of the total sodium current in denervated fibers. Kinetics of the combined currents (i.e., in the absence of TTX) were measured; both activation (m) and inactivation (h) parameters were shifted by approximately -10 mV, whereas neither activation nor inactivation time constants were affected. More extensive investigations of the properties of TTX-resistant currents in both developing and denervated muscle are needed to determine the physiological function of these currents.

7. PATCH-CLAMP STUDIES OF SODIUM CHANNELS IN DEVELOPING RAT MUSCLE

Patch-clamp studies of whole-cell sodium currents in rat myotube preparations have produced data that support opposing hypotheses for the shift in TTX sensitivity during development. Frelin et al.[44] showed a dose–response curve that was fit with a single binding affinity of TTX for its receptor. From this data, it was proposed that sodium channels underwent a gradual shift in TTX affinity during development. Gonoi et al.,[8] using the same technique, found a dose–response relationship suggesting two populations of toxin receptors with different affinities for TTX, suggesting that a shift in the proportion of channel types occurs with development. In an abbreviated repetition of these whole-cell clamp experiments, Weiss and Horn[9,10] obtained results consistent with the latter hypothesis. One population of channels was blocked with a K_d of about 10 nM and the other with a K_d greater than 1 μM. The study then proceeded to look for differences in the single-channel properties of the two channels.

Single-channel currents were measured in outside-out patches from primary cultures of myoblasts and myotubes. The bath solution was (mM): 160 NaCl, 2 CaCl$_2$, 1 MgCl$_2$, 10 HEPES, 5 glucose at pH 7.3. The pipette solution was (mM): 140 CsF, 10 NaCl, 5 EGTA, 10 HEPES at pH 7.4. Experiments were performed at 9.5°C. Unless otherwise indicated, patches were held at −90 to −110 mV and test pulses, to −40 mV, were preceded by a prepulse 20 mV negative to the holding potential.

Figure 1 shows selected single-channel records from excised patches from a myoblast (A) and a myotube (B) in the absence and presence of 156 nM TTX. The records in the top half of Fig. 1 reveal single-channel currents with at least two amplitudes. At this test voltage, −40 mV, the large events were approximately 1.4 pA, and the small events (arrowheads) were approximately 1.0 pA. Both small- and large-amplitude channels were seen in seven out of ten patches from myoblasts and seven out of eight patches from myotubes. One patch from a myotube contained no small events and three patches from myoblasts contained no large events. TTX (5–500 nM) reversibly reduced the frequency of the single-channel openings. In the lower half of Fig. 1, 156 nM TTX has preferentially eliminated large amplitude channels, showing that TTX-sensitive channels have a higher open-channel conductance than TTX-resistant channels. This dose of TTX was calculated to block greater than 94% of the TTX-sensitive and less than 13% of the TTX-resistant sodium channels from the equation:

$$I_{na}/I_{na,max} = K_d/(K_d + [TTX])$$

The K_ds were assumed to be 10 nM and 1 μM.

The distribution of channel amplitudes was bimodal as shown by the amplitude histogram of one experiment in Fig. 2. The data were fit by the weighted sum of two Gaussian densities that yielded mean amplitudes of the large (μ_L) and small (μ_s) events, their standard deviations (σ_L and σ_s, respectively), and the fraction of large events (ω_L). In this experiment, when 156 nM TTX was added to the bath, ω_L changed from 0.53 ± 0.05 to 0.12 ± 0.05. TTX did not shift the mean current amplitudes or significantly affect their standard deviations. In cases where only one population of channel was observed, distributions were well fit by single Gaussian densities comparable to one of the two peaks in Fig. 2.

Mean amplitudes from Gaussian fits like those shown in Fig. 2 were plotted to yield the single-channel current–voltage ($I–V$) relation shown in Fig. 3. The graph points for each type of channel were fit to straight lines by weighted linear regression, and the two slopes were significantly different. The slope conductance for TTX-sensitive channels was 11.7 pS and 9.0 pS for TTX-resistant channels. These values are comparable to

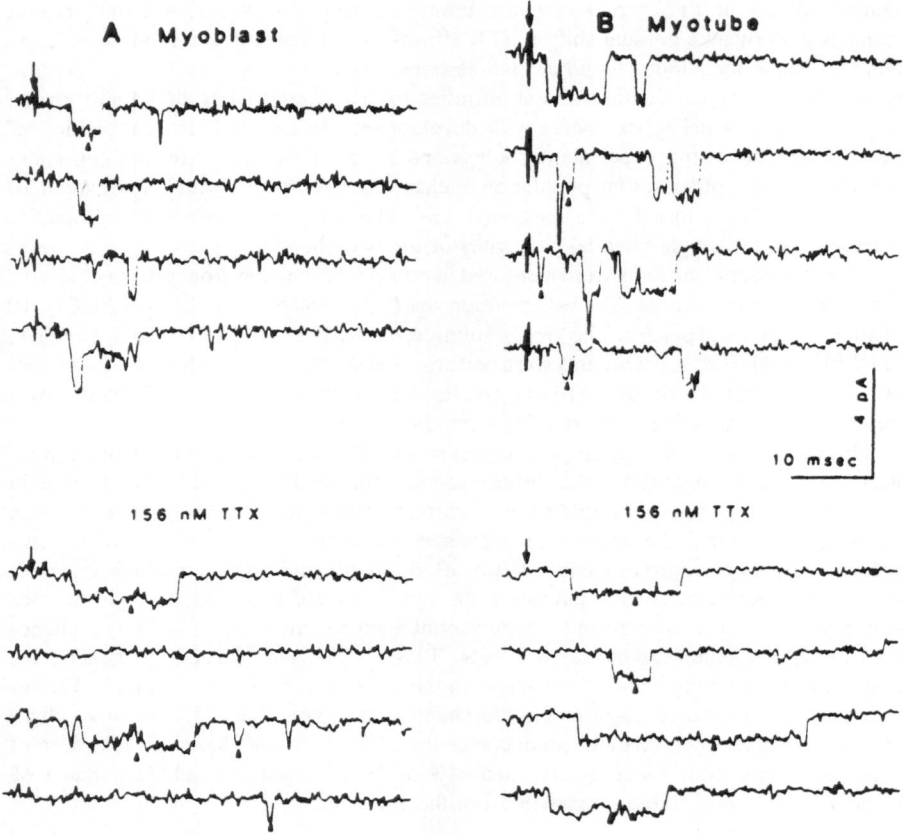

Figure 1. Single-channel currents from a myoblast (A) and a myotube (B) in the absence and presence of 156-nM TTX. The test voltage was -40 mV; the downward arrow marks its beginning. The holding and prepulse potentials were: (A) $V_H = -110$ and $V_{pre} = -130$ mV; (B) $V_H = -100$ and $V_{pre} = -120$ mV. Both large- and small-sized currents (the latter indicated by upward arrowheads) were seen in both patches. The proportion of large events was estimated to be 0.62 and 0.86 for the myoblast and myotube patch, respectively, at this membrane potential. The addition of TTX reduced the frequency of all currents, and the few openings had predominantly small-sized currents. The estimate of the total number of channels was (A) three and (B) nine. Experiments numbers: (A) T42, (B) T33. (Modified from Ref. 9).

those obtained from 13 other experiments where the linear regression of all data points gave slope conductances of 12.1 ± 1.3 pS (95% confidence interval) and 9.8 ± 0.6 pS for large- and small-amplitude events, respectively. Within the voltage range investigated, the individual means μ_L and μ_s never overlapped. The question of whether TTX-resistant sodium channels are a second open state of TTX-sensitive channels, can be addressed by two lines of evidence. First, in one of eight myotube patches only large events were seen, and in three of ten myoblast patches only small events were seen. The ratio of large to small events varied widely in patches from cells in the same culture dish

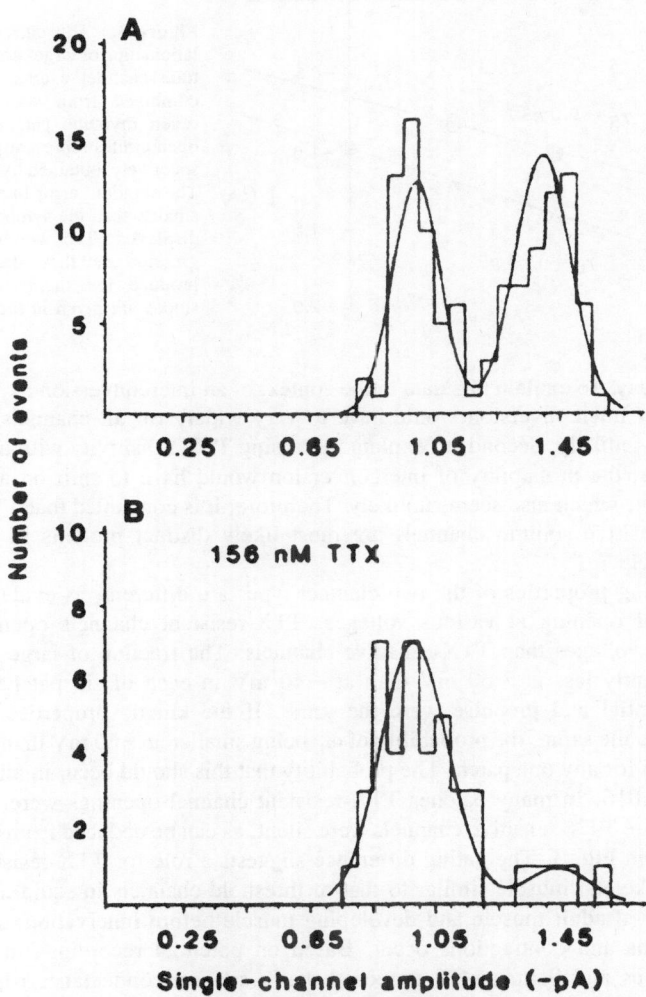

Figure 2. Effect of TTX on the amplitude distribution of individual current amplitudes. The currents were recorded from a 3-day-old myoblast at voltages: $V_H = -90$, $V_{pre} = -110$, $V_{test} = -40$ mV. Two-hundred twenty-five records were obtained before (A) and after (B) addition of 156 nM TTX. Each amplitude is the mean current from a single event. The five parameters for the double-Gaussian curves were estimated by maximum likelihood from the individual amplitudes. In the absence of TTX, $\mu_S = 0.99$ pA, $\sigma_S = 0.08$ pA, $\mu_L = 1.39$ pA, $\sigma_L = 0.09$ pA, and $\omega_L = 0.53$. In the presence of TTX, $\mu_S = 0.98$ pA, $\sigma_S = 0.11$ pA, $\mu_9 = 1.43$ pA, $\sigma_L = 0.5$ pA, and $\omega_L = 0.12$. Experiment number: T35. (Modified from Ref. 9).

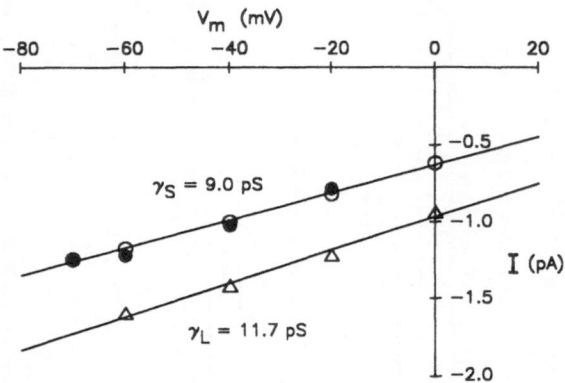

Figure 3. The current–voltage relationships of large- and small-amplitude channel events. The data are combined from six myoblast and seven myotube patches. Filled and open circles represent μ_L and μ_S, respectively, obtained by statistical fits. The standard error bars were usually smaller than the symbols and are not displayed. The weighted linear regression used these standard errors to produce the theoretical lines; the slopes are given in the text.

on the same day. To explain this data in the context of an interconversion hypothesis, the probability of interconversion would have to vary widely for all channels in a patch, which seems unlikely. Second, to explain increasing TTX sensitivity with age in developing muscle, the probability of interconversion would have to shift on a continuous basis with age, which also seems unlikely. Therefore, it is concluded that TTX-resistant and TTX-sensitive sodium channels are most likely distinct proteins in mammalian skeletal muscle.

The gating properties of the two channel types are different, as evidenced by the probability of opening at various voltages. TTX-resistant channels opened at more negative test voltages than TTX-sensitive channels. The fraction of large events, ω_L, was significantly less at -60 mV than at -40 mV in each of six patches when the holding potential and prepulse were the same. If the kinetic properties of the two channels were the same, the probability of ω_L being smaller at -60 mV than at -40 mV should be 0.5 for any one patch. The probability that this should occur in all six patches is 0.5^6, or 0.016. In many patches TTX-resistant channel openings were observed at -70 mV where TTX-sensitive channels were silent, as can be deduced from the leftmost graph points in Fig. 3. The gating difference suggests a role for TTX-resistant sodium channels in skeletal muscle similar to that of threshold channels in squid axons.[11] In both denervated adult muscle and developing muscle before innervation, spontaneous depolarizations and contractions occur. Based on potential recordings in denervated muscle, Purves and Sakmann[49] proposed that a sodium conductance triggers spontaneous action potentials in denervated muscle. Further studies of the TTX-resistant channels will determine if they have the properties required to generate the triggering conductance.

To correlate channel properties to cell function, the kinetic data need to be obtained under physiological conditions. The advent of the loose patch voltage clamp technique[50] makes such measurements of macroscopic currents feasible. Membrane patches can be voltage clamped without exposing the cytoplasmic side of the membrane to artificial solutions as in the vaseline gap, and with much greater ease and convenience than with the multiple microelectrode techniques.

Figure 4 shows loose patch records obtained from a 12-day denervated EDL muscle

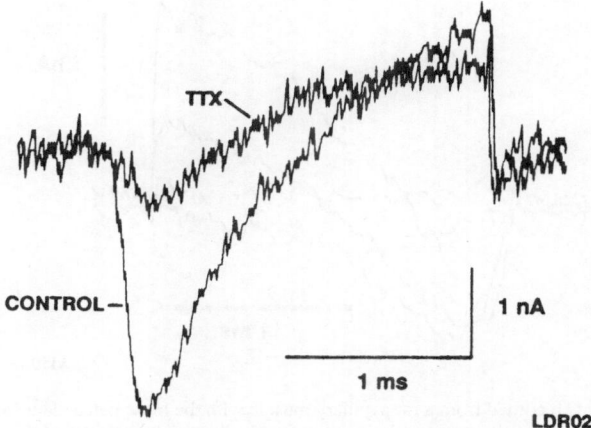

Figure 4. Effect of 150-nM TTX on sodium currents recorded with the loose patch clamp from a 12-day denervated rat EDL muscle at 24°C. The patch was held at -30 mV and the pre- and test pulses were to -60 and 50 mV, respectively, relative to the intrinsic resting potential of the muscle. The prepulse had a 50 msec duration. The mean resting potential of 15 fibers on the surface of the muscle was -83.4 ± 8.3 mV (SD). The solution in the bath and pipette was composed of (mM): 100 NaCl, 35 Na-methanesulfonate, 2 KCl, 2 CaCl$_2$, 1.8 MgCl$_3$, 10 HEPES, 11 glucose (at pH 7.3). The traces were filtered at 10 kHz. Experiment number: LDR02.

from rat in the absence and presence of 150 nM TTX in the bath and patch pipette. The test voltage was the same for both traces and was selected to produce maximum inward current. The records were obtained from different patches, but are representative of the average effect of TTX observed in several patches. The mean is reasonably close to the estimate of 25–30% in previous studies.[6] The kinetic parameters of the TTX-resistant current will be investigated in the presence of sufficient TTX to block TTX-sensitive channels. The data can then be correlated to the conditions that promote spontaneous activity.

The loose patch technique may also be used to voltage clamp syncytial preparations, such as heart muscle.[51] Other voltage clamp methods do not control membrane potential adequately to record sodium currents in ventricular muscle preparations (for review, see Ref. 52). However, because the loose patch technique records from a small area of surface membrane, it can command the membrane potential rapidly. Figure 5 shows a family of single sweep records recorded from a right ventricular papillary muscle from a rat. Test voltages were 25–100 mV depolarizations from the intrinsic resting potential of the cell. (Additional experimental details are provided in the figure legend.) The traces appear to have the expected characteristics of sodium currents seen in other tissues that are under good voltage control. The membrane potential of this cell was not recorded but, based on microelectrode measurements from other experiments, is assumed to be about -75 mV.

With the loose patch technique, it is possible for the first time to examine the kinetics of sodium currents in intact ventricular muscle. Comparison of kinetic parameters obtained from denervated skeletal muscle in the presence of TTX and heart muscle

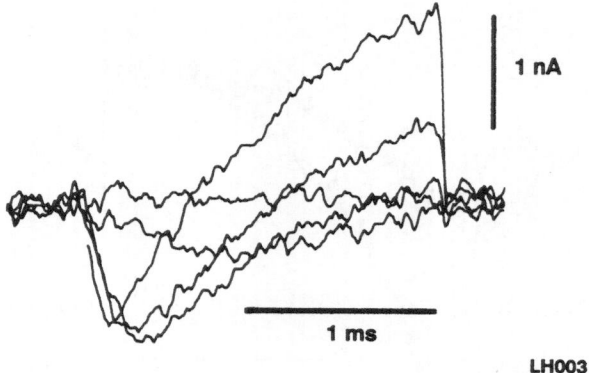

Figure 5.   Currents recorded from a rat papillary muscle with the loose patch clamp at 26°C. There was no holding potential. All voltages are relative to the intrinsic resting potential of the cell. The 50 msec prepulse was −37.5 mV. The test voltages represented are depolarizations of 25, 37.5, 62.5, 75, and 100 mV. The diameter of the patch pipette tip was 13 μm. The calibration bars are 0.75 mA/cm² and 1.5 msec. The traces were filtered at 3 kHz. Experiment number: LR003P4.

will establish if TTX-resistant sodium channels behave equivalently. Of particular interest is a comparison of the voltage range in which the currents activate versus steady-state inactivation (i.e., the existence of "window" currents). Eventually, with comparisons of kinetic data, channel genomes, and the biochemistry of local membrane environments, the question of equivalent function and structure of channels in different cells will be answered.

ACKNOWLEDGMENTS.   I thank Drs. B. Ribalet and J. Talvenheimo for their helpful comments on the manuscript. This work was supported by grants from the Los Angeles affiliate of the American Heart Association (875G1-1), the Dr. Louis Sklarow Memorial Fund (P880125), and the UCLA Laubisch Fund.

REFERENCES

1. Dudel, J., Peper, K., Rudel, R., and Trautwein, W., 1967, The effect of tetrodotoxin on the membrane current in cardiac muscle (Purkinje fibers), *Pflüg. Arch.* **295:** 213–226.
2. Carmeliet, E., and Vereecke, J., 1969, Adrenaline and the plateau phase of the cardiac action potential, *Pflüg. Arch.* **313:** 300–315.
3. Harris, J. B., and Thesleff, S., 1971, Studies on tetrodotoxin-resistant action potentials in denervated skeletal muscle, *Acta Physiol. Scand.* **83:** 382–388.
4. Redfern, P., and Thesleff, S., 1971, Action potential generation in denervated skeletal muscle. II. The action of tetrodotoxin, *Act. Physiol. Scand.* **82:** 70–78.
5. Harris, J. B., and Marshall, M., 1973, Tetrodotoxin-resistant action potentials in newborn rat muscle, *Nature New Biol.* **243:** 191–192.
6. Pappone, P. A., 1980, Voltage-clamp experiments in normal and denervated mammalian skeletal muscle fibers, *J. Physiol.* **306:** 377–410.
7. Bean, B. P., Cohen, C. J., and Tsien, R. W., 1982, Block of cardiac sodium channels by tetrodotox-

in and lidocaine: Sodium current and Vmax experiments, in: *Normal and Abnormal Conduction in the Heart* (A. Paes de Carvalho, B. F. Hoffman, and M. Lieberman, Eds.) Futura, Mt. Kisco, N.Y., pp. 189–206.

8. Gonoi, T., Sherman, S. J., and Catterall, W. A., 1985, Voltage clamp analysis of tetrodotoxin-sensitive and -insensitive sodium channels in rat muscle cells developing *in vitro*, *J. Neurosci.* **5:** 2559–2564.

9. Weiss, R. E., and Horn, R., 1986, Functional differences between two classes of sodium channels in developing rat skeletal muscle, *Science* **233:** 361–364.

10. Weiss, R. E., and Horn, R., 1986, Single-channel studies of TTX-sensitive and TTX-resistant sodium channels in developing rat muscle reveal different open-channel properties, *Ann. N.Y. Acad. Sci.* **479:** 152–161.

11. Gilly, W. F., and Armstrong, C. M., 1984, Threshold channels: A novel type of sodium channel in squid giant axon, *Nature* **309:** 448–450.

12. Kojima, H., Yawo, H., and Kuno, M., 1985, A low-voltage-activated Na channel in the cockroach giant axon, *J. Physiol. Soc. Japan.* **47:** 25a.

13. Llinas, R., and Sugimori, M., 1980, Electrophysiological properties of *in vitro* Purkinje cell dendrites in mammalian cerebellar slices, *J. Physiol.* **305:** 197–213.

14. French, C. R., and Gage, P. W., 1985, A threshold sodium channel in pyramidal cells in rat hippocampus, *Neurosci. Lett.* **56:** 289–294.

15. Goldman, L., and Hahin, R., 1978, Initial conditions and the kinetics of the sodium conductance in *Myxicola* giant axons, *J. Gen. Physiol.* **72:** 879–898.

16. Nagy, K., Kiss, T., and Hof, D., 1983, Single Na channels in mouse neuroblastoma cell membrane, *Pflüg. Arch.* **399:** 302–308.

17. Campbell, D. T., and Hille, B., 1976, Kinetic and pharmacological properties of the sodium channel of frog skeletal muscle, *J. Gen. Physiol.* **67:** 309–323.

18. Adrian, R. H., and Marshall, M. W., 1977, Sodium currents in mammalian muscle, *J. Physiol.* **268:** 223–250.

19. Caille, J., Ildefonse, M., and Rougier, O., 1978, Existence of a sodium current in the tubular membrane of frog twitch muscle fiber: Its possible role in the activation of contraction, *Pflüg. Arch.* **374:** 167–177.

20. Hille, B., and Campbell, D. T., 1976, An improved vaseline gap voltage clamp for skeletal muscle fibers, *J. Gen. Physiol.* **67:** 265–293.

21. Heiny, J. A., and Vergara, J., 1982, Optical signals from surface and T-system membranes in skeletal muscle fibers, *J. Gen. Physiol.* **80:** 203–230.

22. Furman, R. E., Tanaka, J. C., Mueller, P., and Barchi, R. L., 1986, Voltage-dependent activation in purified reconstituted sodium channels from rabbit T-tubular membranes, *Proc. Natl. Acad. Sci. USA* **83:** 488–492.

23. Almers, W., and Levinson, S. R., 1975, Tetrodotoxin binding to normal and depolarized frog muscle and the conductance of a single sodium channel, *J. Physiol.* **247:** 483–509.

24. Catterall, W. A., 1979, Binding of scorpion toxin to receptor sites associated with sodium channels in frog muscle, *J. Gen. Physiol.* **74:** 375–391.

25. Jaimovich, E., Chicheporte, R., Lombet, A., Lazdunski, M., Ildefonse, M., and Rougier, O., 1983, Differences in the properties of Na$^+$ channels in muscle surface and T-tubular membranes revealed by tetrodotoxin derivatives, *Pflüg. Arch.* **397:** 1–5.

26. Barhanin, J., Ildefonse, M., Rougier, O., Sampaio, S. V., Giglio, J. R., and Lazdunski, M., 1984, Tityus γ-toxin, a high-affinity effector of the Na$^+$ channel in muscle, with a selectivity for channels in the surface membrane. *Pflüg. Arch.* **400:** 22–27.

27. Attwell, K., Cohen, I., Eisner, D., Ohba, M., and Ojeda, C., 1979, The steady-state TTX-sensitive ("window") sodium current in cardiac Purkinje fibers, *Pflüg. Arch.* **379:** 137–142.

28. Colatsky, T. J., 1980, Voltage clamp measurement of sodium channel properties in rabbit cardiac Purkinje fibers, *J. Physiol.* **305:** 215–234.

29. Patlak, J. B., and Ortiz, M., 1985, Slow currents through single sodium channels of the adult rat heart, *J. Gen. Physiol.* **86:** 89–104.

30. Chiu, S. Y., 1977, Inactivation of sodium channels: Second-order kinetics in myelinated nerve, *J. Physiol.* **273:** 573–596.
31. Cachelin, A. B., DePeyer, J. E., Kokubun, S., and Reuter, H., 1983, Sodium channels in cultured cardiac cells, *J. Physiol.* **340:** 389–401.
32. Kunze, D. L., Lacerda, A. E., Wilson, D. L., and Brown, A. M., 1985, Cardiac Na currents and the inactivating, reopening, and waiting properties of single cardiac Na channels, *J. Gen. Physiol.* **86:** 691–719.
33. Scanley, B. E., and Fozzard, H. A., 1987, Low-conductance sodium channels in canine cardiac Purkinje cells, *Biophys. J.* **52:** 489–495.
34. Moczydlowski, E., Uehara, A., Guo, X., and Heiny, J., 1986, Isochannels and blocking modes of voltage-dependent sodium channels, *Ann. N.Y. Acad. Sci.* **479:** 269–292.
35. Matteson, D. R., and Armstrong, C. M., 1982, Evidence for a population of sleepy sodium channels in squid axon at low temperature, *J. Gen. Physiol.* **79:** 739–758.
36. Noda, M., Shimizu, S., Tanaka, T., Takai, T., Kayano, T., Ikeda, T., Takahashi, H., Nakayama, H., Kanaoka, Y., Minamino, N., Kangawa, K., Matsuo, H., Raftery, M. A., Hirose, T., Inayama, S., Hayashida, H., Miyata, T., and Numa, S., 1984, Primary structure of *Electrophorus electricus* sodium channel deduced from cDNA sequence, *Nature* **312:** 121–127.
37. Noda, M., Ikeda, T., Kayano, T., Suzuki, H., Takeshima, H., Kurasaki, M., Tadahashi, H., and Numa, S., 1986, Existence of distinct sodium channel RNAs in rat brain, *Nature* **320:** 189–192.
38. Kayano, T., Noda, M., Flockerzi, V., Takahashi, H., and Numa, S., 1988, Primary structure of rat brain sodium channel III deduced from the cDNA sequence, *FEBS Lett.* **228:** 187–194.
39. Auld, V. J., Goldin, A. L., Krafte, D. S., Marshall, J., Dunn, J. M., Catterall, W. M., Lester, H. A., Davidson, N., and Dunn, R. J., 1988, A rat brain Na channel α-subunit with novel gating properties, *Neuron* **1:** 449–461.
40. Stuhmer, W., Methfessel, C., Sakmann, B., Noda, M., and Numa, S., 1987, Patch-clamp characterization of sodium channels expressed from rat brain cDNA, *Eur. Biophys. J.* **14:** 131–138.
41. DeCino, P., and Kidokoro, Y., 1985, Development and subsequent neural tube effects on the excitability of cultured *Xenopus* myocytes, *J. Neurosci.* **5:** 1471–1482.
42. Kidoro, Y., 1973, Development of action potentials in a clonal rat skeletal muscle cell line, *Nature New Biol.* **241:** 158–159.
43. Frelin, C., Vigne, P., and Lazdunski, M., 1983, Na$^+$ channels with high- and low-affinity tetrodotoxin-binding sites in the mammalian skeletal muscle: Difference in functional properties and sequential appearance during rat skeletal myogenesis, *J. Biol. Chem.* **258:** 7256–7259.
44. Frelin, C., Vijverberg, H. P. M., Romey, G., Vigne, P., and Lazdunski, M., 1984, Different functional states of tetrodotoxin-sensitive and tetrodotoxin-resistant Na$^+$ channels occur during the *in vitro* development of rat skeletal muscle, *Pflüg. Arch.* **402:** 121–128.
45. Sherman, S. J., Lawrence, J. C., Messner, D. J., Jacoby, K., and Catterall, W. A., 1983, Tetrodotoxin-sensitive sodium channels in rat muscle cells developing *in vitro*, *J. Biol. Chem.* **258:** 2488–2495.
46. Strichartz, G., Bar-Sagi, D., and Prives, J., 1983, Differential expression of sodium channel activities during the development of chick skeletal muscle cells in culture, *J. Gen. Physiol.* **82:** 365–384.
47. Haimovich, B., Tanaka, J. C., and Barchi, R. L., 1986, Developmental appearance of sodium channel subtypes in rat skeletal muscle cultures, *J. Neurochem.* **47:** 1148–1153.
48. Sherman, S. J., and Catterall, W. A., 1982, Biphasic regulation of development of the high-affinity saxitoxin receptor by innervation in rat skeletal muscle, *J. Gen. Physiol.* **80:** 753–768.
49. Purves, D., and Sakmann, B., 1974, Membrane properties underlying spontaneous activity of denervated muscle fibers, *J. Physiol.* **239:** 125–153.
50. Stuhmer, W., Roberts, W. M., and Almers, W., 1983, The loose patch clamp, in: *Single-Channel Recording* (B. Sakmann and E. Neher, Eds.), Plenum, New York, pp. 123–132.
51. Weiss, R. E., 1988, Macroscopic Na$^+$ currents in intact ventricular heart muscle measured with the loose patch voltage clamp method, *Biophys. J.* **53:** 423a.
52. Johnson, E. A., and Lieberman, M., 1971, Heart: Excitation and contraction, *Rev. Physiol.* **33:** 479–532.

Chapter 18

Sodium Pump in T-Tubules of Frog Muscle Fibers

R. A. Venosa

1. INTRODUCTION

It is well known that in the cytosol of most cells the high concentration of K^+ ($[K^+]_i$) and the relatively low concentration of Na^+ ($[Na^+]_i$) are kept constant in spite of their electrochemical gradients, which promotes the loss of K^+ and the gain of Na^+. The steadiness of $[K^+]_i$ and $[Na^+]_i$ is maintained by a metabolic energy-dependent active transport process first proposed by Dean[1] in skeletal muscle and known as the Na^+ pump (Dean coined the name) or more properly as the Na^+/K^+ pump.

Since then, the active transport of Na^+ and K^+ have been extensively studied, particularly in red blood cells, and nerve and muscle fibers (for a recent review, see Ref. 2). Through the years it became clear that the work of translocation of Na^+ from the inside to the outside of the cell and K^+ in the opposite direction was performed by an Na^+/K^+ ATPase located in the cell membrane.[3,4] This enzyme is a protein large enough to be accessible from either side of the membrane.

External ouabain and other cardiosteroids specifically inhibit the enzyme and thereby the active fluxes of Na^+ and K^+. In frog skeletal muscle fibers, about half of the Na^+ efflux is active and maximally blocked by 30 μM ouabain (or strophanthidin) in a practically irreversible manner with an apparent dissociation constant (K_d) of the order of 0.2 μM.[5] Such very poor reversibility is an advantage in binding studies (Figs. 1 and 2).

R. A. VENOSA • Cátedra de Fisiología y Biofísica, Facultad de Ciencias Médicas, Universidad Nacional de La Plata, 1900 La Plata, Argentina.

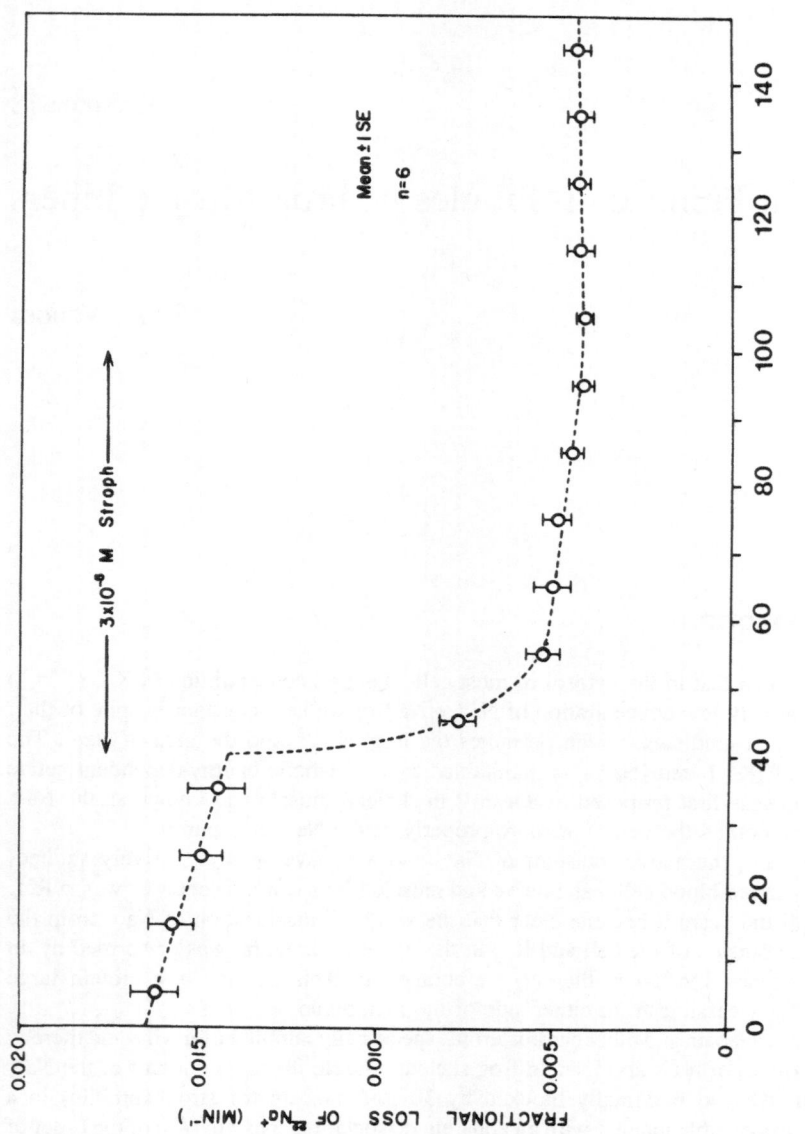

Figure 1. Virtual irreversibility of the inhibition of the active Na⁺ transport by 30 μM strophanthidin in frog sartorius muscles (*R. pipiens*). Mean (± 1 SEM) from six experiments.

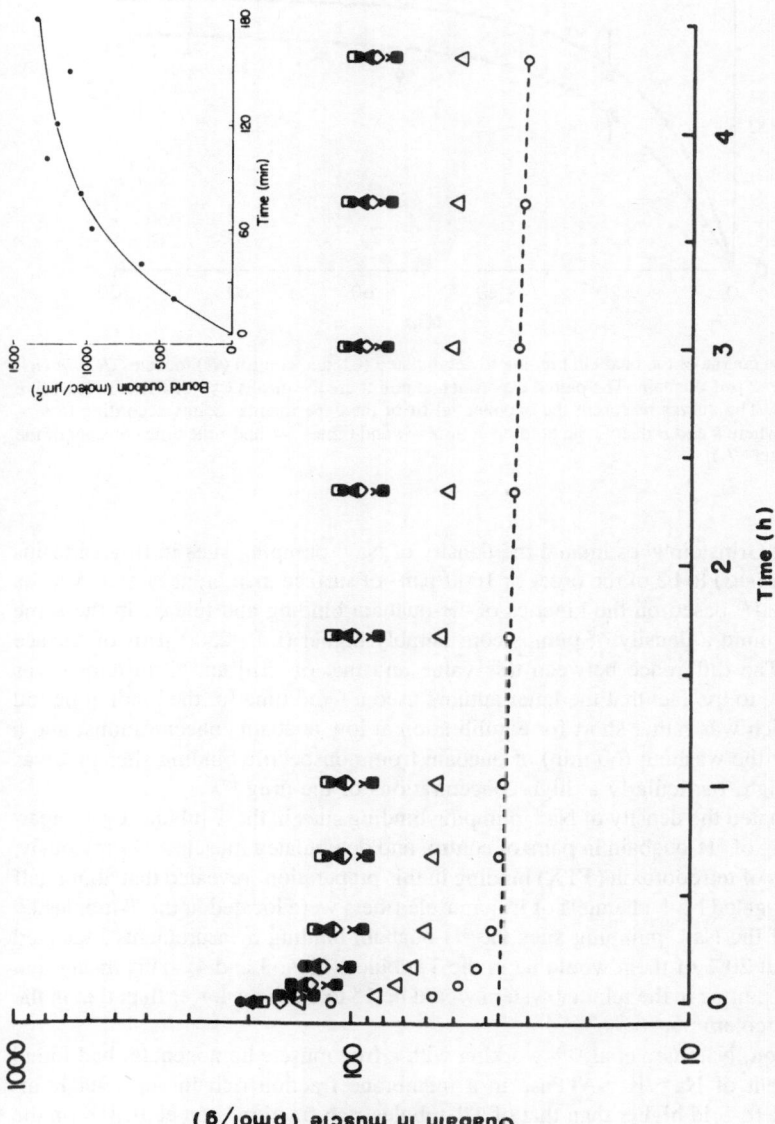

Figure 2. Semilog plot of ouabain washout from eight sartorius muscles (*R. pipiens*) previously exposed to Ringer solution containing 0.36 u*M* labeled (^{3}H) ouabain, each for a different period of time between 20 (○) and 180 (□) min. The dashed line represents the slow single exponential component of the release of the drug. Its extrapolation to time = 0 is a measure of the amount of ouabain bound to specific receptors at the end of the loading period. The inset shows the extrapolated values as a function of the exposure time to ^{3}H-ouabain. (Modified from Ref. 7.)

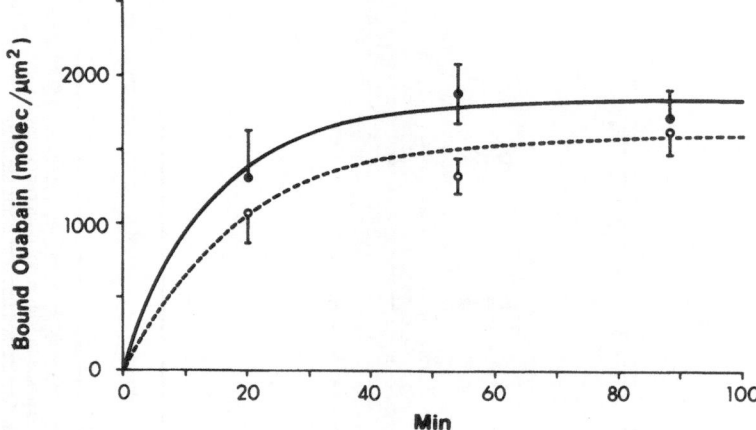

Figure 3. Time course of the ouabain binding to detubulated (○) and control (●) muscles (*R. pipiens*) in the presence of 2 μM ouabain. The paired experimental points are the means (± 1 SEM) of at least five pairs of muscles. The curves represent the exponential fit of the experimental points according to $b = B[(1 - \exp(t/\tau)]$ where b and B denote the binding at time $= t$ and time $= \infty$, and τ the time constant of the uptake. (From Ref. 7.)

Erlij and Grinstein[6] estimated the density of Na^+ pumping sites in frog sartorius muscle (*R. pipiens*) to be of the order of $1600/\mu m^2$ of surface membrane. Later, Venosa and Horowicz,[7] based on the kinetics of ^3H-ouabain binding and release in the same preparation, found a density of pumps considerably higher (i.e., $2500/\mu m^2$ of surface membrane). The difference between this value and that of Erlij and Grinstein[6] was apparently due to the fact that the latter authors used a fixed time for the binding period (50 min), which was rather short for equilibration at low ouabain concentrations, and a fixed time for the washout (60 min) of ouabain from nonspecific binding sites that was not long enough, particularly at high concentrations of the drug.[7]

We estimated the density of Na^+ pumping binding sites in the T-tubules by comparing the binding of ^3H-ouabain in pairs of control and detubulated muscles.[8] Previously, measurements of tetrodotoxin (TTX) binding in this preparation, revealed that about half of the voltage-gated Na^+ channels of frog muscle fibers were located in the T-tubules.[9] In the case of the Na^+ pumping sites the ^3H-ouabain-binding measurements indicated that only about 20% of them would be in the T-tubules (Figs. 3 and 4). This meant that the density of pumps in the tubular system would be 15 to 20 times lower than that in the superficial sarcolemma.

In addition, Narahara et al.[10] working with a frog muscle homogenate, had found that the content of Na^+/K^+-ATPase in a membrane fraction rich in superficial sarcolemma was 14-fold higher than that of a T-tubules-rich fraction. Lau et al.,[11] on the other hand, using vesicles formed in a T-tubule rich membrane fraction from rabbit skeletal muscle reported a ouabain binding capacity of 37 pmol/mg of membrane protein, which would correspond to an estimated density of the order of 180 sites/μm^2

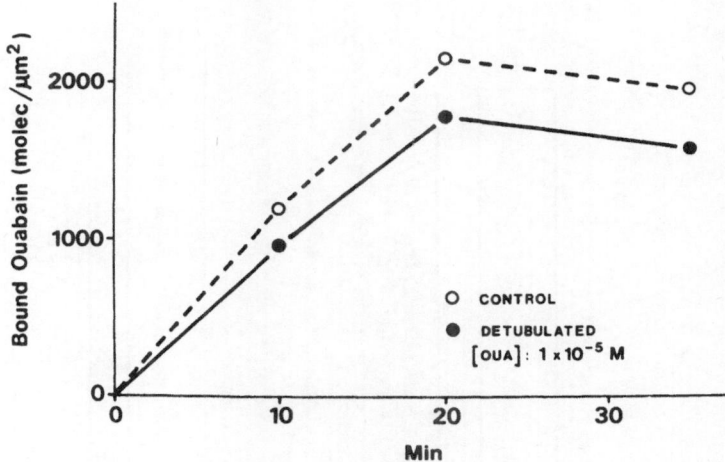

Figure 4. Ouabain binding to paired control (○) and detubulated (●) sartorius muscles from *L. ocellatus*.

of tubular membrane. Measurements in rat muscle yield a binding capacity of 310–721 pmol/g wet wt.[2] or a binding site density of 1430–3330/μm^2 of surface membrane (1300 cm^2/g).[12] On the assumption that rabbit muscle has a binding capacity similar to that of rat muscle, it means that 20 to 50% of the sites per unit area of superficial sarcolemma would be located in the T-tubules opening in that portion of the membrane.

Moreover, Seiler and Fleischer[13] found a Na$^+$/K$^+$-ATPase activity in plasma membrane vesicles from rabbit muscle about ten times larger than that reported by Lau et al.[14] in a T-tubule rich membrane fraction from the same origin.

Up to this point it would therefore appear that the density of Na$^+$ pumps in the T-tubules is substantially lower than in the surface membrane, and that the ouabain/TTX binding ratio is higher in the surface than in the tubules. In fact, Moczydlowski and Latorre[15] used this criterion to identify two membrane fractions: the one with the lower ouabain/saxitoxin ratio as rich in T-tubules, and the one with the higher ratio as surface membrane.

More recently, however, Hidalgo et al.[16] reported a Na$^+$/K$^+$-ATPase activity in the tubules comparable to that found in the surface membrane and Jaimovich et al.[17] a ouabain-binding capacity of 215 pmol/mg of protein in vesicles of T-tubule membranes and of 163 pmol/mg of protein in a fraction enriched in surface membranes from frog muscles (*Caudiverbera caudiverbera*). The density of pump sites can be estimated from their data by assuming that 1 mg of protein corresponds to about 0.27 m^2 of membrane (thickness = 7.5 nm; 50% protein; density = 1 g/cm^3) and therefore the densities of ouabain molecules (pump sites) per μm^2 would be 480 for the tubular membrane and 364 or 654 for the surface membrane depending on whether or not the folds and caveolae of the sarcolemma are taken into account.[18] If, on the average, there are 4 μm^2 of tubular membrane per μm^2 of surface membrane, then there should be (480 × 4) + 364

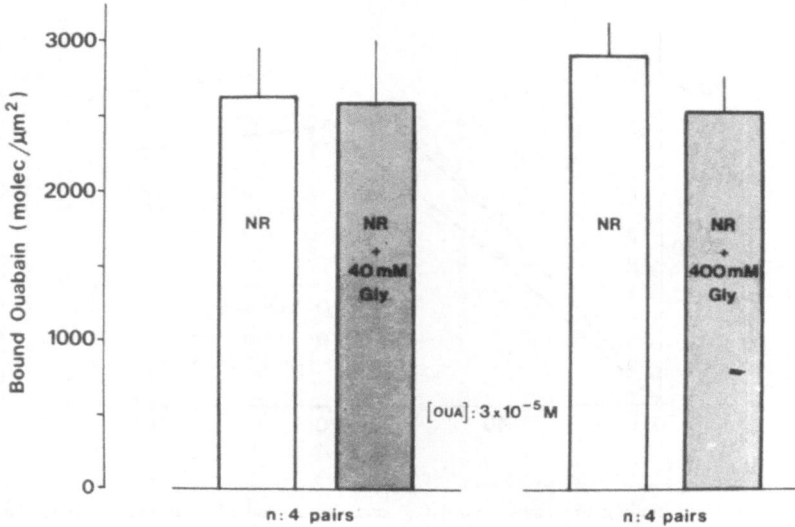

Figure 5. Ouabain binding to paired sartorius muscles (*R. pipiens*) exposed to normal Ringer plus 30 μ*M* ouabain in the absence and in the presence of 40 m*M* and 400 m*M* glycerol. Bars = means + 1 SD. (Venosa and Horowicz, unpublished.)

= 2284, or (480 × 4) + 654 = 2574 pump sites per μm² of surface membrane and its associated T-tubules. These figures are quite close to the value found by Venosa and Horowicz[7] in intact muscles (2500 sites/μm²).

Detubulation produced by glycerol treatment[8] disconnects 90% of the T-tubules from the external medium.[19] On the basis of the data of Jaimovich et al.,[17] after detubulation there should be expected an 80% reduction in ouabain binding instead of the above mentioned 20% observed by Venosa and Horowicz.[7]

Although this discrepancy between the data from isolated membrane fractions and intact muscle fibers could be due to several factors, two likely possibilities are apparent. On the one hand, if the binding in intact muscle fibers were mostly confined to surface membrane it would mean that the agreement between the data of Jaimovich et al.[17] and Venosa and Horowicz[7] on the magnitude of the total binding is fortuitous. On the other hand, if the binding sites were homogeneously distributed, detubulation by glycerol osmotic shock should, somehow, produce a binding increase of its own which roughly cancels out the expected reduction due to the disconnection of the T-tubules from the surface. At a ouabain concentration of 2 μ*M* no sign of nonspecific binding in detubulated muscles was found.[7] In addition, the binding is not significantly affected by the presence of either 40 m*M* or 400 m*M* glycerol Ringer (see Fig. 5). This seems to rule out a chemical effect of glycerol on ouabain binding.

The study of Na⁺ movements and ouabain binding under anisotonic conditions have recently provided some clues on this subject. Some of those results and their possible relation with the data from detubulated muscles are reviewed next.

2. THE HYPOTONIC EFFECT

It is generally accepted that the active Na^+ transport in most cells, including skeletal muscle fibers, is a function of $[Na]_i$. This suggests that a decrease in $[Na^+]_i$ should be followed by a fall in the activity of the Na^+ pump. It would be expected, then, that a reduction in the osmolarity of the external medium should produce a decrease of the active Na^+ extrusion. Quite on the contrary, 10 years ago I found that a reduction of the tonicity of the external medium promoted an increase of the Na^+ efflux, which is completely blocked by strophanthidin (or ouabain) (Fig. 6), not due to depolarization and independent of external Na^+.[20] This stimulation of the Na^+ pump by hypotonicity is quickly reversible and the lower the osmolarity, the larger the stimulation. In Fig. 7 it can be seen that a reduction in osmotic pressure of the external medium to one-half ($\pi = 0.5$) of its normal value ($\pi = 1$) produces an increase close to threefold in the active transport of Na^+. A 2.5-fold increase in active K^+ influx (not shown) was also observed in $\pi = 0.5$ (in $\pi = 1$, half of the osmolarity was provided by sucrose, replacing NaCl; different π's were obtained by changing the sucrose content).

Na^+ and K^+ permeabilities (P_{Na}, P_K), on the other hand, are reduced by 36 and 20% respectively in $\pi = 0.5$ (Venosa, unpublished). The fall in P_K precludes the possibility that the stimulation of the active Na^+/K^+ transport was due to an increased K^+ leak, which would raise $[K^+]_0$ and thereby the activity of the pump.

The author is not aware of a similar behavior of the ouabain-sensitive Na^+/K^+

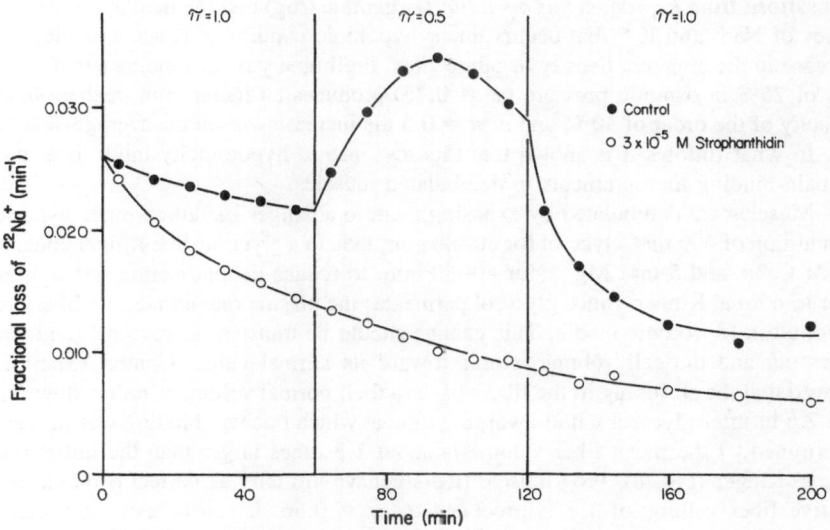

Figure 6. Effect of the reduction of the osmolarity to one-half ($\pi = 0.5$) of its normal value ($\pi = 1$) on the fractional loss of $^{22}Na^+$ from a pair of sartorii one of them in the presence of 30 μM strophanthidin from time zero to the end of the experiment (O) and the other in glycoside free media (●). The hypotonic response is absent when the Na^+ pump is inhibited. (From Ref. 20.)

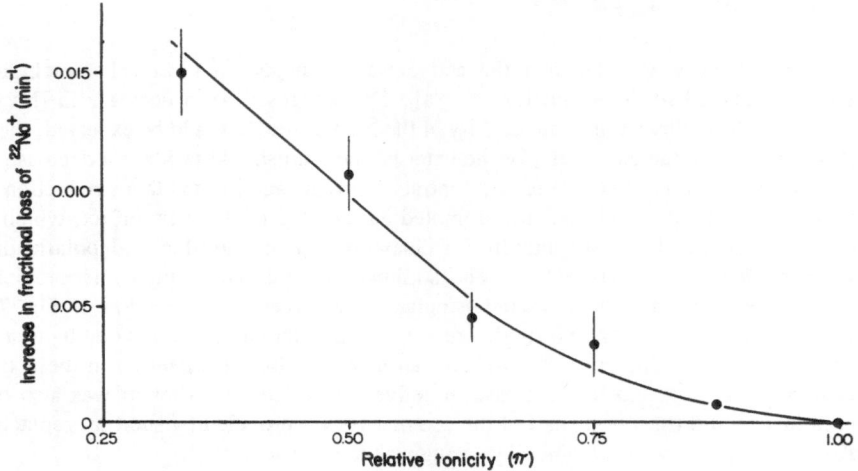

Figure 7.   Increase of $^{22}Na^+$ efflux as a function of tonicity. In $\tau = 1$ the strophanthidin (ouabain)-sensitive portion of the efflux (expressed as fractional loss) was about $0.006\ min^{-1}$, which means that in $\pi = 0.5$ the active Na^+ efflux was roughly three times higher than in $\pi = 1$. Means: $+ 1$ SEM. (From Ref. 20.)

transport in other cell types exposed to hypotonic media. More recently, it was found in frog sartorii from *Leptodactylus ocellatus* (Argentine frog) that the increase in the active fluxes of Na^+ and K^+ that occurs under hypotonic conditions is accompanied by an increase in the apparent density of pump sites. Preliminary results indicate that a reduction of 25% in osmotic pressure ($\pi = 0.75$) produces an increase in ouabain-binding capacity of the order of 30%, and in $\pi = 0.5$ the increase was on the average 40% (Fig. 10). In what follows it is shown that this response to hypotonicity might bear on the ouabain-binding measurements in detubulated muscles.

Muscles are detubulated by exposing them to a Ringer's solution made hypertonic by addition of 400 mM glycerol for 60–80 min, then to a glycerol-free Ringer containing 5 mM Ca^{2+} and 5 mM Mg^{2+} for 60–80 min, to reduce depolarization,[21] and thereafter to normal Ringer. Since glycerol permeates the plasma membrane, the fibers swell upon return to isotonic media. This change should be transient as glycerol (and water) flows out and the cell volume returns toward its normal value. Control experiments showed that the shrinking of the fibers back to their normal volume is rather slow. Thus, 2 to 2.5 hr after glycerol withdrawal, the time at which ouabain binding was previously determined,[7] the mean fiber volume is about 1.3 times larger than the initial one in normal Ringer (Fig. 8). Frog muscle fibers behave virtually as perfect osmometers. A relative fiber volume of 1.3 is produced by $\pi = 0.66$. It can be estimated that this osmotic pressure increases the ouabain-binding capacity by about 37%. This means that in detubulated muscles the observed modest reduction in the number of pump sites to about 80% of the controls in muscles from *R. pipiens* and *L. ocellatus* (Figs. 3 and 4) would probably represent the sum of two opposite effects: (i) an increase of about 35%

promoted by the swelling of the fibers; (ii) a decrease due to the disconnection of the T-tubules from the surface.

In view of the effect of swelling on ouabain binding it seems appropriate to reevaluate the data from detubulated muscles. Because hidden in the reduction of the binding capacity of detubulated muscles to about 80% of the controls there would be a 35% increase due to the volume increase, the percentage of sites left after detubulation would be 80/1.35 = 59. In detubulated fibers up to 10% of the tubules might remain connected to the surface.[19] Therefore, assuming as we did above that there are 4 μm^2 of tubular membrane per μm^2 of superficial sarcolemma, it can be estimated that the actual binding would approximately be 42% for the surface and 58% for the tubules, and, accordingly, the density of pump sites in the surface membrane of intact muscle fibers would be two to three times higher than in the tubular membrane. This distribution of Na^+ pump sites resembles that of the voltage-gated Na^+ channels determined in the same preparation as well as in isolated membrane fractions.[9,17]

The opposite effect of glycerol and detubulation may explain in part early results of Venosa and Horowicz.[22] They found that 1 to 1.5 hr after glycerol withdrawal (when the fibers were maximally swollen) the active Na^+ transport was increased, while later on (6–8 hr) when the fiber volume had presumably diminished or returned to its normal value, the active Na^+ efflux was reduced by about 50%.

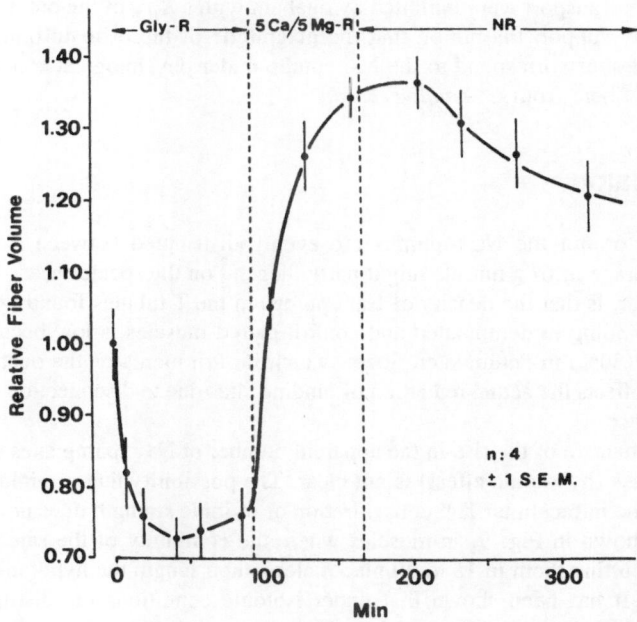

Figure 8. Time course of fiber volume changes in four sartorii from *L. ocellatus* during glycerol treatment. The fiber volumes were estimated by subtracting the extracellular space from the muscle weight. The values from each muscle were normalized with respect to that in normal Ringer and averaged. Means: ± 1 SEM.

It is clear that although the differences between the data from isolated membrane fractions and intact muscle are narrowed by considering the swelling component, and possibly the presence of remnants of tubular membrane accessible to ouabain in glycerol-treated muscles, they are still far from being insignificant.

Another difference between the data from intact muscle and those from isolated membrane fractions refers to the dissociation constant (K_d) of the ouabain binding to the receptor. In intact sartorius muscle from *R. pipiens* it was found to be 0.22 μM,[7] a value practically identical to that required for 50% inhibition ($K_{0.5}$) of the ouabain-sensitive Na$^+$ efflux in the same preparation.[5] In muscles from *L. ocellatus* the K_d for the ouabain-receptor interaction is somewhat higher (1–5 μM). In isolated membrane fractions from *C. caudiverbera* on the other hand, Jaimovich et al.[17] reported K_ds of 10.4 and 9 nM for T-tubule and surface membranes, respectively. A ouabain concentration of 10 nM does not have any appreciable effect on the ouabain-sensitive Na$^+$ efflux in frog muscle from at least three species (*R. pipiens*, *R. temporaria*, and *L. ocellatus*). It would be interesting, therefore, to determine the $K_{0.5}$ of the ouabain-sensitive fraction of the Na$^+$ efflux in muscle fibers from *C. caudiverbera*. If the $K_{0.5}$ for the inhibition of active Na$^+$ transport were similar to those found in other frog species it would suggest that the high-affinity receptors to which ouabain binds under their experimental conditions may be other than those related to the pumping sites which maintain Na$^+$ far from equilibrium across the sarcolemma under physiological conditions. If, on the contrary, the active Na$^+$ transport were inhibited by ouabain with a $K_{0.5}$ of the order of 10 nM, it would strongly support the notion that the magnitude of the ouabain binding and the distribution of sites correspond to the Na$^+$ pump under physiological conditions in the intact muscle fibers from *C. caudiverbera*.

3. CONCLUSION

Whether or not the Na$^+$ pumps are evenly distributed between T-tubules and surface membrane in frog muscle might partly depend on the species used. What seems likely, however, is that the density of Na$^+$ pumps in the T-tubules found by comparing the ouabain binding in detubulated and control-paired muscles is low because the volume increase (30%) in detubulated fibers, which in turn increases the ouabain binding and partially offsets the actual reduction of binding sites due to disconection of T-tubules from the surface.

The mechanism of the rise in the apparent number of Na$^+$ pump sites produced by volume increase (hypotonic effect) is not clear. The possibility that it might be due to a reduction of the intracellular K$^+$ concentration or of ionic strength does not seem likely because, as shown in Fig. 9, in muscles where the continuity of the sarcolemma was disrupted by cutting them in 15 to 20 places along their length the hypotonic effect was not detected. It has been shown that under isotonic conditions the disruption of the sarcolemma per se does not alter ouabain binding.

It would not be unreasonable to assume that under hypotonic conditions and after glycerol treatment the swelling of the fibers could activate latent pumping sites. It is tempting to speculate that the tension in the sarcolemma might modulate the number of active pumps.

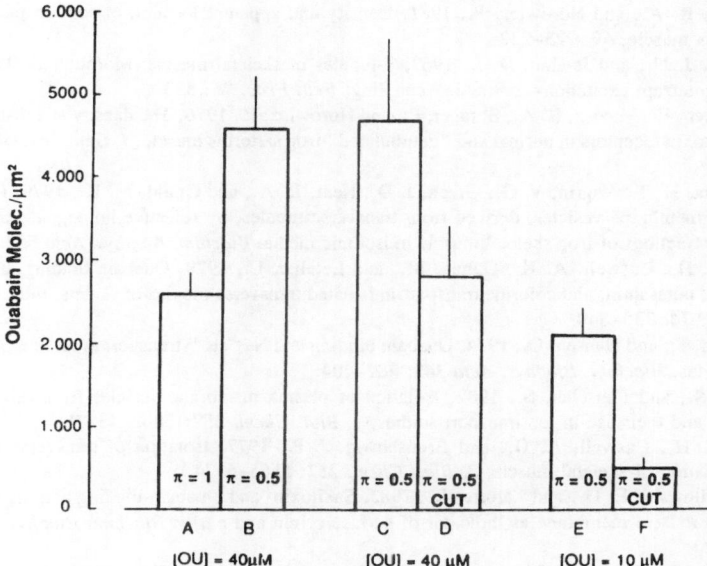

Figure 9. Effect of hypotonicity ($\pi = 0.5$) on ouabain binding (molecules/μm^2) in intact and membrane-disrupted (cut fibers) sartorii from *L. ocellatus*. A–B: eight intact muscles equilibrated in $\pi = 1$ (A) and their paired companions in $\pi = 0.5$ (B) all exposed to 40 μM ouabain. C–D: three intact muscles equilibrated in $\pi = 0.5$ (C) and their cut paired companions (D) all exposed to 40 μM ouabain. E–F: four paired muscles, same conditions as in C–D except that they were exposed to 10 μM ouabain. The increase in ouabain binding promoted by $\pi = 0.5$ in intact muscles is not apparent in membrane-disrupted muscles. Exposure time: 30 min. Bars: means + 1 SEM.

ACKNOWLEDGMENTS. This work was supported by grants from Consejo Nacional de Investigaciones Cientificas y Tecnicas (CONICET) and CIC (Provincia de Buenos Aires).

REFERENCES

1. Dean, R. B., 1941, Theories of electrolyte equilibrium in muscle, *Cold Spr. Harb. Symp. Quant. Biol.* **3**: 331.
2. Clausen T., 1986, Regulation of active Na^+/K^+ transport in skeletal muscle, *Phys. Rev.* **66**:542–580.
3. Skou, J. C., 1957, The influence of some cations on an adenosine triphosphatase from peripheral nerves, *Biochim. Biophys. Acta* **23**: 394–401.
4. Skou, J. C., 1960, Further investigations of Mg^{2+} + Na^+ = activated adenosinetriphosphatase, possibly related to the active, linked transport of Na^+ and K^+ across the nerve membrane, *Biochim. Biophys. Acta* **42**: 6–23.
5. Horowicz, P., Taylor, J. W., and Waggoner, D. M., 1970, Fractionation of sodium efflux in frog sartorius muscle by strophanthidin and removal of external sodium, *J. Gen. Physiol.* **55**: 401–425.
6. Erlij, D., and Grinstein, S., 1976, The number of sodium ions pumping sites in skeletal muscle and its modification by insulin, *J. Physiol.* **259**: 13–31.

7. Venosa, R. A., and Horowicz, P., 1981, Density and apparent location of sodium pump in frog sartorius muscle, *59:* 225–232.
8. Howell, J. N., and Jenden, D. J., 1967, T-tubules of skeletal muscle: Morphological alterations which interrupt excitation–contraction coupling, *Fed. Proc.* **26:** 553.
9. Jaimovich, E., Venosa, R. A., Shrager, P., and Horowicz, P., 1976, The density and distribution of tetrodotoxin receptors in normal and "detubulated" frog sartorius muscle, *J. Gen. Physiol.* **67:** 399–416.
10. Narahara, H. T., Vogrin, V. G., Green, J. D., Kent, R. A., and Gould, M. K., 1979, Isolation of plasma membrane vesicles, derived from transverse tubules, by selective homogenization of subcellular fractions of frog skeletal muscle in isotonic media, *Biochim. Biophys. Acta* **552:** 247–261.
11. Lau, Y. H., Caswell, A. H., Garcia, M., and Letelier, L., 1979, Ouabain binding and coupled sodium, potassium, and chloride transport in isolated transverse tubules of skeletal muscle, *J. Gen. Physiol.* **74:** 335–349.
12. Clausen, T., and Hansen, O., 1974, Ouabain binding and Na^+/K^+ transport in rat muscle cells and adipocytes, *Biochim. Biophys. Acta* **345:** 387–404.
13. Seiler, S., and Fleischer, S., 1982, Isolation of plasma membrane vesicles from rabbit skeletal muscle and their use in ion transport studies, *J. Biol. Chem.* **257:**13862–13871.
14. Lau, Y. H., Caswell, A. H., and Brunschwig, J. P., 1977, Isolation of transverse tubules by fractionation of skeletal muscle, *J. Biol. Chem.* **252:** 5565–6574.
15. Moczydlowski, E. G., and Latorre, R., 1983, Saxitoxin- and Ouabain-binding activity of isolated skeletal muscle membrane as indicator of surface origin and purity, *Biochim. Biophys. Acta* **732:** 412–420.
16. Hidalgo, C., Parra, C., Riquelme, G., and Jaimovich, E., 1986, Transverse tubules from frog skeletal muscle: Purification and properties of vesicles sealed with the inside-out orientation, *Biochim. Biophys. Acta* **855:** 79–88.
17. Jaimovich, E., Donoso, P., Liberona, J. L., and Hidalgo, C., 1986, Ion pathways in transverse tubules: Quantification of receptors in membranes isolated from frog and rabbit skeletal muscle, *Biochim. Biophys. Acta* **855:** 89–98.
18. Dulhunty, A. F., and Franzini-Armstrong, C., 1975, The relative contribution of the folds and caveolae to the surface membrane of frog skeletal muscle fibers at different sarcomere lengths, *J. Physiol.* **250:** 513–539.
19. Franzini-Armstrong, C., Venosa, R. A., and Horowicz, P., 1973, Morphology and accessibility of the transverse tubular system in frog sartorius muscle after glycerol treatment, *J. Membrane Biol.* **14:** 197–212.
20. Venosa, R. A., 1978, Stimulation of the Na^+ pump by hypotonic solutions in skeletal muscle, *Biochim. Biophys. Acta* **510:** 378–383.
21. Eisenberg, R. S., Howell, J. N., and Vaughan, P. C., 1971, The maintainance of the resting potentials in glycerol-treated muscle fibers, *J. Physiol.* **215:** 95–102.
22. Venosa, R. A., and Horowicz, P., 1973, Effects on sodium efflux of treating frog sartorius muscle with hypertonic glycerol solutions, *J. Membrane Biol.* **14:** 33–56.

B. Calcium Channels in T-Tubule

Excitation–Contraction Coupling in Barnacle Muscle Fibers: Does Calcium Entry Trigger Contraction Directly?

Mario Luxoro, Verónica Nassar-Gentina, and Eduardo Rojas

1. INTRODUCTION

It is well known that mechanical activation of muscle from both vertebrates and invertebrates is triggered by an elevation of the concentration of free-calcium in the sarcoplasm. In skeletal muscle fibers from vertebrates this ion is actively transported into the sarcoplasmic reticulum (SR) and released at the level of the terminal cisternae of the SR following the regenerative depolarization of the transverse tubular membrane.[1] However, the question of the origin of the calcium required for mechanical activation in invertebrate skeletal muscle has not been yet elucidated.[2]

In our early studies of the Ca^{2+} entry in response to depolarizing voltage clamp pulses in barnacle skeletal muscle fibers, we proposed that the source of Ca^{2+} for mechanical activation was the external medium.[3,4] Our conclusion was based on two fundamental properties of excitation–contraction (E-C) coupling in these fibers. First, we observed that the early time course (< 200 msec) of the twitch and the Ca^{2+} influx (calculated from the simultaneously recorded inward Ca^{2+} current) had an almost identical time course.[3,4] Second, we observed that the dependence on membrane potential of the maximum tension during a twitch and the Ca^{2+} influx (calculated from

MARIO LUXORO and VERÓNICA NASSAR-GENTINA • Laboratorio de Fisiología Celular, Facultad de Ciencias y Facultad de Medicina, Universidad de Chile, Santiago, Chile. EDUARDO ROJAS • Laboratory of Cell Biology and Genetics, National Institute of Diabetes and Digestive and Kidney Diseases, National Institutes of Health, Bethesda, Maryland 20892.

the corresponding inward current) were rather similar for both processes.[3,4] Consistent with these results were our observations that maximum tension during the twitches elicited by depolarizing voltage clamp pulses decreases rapidly and steadily following the removal of the external calcium ions.[5,6] In contrast, Ca^{2+} removal from the external solution does not impair twitch generation in vertebrate skeletal muscle.[7] These results indicate that mechanical activation evoked by membrane depolarization in the barnacle skeletal muscle requires Ca^{2+} entry from the external medium, and, at least in this respect, this invertebrate muscle is quite different from vertebrate skeletal muscle.

The ability of the SR to accumulate Ca^{2+} has been firmly established and, recently, the putative Ca^{2+}-release channel from highly purified SR vesicles has been incorporated into bilayer membranes (see Chapter 33, this volume). These two features suggest strongly that in vertebrate muscle the SR is the intracellular Ca^{2+} store involved in E-C coupling. Although crustacean skeletal muscles have a well-developed SR system,[8,9] the question of the role played by the SR in E-C coupling in the barnacle muscle remains to be elucidated.[10,11]

In this chapter, we summarize the results of our experiments focused on the question: Is the SR the source for the Ca^{2+} required for the activation of the contractile machinery? Using different approaches we arrived at the conclusion that in barnacle muscle the SR is an important, but not the only, source of the Ca^{2+} required for mechanical activation. Thus, our results suggest that, under physiological conditions, the elevation of $[Ca^{2+}]_i$ following Ca^{2+} entry may trigger the synthesis of inositol 1,4,5-trisphosphate (IP_3), which in turn may induce Ca^{2+} release from the SR. Under particular conditions, external Ca^{2+} may directly activate the contractile machinery.

2. METHODOLOGICAL CONSIDERATIONS

The experiments were performed using single giant muscle fibers from *Megabalanus psittacus* (Darwin), quite abundant along the Chilean seashore. The dissection of the muscle fibers, the chamber used, and the recording set-up have been described previously.[5,12] Briefly, a single muscle fiber was mounted horizontally in a voltage-clamp chamber with its tendon attached to a force transducer and with its other end cannulated. A platinized steel needle (200 μm in diameter) was introduced longitudinally and used as the current electrode. To control the internal composition we used two methods. For voltage clamp experiments we kept the surface membranes intact and used the needle to introduce solutions into the fiber. For biochemical protocols we used giant muscle fibers that had been cut longitudinally several times to expose as much as possible the myofibrils and to facilitate exchange by diffusion of externally applied solutions.

We used the open fibers for three different experimental protocols: First, to determine the generation of tension as a function of the pCa of the external solution. Second, to determine phosphoinositide turnover; for these studies we incubated each open fiber with tritiated inositol and measured the formation of phosphatidylinositol (PI), phosphatidylinositol 4-phosphate (PIP), and phosphatidylinositol 4,5-bisphosphate (PIP_2); details on the methods used have been described previoulsy.[11] Third, we used the protein aequorin to measure IP_3-evoked release of calcium from internal stores. The light was detected using an ultralow dark current photomultiplier.[11] Control experiments without fibers showed that the IP_3 used was virtually free of Ca^{2+}.

3. RESULTS AND DISCUSSION

3.1. Effects of Barium and Strontium on Mechanical Activation

Muscle fibers internally perfused with a Cs^+ solution (mM: 170 Cs-acetate, 30 tetraethylammonium (TEA)-Cl, 440 sucrose, 10 tris-acetate, pH = 7.4) and immersed in artificial seawater (Ca^{2+}-ASW; mM: 440 NaCl, 50 $MgCl_2$, 10 $CaCl_2$, 5 trisCl, pH = 7.4) were subjected to a series of depolarizing voltage clamp pulses (50 msec in duration) while ionic currents, the time integral of the inward currents, and peak tension were recorded.

Replacement of Ca^{2+} in the external ASW by Ba^{2+} or Sr^{2+}, so that inward currents were carried by Ba^{2+} or Sr^{2+} (see Fig. 1B), reversibly abolished the development of tension. The effects of replacing the Ca^{2+}-ASW by a modified ASW in which either Mg^{2+}, Ba^{2+}, or Sr^{2+} was used in place of Ca^{2+} are shown in Fig. 1A. Switching from Ca^{2+}-ASW to 0 Ca^{2+}-ASW (with either Mg^{2+}, Ba^{2+}, or Sr^{2+} in place of Ca^{2+}) was followed by a cessation of the twitches within a few minutes.

It should be pointed out that with Mg^{2+} in place of Ca^{2+} no inward currents were recorded. Furthermore, recovery of the size of the peak tension during the twitches was faster when switching from Mg^{2+} back to Ca^{2+} than when switching from Ba^{2+} or Sr^{2+} (Fig. 1A).

When external Ca^{2+} is replaced by Ba^{2+}, contraction is also prevented. Indeed, Fig. 1B, record 4, shows that although the inward current recorded with the fiber in Ba^{2+}-ASW is similar in magnitude to that in record 3 made with the fiber in Ca^{2+}-ASW, no tension was developed. The rate of recovery of the tension after switching back to 10 mM Ca^{2+}-ASW was drastically reduced in those fibers which were stimulated in the presence of Ba^{2+}-ASW. This is illustrated in Fig. 1B (compare records 3 and 5). Record 5 was made 20 min after replacing Ba^{2+}-ASW by Ca^{2+}-ASW. Although after this time the inward current had completely recovered, the tension was about one-half of that in the control record. Tail currents (i.e., currents after the pulse) in muscle fibers that have been immersed in Ba^{2+}-ASW (or Sr^{2+}) occasionally deactivate extremely slowly (Fig. 1B, record 4).

These results indicate that, although Ba^{2+} and Sr^{2+} can substitute for Ca^{2+} as charge carriers through the Ca^{2+}-channel, once inside the muscle neither Ba^{2+} nor Sr^{2+} can replace Ca^{2+} to trigger the series of events leading to development of tension. On the contrary, Ba^{2+} inside the fiber reduces the rate of recovery of peak tension during the twitch.

Recovery of the twitch tension after exposure of the fiber to Ba^{2+} or Sr^{2+}-ASW could be accelerated if the concentration of Ca^{2+} is augmented. These data suggest that Ba^{2+} (and Sr^{2+}) compete with Ca^{2+} for an internal-specific Ca^{2+} receptor site. We postulate the existence of an internal-specific site based on our observation that exposure of resting, nonstimulated fibers to Ba^{2+}-ASW or Sr^{2+}-ASW per se does not lead to inhibition of the twitch tension after switching back to Ca^{2+}-ASW. Inhibition is ob-

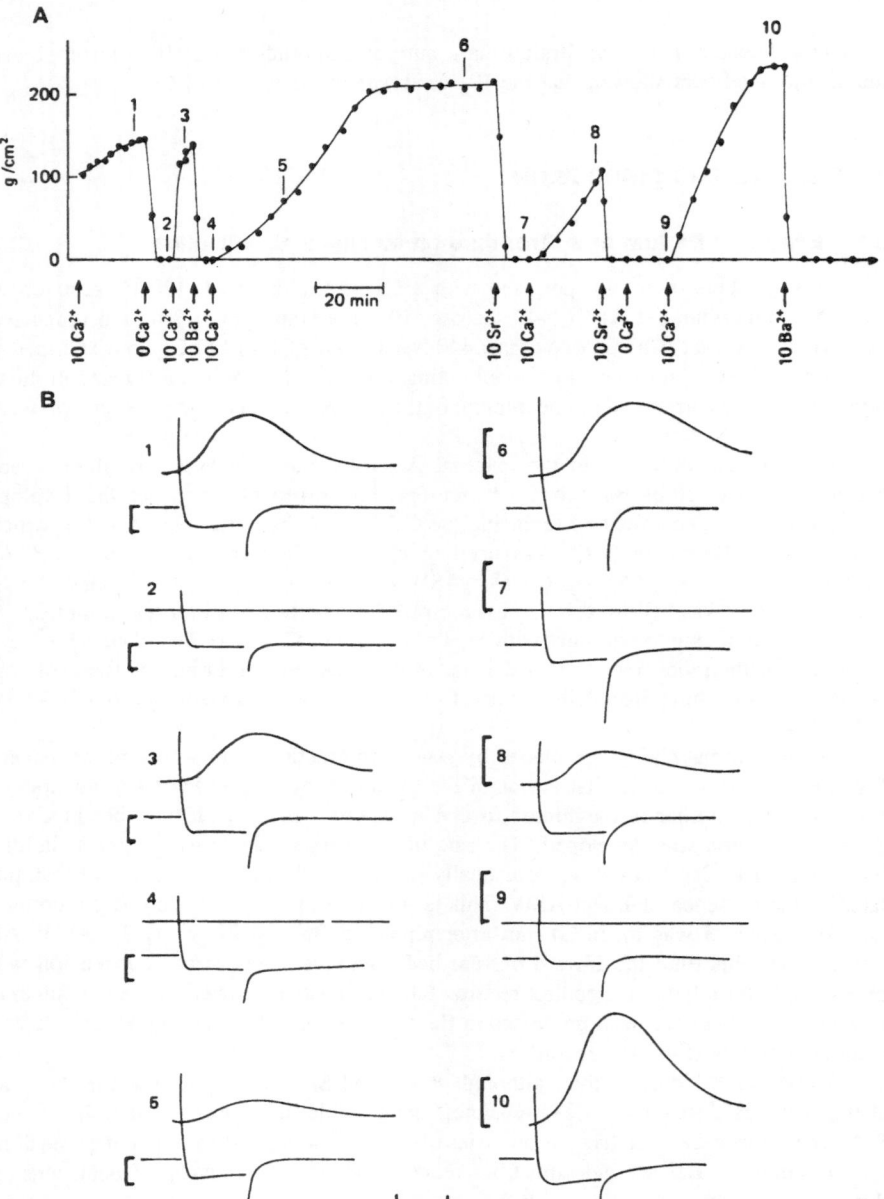

Figure 1. Peak tension during twitches recorded under different external conditions (A) and the temporal relationship between inward current and development of tension (B). The arrows in Fig. 1A indicate the time at which the external ASW was changed. The numbers on top of the figure indicate the time at which the records of tension and currents, depicted in Fig. 1B, were made. Internal solution (mM): 170 Cs-acetate, 30 TEACl, 440 sucrose, 10 tris-Cl, pH = 7.4. Holding potential −35 mV; voltage clamp pulse 50 mV, 50 msec. (B) Records of tension and currents from part (A). Vertical calibrations: Left side = 2 mA/cm^2; right side = 100 g/cm^2. Horizontal calibration: 40 msec (for the currents traces), 80 msec (for the tension traces). (From Ref. 5, with permission.)

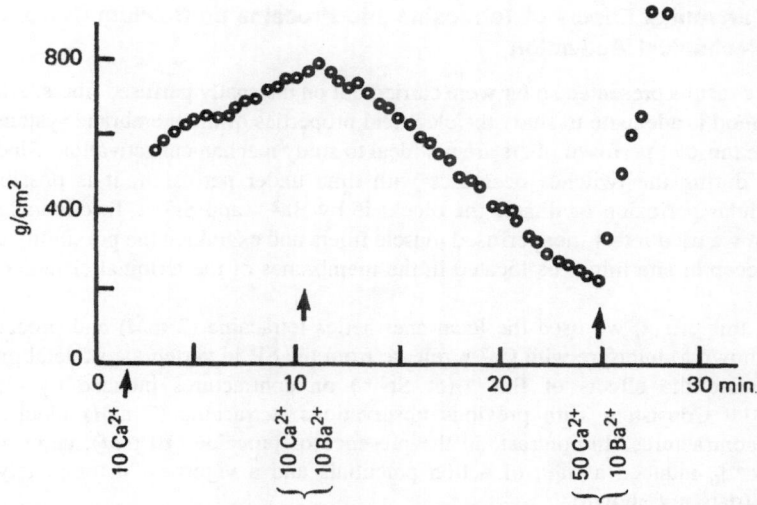

Figure 2. Effect of high external concentration of Ca^{2+} on the inhibition caused by Ba^{2+}. Peak tension was recorded in different ASW. First in 10 mM Ca^{2+}-ASW, then in 10 mM Ca^{2+} + 10 mM Ba^{2+}-ASW and, finally, in 50 nM Ca^{2+} + 10 mM Ba^{2-}-ASW. The inhibition of tension observed in 10 nM Ca^{2+} + 10 mM Ba^{2+} is abolished by 50 mM Ca^{2+} in the presence of 10 mM Ba^{2+}. The tension increased rapidly, the amplifier saturated, and, finally, the muscle fiber disrupted. Holding potential: -52 mV; voltage pulses: 45 mV, 40 msec. (From Ref. 5, with permission.)

served only if Ba^{2+} or Sr^{2+} currents are elicited by stimulation of the fiber.[5] To examine the possibility of a competitive inhibition by Ba^{2+} at an intracellular Ca^{2+} site, we tested the effects of different $[Ba^{2+}]_0$ at different $[Ca^{2+}]_0$. Figure 2 illustrates the results of one experiment in which Ba^{2+} (10 mM) was applied in the presence of Ca^{2+} (10 mM). Under these conditions the onset of the Ba^{2+} blockade of the twitch tension was slower than that measured in the absence of Ca^{2+}. The blockade by Ba^{2+} could be reversed by increasing $[Ca^{2+}]_0$ from 10 to 50 mM.

Application of Sr^{2+} (5 mM) in the presence of 10 mM $[Ca^{2+}]_0$ blocked the twitches elicited by depolarizing voltage clamp pulses. Increasing the $[Sr^{2+}]_0$ from 5 to 30 mM, in the absence of Ca^{2+}, sometimes caused generation of tension. Whenever this happened, it occurred with a delay of several seconds and always associated with long-lasting tails of inward currents. Considering the fact that troponin presents some affinity for Sr^{2+}, although much less than for calcium,[13] a large $[Sr^{2+}]_i$ will be needed to induce tension.

The results presented so far suggest that both Ba^{2+} and Sr^{2+} compete with Ca^{2+} for an intracellular Ca^{2+} receptor site. We propose that the interaction between Ca^{2+} and this receptor is the first intracellular and most significant event of the excitation–contraction coupling cycle in barnacle skeletal muscle.

An obvious candidate for this Ca^{2+}-receptor is troponin. However, there is no affinity of troponin for barium.[13] Thus, the competitive inhibition by Ba^{2+} or Sr^{2+} must occur at a different Ca^{2+} receptor site.

3.2. Differential Effects of Tetracaine and Procaine on Calcium Currents and Mechanical Activation

The results presented so far were carried out on internally perfused fibers. Although this method is adequate to study the electrical properties of the membrane system of the barnacle muscle, perfused fibers are not ideal to study mechanical activation. Since peak tension during the twitches decreases with time under perfusion, it is possible that intracellular perfusion facilitates the blockade by Ba^{2+} and Sr^{2+}. To circumvent this problem we used intact, nonperfused muscle fibers and examined the possibility that the Ca^{2+} receptor site might be located in the membranes of the terminal cisternae of the SR.

To this effect, we used the local anesthetics tetracaine (2 mM) and procaine (10 mM), known to interfere with Ca^{2+} release from the SR in vertebrate skeletal muscles, and studied the effects of Ba^{2+} (or Sr^{2+}) on contractures induced by elevating $[K^+]_0$.[14] Consistent with previous observations, tetracaine (2 mM) blocked high $[K^+]_0$ contractures. In contrast, in the presence of procaine (10 mM), application of high $[K^+]_0$ induced a train of action potentials and a vigorous, incompletely fused tetanus (data not shown).

In an attempt to identify the site of action of barium and strontium, we studied the effects of tetracaine and procaine in two preparations—the open and the internally perfused fiber. In the open fiber, contractures were elicited by the application of a sudden increase in $[Ca^{2+}]_0$ from 0.1 (pCa = 7) to 6.3 μM (pCa 5.2; Ca^{2+} buffered with 20 mM EGTA). Addition of tetracaine or procaine did not affect maximal tension developed during a high $[Ca^{2+}]_0$-evoked contracture (Fig. 3). Thus, tetracaine and

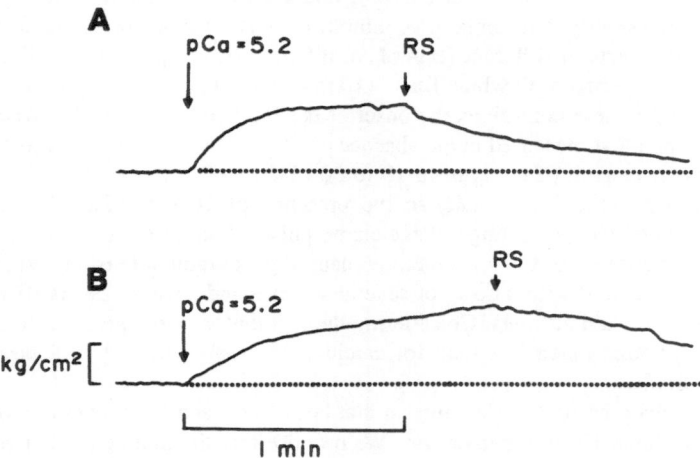

Figure 3. Effect of tetracaine on development of tension evoked by exposure to a high calcium solution in open fibers. (A) the control experiment in the absence of tetracaine (B) the effect of 2 mM tetracaine applied 30 min before decreasing the pCa from 7 to 5.2. The arrows indicate the time at which solutions were changed. RS = relaxing solution.

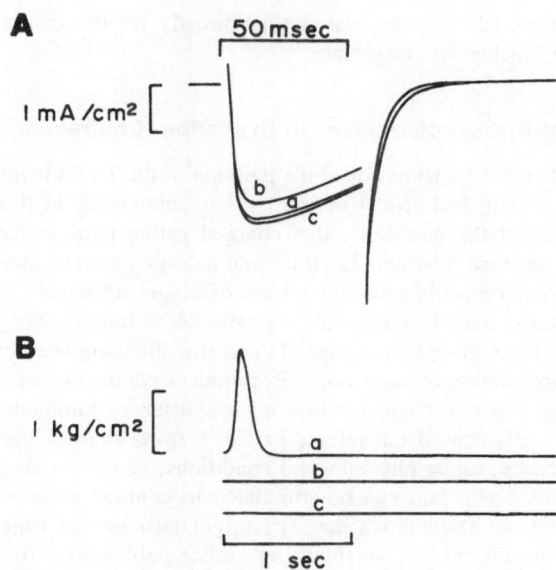

Figure 4. Effect of tetracaine on Ca^{2+} currents and on twitch tension. (A) Ca^{2+} currents: (a) control; (b) in the presence of tetracaine (2 mM); (c) in the presence of tetracaine and 15 mM Ca^{2+}. (B) Traces of the development of tension: (a) control; (b) in the presence of tetracaine; (c) with tetracaine in the presence of 15 mM Ca^{2+}. Traces (b) and (c) represent the resting tension of the fiber related to pulses (b) and (c). Internal solution (mM): 180 Cs-acetate, 20 TEACl, 5 tris-acetate, 400 sucrose at pH 7.2.

procaine do not affect mechanical activation induced by direct application of Ca^{2+} to the contractile machinery.

To determine whether or not these drugs affect membrane ionic currents in the barnacle muscle, we studied the effect of tetracaine and procaine measured under voltage clamp conditions. Consistent with previously reported results, we found that tetracaine and procaine are potent blockers of potassium outward currents in barnacle muscle. Inward currents recorded with Cs^+ in place of K^+ inside the fibers were marginally larger in the presence of procaine and 10 to 25% smaller in the presence of tetracaine. As shown in Fig. 4, mechanical activation was completely blocked by tetracaine.

To rule out the possibility that the blockade of the twitch induced by tetracaine was due to the small inhibition of the inward current observed above, the external concentration of Ca^{2+} was increased in order to restore the size of the calcium current to that of the control experiment. While increasing $[Ca^{2+}]_0$ from 10 to 15 mM was sufficient to restore the size of the inward calcium current, the twitch remained blocked (Fig. 4). As in the case of the inhibition caused by Ba^{2+} or Sr^{2+}, tetracaine blocked mechanical activation at an intracellular site located between the site of Ca^{2+} entry and troponin.

Muscle fibers which were exposed to tetracaine for an extended period (> 120 min) generated delayed long-lasting contractures only associated with long-lasting tails of current in response to depolarizing voltage clamp pulses. Probably, under this circumstance, the enhanced Ca^{2+} entry was acting directly on the contractile machinery bypassing the site blocked by tetracaine.

3.3. The Role of Phosphoinositides in Excitation–Contraction Coupling

The nature of signal transmission at the junction of the T-tubule membrane and the SR has not been established. Most of the experimental work in this area has been interpreted in terms of the possibility that charged gating particles control the Ca^{2+} release from the SR (see Chapter 25, this volume). In contrast, the possibility that membrane depolarization could cause the release of trigger substances from the T-tubule membrane has been given substance only recently.[15,16] Indeed, Vergara et al,[15] by demonstrating that IP_3 was rapidly produced in electrically stimulated muscle fibers and that intracellular application of exogenous IP_3 produces contracture of the muscle fiber, presented evidence that this could be the trigger substance. Simultaneously, Volpe et al.[16] showed that IP_3 caused the release of Ca^{2+} from internal stores in vertebrate muscle fibers. Because, under physiological conditions, vertebrate skeletal muscle neither gains nor loses Ca^{2+} and can be stimulated to contract even in the absence of external calcium,[7] we studied whether IP_3 might also be the trigger substance in barnacle skeletal muscle which, as shown in earlier publications from this laboratory,[3,5] has an absolute requirement for external Ca^{2+}.

Figure 5 depicts several records of contractures induced by consecutive microinjections by exogenous IP_3. A large dose (upper record, left side) induced a vigorous contracture (peak tension about 1.25 kg/cm^2 for 8 nmol of IP_3 injected). The amplitude of the tension decreased with consecutive applications at a high concentration (80 μM). In contrast, consecutive applications at a lower concentration (2 μM) induced contractures of similar shape and amplitude.[11] Since barnacle muscle fibers are equipped with a very effective Ca^{2+} transport system which, together with the SR,[10] participates in the maintenance of $[Ca^{2+}]_i$, the decay in the maximum amplitude of the force developed in consecutive IP_3-induced contractures is probably due to depletion of Ca^{2+} stores. Exposure of the fiber to high K^+-ASW depolarized the membrane and induced Ca^{2+} entry through voltage-gated Ca^{2+}-channels presumably present in the transverse tubular membrane.[12] The onset of the contracture in response to Ca^{2+} entry was substantially delayed (Fig. 5, second chart record from the top). Since the intracellular Ca^{2+} stores had been depleted by three applications of IP_3 (Fig. 5, top record), we interpreted this result to indicate that in those fibers in which the IP_3-sensitive Ca^{2+} pool had been depleted, the Ca^{2+} permeating the T-tubule membrane acted directly on the Ca^{2+} sites known to be present in the troponin molecules.

Incubation of the fiber in ASW containing 20 mM Cd^{2+}, a potent blocker of voltage-gated Ca^{2+}-channels, potentiated the effects of IP_3 (Fig. 5, third record from top).

Successive applications of IP_3 to cut open barnacle muscle fibers that had been previously incubated for 3 to 5 min in a medium containing ATP and Mg^{2+} ($[Ca^{2+}]$ adjusted to 1 μM) always elicited a transient increase in Ca^{2+}. The size of the response (i.e., pmol Ca^{2+}/g of muscle) depended on $[IP_3]$ as illustrated in Fig. 6. Regardless of the concentration applied, IP_3 induced a rapid elevation of $[Ca^{2+}]$ to a peak value. This peak was followed by a rapid and then by a slow decrease toward the basal level. The complex time course of the Ca^{2+} transients shown in Fig. 6 suggest that more than one

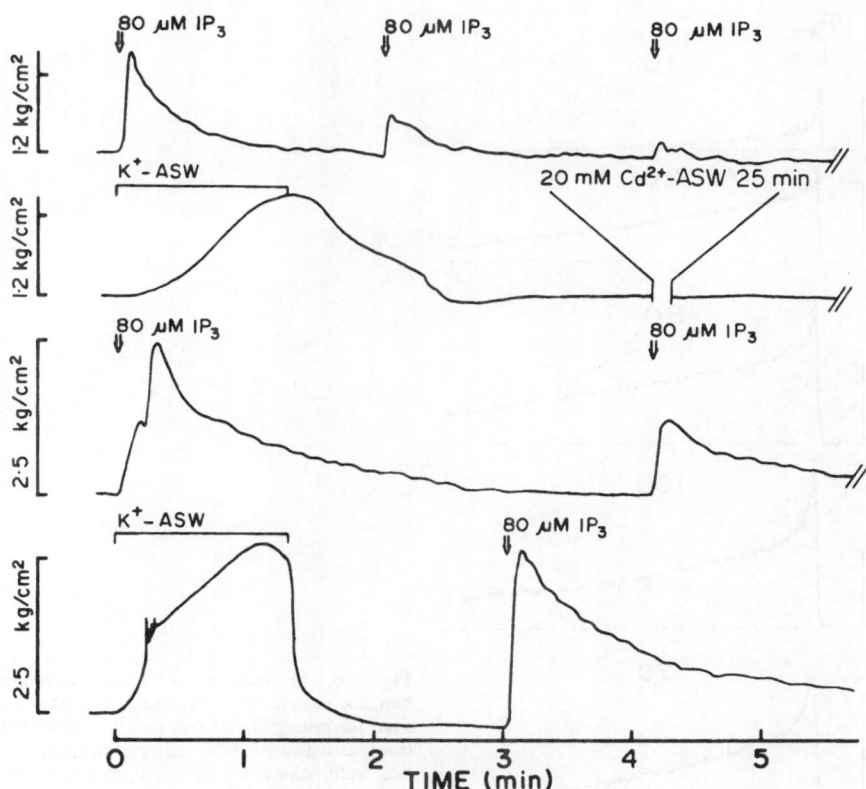

Figure 5. IP$_3$-evoked Ca^{2+} unloading of internal stores and high K$^+$-ASW-induced Ca^{2+} entry and reloading. The fiber was bathed in ASW and perfused with an internal medium containing 0.5 mM Mg^{2+}. Four segments of a continuous chart record are shown. The record on top of the figure shows contractures evoked by three consecutive short perfusion with IP$_3$ (2sec, 100 μl, 80 μM). After the third application of IP$_3$ the internal stores were reloaded depolarizing the muscle fiber with K$^+$-ASW as indicated by the horizontal bar (second row). External ASW was replaced by ASW containing 20 mM CdCl$_2$. In the presence of Cd^{2+} (third row) the fiber was perfused again with IP$_3$. At the end of the third row, Cd^{2+} was removed. Finally (bottom row), after reloading once more the SR, a short perfusion was made. The marks (//) shown at the end of each row indicate that the record continues below without time lapse between the segments. (From Ref. 11, with permission.)

Ca^{2+} pool may be involved.[11] [IP$_3$] for each calcium transient is indicated by the numbers above each record and was increased from 10 (top record) to 320 μM (bottom record). The Ca^{2+} released during each exposure to IP$_3$ ranged from 7.8 nmol/g released with 10 μM IP$_3$ to 19 nmol/g released with 320 μM IP$_3$. Saturation of the response was achieved with concentrations above 80 μM.

In an attempt to identify the signal triggering IP$_3$ production, we measured the effects of [Ca^{2+}] on the incorporation of [^{3}H]inositol to PI, PIP, and PIP$_2$. Incubation of the open fibers was done in the medium used for intracellular perfusion,[11] in which

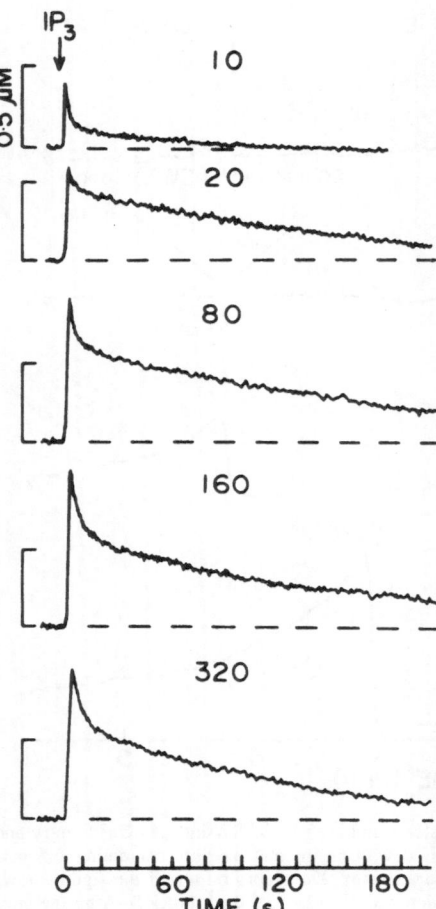

Figure 6. IP$_3$-evoked Ca^{2+} release in an open barnacle muscle fiber. The same open fiber was used for testing IP$_3$ at five different concentrations, as indicated above each record (μM). Vertical calibrations: Change in light output after addition of Ca^{2+} to increase the concentration in the reaction chamber from 0.1 to 0.5 μM. Peak values of [Ca^{2+}] as a function of increasing [IP$_3$]: 0.4, 0.6, 0.9, 1.0, and 1.0 μM for the early component; 0.1, 0.4, 0.5, 0.5, and 0.5 μM for the late component.

[Ca^{2+}] was controlled with the Ca^{2+} buffer EGTA (20 mM). Under these conditions we could show that there is a substantial and rapid incorporation of [^{3}H]inositol to PI, PIP, and PIP$_2$ in barnacle muscle fibers. Furthermore, in the presence of tetracaine (2 mM), [^{3}H]inositol incorporation to PIP was substantially augmented, perhaps indicating an inhibition of the formation of PIP$_2$ from PIP.

Assuming that IP$_3$ is indeed the chemical messenger coupling the electrical events at the level of the T-tubular membrane to the calcium release channel known to be present in the SR membrane (see Chapter 33, this volume), one could predict that tetracaine would be unable to block IP$_3$-induced contractures as those illustrated in Fig. 5. Indeed, the response of fibers microinjected with small doses of IP$_3$ (volume injected = 1 μl; [IP$_3$] = 1 mM) was not affected by preincubation (30 min) in the presence of tetracaine. In contrast, caffeine-evoked contractures (10 mM) under the same conditions

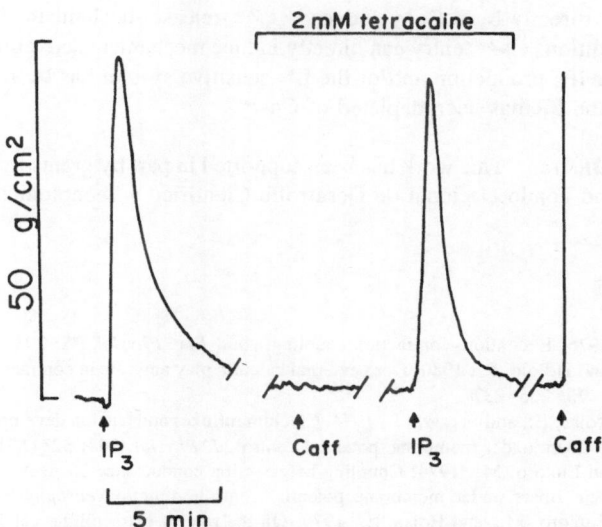

Figure 7. Differential effects of tetracaine on IP_3- and caffeine-evoked contractures. Traces represent tension as a function of time. Fiber bathed in ASW containing 10 mM Ca^{2+}. IP_3 was applied intracellularly by microinjection as indicated (1 μl of a Cs-acetate solution (mM): 180 Cs-acetate, 20 TEACl, 5 tris-acetate, 400 sucrose at pH 7.2 containing 1 mM IP_3). Cut marks (//) represent a resting period of about 30 min. During the second application of caffeine (10 mM) the fiber broke.

were blocked by tetracaine (Fig. 7). The blockade was readily reversed after the removal of tetracaine.[17]

To determine a causal relationship between Ca^{2+} entry and IP_3 production, open fibers were incubated in the presence of [^{3}H]inositol added to the intracellular solution with [Ca^{2+}] buffered at two different pCa values, namely, 7.2 and 5.6. After an incubation period of 30 min, the supernatant was adsorbed in a formic anionic exchange column in order to separate the inositol mono- and polyphosphates so produced.[18] Our preliminary data show that the production of IP_3 is definitely enhanced at pCa 5.6.

4. CONCLUDING REMARKS

Our results show clearly that, under physiological conditions, Ca^{2+} entry from the extracellular medium is the first step in excitation–contraction coupling in barnacle skeletal muscle. The depolarization of the membrane at the level of the neuromuscular junction propagates electrotonically on the surface sarcolemma, clefts, and transverse tubular membranes. This depolarization activates voltage-gated Ca^{2+}-channels which are known to be present in the membrane systems of the barnacle muscle. The entry of Ca^{2+} causes a rapid and localized increase in [Ca^{2+}]$_i$ near the relevant membranes. The transient elevation of [Ca^{2+}]$_i$ in turn may trigger the release of extra Ca^{2+} from the SR either via an IP_3-sensitive Ca^{2+} release channel, which is presumably located in the SR membranes, or directly by a Ca^{2+}-induced Ca^{2+}-release mechanism. Under two experimental conditions Ca^{2+} entry can directly induce mechanical activation: First, when the intracellular IP_3 production and/or the IP_3-sensitive system has been blocked, and, second, when the SR has been depleted of Ca^{2+}.

ACKNOWLEDGMENTS.   This work has been supported in part by grants from the University of Chile and Fondo Nacional de Desarrollo Cientifico y Tecnologico de Chile.

REFERENCES

1. Ebashi, S., 1976, Excitation–contraction coupling, *Ann. Rev. Physiol.* **38:** 293–313.
2. Caputo, C., and DiPolo, R., 1980, Does external calcium play any role in contractile activation?, *J. Gen. Physiol.* **75:** 235–237.
3. Atwater, I., Rojas, E., and Vergara, J., 1974, Calcium influxes and tension development in perfused single muscle fibers under membrane potential control, *J. Physiol.* **243:** 523–551.
4. Rojas, E., and Luxoro, M., 1974, Coupling between ion conductance changes and contraction in barnacle muscle fibers under membrane potential control, *Actual. Neurophysiol.* **10:** 159–169.
5. Hidalgo, J., Luxoro, M., and Rojas, E., 1979, On the role of extracellular calcium in triggering contraction in muscle fibers from barnacle under membrane potential control, *J. Physiol.* **288:** 313–330.
6. Bacigalupo, J., Luxoro, M., Rissetti, S., and Vergara, C., 1979, Extracellular space and diffusion barriers in muscle fibers from *Megabalanus psittacus* (Darwin), *J. Physiol.* **288:** 301–312.
7. Armstrong, C. M., Bezanilla, F. M., and Horowicz, P., 1972, Twitches in the presence of EGTA, *Biochim. Biophys. Acta,* **267:** 605–608.
8. Franzini-Armstrong, C., Eastwood, A. B., and Peachey, L. D., 1986, Shape and disposition of clefts, tubules, and sarcoplasmic reticulum in long and short sarcomere fibers of crab and crayfish, *Cell Tissue Res.* **244**(1): 9–19.
9. Crowe, L. M., and Baskin, R. J., 1981, Activation of the contractile system in crustacean muscle: Ultrastructural evidence for the role of the T-system, *Tissue Cell.* **13:** 153–164.
10. Garcia, A. M., Lennon, A. M., and Hidalgo, C., 1975, Sarcoplasmic reticulum from barnacle muscle: Composition and calcium uptake properties, *FEBS Lett.* **58:** 344–348.
11. Rojas, E., Nassar-Gentina, V., Luxoro, M., Pollard, M. E., and Carrasco, M. A., 1987, Inositol 1,4,5-trisphosphate-induced calcium release from the sarcoplasmic reticulum and contraction in crustacean muscle, *Can. J. Physiol. Pharmacol.* **65:** 672–680.
12. Keynes, R. D., Rojas, E., Taylor, R. E., and Vergara, J., 1973, Calcium and potassium systems of a giant barnacle muscle fiber under membrane potential control, *J. Physiol.* **229:** 409–455.
13. Fuchs, F., 1971, Ion exchange properties of the calcium receptor site of troponin, *Biochim. Biophys. Acta* **245:** 221–229.
14. Luxoro, M., and Nassar-Gentina, V., 1984, Potassium-induced depolarizations and generation of tension in barnacle muscle fibers: Effects of external calcium, strontium, and barium, *Quart. J. Exp. Physiol.* **69:** 235–243.
15. Vergara, J., Tsien, R. Y., and Delay, M., 1985, Inositol 1,4,5-trisphosphate: A possible chemical link in excitation–contraction coupling in muscle, *Proc. Natl. Acad. Sci. USA* **82:** 6352–6356.
16. Volpe, P., Salviati, G., Di Virgilio, and F. Pozzan, T., 1985, Inositol 1,4,5-trisphosphate induces calcium release from sarcoplasmic reticulum of skeletal muscle, *Nature* **310:** 347–349.
17. Nassar, V., Rojas, E., Carrasco, M. A., and Luxoro, M., 1987, Acoplamiento excitacion–contraccion en musculo de crustaceo, *Arch. Biol. Med. Exp.* **20:** R154.
18. Downes, C. P., and Michell, R. H., 1981, The polyphosphoinositide phosphodiesterase of erythrocyte membranes, *Biochem. J.* **198:** 133–140.

Chapter 20

Biochemical Structure of the Dihydropyridine Receptor

Jane A. Talvenheimo, Shu-Rong Wen, Kyung Sook Kim, and Anthony H. Caswell

1. INTRODUCTION

When radiolabeled dihydropyridine compounds first became available, they opened up a wide area of investigation for researchers interested in the biochemical structure of voltage-dependent Ca^{2+}-channels. Dihydropyridines (DHPs) were seen to bind to a high-affinity site closely associated with an important class of Ca^{2+}-channels (designated DHP-sensitive Ca^{2+}-channels), and provided the first tools for identifying and extracting the Ca^{2+}-channel protein from membranes. Investigations of the DHP receptor/Ca^{2+}-channel protein have focused on three different approaches for identifying the high-affinity DHP-binding protein: (i) radiation inactivation of the binding activity to determine the target size of the receptor, (ii) solubilization and purification of the binding activity from membranes, and (iii) photoaffinity labeling of the drug receptor in intact membranes. Nearly all of these studies have been performed using skeletal muscle transverse tubule membranes (T-tubules), a well-characterized membrane fraction[1] as a source of DHP receptors. T-tubule membranes contain the highest density of DHP binding sites (from 10 to 100 pmol of binding sites per mg membrane protein) of any tissue surveyed for DHP-binding activity. The goal of this chapter is to briefly review

JANE A. TALVENHEIMO, KYUNG SOOK KIM, and ANTHONY H. CASWELL • Department of Pharmacology, University of Miami School of Medicine, Miami, Florida 33101. SHU-RONG WEN • Department of Pharmacology, University of Miami School of Medicine, Miami, Florida 33101; and Department of Pharmacology, Beijing Medical University, Beijing, People's Republic of China.

what is known about the protein structure of the T-tubule DHP receptor/Ca^{2+}-channel, to present some data on the purification and reconstitution of the DHP receptor, and to highlight some of the unanswered questions regarding the structure and possible functional role of this protein.

2. TARGET SIZE ANALYSIS

Target size analysis provides an estimate of the size of a receptor by measuring the dose of high-energy electron (1.5–10 MeV) radiation required to inactivate a particular function, such as specific ligand-binding activity. Estimates of the DHP receptor size obtained by this method have varied somewhat, depending on the particular DHP used to measure receptor function. Radiation inactivation of nitrendipine binding to T-tubule membranes gave a size of 210 kDa,[2] while inactivation of nimodipine and PN200-110 binding yielded estimates of 178 and 136 kDa, respectively.[3] Since these DHPs bind to the same site, it is not clear why the target size of the drug-binding site varies with the drug used. Nevertheless, these data provided one of the first indications that the DHP-binding protein is comparable in size to the voltage-dependent Na^+ channel, and is large enough to span the membrane.

3. SOLUBILIZATION AND PURIFICATION OF THE DHP RECEPTOR

Several laboratories have solubilized the DHP receptor from brain and skeletal muscle using digitonin or CHAPS,[4–6] and have successfully purified the binding protein several-hundred-fold from T-tubules. Based on the observation by Goll et al.[3] that the solubilized receptor can be absorbed to lectin columns, Curtis and Catterall[7] developed a purification procedure in which the receptor is labeled with ^3H-nitrendipine, solubilized in digitonin, and then purified by wheat germ agglutinin-Sepharose (WGA-Sepharose) chromatography, anion exchange chromatography, and sucrose gradient sedimentation. Borsotto et al.[8,9] developed a similar method, in which the DHP-labeled receptor was solubilized with CHAPS and purified by gel filtration, WGA-Sepharose, and anion exchange chromatography. Both of these methods, with some modifications, are still widely used to isolate the receptor protein.

Analysis of the protein composition of the purified DHP receptor on SDS polyacrylamide gels initially revealed the presence of at least three polypeptides, although there was some disagreement among different laboratories about the identity of the receptor subunits. Curtis and Catterall[7] identified three polypeptides with relative molecular sizes (M_r) of 160 kDa, 51 kDa, and 33 kDa, which were designated the α-, β-, and γ-subunits, respectively. Upon reduction of disulfide bonds, the M_r of the α-subunit was reported to shift from 160 kDa to 130 kDa. Borsotto et al.[8] also found a large polypeptide of about 140 kDa (after reduction) and a 33 kDa peptide, but did not detect a 51 kDa peptide corresponding to the β-subunit in their purified receptor preparation. In addition, they described the appearance of a 32 kDa peptide, distinct from the 33 kDa peptide, following disulfide bond reduction. This smaller peptide, with an estimated M_r ranging from 27 to 32 kDa, has since been designated the δ-subunit. Flockerzi et al.[10] identified polypeptides with M_r = 142, 122, 56, 31, 26, and 22 kDa as the major components of purified DHP receptor samples.

Using a modification of the method published by Curtis and Catterall[7] we have also purified the rabbit T-tubule DHP receptor.[11] Figure 1 shows the polypeptide composition of highly purified receptor analyzed by SDS polyacrylamide gel electrophoresis. Lanes 1–4 contain aliquots from four consecutive sucrose gradient fractions containing the peak of DHP-binding activity. The major peptide components that comigrated with the peak of DHP-binding activity in this sample have M_r = 170, 140, 55, 33, 27, and 24 kDa. In this experiment and in the studies summarized above, it must be emphasized that the only criterion for identifying polypeptides as "subunits" of the DHP receptor/Ca^{2+}-channel protein is that these polypeptides copurify with DHP-binding activity. Initially it was not known which of the proteins in these samples actually bound the DHP drugs, nor was it known whether some of the polypeptides were contaminants or proteolytic fragments of larger proteins.

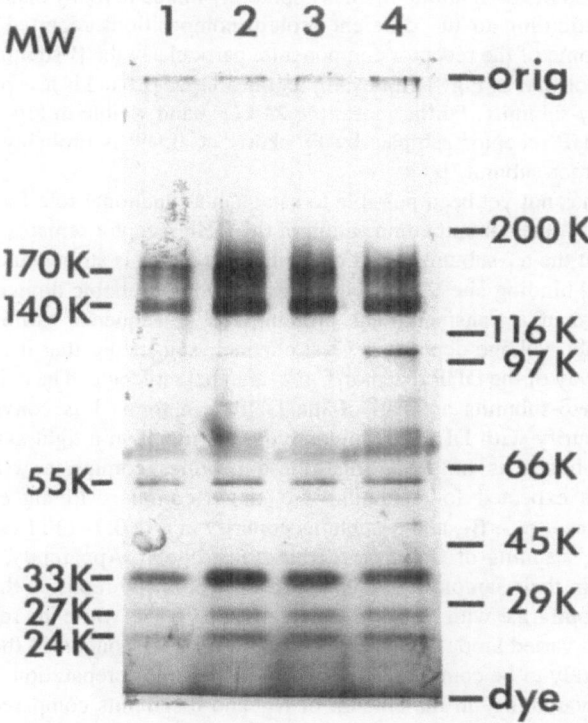

Figure 1. Purified rabbit skeletal muscle DHP receptor analyzed by SDS polyacrylamide gel electrophoresis. Four consecutive sucrose gradient fractions containing the peak of DHP-binding activity from the final step of purification were lyophilized, dissolved in sample buffer containing 20 mM dithiothreitol, and run on a 4–15% gradient gel as described in Talvenheimo et al.[11] Each lane contains 2 µg of protein. Lane 2 contained the highest specific DHP-binding activity. The gel was silver-stained to visualize protein. Molecular weight markers were myosin, β-galactosidase, phosphorylase B, bovine serum albumin, egg albumin, and carbonic anhydrase.

There are several reasons for the discrepancies in the protein composition reported for the purified DHP receptor. Part of the apparent disagreement is simply due to the inherent variability in estimating molecular weight by SDS gel electrophoresis. For example, it is now clear that the 170 and 140 kDa peptides in our samples are equivalent to the 160 and 130 kDa peptides of Curtis and Catterall,[7] and to the 140 and 122 kDa peptides identified by Flockerzi et al.[10] A more important factor underlying the discrepancies in subunit composition is that the α-subunit, which migrates as a diffuse band in polyacrylamide gels, was originally thought to consist of a single polypeptide, but was subsequently shown to contain two distinct polypeptides, α_1 and α_2. These are shown clearly separated in Fig. 1. The α_1-subunit has a M_r of 165–170 kDa, which does not shift under reducing conditions, and contains the binding sites for all three classes of drug which interact with DHP-sensitive Ca^{2+}-channels: the DHP drugs, phenylalkylamines (verapamil), and benzothiazepines (diltiazem). The α_1-subunit has also been completely sequenced. The α_2-subunit has a M_r of 170 kDa before reduction, but shifts to 135–140 kDa after disulfide reduction, giving rise to a smaller 27–30 kDa subunit, designated the δ-subunit, which is apparently linked to α_2 by disulfide bonds. A third factor contributing to the different protein compositions reported for the DHP receptor is that some of the receptor components, particularly the β-subunit,[7] are quite sensitive to proteolysis. In Fig. 1, for example, the 55 kDa β-band is less prominent than the α_1-, α_2-, or γ-subunits. Furthermore, the 24 kDa band visible in Fig. 1, which was also found in DHP receptor samples by Flockerzi et al.[10] is probably a proteolytic fragment of a larger subunit.[12]

Because it has not yet been possible to establish a functional role for the α_2, β, γ, and δ polypeptides, the subunit composition of the DHP receptor remains controversial. The evidence that the α_1-subunit is part of the DHP receptor is strong since this subunit contains the drug binding sites. It also contains large hydrophobic domains, consistent with the structure of a transmembrane protein, and its sequence exhibits significant homology with the voltage-dependent Na^+ channel, suggesting that it forms the ion-conducting pathway of the DHP receptor/Ca^{2+}-channel molecule. The evidence that the α_2-, β-, γ-, and δ-subunits are part of the DHP receptor is less convincing: These polypeptides copurify with DHP-binding activity and maintain a tight association with the α_1-subunit, but it is not clear that these peptides comigrate with a constant stoichiometry, as expected for subunits of a single complex. Leung et al.[13] have reported that the α_1-, α_2-, β-, and γ-subunits copurify in a $1:0.79:1:1$ ratio (estimated by densitometric scanning of Coomassie Blue-stained gels). Apparently, the δ-subunit was not present in their samples. In contrast, Sieber et al.[12] reported that the α_1, β-, and γ-subunits comigrate with a stoichiometry of $1:1.7:1.4$, while the relative amount of α_2-/δ-subunits varied among different preparations. They concluded that the α_2- and δ-subunits are likely to be contaminants in the DHP receptor preparation. Similarly, we have also noted variability in the amount of α_2- and δ-subunits compared to the other peptides (Talvenheimo and Wen, unpublished data).

Since the structure of the DHP receptor/Ca^{2+}-channel complex, like that of the Na^+ channel, has probably been highly conserved during evolution, information about the structure of this channel from species other than rabbit would be useful for predicting which polypeptides are most likely to represent receptor subunits. Borsotto et al.[9]

partially purified DHP receptor from chick and frog skeletal muscle and identified a large subunit with $M_r = 135–141$ kDa in both species. The DHP receptor from chick heart, purified by Cooper et al.,[14] contains prominent polypeptides of 170, 140, 32, and 29 kDa, presumably analogous to the α_1-, α_2-, γ-, and δ-subunits found in rabbit skeletal muscle. The chick heart receptor α_1, α_2, and δ polypeptides were also recognized by monoclonal antibodies raised against the corresponding rabbit skeletal muscle peptides. When we purified the DHP receptor from frog skeletal muscle T-tubules, we found a polypeptide composition differing somewhat from the results we obtain for rabbit skeletal muscle. As shown in Fig. 2, the major polypeptides in the purified frog DHP receptor are 177, 152, 105, 96, 67, 47, and 23 kDa. It seems likely that the 177 and 152 kDa peptides correspond to α_1 and α_2, respectively. Whether any of the other peptides correspond to subunits identified in rabbit muscle is not clear. The α_1- and α_2-subunits have also been identified by immunochemical methods or by affinity labeling in rat brain,[15] guinea pig heart,[16] guinea pig skeletal muscle,[17] and hamster heart,[18] indicating that the association of α_1 and α_2 is highly conserved. The functional significance of this association, however, along with the role of the β-, γ-, and δ-subunits, is likely to remain controversial until it is proved that these peptides either are or are not required for DHP receptor/Ca^{2+}-channel function. The evidence implicating each of the polypeptides in the DHP receptor/Ca^{2+}-channel structure is reviewed below.

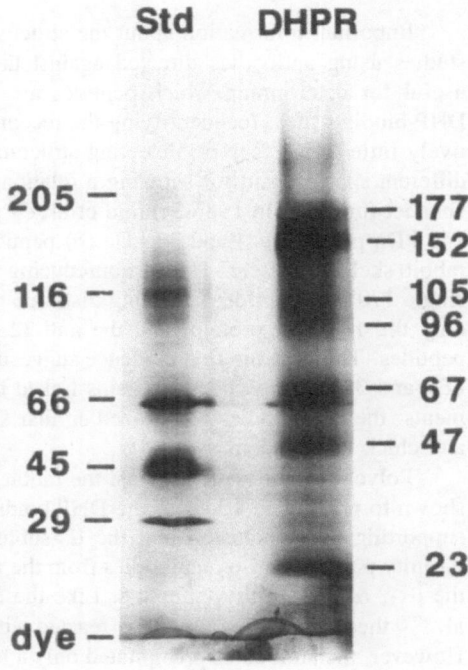

Figure 2. Polypeptide composition of purified frog skeletal muscle DHP receptor. Frog T-tubule membranes were isolated from frog skeletal muscle as described by Hidalgo et al.[61] The frog DHP receptor was purified as described for the rabbit DHP receptor in Talvenheimo et al.[11] The fraction containing the peak of DHP-binding activity from the final sucrose gradient step was lyophilized, dissolved in sample buffer containing 20 mM dithiothreitol, and an aliquot containing 10 µg of protein was run on a 4–15% gradient gel. The gel was silver-stained to visualize protein. The molecular weight markers were identical to those used in Fig. 1.

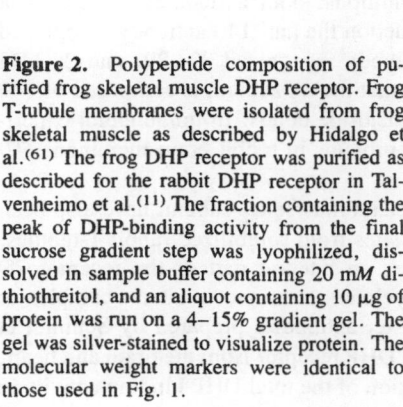

4. PHOTOAFFINITY LABELING

Using photoreactive derivatives of DHPs, several groups have labeled the DHP receptor in intact membranes. In some of the early studies, the labeled peptides ranged in molecular size from 145 kDa[19] and 42 kDa[20] to 32–33 kDa.[21,22] Although some of these results can be reconciled with the subunit composition of the purified receptor, more recent studies suggest that smaller labeled peptides were proteolytic fragments of a larger DHP-binding peptide. The recent studies, using ^{3}H-azidopine to covalently label the DHP receptor both in intact membranes and in purified receptor samples, have confirmed and extended the results of Ferry et al.,[19] showing that a single large peptide with M_r = 145–165 kDa is specifically labeled by DHPs.[16–18,23–25] The identity of this labeled peptide as the α_1-subunit was confirmed by the fact that it comigrates with α_1; its size is unchanged by disulfide reduction; it is phosphorylated under conditions where only α_1, not α_2, is phosphorylated; and it can be separated completely from α_2 under denaturing conditions. The DHP receptor has also been labeled covalently with diltiazem[23] and with a photoreactive verapamil derivative.[26] In both cases the only polypeptide labeled specifically by these drugs was identified as the α_1 subunit. Together these data indicate that the binding sites for all three classes of drug reside on the α_1 subunit.

5. IMMUNOCHEMICAL STUDIES OF THE DHP RECEPTOR COMPOSITION

Important information about the structure of the DHP receptor has emerged from studies using antibodies directed against the receptor protein. Antibodies have been useful for determining which peptides are recognized by antibodies that precipitate DHP-binding sites, for identifying the receptor components in tissues that contain relatively little DHP receptor, detecting structural similarities among DHP receptors from different species, and for showing a relationship between specific peptides and Ca^{2+}-channel function. In 1986 Schmid et al.[27] prepared polyclonal antibodies against the 140 kDa peptide (α_2) and 32 kDa (δ) peptide found in a DHP receptor purified from rabbit skeletal muscle. Under nonreducing conditions, both antibodies recognized a single 170 kDa peptide; following disulfide reduction the anti-140 antibody recognized only the 140 kDa protein, and the anti-32 antibody recognized 32, 29, and 26 kDa peptides. This was the first evidence suggesting that the α_2-subunit is composed of 140 kDa and 32 kDa polypeptide chains linked by disulfide bonds. In immunoblot experiments, these antibodies recognized similar size proteins in rabbit brain membranes[27] and chick heart membranes.[14]

Polyclonal antibodies against the rabbit skeletal muscle α_2-subunit have also been shown to precipitate 100% of the DHP-binding sites from solubilized rabbit T-tubules, supporting the conclusion that the α_2-subunit is an integral part of the receptor.[28] Affinity-purified anti-α_2 antibodies from the polyclonal serum precipitated a complex of the α_1-, α_2-, β-, and γ-subunits. Like the anti-α_2 antibodies prepared by Schmidt et al.,[27] these antibodies also cross-reacted with a DHP receptor from the brain and heart. However, the antibodies precipitated only a fraction of the total DHP-binding activity in brain and heart membranes, suggesting that some of the antibodies in this polyclonal serum recognize epitopes on the muscle receptor which are not present on the brain and heart receptor. Norman et al.,[29] using monoclonal anti-α_2 antibodies prepared against the rabbit receptor, also found cross-reactivity with a similar protein in rabbit heart and brain, and with rat, mouse, and frog skeletal muscle. The monoclonal antibodies prepared by Vandaele et al.[30] precipitated desmethoxyverapamil-binding activity in addition to DHP-binding sites, and cross-reacted with brain, heart, and smooth muscle in immunoblot experiments. Although the subunit specifity of these antibodies was not determined, the antibodies precipitated α_1-, α_2-, γ-, and δ-subunits, supporting the conclusion that these polypeptides are tightly associated.

Monoclonal antibodies specific for the α_1 subunit were first prepared by Leung et al.[13] and by Morton and Froehner.[31] These antibodies, like the anti-α_2 antibodies, precipitate DHP-binding sites, and were used in immunoblot experiments to provide some of the first definitive evidence that α_1 and α_2 are two distinct proteins. In immunoblots of purified DHP receptor these antibodies recognized a 170–200 kDa protein that was not affected by reduction of disulfide bonds. The antibodies did not recognize the 140 kDa α_2-subunit. In the same series of experiments, the α_1- and α_2-subunits were further distinguished by the fact that the α_2-subunit binds ^{125}I-wheat germ agglutinin, while the α_1-subunit does not. In addition, the protein recognized by the anti-α_1 antibody could be labeled covalently by photoreactive DHPs,[25] while α_2 is not labeled by DHPs. Anti-α_1 antibodies do not recognize α_2 in immunoblots; conversely, anti-α_2 antibodies do not recognize α_1, indicating that the two subunits do not share common antigenic sites. However, the fact that both anti-α_1 and anti-α_2 antibodies precipitate DHP-binding activity, which resides on the α_1-subunit, and that both antibodies precipitate the α_1 and α_2 polypeptides from solubilized receptor preparations, supports the idea that α_1 and α_2, though structurally distinct, are normally tightly associated in a noncovalent complex.

A monoclonal antibody against the 52 kDa β-subunit is also able to precipitate DHP-binding sites from T-tubules, providing direct evidence that the β-subunit, like the α_2-subunit, is tightly associated with the DHP-binding α_1-subunit, and is not a contaminant in this preparation.[32] The anti-β antibody did not recognize α_1 in immunoblots. Furthermore, different peptide maps were obtained from the β- and α_1-subunits following limited proteolysis, indicating that the β-subunit is not a proteolytic fragment of the α_1-subunit. Considering the variability in the amount of the β-subunit detected in purified DHP receptor preparations from different laboratories, this immunological evidence for the association of the β-subunit with DHP-binding activity is particularly important.

Two monoclonal antibodies have been reported to exhibit functional effects on DHP-sensitive Ca^{2+}-channels. Malouf et al.,[33] by selecting antibodies specific for rabbit T-tubule membrane components, identified one antibody that appeared to increase the open probability of DHP-sensitive Ca^{2+}-channels incorporated into planar lipid bilayers. This antibody precipitated T-tubule proteins with $M_r = 175, 90, 55$, and 34 kDa. While the sizes of the 175, 55, and 34 kDa peptides are consistent with the sizes of DHP receptor subunits, the identities of the peptides were not confirmed, nor was the specificity of this antibody for the DHP receptor tested by precipitation of DHP-binding activity. With more complete characterization of the antibody, it may be possible to establish a functional relationship between one of these peptides and the Ca^{2+}-channel.

The anti-α_1 antibody developed by Morton et al.,[34] which has been shown to precipitate DHP-binding activity and is specific for the α_1-subunit, inhibits slow Ca^{2+} current in the BC3H1 muscle cell line.[31] This effect implies that the α_1-subunit of the DHP receptor is in fact part of the DHP-sensitive Ca^{2+}-channel. Combined with reconstitution experiments and the sequence data showing structural homology between the α_1-subunit and voltage-dependent Na^+-channels, this result provides one of the most compelling arguments for the ability of the DHP receptor to function as a voltage-dependent Ca^{2+}-channel.

6. PHOSPHORYLATION OF THE DHP RECEPTOR

Because cAMP-dependent protein kinase increases Ca^{2+}-currents in cardiac cells,[35-38] there has been considerable interest in determining whether the protein subunits of the DHP receptor are phosphorylated, and whether this is correlated with functional effects. Phosphorylation of the skeletal muscle DHP receptor was first investigated by Curtis and Catterall,[39] who showed that the β-subunit and one of the two α-subunits are phosphorylated in soluble form by cAMP-dependent protein kinase. In intact T-tubule membranes, only the β-subunit was significantly phosphorylated by exogenous cAMP-dependent protein kinase. Hosey et al.[40] also showed that one of the two α-subunits, subsequently determined to be the α_1-subunit (at this time the existence of two separate peptides, α_1 and α_2, was not fully appreciated), is phosphorylated by cAMP-dependent kinase and by Ca^{2+}/calmodulin-dependent kinase. However, these investigators concluded that the phosphorylated 54 kDa peptide that appeared in their preparations was not a separate β-subunit, but a proteolytic fragment of the α-subunit. Nastaínczyk et al.[41] resolved this issue by showing that the phosphorylated 165 kDa and 55 kDa subunits give rise to different phosphorylated tryptic fragments; the labeled 55 kDa subunit is therefore not a proteolytic fragment of the labeled α_1-subunit.

The first report of a functional effect of phosphorylation was based on reconstitution experiments in which the purified DHP receptor was incorporated into lipid bilayers and single Ca^{2+}-channel currents were measured.[42] The purified reconstituted DHP receptor sample contained single channels with the properties expected for DHP-sensitive Ca^{2+}-channels. When the catalytic subunit of cAMP-dependent kinase was added to the experimental solution, the open probability of the DHP-sensitive channel increased significantly. Using ^{32}P to label the purified DHP receptor, these investigators detected phosphorylation of 140 kDa (α_1) and 55 kDa (β) peptides, suggesting a correlation between the phosphorylation state of one of these two peptides and the open probability of the Ca^{2+}-channel. To determine which subunit, α_1 or β, is more likely to mediate the phosphorylation-induced increase in channel activity, Nastaínczyk et al.[41] compared the rates of phosphorylation of the two subunits. They showed that α_1 was preferentially phosphorylated by cAMP-dependent kinase, and proposed that phosphorylation of the α_1-subunit regulates Ca^{2+}-channel activity. These results must be interpreted with some caution, however, because it is not known whether the sites phosphorylated on the soluble receptor by exogenous kinase are identical to those phosphorylated *in vivo*.

Phosphorylation of the receptor by an intrinsic kinase activity in T-tubules was first

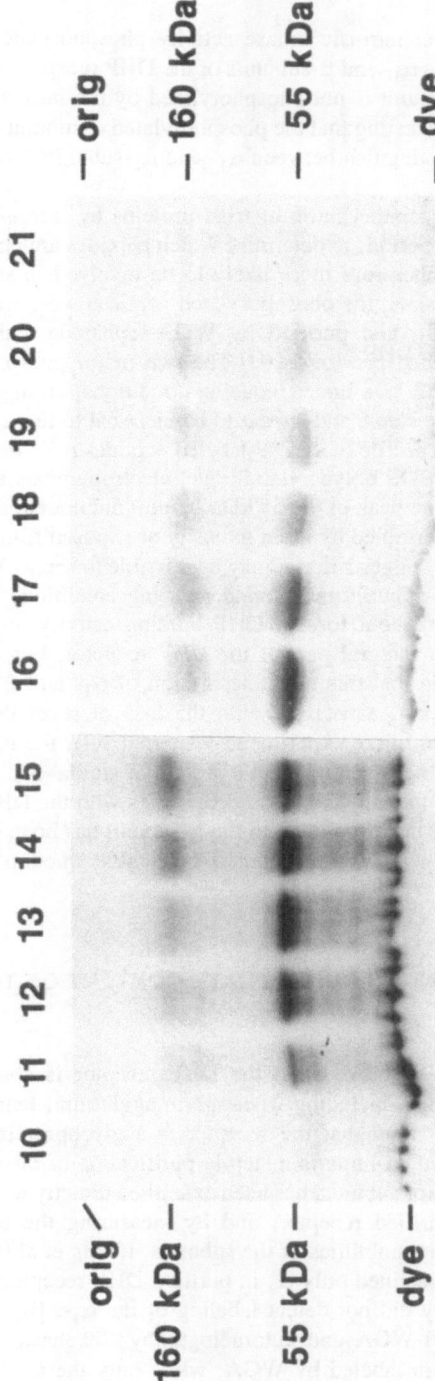

Figure 3. Autoradiogram of ^{32}P-labeled DHP receptor eluted from DEAE-Sephadex. Rabbit skeletal muscle T-tubule membranes were purified according to Lau et al.[62] The intact membranes were labeled by endogenous kinase activity in a reaction mixture containing 1.6 mg membrane protein, 125 mM sucrose, 80 mM tris-MOPS, 20 mM tris-HCl, pH 7, 2 mM EGTA, 10 mM MgSO$_4$. The reaction was initiated by adding [^{32}P]ATP (100 μM final concentration); the sample was incubated for 1 min at 25°C; then the reaction was terminated by adding two volumes of 50 mM NaF, 50 mM Na phosphate, pH 7, 20 mM EDTA. The ^{32}P-labeled membranes were mixed with 19 mg of unlabeled membranes and solubilized in digitonin. The DHP receptor was labeled with 2 mM ^{3}H-PN200-110 and purified by WGA-Sepharose and DEAE-Sephadex chromatography as described by Talvenheimo et al.,[11] with 20 mM NaF added to all buffers. The DEAE-Sephadex column was eluted with a linear 15–200 mM NaCl gradient. Aliquots (0.4 mL) from the fractions containing the peak of DHP receptor were concentrated and desalted by filtration through Centricon filters (10 kDa cut-off), then lyophilized, dissolved in sample buffer containing 20 mM dithiothreitol, and run on 4–15% gradient gels. The gels were dried prior to autoradiography.

described by Imagawa et al.[43] The intrinsic kinase activity phosphorylates several peptides in intact triads, including the α_1- and β-subunits of the DHP receptor. Imagawa et al.[43] also showed that the α_2-subunit is not phosphorylated by the intrinsic kinase activity or by exogenous kinases, suggesting that the phosphorylated α-subunit observed in earlier studies, which failed to distinguish between α_1- and α_2-subunits, was in fact the α_1-subunit.

We have also examined the phosphorylation of triad proteins by intrinsic kinase activity, but using shorter incubation periods to determine which peptides are phosphorylated on a short time-scale and are therefore more likely to be involved in short-term regulatory events. For these experiments, the phosphorylated peptides were solubilized from the membranes with digitonin, and purified by WGA-sepharose and DEAE-Sephadex chromatography as described previously.[11] The two major phosphorylated peptides, with M_r = 170 kDa and 55 kDa under reducing conditions, comigrate with DHP-binding activity in both of these steps, and appear to be identical to the α_1- and β-subunits. However, when the elution profile of the ^{32}P-labeled peptides from the DEAE-Sephadex column was analyzed by SDS polyacrylamide gel electrophoresis and autoradiography (Fig. 3), we found that the peak of the 55 kDa subunit did not coincide with the peak of the 170 kDa subunit (determined by grain intensity of exposed film). This is an intriguing observation, because it suggests that it may be possible to separate the 170 kDa α_1-subunit from the 55 kDa β-subunit under nondenaturing conditions. If the β peptide can be separated from α_1 without loss of DHP binding activity, this would suggest that the β-subunit is not an integral part of the DHP receptor, but a closely associated protein. It is also possible that this slight separation of α_1- and β-subunits represents denaturation of the receptor, associated with the loss of reversible DHP-binding activity; this will be tested in future experiments. Alternatively, the labeled 55 kDa peptide may not be identical to the β-subunit, but a protein of similar M_r. This last possibility seems very unlikely since the labeled peptide copurifies with the DHP receptor, and phosphorylation of a peptide corresponding to the β-subunit has been observed in highly purified receptor samples, which do not contain detectable amounts of other proteins in this size range.

7. HYDROPHOBIC DOMAINS AND CARBOHYDRATE CONTENT OF THE DHP RECEPTOR

In 1983 Glossmann and Ferry[5] showed that the DHP receptor is absorbed to several types of lectin affinity columns, including wheat germ agglutinin, lentil lectin, and Concanavalin A (Con A), indicating that the receptor is a glycoprotein. Lectin affinity chromatography has remained an important step in purification of the receptor. Glycosylation of the individual receptor subunits has been examined directly using lectin binding to Western blots of the purified receptor, and by measuring the effects of deglycosylation on the electrophoretic mobilities of the subunits. Leung et al.[13] found that WGA conjugated to peroxidase stained only α_2 in purified DHP receptor samples blotted on nitrocellulose paper. They did not detect labeling of the α_1-, β-, γ-, or δ-subunits by this method. Using ^{125}I-WGA and autoradiography, Takahashi et al.[24] found that the α_2-, γ-, δ-subunits are labeled by WGA, while only the α_2-subunit is labeled by ^{125}I-Con A. Both groups have shown that the α_2-subunit binds strongly to WGA-Sepharose, even in conditions where the α_1-subunit can be eluted. When the purified receptor was deglycosylated with neuraminidase and endoglycosidase F,[24] the electrophoretic mobilities of the α_2- and γ-subunits increased, while that of the α_1-subunit was unchanged. It is clear from these results that the α_2-, γ-, and δ-subunits are glycosylated, but the status of α_1 is ambiguous. While the α_1-subunit does not appear to be labeled in Western blots by WGA or Con A, one study has suggested that the electrophoretic mobility of α_1 is altered by endoglycosidase F treatment.[44] Although this result may have been caused by proteolytic cleavage, it is possible that α_1 is glycosylated, but to a much lesser extent than α_2. The heavy glycosylation of α_2, along with the apparent lack of glycosylation of α_1 (or at least much less extensive glycosylation), supports the conclusion that α_1 and α_2 are very closely associated, since both subunits are retained by lectin affinity columns that appear to interact primarily with the α_2-subunit. The functional significance of the glycosylation is not known.

Hydrophobic regions of the DHP receptor subunits have been detected by covalent labeling with a hydrophobic probe, 3-(trifluoromethyl)-3-(m-[125I]iodophenyl)diazirine ([125I]TID). [125I]TID preferentially labels the α_1- and γ-subunits of the purified DHP receptor.[24] This result has since been substantiated for the α_1-subunit by hydrophobicity analysis of the amino acid sequence.[45] The α_2-subunit was not significantly labeled by [125I]TID, confirming the structural difference between the α_1- and α_2-subunits.

8. RECONSTITUTION OF DHP-SENSITIVE CA²⁺-CHANNEL FUNCTION FROM THE PURIFIED DHP RECEPTOR

Because skeletal muscle appears to contain far more high-affinity DHP-binding sites than functional Ca^{2+}-channels, and these receptor sites have a higher apparent affinity for DHPs than the DHP-sensitive Ca^{2+}-channels in skeletal muscle,[46,47] there has been considerable controversy over whether the DHP receptor functions as a Ca^{2+}-channel. This issue has been addressed by reconstituting the purified receptor into lipid vesicles or planar lipid bilayers. When the isolated receptor was reconstituted into lipid vesicles, Curtis and Catterall[48] were able to measure channel-mediated isotope transport into the vesicles. This transport was blocked by DHPs and verapamil, and by inorganic Ca^{2+}-channel blockers such as La^{3+}, as predicted for ion transport through a DHP-sensitive Ca^{2+}-channel. A quantitative estimate of the number of DHP receptors present per vesicle and the initial rate of channel-mediated Ca^{2+} transport indicated that only 2–3% of the DHP receptors were functioning as ion channels, which is consistent with the idea that most of the DHP receptors in skeletal muscle do not normally conduct ions. The single-channel properties of a purified DHP receptor were first studied by incorporating the protein into bilayers formed on the tip of a patch pipette.[42] These authors identified divalent cation-selective channels with a single-channel conductance of 20 pS in purified DHP receptor preparations. The channels were activated by the DHP agonist Bay K 8644 and by the addition of cAMP-dependent protein kinase, and were inhibited by the verapamil derivative D600.

We used a similar approach to study single channels in purified DHP receptor

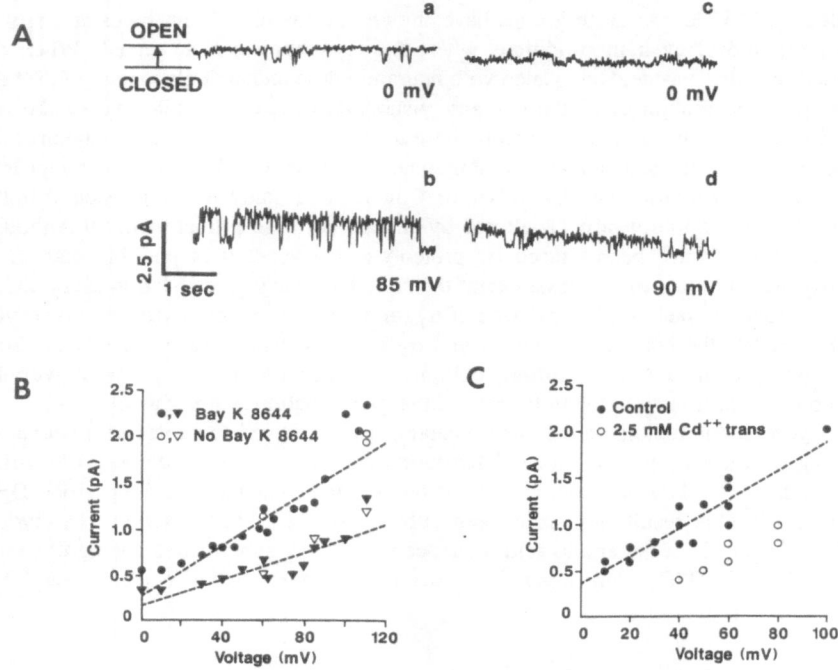

Figure 4. (A) Purified DHP receptor incorporated into planar lipid bilayers exhibits two conductance levels. DHP receptor purified to a specific binding activity of 1800 pmol/mg was incorporated into planar lipid bilayers as described in Talvenheimo et al.[11] The *cis* compartment contained 80 mM BaCl$_2$, the *trans* compartment contained 50 mM NaCl. Bay K 8644 (6 μM final concentration) was added to both sides of the bilayer. Single-channel currents were recorded at the indicated holding potentials. Records (a) and (b) were selected to show the larger conductance level; records (c) and (d), obtained from the same DHP receptor preparation, were selected to show the smaller conductance level. All records were filtered at 50 Hz. (B) Current–voltage relationships for purified DHP receptor. Single-channel current amplitudes are plotted versus holding potential for both the large conductance level (circles) and the small conductance level (triangles) exhibited by the purified DHP receptor preparation (1800 pmol/mg). Ionic conditions were exactly as described for (A). Open symbols ($\bigcirc$, $\triangledown$) indicate current levels measured in the absence of Bay K 8644, closed symbols ($\bullet$, $\blacktriangledown$) indicated current levels measured in the presence of 6 μM Bay K 8644 (both sides of the bilayer). The slope conductances, calculated by least square fit to the data between + 20 mV and + 100 mV, are 16 ± 3 pS for the larger current level and 7 ± 1 pS for the smaller current level. (C) Cadmium block of the 16 pS ca^{2+}-channel. Single-channel current amplitudes were measured as a function of membrane potential in the absence ($\bullet$) and presence ($\bigcirc$) of 2.5 mM cadmium (*trans* side only). Ionic conditions and DHP receptor preparation were identical to those used in (A). 6 μM Bay K 8644 was present on both sides of the bilayer. (This figure is adapted from Talvenheimo et al.[11])

samples. The purified receptor was reconstituted into lipid vesicles, then incorporated into planar lipid bilayers. We found that this preparation contains divalent cation-selective channels with two distinct conductance levels of 7 pS and 16 pS.[11] Figure 4A shows single-channel current records for each conductance level at two membrane potentials (see figure legend for details). The current–voltage relationships are shown in Fig. 4B. Both conductance levels were observed in the absence, as well as in the presence, of the DHP activator Bay K 8644 (Fig. 4B, open symbols). To characterize the 16 pS channel further, we tested its sensitivity to the inorganic Ca^{2+}-channel blocker Cd^{2+} and to the DHP antagonist nifedipine. When Cd^{2+} was added to one side of a bilayer containing only the 16 pS conductance level, it reduced the apparent single-channel conductance by about 40% (Fig. 4C). This is consistent with the fact that Cd^{2+} blocking events are too brief to be resolved in our recording system, giving rise to an apparent current reduction. The 16 pS conductance level is also blocked by nifedipine, as expected for a DHP-sensitive Ca^{2+}-channel. Figure 5A shows single-channel currents recorded before and after the addition of nifedipine. Although bursts of activity still occur, nifedipine clearly reduces the fraction of time the channel is open. Current amplitude histograms measured before and after the addition of nifedipine are shown in Fig. 5B. The fraction of time spent in the open state, which is proportional to the relative area under the current amplitude peak centered at 1 pA, is reduced from 0.46 to 0.15 by 2 μM nifedipine. We concluded from these experiments that the purified DHP receptor preparation does contain DHP-sensitive Ca^{2+}-channels, which are very similar to those observed in intact T-tubule membranes.[49]

Although we did not test the DHP sensitivity of the 7 pS channel in the purified receptor preparation, the occurrence of this conductance level in the absence of the 16 pS conductance level suggested that it might represent a second type of Ca^{2+}-channel that copurifies with the DHP receptor. Pelzer et al.[50] identified a similar conductance level in their purified receptor samples and found that it was not sensitive to DHPs. Since there is considerable evidence for at least two components of Ca^{2+} current in skeletal muscle,[51–53] including one that is insensitive to DHPs, the presence of two distinct types of Ca^{2+}-channel in these preparations would not be surprising. The observation does serve, however, to underscore one of the limitations of this approach to reconstituting channel function. Because observations are limited to single molecules, it is not possible to assess the purity of a preparation using this method or to attribute the observed channel activity to the major peptide component in a sample without corroborative evidence. For example, Morton et al.[34] demonstrated recently that a monoclonal antibody against the α_1-subunit is able to inhibit both whole-cell Ca^{2+} current and single Ca^{2+}-channels in cultured muscle cells. This type of evidence, if extended to purified DHP receptor samples, would support a functional connection between the observed single-channel activity and the polypeptide subunits in the purified sample.

An alternative explanation for the presence of two current levels in purified receptor samples is that the DHP-sensitive Ca^{2+}-channel exhibits multiple conductance levels. In support of this interpretation, Ma and Coronado[54] have described the occurrence of three conductance levels, 3 pS, 9 pS, and 12 pS in bilayers containing DHP-sensitive Ca^{2+}-channels. All three conductance levels were sensitive to DHPs, and the frequency of occurrence of the 3 pS and 9 pS events was reciprocally related to the frequency of the 12 pS events, suggesting they might be subconductance states of the same channel molecule. If this is the case, then the 7 and 16 pS current levels we observe in the purified receptor preparation may be identical to the 9 and 12 pS levels described by Ma and Coronado.[54] Another group[55] studying purified DHP receptors reconstituted into planar lipid bilayers has also described the occurrence of two conductance levels, 12–14 pS and 22 pS, which appeared to be statistically coupled to each other. Whether the 22 pS level is related to the 20 pS events reported by Flockerzi et al.[42] is unclear. Smith et

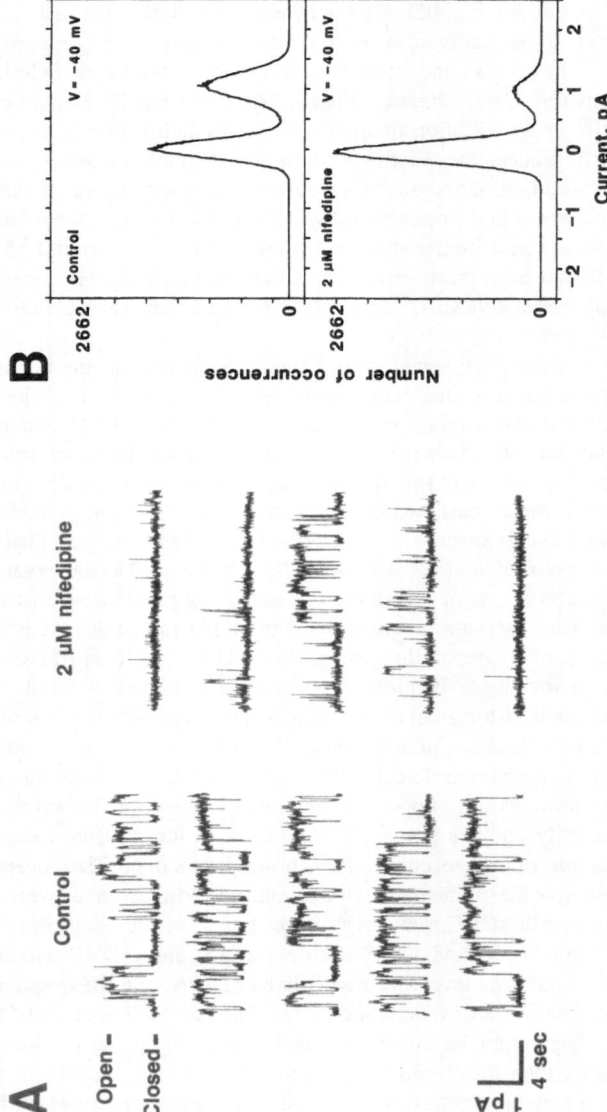

Figure 5. Inhibition of the 16 pS Ca^{2+}-channel by nifedipine. (A) Single-channel records before and after the addition of 2 μM nifedipine. DHP receptor purified to a specific binding activity of 2500 pmol/mg was reconstituted into a planar lipid bilayer. The *cis* and *trans* compartments contained 100 mM BaCl$_2$, and 7 μM Bay K 8644 was present on both sides of the bilayer. The membrane potential was held at −40 mV. After control records were obtained, nifedipine was added to both sides of the bilayer to a final concentration of 2 μM, and recording was resumed 5 min after the drug addition. The records shown were filtered at 100 Hz. (B) current amplitude histograms before and after the addition of nifedipine. Current amplitudes from the experiment in (A) were measured and are shown plotted in a frequency histogram. Fractional open and closed times were determined by measuring the area under each amplitude peak. (This figure is adapted from Talvenheimo et al.[11])

al.[55] report that 22 pS events were relatively infrequent, while Flockerzi et al.[42] reported this conductance level to be the predominant type of activity in their preparations. The major conductance level we have observed in purified receptor samples is 16 pS, close to that reported by Smith et al.,[55] but in intact T-tubules we have observed a divalent cation-selective channel with a conductance of about 25 pS.[11] Whether these represent different channel molecules remains open to question. We cannot exclude the possibility that the DHP-sensitive Ca^{2+}-channel exhibits different conductance levels that can interconvert, as suggested by Ma and Coronado. Some as yet unidentified factors, which may or may not be preserved during purification, could control this interconversion. It should be emphasized that the occurrence of multiple-conductance states of a single Ca^{2+}-channel and the presence of different types of Ca^{2+}-channel are not mutually exclusive possibilities. If multiple-subconductance states of one channel type coexist with different channel types, this, combined with differences in experimental conditions, may account for the diversity of the results obtained in reconstitution studies. This issue may be resolved unequivocably when DHP-sensitive Ca^{2+}-channels are produced by cDNA cloning and are studied in a membrane that does not contain other Ca^{2+}-channel types.

9. MOLECULAR CLONING OF THE DHP RECEPTOR α_1-SUBUNIT

The DHP-binding subunit of the DHP subunit, α_1, has been sequenced by methods similar to those used to sequence the nicotinic acetylcholine receptor and the voltage-dependent Na^+-channel. Tanabe et al.[45] purified the α_1 polypeptide by HPLC, treated it with trypsin, and then purified and sequenced several of the tryptic fragments. Based on the partial amino acid sequence for one of the fragments, oligonucleotide probes were synthesized to screen a cDNA libary for cDNAs encoding the DHP receptor. They were able to obtain a full-length cDNA for the DHP receptor. The receptor cDNA codes for a 1,873-amino-acid polypeptide with a calculated M_r of 212,018. Analysis of the DHP receptor sequence revealed the presence of four internal repeat regions within the polypeptide that exhibit considerable homology with each other, a feature also present in voltage-dependent Na^+-channels. These four internal repeats also display significant sequence homology to the corresponding regions in the rat brain type II Na^+-channel.[56] Each of these regions contains five hydrophobic segments and one positively charged segment, like the corresponding regions in the Na^+-channel. The structural similarities between the α_1 polypeptide of the DHP receptor and the Na^+-channel led Tanabe et al.[45] to propose that α_1, in addition to containing the drug-binding sites that modulate Ca^{2+}-channel function, also contains the ion-conducting channel, and that the DHP receptor gene shares a common ancestor with the Na^+-channel gene. This, of course, raises the obvious question, Is the α_1-subunit alone sufficient to form a DHP-sensitive voltage-dependent Ca^{2+}-channel, or are the other polypeptides required for function? Functional DHP-sensitive Ca^{2+}-channels have not yet been expressed from mRNA encoding the α_1-subunit, but it is not known whether this is due to a problem with the expression of the protein or to a requirement for additional receptor subunits.

The receptor subunits can be separated from one another by conventional methods, but this requires denaturing conditions because the subunits are normally so tightly associated. There have been no reports of receptor or channel function reconstituted from subunits purified to homogeneity by these methods. Based on the accumulated evidence from a number of laboratories, Catterall et al.[57] recently proposed a model for the structure of the DHP receptor/Ca^{2+}-channel complex in which the α_1-subunit forms the transmembrane ion-conducting pore. The α_1-subunit is associated with α_2-, β-, and γ-subunits. A δ-subunit is linked by disulfide bonds to the α_2-subunit. The α_2-, γ-, and δ-subunits are all glycosylated and therefore are predicted to be exposed at the external membrane surface. The β-subunit must be accessible from the interior of the cell because it is phosphorylated in the intact membrane, but there is no data to indicate that it spans the membrane. There is no information on the three-dimensional structure of the receptor, the folding patterns of the individual polypeptide chains within the membrane, or the nature of the subunit interactions. The stoichiometry of the subunit association (one of each subunit) is based on densitometric scans of SDS polyacrylamide gels containing the highly purified receptor[32] and on the hydrodynamic estimate of the size of cardiac DHP receptor complexes.[58] This model may be simplified when more information is obtained regarding the functional role of each of the constituents.

Although a primary subunit (and possibly the only required subunit) of the DHP receptor has been sequenced, there are still several gaps in our knowledge about the structure of this receptor/channel complex. If the α_1-subunit alone is sufficient to form a DHP-sensitive Ca^{2+}-channel, what functions do the other polypeptides serve? Two possible functions that may be served by the other subunits are the formation of structural associations between the DHP receptor and other components of the triad junction, and the regulation of channel activity. Electrophysiological data, cell fractionation studies, and immunochemical data all indicate that the DHP receptor is localized to the T-tubule membrane in skeletal muscle, but it is not known whether the receptor interacts specifically with other proteins in the T-tubule membrane or in the junctional sarcoplasmic reticulum (SR). The α_2-, β-, γ-, and δ-subunits may be required for anchoring the receptor protein, or for mediating interactions between the receptor and other cellular constituents. Alternatively, one or more of these subunits may be required for intracellular modulation of channel activity. It has become apparent that in addition to phosphorylation, Ca^{2+}-channels are modulated directly by guanine nucleotide-binding proteins[59] and may also be regulated by inositol polyphosphates. They therefore must contain a number of unidentified regulatory sites.

Do all DHP receptors function as Ca^{2+}-channels? The fact that the α_1-subunit contains the DHP-binding site and is structurally homologous to voltage-dependent Na^+-channels strongly suggests that all DHP receptors are capable of functioning as Ca^{2+}-channels. If only a small fraction of the DHP receptors actually conduct ions at any given time, as is suggested by physiological data, there must be factors that determine whether the receptors will conduct ions, but these have not been identified. Furthermore, the conditions that govern the appearance of multiple-conductance states of the DHP-sensitive channel have not been defined. If the primary function of the DHP receptor is to sense membrane potential across the T-tubule membrane as suggested by Rios and Brum,[60] and not to conduct ions at all, then a critical unsolved problem is the mechanism by which membrane potential changes are conveyed from the receptor to the SR. We have demonstrated an indirect association between the α_1-subunit and the SR

Ca^{2+} release channel (Chapter 23, this volume), but the functional significance of this interaction is still unknown.

ACKNOWLEDGMENTS. This work was supported by NIH Grants R01 HL 36029 (J.A.T.) and AM 21601 (A.H.C.), and by grants from the Muscular Dystrophy Association (J.A.T. and A.H.C.)

REFERENCES

1. Hidalgo, C., 1986, Isolation of muscle membranes containing functional ionic channels, in: *Ionic Channels in Cells and Model Systems* (R. Latorre, Ed.), pp. 101–125, Plenum, New York.
2. Norman, R. I., Borsotto, M., Fosset, M., Lazdunski, M., and Ellory, J., 1983, Determination of the molecular size of the nitrendipine-sensitive Ca^{2+}-channel by radiation inactivation, *Biochem. Biophys. Res. Comm.* **111:** 121–127.
3. Goll, A., Ferry, D., and Glossmann, H., 1983, Target size analysis of skeletal muscle Ca^{2+}-channels: Positive allosteric heterotropic regulation by D-*cis*-diltiazem is associated with apparent channel oligomer dissociation, *FEBS Lett.* **157:** 63–69.
4. Curtis, B. M., and Catterall, W. A., 1983, Solubilization of the calcium antagonist receptor from rat brain, *J. Biol. Chem.* **258:** 7280–7283.
5. Glossmann, H., and Ferry, D. R., 1983, Solubilization and partial purification of putative calcium channels labelled with [^{3}H]-nimodipine, *Naunyn-Schmied. Arch. Pharmacol.* **323:** 279–291.
6. Borsotto, M., Norman, R. I., Fosset, M., and Lazdunski, M., 1984, Solubilization of the nitrendipine receptor from skeletal muscle transverse tubule membranes, *Eur. J. Biochem.* **142:** 449–455.
7. Curtis, B. M., and Catterall, W. A., 1984, Purification of the calcium antagonist receptor of the voltage-sensitive calcium channel from skeletal muscle transverse tubules, *Biochemistry* **23:** 2113–2118.
8. Borsotto, M., Barhanin, J., Norman, R. I., and Lazdunski, M., 1984, Purification of the dihydropyridine receptor of the voltage-dependent Ca^{2+}-channel from skeletal muscle transverse tubules using (+)[^{3}H]PN200-110, *Biochem. Biophys. Res. Comm.* **122:** 1357–1366.
9. Borsotto, M., Barhanin, J., Fosset, M., and Lazdunski, M., 1985, The 1,4-dihydropyridine receptor associated with the skeletal muscle voltage-dependent Ca^{2+}-channel, *J. Biol. Chem.* **260:** 14255–14263.
10. Flockerzi, V., Oeken, H.-J., and Hofmann, F., 1986, Purification of a functional receptor for calcium-channel blockers from rabbit skeletal muscle microsomes, *Eur. J. Biochem.* **161:** 217–224.
11. Talvenheimo, J. A., Worley III, J. F., and Nelson, M. T., 1987, Heterogeneity of calcium channels from a purified dihydropyridine receptor preparation, *Biophys. J.* **52:** 891–899.
12. Sieber, M., Nastainczyk, W., Zubor, V., Wernet, W., and Hofmann, R., 1987, The 165 KDa peptide of the purified skeletal muscle dihydropyridine receptor contains the known regulatory sites of the calcium channel, *Eur. J. Biochem.* **167:** 117–122.
13. Leung, A. T., Imagawa, T., and Campbell, K. P., 1987, Structural characterization of the 1,4-dihydropyridine receptor of the voltage-dependent Ca^{2+}-channel from rabbit skeletal muscle, *J. Biol. Chem.* **262:** 7943–7946.
14. Cooper, C. L., Vandaele, S., Barhanin, J., Fosset, M., Lazdunski, M., and Hosey, M. M., 1987, Purification and characterization of the dihydropyridine-sensitive voltage-dependent calcium channel from cardiac tissue, *J. Biol. Chem.* **262:** 509–512.
15. Takahashi, M., and Catterall, W. A., 1987, Identification of an α-subunit of dihydropyridine-sensitive brain calcium channels, *Science* **236:** 88–91.
16. Ferry, D. R., Goll, A., and Glossmann, H., 1987, Photoaffinity labelling of the cardiac calcium channel. (−)-[^{3}H]azidopine labels a 165 kDa polypeptide, and evidence against a [^{3}H]-1,4-di-

hydropyridine-isothiocyanate being a calcium channel-specific affinity ligand, *Biochem. J.* **243:** 127–135.

17. Vaghy, P. L., Striessnig, J., Miwa, K., Knaus, H.-G., Itagaki, K., McKenna, E., Glossmann, H., and Schwartz, A., 1987, Identification of a novel 1,4-dihydropyridine- and phenylakylamine-binding polypeptide in calcium channel preparations, *J. Biol. Chem.* **262:** 14337–14342.

18. Kuo, T. H., Johnson, D. F., Tsang, W., and Wiener, J., 1987, Photoaffinity labeling of the calcium channel antagonist receptor in the heart of the cardiomyopathic hamster, *Biochem. Biophys. Res. Comm.* **148:** 926–933.

19. Ferry, D. R., Kampf, K., Goll, A., and Glossmann, H., 1985, Subunit composition of skeletal muscle transverse tubule calcium channels evaluated with the 1,4-dihydropyridine photoaffinity probe ^{3}H-azidopine, *EMBO J.* **4:** 1933–1940.

20. Horne, P., Triggle, D. J., and Venter, J. C., 1984, Nitrendipine and isoproterenol induce phosphorylation of a 42 K dalton protein that comigrates with affinity-labeled calcium channel regulatory subunit, *Biochem. Biophys. Res. Comm.* **121:** 890–898.

21. Sarmiento, J. G., Epstein, P. M., Smilowitz, H., Chester, P. N., Wehinger, E., and Janis, R. A., 1985, Photoaffinity labeling of the 1,4-dihydropyridine Ca^{2+}-channel binding site in cardiac, skeletal, and smooth muscle membranes, *Fed. Proc.* **44:** 164.

22. Campbell, K. P., Lipschutz, G. M., and Denney, G. H., 1984, Direct photoaffinity labeling of the high-affinty nitrendipine-binding site in subcellular membrane fractions isolated from canine myocardium. *J. Biol. Chem.* **259:** 5384–5387.

23. Galizzi, D.-P., Borsotto, M., Barhanin, J., Fosset, M., and Lazdunski, M., 1986, Characterization and photoaffinity labeling of receptor sites for the Ca^{2+}-channel inhibitors D-*cis*-diltiazem, (+/−)-bepridil, desmethoxyverapamil, and (+)-PN200-110 in skeletal muscle transverse tubule membranes, *J. Biol. Chem.* **261:** 1393–1397.

24. Takahashi, M., Seagar, M. J., Jones, J. F., Reber, B. F. X, and Catterall, W. A., 1987, Subunit structure of dihydropyridine-sensitive calcium channels from skeletal muscle, *Proc. Natl. Acad. Sci. USA* **84:** 5478–5482.

25. Sharp, A. H., Imagawa, T., Leung, A. T., and Campbell, K. P., 1987, Identification and characterization of the dihydropyridine-binding subunit of the skeletal muscle dihydropyridine receptor, *J. Biol. Chem.* **262:** 12309–12315.

26. Striessnig, J., Moosburger, K., Goll, D., Ferry, D. R., and Glossmann, H., 1986, Stereoselective photoaffinity labeling of the purified 1,4-dihydropyridine receptor of the voltage-dependent calcium channel, *Eur. J. Biochem.* **161:** 217–224.

27. Schmid, A., Barhanin, J., Coppola, T., Borsotto, M., and Lazdunski, M., 1986, Immunochemical analysis of subunit structures of 1,4-dihydropyridine receptors associated with voltage-dependent Ca^{2+}-channels in skeletal, cardiac, and smooth muscles, *Biochemistry* **25:** 3492–3495.

28. Takahashi, M., and Catterall, W. A., 1987, Dihydropyridine-sensitive calcium channels in cardiac and skeletal muscle membranes: Studies with antibodies against the α-subunits, *Biochemistry* **26:** 5518–5526.

29. Norman, R. I., Burgess, A. J., Allen, E., and Harrison, T. M., 1987, Monoclonal antibodies against the 1,4-dihydropyridine receptor associated with voltage-sensitive Ca^{2+}-channels detect similar polypeptides from a variety of tissues and species, *FEBS Lett.* **212:** 127–132.

30. Vandaele, S., Fosset, M., Galizzi, J.-P., and Lazdunski, M., 1987, Monoclonal antibodies that coimmunoprecipitate the 1,4-dihhydropyridine and phenylalkylamine receptors and reveal the Ca^{2+}-channel structure, *Biochemistry* **26:** 5–9.

31. Morton, M. E., and Froehner, S. C., 1987, Monoclonal antibody identifies a 200 kDa subunit of the dihydropyridine-sensitive calcium channel, *J. Biol. Chem.* **262:** 11904–11907.

32. Leung, A. T., Imagawa, T., Block, B., Franzini-Armstrong, C., and Campbell, K. P., 1988, Biochemical and ultrastructural characterization of the 1,4-dihydropyridine receptor from rabbit skeletal muscle, *J. Biol. Chem.* **263:** 994–1001.

33. Malouf, N. N., Coronado, R., McMahon, D., Meissner, G., and Gillespie, G. Y., 1987, Monoclonal antibody specific for the transverse tubular membrane of skeletal muscle activates the dihydropyridine-sensitive Ca^{2+}-channel, *Proc. Natl. Acad. Sci. USA* **84:** 5019–5023.

34. Morton, M. E., Caffrey, J. M., Brown, A. M., and Froehner, S. C., 1988, Monoclonal antibody to the α_1-subunit of the dihydropyridine-binding complex inhibits calcium currents in BC3H1 myocytes, *J. Biol. Chem.* **263:** 613–616.

35. Osterrieder, W., Brum, G., Hescheler, J., Trautwein, W., Flockerzi, V., and Hofmann, F., 1982, Injection of subunits of cyclic AMP-dependent protein kinase into cardiac myocytes modulates Ca^{2+} current, *Nature* **298:** 576–578.

36. Cachelin, A. B., de Peyer, J. E., and Kokubun, S., and Reuter, H., 1983, Ca^{2+}-channel modulation by 8-bromocyclic AMP in cultured heart cells, *Nature* **304:** 462–464.

37. Brum, G., Flockerzi, V., Hofmann, F., Osterrieder, W., and Trautwein, W., 1983, Injection of catalytic subunit of cAMP-dependent protein kinase into isolated cardiac myocytes, *Nature* **298:** 576–578.

38. Bean, B. P., Nowycky, M. C., and Tsien, R. W., 1984, β-adrenergic modulation of calcium channels in frog ventricular heart cells, *Nature* **307:** 371–375.

39. Curtis, B. M., and Catterall, W. A., 1985, Phosphorylation of the calcium antagonist receptor of the voltage-sensitive calcium channel by cAMP-dependent protein kinase, *Proc. Natl. Acad. Sci. USA* **82:** 2528–2532.

40. Hosey, M. M., Borsotto, M., and Lazdunski, M., 1986, Phosphorylation and dephosphorylation of dihydropyridine-sensitive voltage-dependent Ca^{2+}-channel in skeletal muscle membranes by cAMP-and Ca^{2+}-dependent processes, *Proc. Natl. Acad. Sci. USA* **83:** 3733–3737.

41. Nastainczyk, W., Rohrkasten, A., Sieber, M., Rudolph, C., Schachtele, C., Marme, D., and Hofmann, F., 1987, Phosphorylation of the purified receptor for calcium-channel blockers by cAMP kinase and protein kinase C, *Eur. J. Biochem.* **169:** 137–142.

42. Flockerzi, V., Oeken, H.-J., Hofmann, F., Pelzer, D., Cavalie, A., and Trautwein, W., 1986, Purified dihydropyridine-binding site from skeletal muscle T-tubules is a functional calcium channel, *Nature* **323:** 66–68.

43. Imagawa, T., Leung, A. T., and Campbell, K. P., 1987, Phosphorylation of the 1,4-dihydropyridine receptor of the voltage-dependent Ca^{2+}-channel by an intrinsic protein kinase in isolated triads from rabbit skeletal muscle, *J. Biol. Chem.* **262:** 8333–8339.

44. Hosey, M. M., Barhanin, J., Schmid, A., Vandaele, S., Ptasienski, J., O'Callahan, C., Cooper, C., and Lazdunski, M., 1987, Photoaffinity labeling and phosphorylation of a 165-kDa peptide associated with dihydropyridine and phenylalklamine-sensitive calcium channels, *Biochem. Biophys. Res. Comm.* **147:** 1137–1145.

45. Tanabe, T., Takeshima, H., Mikami, A., Flockerzi, V., Takahashi, H., Kangawa, K., Kojima, M., Matsuo, H., Hirose, T., and Numa, S., 1987, Primary structure of the receptor for calcium-channel blockers from skeletal muscle, *Nature* **328:** 313–318.

46. Almers, W., Fink, R., and Palade, P. T., 1981, Calcium depletion in frog muscle tubules: The decline of calcium current under maintained depolarization, *J. Physiol.* **312:** 177–207.

47. Schwartz, L. M., McCleskey, E. W., and Almers, W., 1985, Dihydropyridine receptors in muscle are voltage-dependent, but most are not functional calcium channels, *Nature* **314:** 747–751.

48. Curtis, B. M., and Catterall, W. A., 1986, Reconstituion of the voltage-sensitive calcium channel purified from skeletal muscle transverse tubules, *Biochemistry* **25:** 3077–3083.

49. Affolter, H., and Coronado, R., 1985, Agonists Bay-K8644 and CGP-28392 open channels reconstituded from skeletal muscle transverse tubules, *Biophys. J.* **48:** 341–347.

50. Pelzer, D., Cavalie, A., Flockerzi, V., Hofmann, F., and Trautwein, W., 1987, Two types of Ca channels from skeletal muscle transverse-tubule membranes in lipid bilayers: Differences in conductance properties, gating kinetics, and chemical modulation, *Pflüg. Arch.* **408**(Suppl. 1): R35.

51. Cota, G., and Stefani, E., 1986, A fast-activated inward calcium current in twitch muscle fibers of the frog (*Rana montezume*), *J. Physiol.* **370:** 151–163.

52. Beam, K. G., Knudson, C. M., and Powell, J. A., 1986, A lethal mutation in mice eliminates the slow calcium current in skeletal muscle cells, *Nature* **320:** 168–170.

53. Cognard, C., Lazdunski, M., and Romey, G., 1986, Different types of Ca^{2+}-channels in mammalian skeletal muscle cells in culture, *Proc. Natl. Acad. Sci. USA* **83:** 517–521.

54. Ma, J., and Coronado, R., 1988, Heterogeneity of conductance states in calcium channels of skeletal muscle, *Biophys. J.* **53:** 387–395.
55. Smith, J. S., McKenna, E. J., Ma, J., Vilven, J., Vaghy, P. L., Schwartz, A., and Coronado, R., 1987, Calcium-channel activity in a purified dihydropyridine-receptor preparation of skeletal muscle, *Biochemistry* **26:** 7182–7188.
56. Noda, M., Ikeda, T., Kayano, T., Suzuki, H., Takeshima, H., Kurasaki, M., Takahashi, H., and Numa, S., 1986, Existence of distinct sodium channel messenger RNAs in rat brain, *Nature* **320:** 188–192.
57. Catterall, W. A., Seagar, M. J., and Takahashi, M., 1988, Molecular properties of dihydropyridine-sensitive calcium channels in skeletal muscle, *J. Biol. Chem.* **263:** 3535–3538.
58. Horne, W. A., Weiland, G. A., and Oswald, R. E., 1986, Solubilization and hydrodynamic characterization of the dihydropyridine receptor from rat ventricular muscle, *J. Biol. Chem.* **261:** 3588–3594.
59. Yatani, A., Codina, J., Imoto, Y., Reeves, J. P., Birnbaumer, L., and Brown, A. M. 1987, A G protein directly regulates mammalian cardiac calcium channels, *Science* **238:** 1288–1292.
60. Rios, E., and Brum, G., 1987, Involvement of dihydropyridine receptors in excitation–contraction coupling in skeletal muscle, *Nature* **325:** 717–720.
61. Hidalgo, C., Parra, C., Riquelme, G., and Jaimovich, E., 1986, Transverse tubules from frog skeletal muscle: Purification and properties of vesicles sealed with the inside-out orientation, *Biochim. Biophys. Acta.* **855:** 79–88.
62. Lau, Y. H., Caswell, A. H., and Brunschwig, J. P., 1977, Isolation of transverse tubules by fractionation of triad junctions of skeletal muscle, *J. Biol. Chem.* **252:** 5565–5574.

Chapter 21

Role of Slow Inward Calcium Current in Excitation–Contraction Coupling

Vincent Jacquemond and Oger Rougier

1. INTRODUCTION

The role of external calcium in the process of excitation–contraction coupling of skeletal muscle fibers had been investigated extensively. Removal of external calcium has been shown to not prevent the contractile activation, but it does have distinct effects on contracture amplitude and duration and on the voltage-dependence of contractile inactivation.[1–5] Furthermore, it has been demonstrated recently that the magnitude and time course of myoplasmic calcium transients elicited by membrane depolarization depend strongly on extracellular calcium.[6]

Calcium influx through slow calcium channels had been supposed to play a role in the development of prolonged contractile responses.[7,8] Two main lines of results, however, have tended to exclude such a possibility: (i) potassium contractures are prolonged when external calcium is replaced by the calcium channel blockers nickel or cobalt[9,10]; and (ii) myoplasmic calcium transients are not reduced during voltage-clamp pulses to large positive potentials at which calcium current approaches its equilibrium potential.[11]

The calcium channel blocking agents from the 1,4-dihydropyridine (DHP) class of compounds act as inhibitors of excitation–contraction coupling in different experimental situations.[12–16] A close relationship between tubular DHP receptors and voltage-

VINCENT JACQUEMOND and OGER ROUGIER • Laboratoire de Physiologie des Elements Excitables, Université Claude Bernard, F-69622 Villeurbanne Cedex, France.

sensing molecules linked to Ca^{2+} release from the sarcoplasmic reticulum (SR) may be implicated in the above effects.[14,17,18]

We report here experiments showing that treatments which affect the slow calcium channel activity also affect the simultaneously recorded contractile responses. Part of these results seems difficult to interpret without taking into account a role of Ca^{2+} influx through slow calcium channels in the coupling process.

2. METHODS

Experiments were performed at 18°C on intact single fibers isolated from the semitendinosus of the frog *Rana esculenta*. Ionic currents and contraction were recorded in the test gap (250 μm in length) of a double sucrose-gap device as previously described.[19] Linear ionic and capacitative currents were not subtracted.

The recording solution used contained (mM): Tetraethylammonium methylsulphonate (TEA-CH_3SO_3), 122; $CaCl_2$, 1.8; 4-aminopyridine, 3; glucose, 5; HEPES, 10, pH 7.4. In some experiments, 1.8 or 5 mM of Ca-$(CH_3SO_3)_2$ was used instead of 1.8 mM of

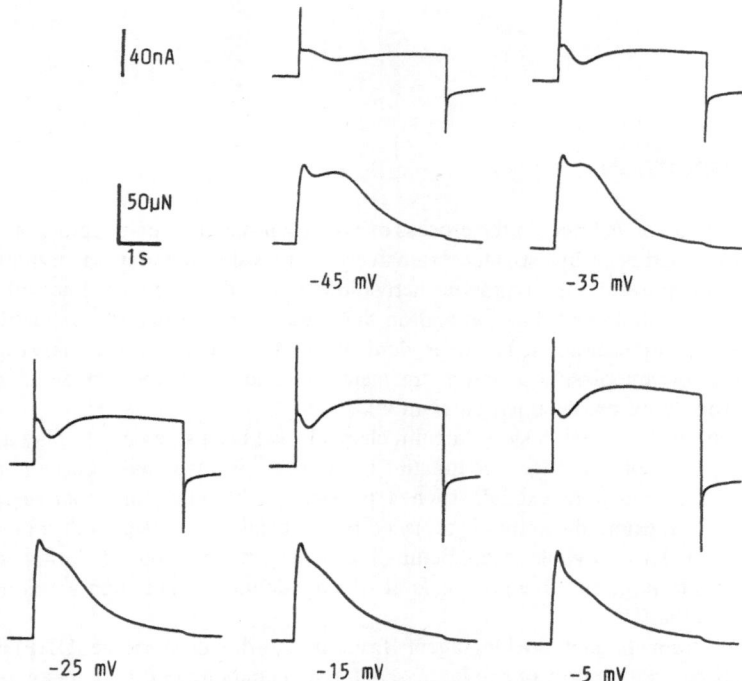

Figure 1. Simultaneous control recordings of membrane current (upper traces) and mechanical tension (lower traces) for step depolarizations from a holding potential of −90 mV to the values respectively indicated under each pair of traces. Successive pulses were delivered at a frequency of 1/min. Notice the clear biphasic time course of the contractile responses for low values of depolarization.

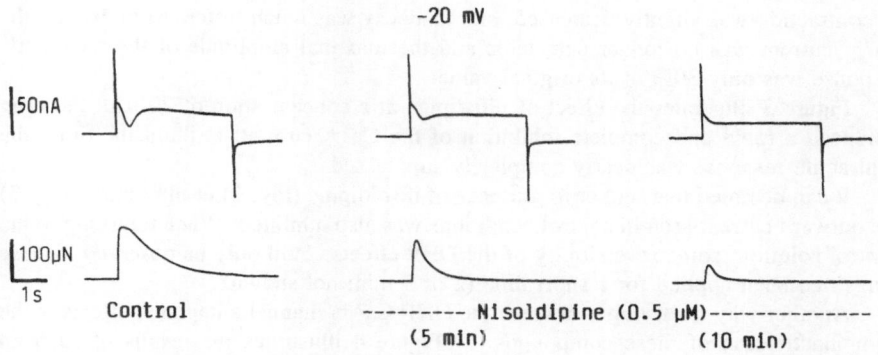

Figure 2.   Effects of nisoldipine (0.5 μM) on membrane current and mechanical tension. Left, control; middle and right, respectively, 5 and 10 min after nisoldipine was applied. Note the simultaneous decrease of Ca^{2+} current and contraction.

$CaCl_2$. Isotonic mannitol was used in the isolating compartments. At the beginning of each experiment, the holding potential (HP) was chosen so that the rheobasic depolarization for contractile threshold was 40 mV. In these conditions, the holding potential can be assumed to have a value of −90 mV. When long-lasting step depolarizations (> 1 sec) were used, this was done at a frequency of 1/min. The experiments with dihydropyridines were performed in the dark. Light inactivation of these compounds was performed with a flashlamp (similar to the method used by Nerbonne et al.).[20]

2.1.   Calcium Current and Contraction in Control Conditions

Figure 1 presents simultaneous recordings of current and contraction for different depolarizing steps in a fiber exposed to the control solution. It should be noted that the time course and amplitude of the inward current are altered by the presence of a remaining outward current whose relative contribution to the whole current varied greatly from one fiber to another. For a pulse to −45 mV, a slight and slow Ca^{2+} current was activated; the contractile response was composed of an early, rapidly rising phase followed by a slower one. Such a clear dissociation between the two phases of contraction was not observed in every tested fiber. When the fiber was depolarized to −35 and then to −25 mV, the calcium current became clearly activated and the time course of the slow contractile phase appeared to be correlated closely with that of the slow Ca^{2+} current. For higher values of depolarization, the Ca^{2+} current activation was faster, and the two contractile phases were less discriminated. These observations led us to suggest that the entry of Ca^{2+} through slow Ca^{2+}-channels might be a relevant signal to control the time course of prolonged contractile responses.

2.2.   Effects of Dihydropyridine Calcium Channel Antagonists

The effect of nisoldipine at a concentration of 0.5 μM is illustrated in Fig. 2. After 5 min in the presence of nisoldipine, Ca^{2+} current was reduced, the maximal amplitude of contraction was slightly decreased, and its decay was much faster. After 10 min the Ca^{2+} current was no longer detectable and the maximal amplitude of the contractile response was only 30% of its original value.

Figure 3 illustrates the effect of nifedipine at a concentration of 20 μM: the drug produced a rapid and complete inhibition of the Ca^{2+} current; at the same time, the contractile response was nearly completely suppressed.

It can be noted that, either in presence of nisoldipine (Fig. 2) or nifedipine (Fig. 3) the outward current seen in control conditions was also inhibited. When returning to the control solution, some reversibility of the DHP effects could only be observed if these drugs had been applied for a short time (2 or 3 min; not shown).

Another way for trying to reverse the DHP Ca^{2+}-channel antagonist effects is the light inactivation of these compounds.[20] Figure 4 illustrates the results of such an experiment: the recordings were obtained with a depolarizing pulse to -30 mV. As already described, nifedipine at a concentration of 0.5 μM produced a progressive inhibitory effect on Ca^{2+} current and contraction. When a single flash was delivered just before stimulating the fiber in presence of nifedipine, it was poorly effective in producing a partial reversibility (not shown). On the other hand, after returning to the control solution, a partial reversibility of the nifedipine effects was obtained after a flash. The outward current inhibition by nifedipine was not reversed and thus, Ca^{2+} current was observed more clearly after returning to control conditions than in control conditions.

From these experiments, it seems that the DHP Ca^{2+}-channel antagonists are potent inhibitors of the whole prolonged contractile responses. To test the effects of these compounds on the contractile responses independent of their effects on Ca^{2+} current, we performed experiments using voltage-clamp steps of 50 msec duration. For such durations, the slow calcium current is not, or is only poorly, activated in control condi-

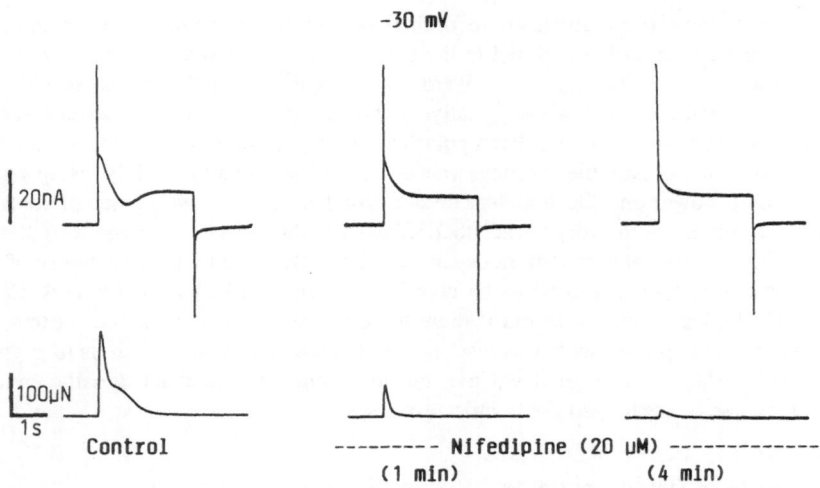

Figure 3. Effects of nifedipine (20 μM) on membrane current and mechanical tension. Left, control: middle and right, respectively, 1 and 4 min after nifedipine was applied. Note that the contraction is almost entirely suppressed.

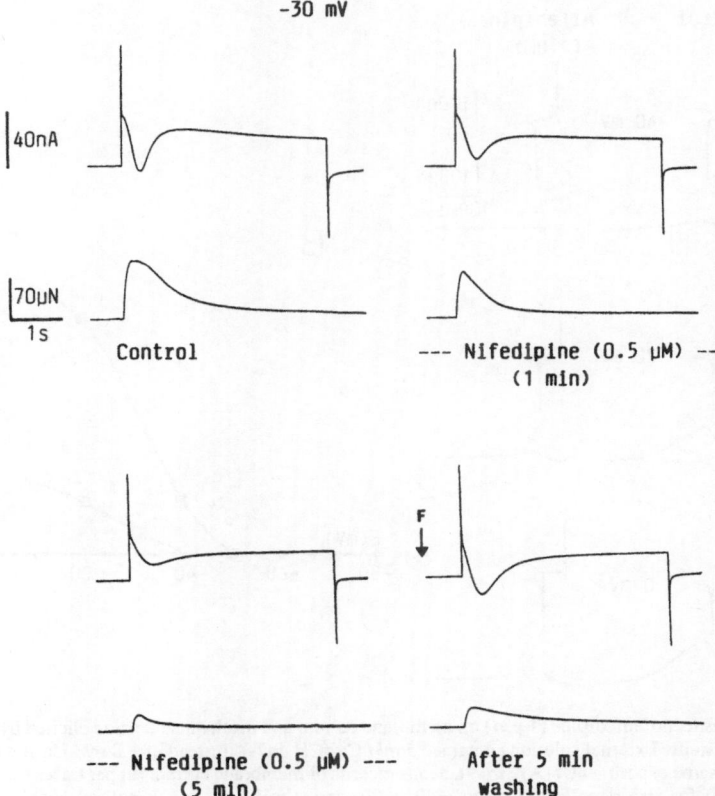

Figure 4.   Photoremoval of nifedipine-induced blockade of Ca^{2+} current and mechanical tension. After washing, single flashes (F) were delivered between each successive depolarizing step. Partial reversibility of the nifedipine-induced blockade of inward current and contractile response can be seen.

tions. Figure 5 illustrates the results of such an experiment: in the presence of 1 μ*M* nifedipine. Currents were not modified, but contraction was dramatically decreased for each potential, as can be seen from the peak tension–voltage relationship (Fig. 5B).

This observation confirms that the effect of DHP Ca^{2+}-channel antagonists on the development of contractile responses is not only related to their Ca^{2+} entry-blocking potency, which agrees with the results of Rios and Brum[14] and Gamboa-Aldeco et al.[15]

2.3.   Effects of (±) Bay K 8644: A DHP Ca^{2+}-Channel Agonist

Figure 6 presents the effects of (±) Bay K 8644 for a test depolarizing step to −30 mV. The amplitude of the Ca^{2+} current was increased considerably in the presence of the drug, and its time to peak was largely decreased. The experiment shows that there is

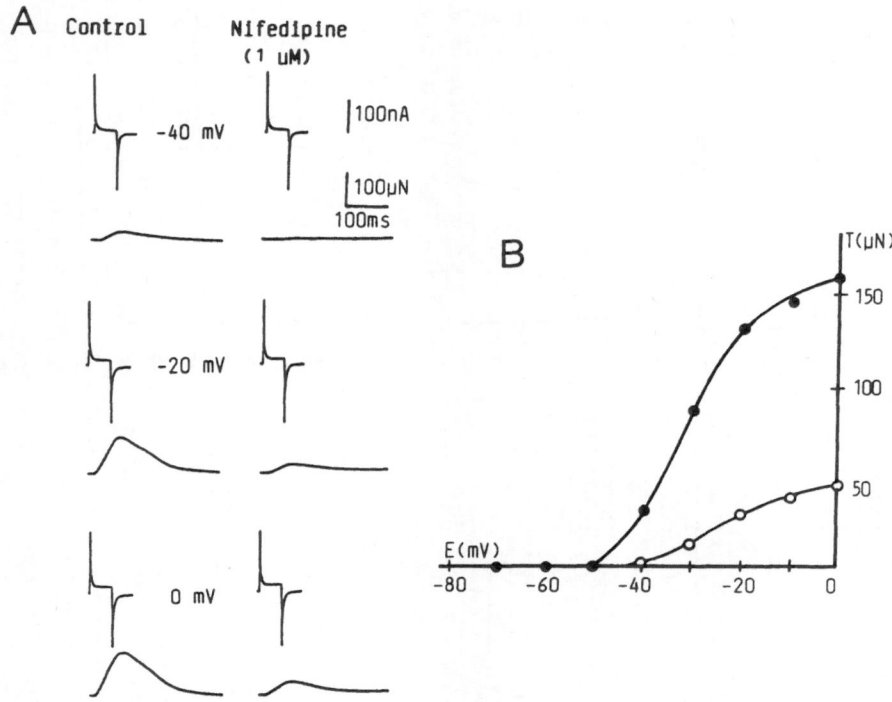

Figure 5. Effect of nifedipine (1 μM) on membrane current and mechanical tension elicited by 50 msec depolarizing steps. External solution contained 5 mM Ca-(CH$_3$SO$_3$)$_2$ instead of 1.8 mM CaCl$_2$. (A and B are from the same experiment.) (A) Simultaneous records of membrane current (upper traces) and tension (lower traces) for step depolarizations to -40, -20, and 0 mV. Left panel, control conditions; right panel, after 12 min exposure to nifedipine. (B) Peak–tension versus voltage relationship in control conditions ($\bullet$) and with nifedipine added ($\bigcirc$).

a close correlation between the evolution of the time course of the Ca^{2+} current and that of the slow phase of contraction. In addition, the maximal amplitude of the two events was increased simultaneously. This effect was difficult to reverse when returning to the control solution. Furthermore, the early phase of contraction was decreased considerably in the presence of the drug. Figure 7 presents the results of an experiment designed to further investigate this effect. The test depolarizing step was to -30 mV. Increasing concentrations of ($\pm$) Bay K 8644 were successively applied to the preparation, each of them during 4 min. It can be seen that the drug produced a dose-dependent inhibitory effect on the early contractile phase (see the arrows), which was maximum at 0.1 μM. For this and higher concentrations of Bay K 8644, the slow phase of contraction was easily discriminated.

In summary, the effects of ($\pm$) Bay K 8644 on the membrane processes controlling the prolonged contractile responses seem to be due to a dual action of this molecule: On

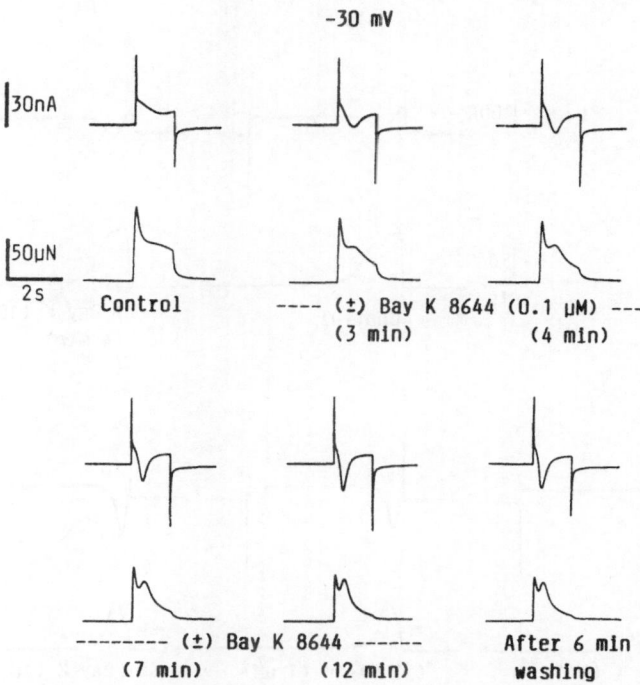

Figure 6. Effects of ($\pm$) Bay K 8644 (0.1 μM) on Ca^{2+} current and mechanical tension. External solution contained 1.8 mM Ca-$(CH_3CO_3)_2$ instead of 1.8 mM $CaCl_2$. Note the progressive decrease of the early phase of contraction and the modification of the slow phase during the increase in Ca^{2+} current.

one hand, it favored the development of the slow phase of contraction; on the other hand, it inhibited the early one.

Because we observed an inhibitory effect of nifedipine on the whole prolonged contractile responses (see Figs. 2, 3, and 4), we tested the effect of Bay K 8644 after having applied nifedipine; if these two drugs have a similar inhibitory action on a membrane process linked to SR Ca-release, the resulting effect of applying Bay K 8644 in these conditions should thus only be due to its Ca^{2+}-channel agonist properties. Results of such an experiment are presented in Figs. 8 and 9.

In control conditions (Fig. 8), for a test depolarizing step to -30 mV, there was not a clear dissociation between the two phases of contraction. Applying nifedipine at a concentration of 0.5 μM produced a complete inhibition of the contractile response. After returning to the control solution and applying several flashes, the inward current inhibition was reversed, and a slight contractile response could be obtained (Fig. 8; lower pairs of recordings). In Fig. 9 the control recordings represent current and contraction after partial photoremoval of nifedipine in the previous experiment (same as in Fig. 8; lower right pair of traces). When ($\pm$) Bay K 8644 was added at a concentration of 1 μM, it produced a parallel increase in the amplitude of the inward current and of the

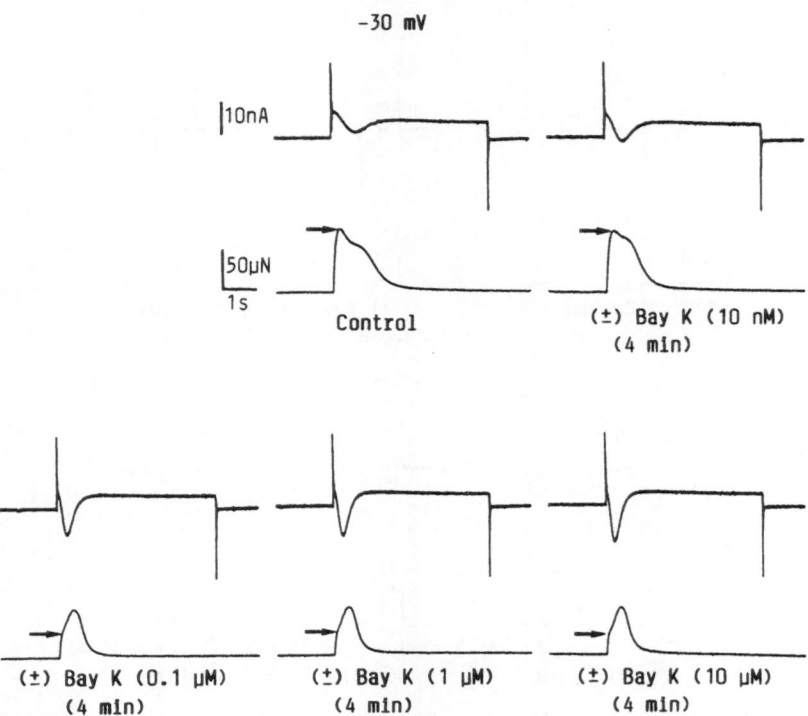

Figure 7. Depressing effect of (±) Bay K 8644 on the early rapid phase of a prolonged contractile response. Increased concentrations of (±) Bay K were successively applied on the fiber during 4 min. Arrows indicate the estimated maximal amplitude of the early contractile phase. Note that the two components of contraction are clearly separated; the slow phase, which becomes higher than the early one, follows the time course of the inward Ca^{2+} current.

slow phase of contraction. These results agree with a clear correlation between Ca^{2+} entry and the development of a slow contractile response.

2.4. Effects of External Ca^{2+} Substitution by the Ca^{2+} Channel Blockers Ni^{2+} or Co^{2+}

The results presented in this section were obtained either in presence of Ni^{2+} or of Co^{2+}. Steady-state comparative records of current and contraction obtained either in control conditions (upper pairs of traces) or 5 min after external Ca^{2+} had been replaced by Ni^{2+} (lower pairs of traces) are presented in Fig. 10 (same fiber as in Fig. 1). For depolarizations to -25 and -5 mV, the maximal amplitude and the area under the contractile responses were larger in presence of Ni^{2+} than in control conditions. For a depolarization to -45 mV the contractile response was smaller in presence of Ni^{2+}, this being due to a shift of the tension activation curve toward more positive potentials.[9] These results agree with previous observations[9,10] and thus indicate that the inward

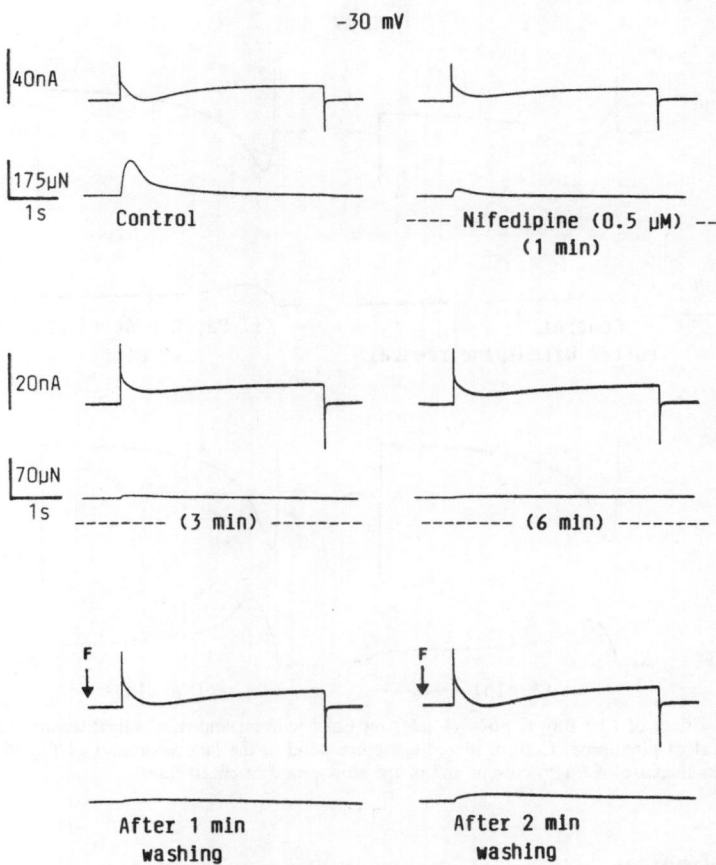

Figure 8. Partial photoremoval of nifedipine-induced blockade of Ca^{2+} current and mechanical tension. On this preparation, nifedipine (0.5 μM) induced a nearly complete blockade of the contraction. After nifedipine removal, single flashes (F) were delivered before each depolarizing step (lower pairs of traces). Note the higher gains for the four last sets of recordings.

Ca^{2+} current is not necessary to maintain the duration of prolonged contractile responses.

It is possible, however, to distinguish the effect of inward Ca^{2+} current suppression from the steady-state effect of Ni^{2+} or Co^{2+}; Fig. 11 illustrates recordings of current and contraction at different times following a Co^{2+} for Ca^{2+} substitution. The test depolarization was to -30 mV. The control recordings were obtained after removal of Bay K 8644 so that the two contractile phases were clearly distinct (same fiber as in Fig. 7). Immediately after the substitution, the inward Ca^{2+} current and the corresponding slow phase of tension were suppressed. During the following stimulations, the contraction increased again, together with a decrease in its rate of relaxation. Attaining steady-state conditions for this effect took between 5 and 10 min.

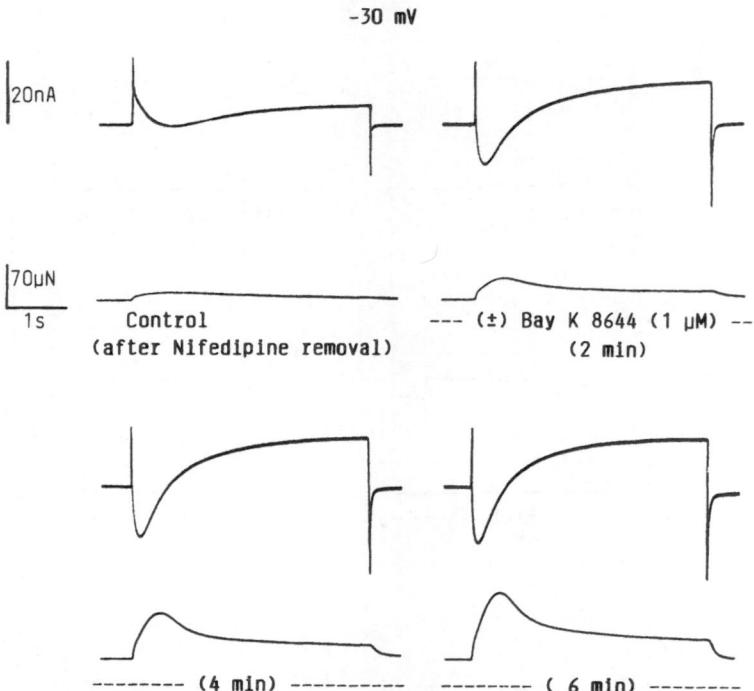

Figure 9. Effect of (±) Bay K 8644 (1 μM) on Ca^{2+} current and mechanical tension after partial photoremoval of nifedipine. Control recordings correspond to the last recordings of Fig. 8. Note the simultaneous increase of Ca^{2+} current and of the slow phase of contraction.

3. DISCUSSION

Our results show that different treatments that modulate the Ca^{2+}-channel activity also affect the simultaneously recorded contractile responses. However, none of the reported effects can be solely and simply interpreted in terms of increased or decreased Ca^{2+} entry through slow Ca^{2+} channels.

In control conditions, the second phase of the contractile responses elicited by long-lasting membrane depolarizations presented a time course that seemed to depend on the kinetics of Ca^{2+} current over a range of membrane potentials from -40 to -20 mV. This observation led us to consider that the inward flux of Ca^{2+} could be important to control the duration of these contractile responses.

Previous studies designed to search for the role of the slow inward Ca^{2+} current have generated controversial hypotheses: (i) Ca^{2+} current may be responsible for a part of the prolonged contractile responses, through a mechanism of Ca^{2+}-induced Ca^{2+} release at the SR level[8,21]; (ii) Ca^{2+} current may have no role in the excitation–contraction coupling process[22]; (iii) Ca^{2+} current may shorten the duration of contractures by causing ionic depletion in the T-tubules.[10]

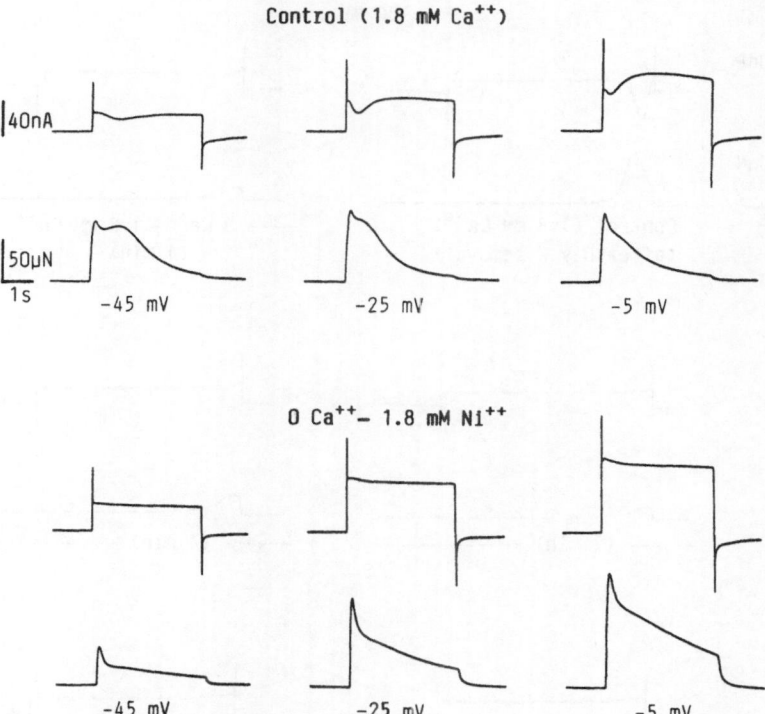

Figure 10. Comparative recordings of membrane current and mechanical tension in control (1.8 mM CaCl₂) and after 5 min (steady action) in 1.8 NiCl₂. Note that in NiCl₂ the calcium inward current is completely suppressed but that the contraction is greater for depolarizations to −25 and −5 mV in NiCl₂ (same fiber as in Fig. 1).

The possibility of a Ca^{2+} current dependent phase of contraction was tested further by studying the role of agents acting on Ca^{2+} channels.

3.1. Experiments with Inorganic Ca^{2+}-Channel Blockers

Substituting the external Ca^{2+} by Ni^{2+} or Co^{2+} produced a rapid simultaneous suppression of Ca^{2+} current and of the slow contractile component, thus confirming the proposal that these two events are highly correlated. However, the steady-state contractile responses often presented a larger amplitude and always a slower decay phase than in the presence of Ca^{2+}. Lorkovic,[23] Caputo,[9] and Lorkovic and Rudel[10] reported similar observations on potassium contractures experiments. Their observations were made at the steady state, so that the initial decrease of the slow Ca^{2+} current dependent contractile phase could not be observed.

The potentiating effect on potassium contractures of Ni and Co led Lorkovic and Rudel[10] to suggest that the presence of external divalent cations is important for the function of membrane mechanisms implicated in excitation–contraction coupling. The

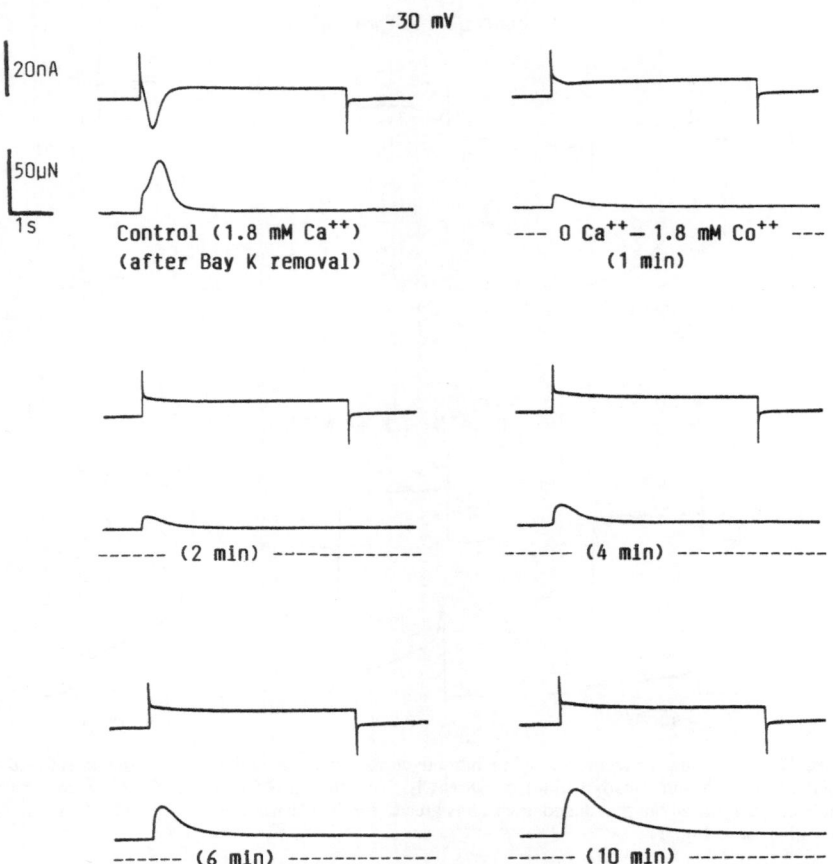

Figure 11. Effects of a Co^{2+} for Ca^{2+} substitution on membrane current and mechanical tension. The control recordings were obtained after removal of ($\pm$) Bay K 8644 (10 μM). Time after substitution is indicated under each pair of recordings. Note the immediate suppression of Ca^{2+} current and of the second phase of contraction, and the progressive increase of the first phase of contraction.

recent finding by Brum et al.[6] (see Chapter 25, this volume), that the tubular "voltage sensor" linked to SR Ca^{2+} release needs the presence of external cations to operate confirms this hypothesis.

In line with this idea, one can suppose that the rapid decay of contraction observed in our experiments when Ca^{2+} current is flowing is due to a tubular calcium depletion.[24] Another possibility to explain the potentiating action of Ni (and Co) might be an increase in the intracellular concentration of IP_3 (a possible link in excitation–contraction coupling) since Ni^{2+} has been proposed to act as an inhibitor of IP_3-phosphatase.[25]

3.2. Experiments in the Dihydropyridine Series

The tested DHP Ca^{2+}-channel antagonists nisoldipine and nifedipine inhibited the slow Ca^{2+} current and the corresponding phase of contraction. Furthermore, the early, rapidly rising phase of contraction was also reduced dramatically, as confirmed in experiments with short depolarizing pulses where the slow Ca^{2+} current was not activated. This is in agreement with other reports showing that DHP Ca^{2+}-channel antagonists have an inhibitory action on charge movements, Ca^{2+} transients, and contractile responses in different experimental conditions,[13–15, 26] and thus that DHP receptors could be closely linked to SR Ca^{2+}-release.

The DHP Ca^{2+}-channel agonist ($\pm$) Bay K 8644 had a more complex action. This compound increased in a parallel manner the amplitude and the kinetics of the Ca^{2+} current and of the slow phase of contraction, so that this slow contractile component looked like the mirror image of the Ca^{2+} current. Like DHP Ca^{2+}-channel antagonists, Bay K 8644 also decreased the first rapidly rising phase of contraction. This dual action of Bay K 8644 led to experimental situations where the two phases of contraction were clearly dissociated (Fig. 6).

These results with DHP compounds indicate that DHP-binding molecules, which have been commonly assumed to correspond to Ca^{2+} channels, are implicated in a complex manner in the process of excitation–contraction coupling. On one hand these molecules may work like voltage-dependent Ca^{2+} channels (ionophores) comparable to the L type found in heart myocytes.[27] So it is not surprising that Ca^{2+} current flowing through these Ca^{2+} channels plays a role in excitation–contraction coupling as observed in heart muscle. This could be by a direct action on contractile proteins as in frog heart,[28] or by a mechanism of Ca^{2+}-induced Ca^{2+}-release, as in mammalian heart,[29] or by any other mechanism (e.g., enzyme activation).

On the other hand these molecules also have the properties of the so-called "voltage sensors" introduced in the work of Hodgkin and Horowicz[30] to designate the voltage-dependent membrane mechanism that couples the tubular membrane depolarization to the SR calcium release.

It is not possible at present to say whether these two functions (Ca^{2+} ionophore and voltage sensor) are properties of the same molecule or of two different ones. The fact that DHP-binding experiments on tubular membrane preparation show a single class of sites[31] can be considered as an indication that these molecules belong at least to the same family of membrane proteins.

Concerning the "voltage sensor," its link with the mechanism of SR Ca^{2+}-release remains a mystery: Is it a mechanical,[32] a chemical,[25] or an ionic[33] coupling? Whatever the mechanism(s), a complete and realistic model will have to take into account all the known properties of excitation–contraction coupling in skeletal muscle including the role of external Na^+ ions[19,34] and also of K^+ ions.

In summary, our results are both consistent with (i) the existence of a Ca^{2+} current dependent phase of contraction, (ii) the possibility that tubular Ca^{2+} depletion favors the inactivation of prolonged contractile responses, and (iii) the existence of a close link between DHP receptors and SR Ca^{2+}-release.

We suggest that these three possibilities do not exclude one another.

4. SUMMARY

Simultaneous recordings of slow inward Ca^{2+} current and contraction of frog twitch muscle fibers have been done in a double sucrose-gap device; the external solution used contained a physiological Ca^{2+} concentration. In control conditions, the contractile responses elicited by long-lasting depolarizing steps to a potential between -40 and -20 mV showed clearly the presence of two components: an initial one which is rapidly activated upon membrane depolarization, and a slower one whose time course seems to be correlated with that of the slow inward Ca^{2+} current.

The Ca^{2+}-channel blockers nifedipine and nisoldipine produced an inhibition of the two components of contraction. Conversely, the Ca^{2+}-channel agonist ($\pm$) Bay K 8644 had a dual effect: It promoted the slow contractile component, and inhibited the fast one.

When external Ca^{2+} was replaced by the Ca^{2+}-channel blockers Co or Ni, a rapid suppression of the slow contractile component could be observed. Then the contraction was potentiated, with an important decrease in the rate of relaxation. These results indicate that the entry of Ca^{2+} through slow Ca^{2+} channels may be important to control the time course of prolonged contractile responses, although not necessary to maintain them.

ACKNOWLEDGMENTS. We thank Drs. Bonvallet, Ildefonse, and Tourneur for comments and suggestions in reading the manuscript, and J. Barbery for typing and graphical work.

REFERENCES

1. Almers, W., Fink, R., and Palade, P. T., 1981, Calcium depletion in frog muscle tubules: The decline of calcium current under maintained depolarization, *J. Physiol.* **312:** 177–207.
2. Luttgau, H. C., and Spiecker, W., 1979, The effects of calcium deprivation upon mechanical and electrophysiological parameters in skeletal muscle fibers of the frog, *J. Physiol.* **296:** 411–429.
3. Cota, G., and Stefani, E., 1981, Effects of external calcium reduction on the kinetics of potassium contractures in frog twitch muscle fibers, *J. Physiol.* **317:** 303–316.
4. Graf, F., and Schatzmann, H. J., 1984, Some effects of removal of external calcium on pig striated muscle. J. Physiol., **349:** 1–13.
5. Bolanos, P., Caputo, C., and Velaz, L., 1986, Effects of calcium, barium, and lanthanum on depolarization–contraction coupling in skeletal muscle fibers of *Rana pipiens, J. Physiol.* **370:** 39–60.
6. Brum, G., Rios, E., and Stefani, E., 1988, Effects of extracellular calcium on calcium movements of excitation–contraction coupling in frog skeletal muscle fibers, *J. Physiol.* **398:** 441–473.
7. Sanchez, J. A., and Stefani, E., 1978, Inward Ca^{2+} current in twitch muscle fibers of the frog, *J. Physiol.* **283:** 197–209.
8. Potreau, D., and Raymond, G., 1980, Calcium-dependent electrical activity and contraction of voltage-clamped frog single muscle fibers, *J. Physiol.* **307:** 9–22.
9. Caputo, C., 1981, Nickel substitution for calcium and the time course of potassium contractures of single muscle fibers, *J. Musc. Res. Cell. Motility* **2:** 167–182.
10. Lorkovic, H., and Rudel, R., 1983, Influence of divalent cations on potassium contracture duration in frog muscle fibers, *Pflüg. Arch.* **398:** 114–119.

11. Brum, G., Stefani, E., and Rios, E., 1987, Simultaneous measurements of Ca^{2+} currents and intracellular Ca^{2+} concentrations in single skeletal muscle fibers of the frog, *Can. J. Physiol. Pharmacol.* **65**: 681–685.

12. Ildefonse, M., Jacquemond, V., Rougier, O., Renaud, J. F., Fosset, M., and Lazdunski, M., 1985, Excitation–contraction coupling in skeletal muscle: Evidence for a role of slow Ca^{2+}-channels using Ca^{2+}-channel activators and inhibitors in the dihydropyridine series, *Biochem. Biophys. Res. Commun.* **129**: 904–909.

13. Rakowski, R. F., Olszewska, E., and Paxson, C., 1987, High-affinity effect of nifedipine on K contracture in skeletal muscle suggests a role for calcium channels in excitation–contraction coupling, *Biophys. J.* **51**: 550a.

14. Rios, E., and Brum, G., 1987, Involvement of dihydropyridine receptors in excitation–contraction coupling in skeletal muscle, *Nature* **325**: 717–720.

15. Gamboa-Aldeco, R., Huerta, M., and Stefani, E., 1988, Effect of Ca^{2+}-channel blockers on K^+ contractures in twitch fibers of the frog (*Rana pipiens*), *J. Physiol.* **397**: 389–399.

16. Jacquemond, V., and Rougier, O., 1988, Nifedipine and Bay K inhibit contraction independently from their action on calcium channels, *Biochem. Biophys. Res. Commun.* **152**: 1002–1007.

17. Lamb, G. D., and Walsh, T., 1987, Calcium currents, charge movement, and dihydropyridine binding in fast- and slow-twitch muscles of rat and rabbit, *J. Physiol.* **393**: 595–617.

18. Brum, G., Fitts, R., Pizarro, G., and Rios, E., 1988, Voltage sensors of the frog skeletal muscle membrane require calcium to function in excitation–contraction coupling, *J. Physiol.* **398**: 475–505.

19. Caille, J., Ildefonse, M., and Rougler, O., 1978, Existence of a sodium current in the tubular membrane of frog twitch muscle fiber: Its possible role in the activation of contraction, *Pflüg. Arch.* **379**: 117–119.

20. Nerbonne, J. M., Richard, S., and Nargeot, J., 1985, Ca^{2+}-channels are unblocked within a few milliseconds after photoconversion of nifedipine, *J. Mol. Cell. Cardiol.* **17**: 511–515.

21. Raymond, G., and Potreau, D., 1981, Effets des anesthesiques locaux (procaine, tetracaine) sur la permeabilite calcique lente et la contraction de la fibre musculaire squelettique de grenouille, *C. r. hebd. Seanc. Acad. Sci. Paris III* **292**: 637–640.

22. Gonzalez-Serratos, H., Valle Aguilera, R., Lathrop, D. A., and Del Carmen Garcia, M., 1982, Slow inward Ca currents have no obvious role in muscle excitation–contraction coupling, *Nature* **298**: 292–294.

23. Lorkovic, 1967, Effects of divalent cations on frog twitch muscles, *Am. J. Physiol.* **212**: 623–628.

24. Almers, W., Fink, R., and Palade, P. T., 1981, Calcium depletion in frog muscle tubules: The decline of calcium current under maintained depolarization, *J. Physiol.* **312**: 177–207.

25. Vergara, J., Tsien, R. Y., and Delay, M., 1985, Inositol 1,4,5-trisphosphate: A possible chemical link in excitation–contraction coupling in skeletal muscle, *Proc. Natl. Acad. Sci. USA* **82**: 6352–6356.

26. Lamb, G. D., 1986, Components of charge movement in rabbit skeletal muscle: The effect of tetracaine and nifedipine, *J. Physiol.* **376**: 85–100.

27. Tsien, R. W., 1987, Calcium currents in heart cells and neurons, in: *Neuromodulation* (L. K. Kaczmarek and I. B. Levitan, Eds.) pp. 206–242, Oxford University Press, Oxford.

28. Vassort, G., and Rougier, O., 1972, Membrane potential and slow inward current dependence of frog cardiac mechanical activity, *Pflüg. Arch.* **331**: 191–203.

29. Fabiato, A., and Fabiato, F., 1977, Calcium release from the sarcoplasmic reticulum, *Circ. Res.* **40**: 119–129.

30. Hodgkin, A. L., and Horowicz, P., 1960, Potassium contractures in single muscle fibers, *J. Physiol.* **153**: 386–403.

31. Fosset, M., Jaimovich, E., Delpont, E., and Lazdunski, M., 1983, [3H]Nitrendipine receptors in skeletal muscle: Properties and preferential localization in transverse tubules, *J. Biol. Chem.* **258**: 6086–6092.

32. Schneider, M. F., and Chandler, W. K., 1973, Voltage-dependent charge movement in skeletal muscle: A possible step in excitation–contraction coupling, *Nature* **242**: 244–246.

33. Mathias, R. T., Levis, R. A., and Eisenberg, R. S., 1980, Electrical models of excitation–contraction coupling and charge movement in skeletal muscle, *J. Gen. Physiol.* **76:** 1–31.
34. Caille, J., Ildefonse, M., Roy, G., and Rougier, O., 1981, Surface and tubular sodium currents in frog twitch muscle fiber: Implication in excitation–contraction coupling, in: *Molecular Aspects of Muscle Function* (E. Varga, A. Kover, T. Kovacs, and L. Kovacs, Eds.), pp. 389–409, Pergamon, Oxford, England.

C. Molecular Architecture of the Triad

Chapter 22

Molecular Architecture of T-SR Junctions: Evidence for a Junctional Complex That Directly Connects the Two Membrane Systems

Clara Franzini-Armstrong, Barbara Block, and Donald G. Ferguson

1. INTRODUCTION

Contraction of all types of muscle fibers is activated by an increase in the cytoplasmic concentration of calcium ions. In the skeletal muscle fibers of vertebrates the calcium required for activation is released rapidly from an internal membrane system, the sarcoplasmic reticulum (SR). In other muscles release from the SR and influx through the surface membrane may contribute in variable proportion to the increase of intracellular calcium needed for myofibrillar activation. Most striated muscle fibers have extensive tubular invaginations of the surface membrane, forming networks called the transverse (T) tubular systems. Individual components are called transverse (T) tubules, even though their orientation is not always transverse to the long axis of the fiber.

CLARA FRANZINI-ARMSTRONG • Department of Anatomy, University of Pennsylvania, Philadelphia, Pennsylvania 19104-6018.   BARBARA BLOCK • Departments of Biology and Anatomy, University of Pennsylvania, Philadelphia, Pennsylvania 19104-6018.   DONALD G. FERGUSON • Department of Physiology and Biophysics, University of Cincinnati, Cincinnati, Ohio 45267.   *Present address for C.F-A.:* Department of Biology, University of Pennsylvania, Philadelphia, Pennsylvania 19104-6018. *Present address for B.B.:* Department of Organismal Biology and Anatomy, University of Chicago, Chicago, Illinois 60637.

Two basic principles of excitation–contraction coupling for skeletal muscle were established approximately 30 years ago in two classical sets of experiments. The so-called "local stimulation" experiments demonstrated that local depolarization of the sarcolemma induced a localized, graded contraction only when it occurred at sites of T-tubule invaginations.[1] From these experiments, and from structural observations showing intimate connections between T-tubule and SR elements[2,3] (Fig. 1), the concept developed that depolarization of the T-tubules is a link in excitation–contraction coupling, and that an indirect interaction between T-tubules and SR results in calcium release from the latter. In the second set of experiments,[4] it was demonstrated that depolarization of the external surface is the first step in excitation–contraction coupling and that release of calcium from the SR is a steep function of the surface membrane (and T-tubule) voltage.

Morphology sets the framework for these events and establishes some of the boundaries within which models of excitation–contraction coupling must operate. This is done in two ways: first, the minimum common structure in all muscle fibers is determined. This must be the structure that underlies the common functional features of all muscle fibers. In modern times, electron microscopy plays a role in the identification and localization of key, biochemically identified molecules. Second, qualitative and quantitative variations in the common structure are established, and then compared with functional variations of one or more key steps in excitation–contraction coupling. Both morphological approaches, the basic and the comparative, have significantly contributed to the understanding of the key components of excitation–contraction coupling, and examples of both approaches will be given.

T-tubules and external surface membranes both interact with the SR at specialized junctions where the membranes of the two separate cell organelles face each other across a narrow junctional gap of approximately 10 nm. Uniformity of the junctional gap width is established by a set of evenly spaced macromolecules, called the junctional feet[5] (Figs. 2 and 3). This chapter deals with the molecular architecture of the T-SR junction and with the functional significance of variations in the frequency and size of the junctional contacts in various fiber types. An initial observation is that junctional feet and a junctional gap of uniform width are present in muscle fibers from the entire animal kingdom, including smooth and cardiac muscle (Figs. 3, 4, 5, and 7). The conclusion is that the feet must play a key role in excitation–contraction coupling. The similarity in the size and basic disposition of feet in muscle fibers from the entire animal kingdom and with different functional properties, also suggest that they may have evolved early in phylogeny and have maintained common characteristics and function. T-tubules, on the other hand, are missing from some muscle fibers, particularly those of very small diameter[6-9] (Fig. 6). The conclusion is that T-tubules are a component of the excitation–contraction coupling pathway, which can be bypassed under the appropriate conditions. It is quite well established that T-tubules are invaginations of the surface membrane and have the function of transmitting the surface depolarization to the fiber interior, thus allowing a more synchronous and rapid initiation of contraction through the cross section of large muscle fibers. In certain fibers the inward spread of depolarization is not needed, and the surface membrane directly interacts with the T-tubules. Other fibers have a mixture of T-SR and surface-SR junctions.

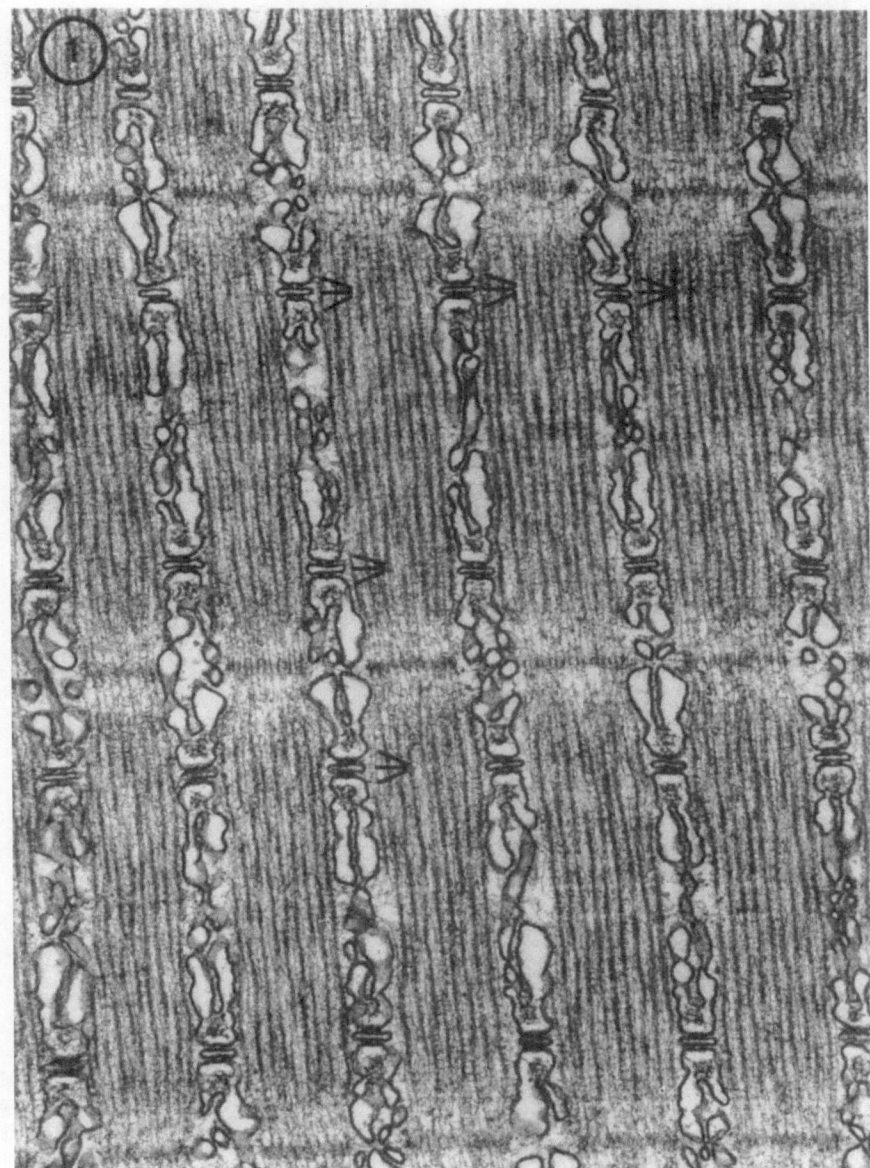

Figure 1. Longitudinal section of fast muscle fiber from swimbladder of toadfish. This is one of the fastest known muscles in vertebrates and it has an extremely orderly disposition of membranes. T-tubule networks are located at A-L junction, and form triads (3 lines) with elements of sarcoplasmic reticulum. ×32,500.

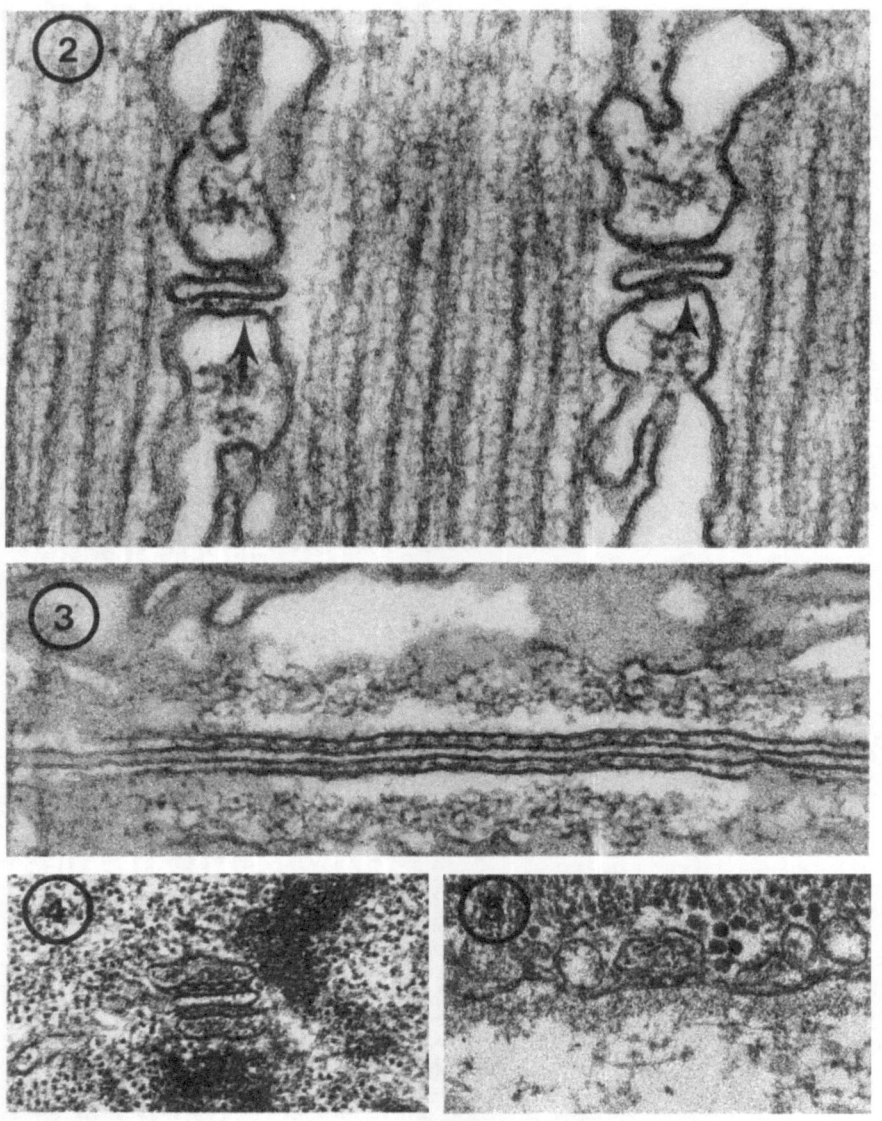

Figure 2. Detail of two triads, same muscle as in Fig. 1. Two junctional feet occupy each junctional gap above and below the T-tubule profile. Each foot appears either as a pillar, with a less dense core (arrowhead), or as an elongated line halfway between the two membranes (arrows). The two appearances derive from different orientations of feet relative to viewer. (From Ref. 40.) ×80,000.

Figure 3. This image is from a plane of section perpendicular to the one in Fig. 2, but still parallel to the fiber axis (or longitudinal). Feet form a row, with a regular spacing of approximately 30 nm. Size of junctional gap is constant and approximately 10 nm. (From Ref. 40.) ×80,000.

Figure 4. Triads in slow tonic fibers of frog have a different shape and orientation than those in twitch fibers. However, size of junctional gap and spacing and appearance of feet are the same. (From Ref. 33.) ×65,000.

Figure 5. In many fibers SR forms junctions with surface membrane (peripheral couplings), which are similar in structure to triads. From a slow tonic fiber in the frog. (From Ref. 41.) ×82,000.

Figure 6.   Some fibers of small diameter (e.g., those in amphioxus and in scallop muscles) have no T-tubules. In this cross section from fast adductor muscle of scallop, SR is in the form of round profiles immediately below plasmalemma. Peripheral SR forms peripheral couplings with surface membrane (arrows). (From Ref. 8.) ×25,000.

Figure 7.   Detail of peripheral coupling from fast adductor of scallop, showing junctional feet. (From Ref. 8.) ×100,000.

2. PARTICLES ASSOCIATED WITH THE SR JUNCTION MEMBRANE

The two areas of SR and of T-tubules (or surface) membranes that participate in the junction [called junctional SR (JSR), junctional T (JT), and junctional surface (JS) membranes] constitute specialized membrane domains which play a major role in excitation–contraction coupling, by virtue of their intrinsic components. Other major functions of T-tubule and SR membrane are relegated to the free, or nonjunctional domains of the two membranes, where a variety of ionic channels, pumps, and their accessory molecules are located (e.g., see Ref. 10). Transition between junctional and free domains of the SR and T-tubule membranes is abrupt, and can be seen clearly in freeze-fracture images.[11]

Two major components of the junctional membranes are recognized by morphological approaches: the feet associated with the JSR membrane, and the tetrads of particles in the JT membranes. The two components, and their relationship to each other, are described and identified below.

In thin sections oriented at right angles to the junctional gap, the feet appear either as 20-nm-wide columns (or pillars), often with a less dense central core, or as a dense line between the two membranes (Fig. 2). The feet apparently touch both SR and T-tubule membranes. In all muscle fibers feet are disposed in rows, at periodic center-to-center distances of approximately 30 nm. At least two, and in many cases more, adjacent rows of feet occupy the junctional gap, and feet in adjacent rows are in register, thus forming a tetragonal arrangement. This is seen particularly well in sections cut in a direction parallel to the plane of the junction and grazing the junctional gap. Figures 8–11 illustrate grazing views of several junctions from a variety of sources, in which the gap is occupied by 2–10 rows of feet. Despite the variation in the number of rows, the spacing and disposition of the feet are obviously the same in the two cases. In this view the apparent shape of the feet is that of small diamonds oriented with the diagonals at a slightly skewed angle relative to the long axis of the T-tubules (Fig. 11).

When a microsomal fraction is obtained by homogenizing muscle fibers, SR and T-tubules are often separated. Although the feet touch both T-tubule and SR membranes, they are components of the SR, because they remain associated with it when SR and T-tubules are separated during fractionation of the muscle fibers.[12] Taking advantage of the fact that feet-bearing vesicles have a high content of calsequestrin and thus a higher density than other SR vesicles, a JSR-enriched fraction can be isolated using sucrose density gradients (Fig. 12).[13]

Images of heavy-SR vesicles prepared by freeze-drying and rotary shadowing[14] offer a different view of the junctional feet, one that is most appropriate to the study of their shape (Figs. 13 and 14). In gently isolated vesicles, feet maintain their disposition in double rows over the surface of the vesicle (Fig. 13). Each foot is composed of four subunits of equal size. The overall shape of the four subunits is approximately spherical, and they aggregate to form a tetrafoil arrangement, with a depression in the center (Fig. 14). Adjacent feet abut corner to corner, with an overlap of one subunit, so that the individual feet are arranged in a skewed position relative to the long axis of the double row. In the *in situ* membranes this coincides with the long axis of T-tubules. Comparison with the structure of a triad in an intact muscle fiber shows that this is the native disposition (Fig. 15).

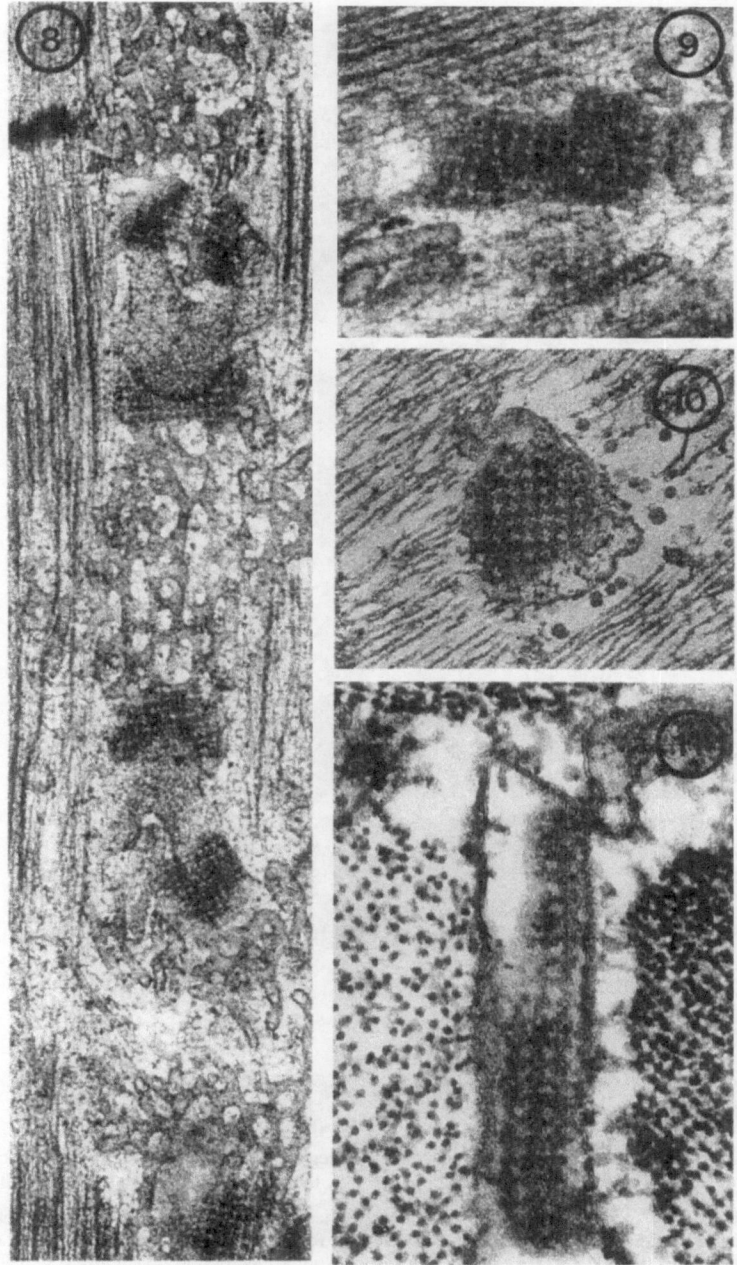

Figure 8–11. Grazing views of junctional gap in a variety of muscle fibers, showing that feet always form a tetragonal arrangement with equal spacings. Where feet form multiple rows, the shape of the junction is round or ovoidal (Figs. 8–10); where feet form 2 to 3 elongated rows, the junction is elongated (Fig. 11). Figures 8 and 9, from a scorpion, ×50,000 and 90,000, respectively. Figure 10 from a tonic fiber in frog, ×70,000 (from Ref. 41). Figure 11 from a twitch fiber in a small fish, ×110,000.

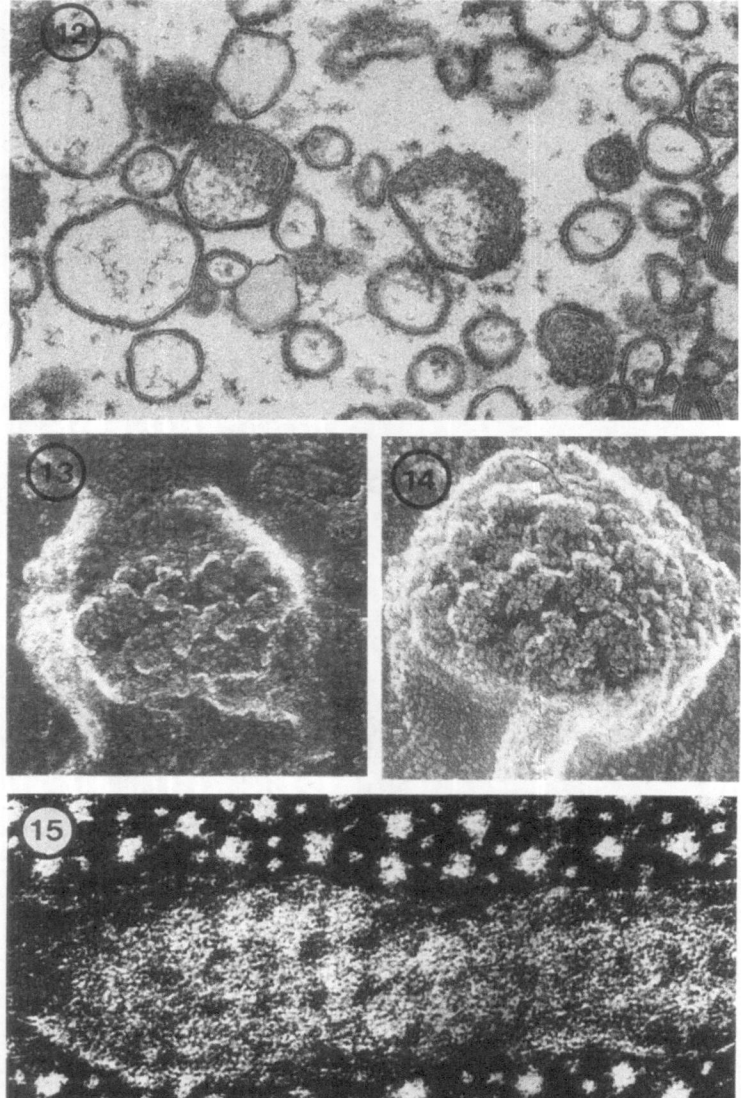

Figure 12. "Heavy SR" fraction isolated from rabbit leg muscle. SR vesicles derived from triads have a visible content of calsequestrin and junctional feet on their surface. (From Ref. 12.) ×110,000.

Figures 13 and 14. Shadowed replicas from freeze-dried heavy-SR vesicles. In Fig. 13, feet maintain their arrangement in double rows. Each foot consists of four subunits, and feet are in contact corner-to-corner, but with a slight skew. A single foot in Fig. 14 shows clearly four apparently spherical subunits surrounding a central depression. (From Ref. 14.) ×230,000 and ×500,000 images, respectively, are printed in negative contrast.

Figure 15. Thin section of a grazing view of a junctional gap, printed in negative contrast. Comparison with Fig. 13 shows that feet are slightly skewed relative to each other, as in isolated vesicles. (From Ref. 14.) ×220,000.

3. THE FOOT PROTEIN AS AN ION CHANNEL

Two recent breakthroughs are responsible for assigning a specific role to the junctional feet. First, a large-molecular-weight polypeptide was identified with a subunit of the junctional feet, and was shown by use of antibodies to be located in the JSR membranes.[15] The large-molecular-weight polypeptide was shown to be an intrinsic component of the SR membrane.[16] Second, a large molecular complex, probably composed of four polypeptides, was purified and identified with complete junctional feet.[17–19] The so-called "foot protein" was shown to bind ryanodine with a high affinity (see Chapter 31, this volume). The isolated foot protein, when reconstituted in a lipid bilayer,[20] forms a channel with a high permeability to calcium and with pharmacological properties similar to those of rapid calcium release from heavy SR.[21] (see Chapter 32, this volume). On the basis of these data the foot protein is thought to be the site of calcium release from the SR during excitation–contraction coupling.

Identification of the isolated protein with the feet is unequivocal and uncontroversial, because it is based on unique structural features. Figures 16 and 17 show images of purified foot proteins from the laboratory of Dr. K. P. Campbell, which were adsorbed on mica, freeze-dried, and rotary shadowed. The isolated molecules are composed of four equal subunits and have a size very close to that of the junctional feet. The four subunits have a spherical shell, but may be hollow because they are filled by uranyl acetate when viewed by negative staining.[20] The *in situ* and the detergent-dissociated protein, however, differ in one significant detail: the former has a central depression, while the latter has a large central bump, occasionally showing evidence of four components. The simplest explanation for this difference is that the central bump is the hydrophobic component of the molecule. When the feet are associated with the SR membrane the central bump is contained within the lipid bilayer and therefore it is not visible. The isolated molecule, on the other hand, adheres to the mica with its hydrophylic region, and presents the central hydrophobic portion to the viewer.

The hydrophobic component of the foot protein in the interior of the JSR membrane is exposed when the membrane is split in the freeze-fracture technique. Figure 18 shows the luminal leaflet of the JSR in a muscle fiber from a fast fish muscle. The exposed hydrophobic interior of the membrane is decorated by periodic bumps which have the same periodicity and dispositions as the junctional feet and a size similar to that of the central bump in the isolated molecule. At high magnification (Fig. 19), the foot protein roots show a fourfold substructure, indicating that the four subunits participate equally in the formation of the intramembrane component of the molecule. Thus, the JSR bumps represent the roots of the foot proteins into the SR membrane. The foot protein forms a channel with two components, both having a fourfold symmetry: the intramembrane component presumably spans the whole thickness of the SR membrane, forming the calcium release channel; the cytoplasmic component (or foot) is quite large, and spans the width of the junctional gap, establishing contact with the T-tubule membrane.

The foot protein constitutes an ionic channel of unusually large dimensions (approximately four times larger than the dihydropyridine receptor). Part of the uniqueness of this molecule is its large hydrophilic domain, which forms the junctional feet. The significance of this will be discussed below.

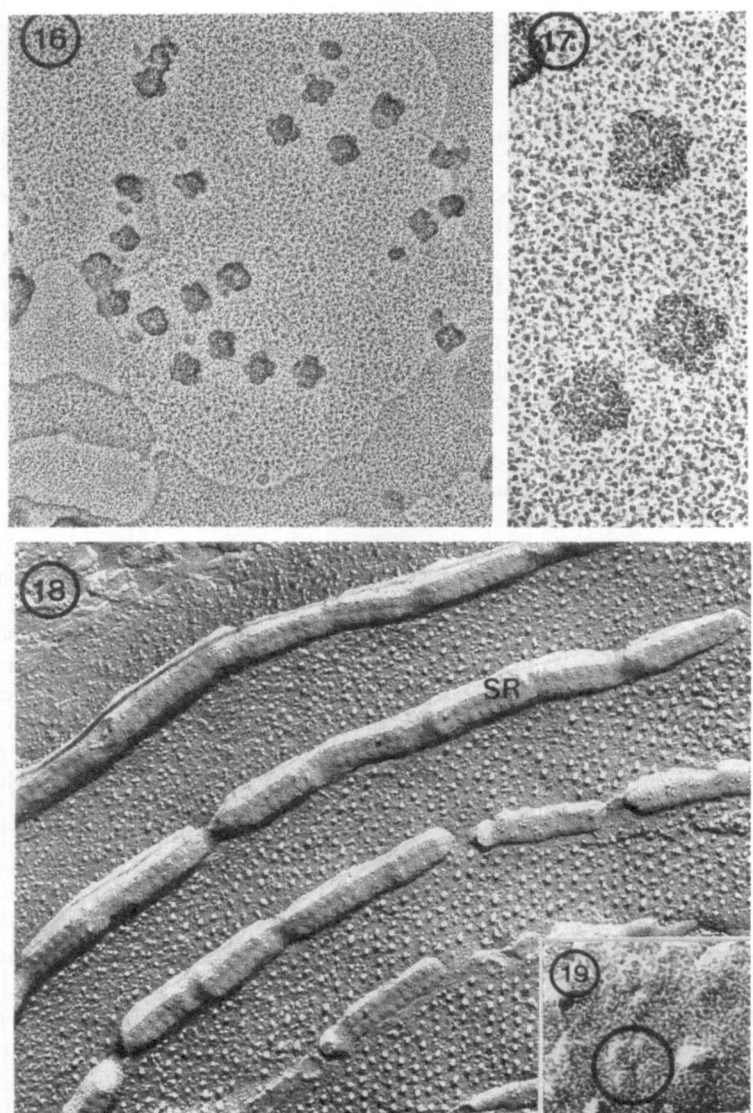

Figures 16 and 17. Isolated foot proteins from the laboratory of K. P. Campbell. Proteins in this pure fraction have the same four subunit composition as *in situ* feet (compare with Figs. 13 and 14). However, the center of the isolated molecule has a large bump instead of a pit. Stereomicrographs (not shown) indicate that central bump is quite tall. See text for interpretation. (From Ref. 24.) ×135,000 and 376,000, respectively.

Figures 18 and 19. Freeze-fractures showing luminal leaflet of junctional SR membrane, immediately below junctional feet, from a fish muscle. Two rows of intramembranous "bumps" mark sites at which feet are attached to membrane and represent hydrophobic portions of foot proteins, forming presumed calcium release channel. In Fig. 19, circled "bump" is composed of four subunits. (From Ref. 24.) ×66,000 and 376,000, respectively.

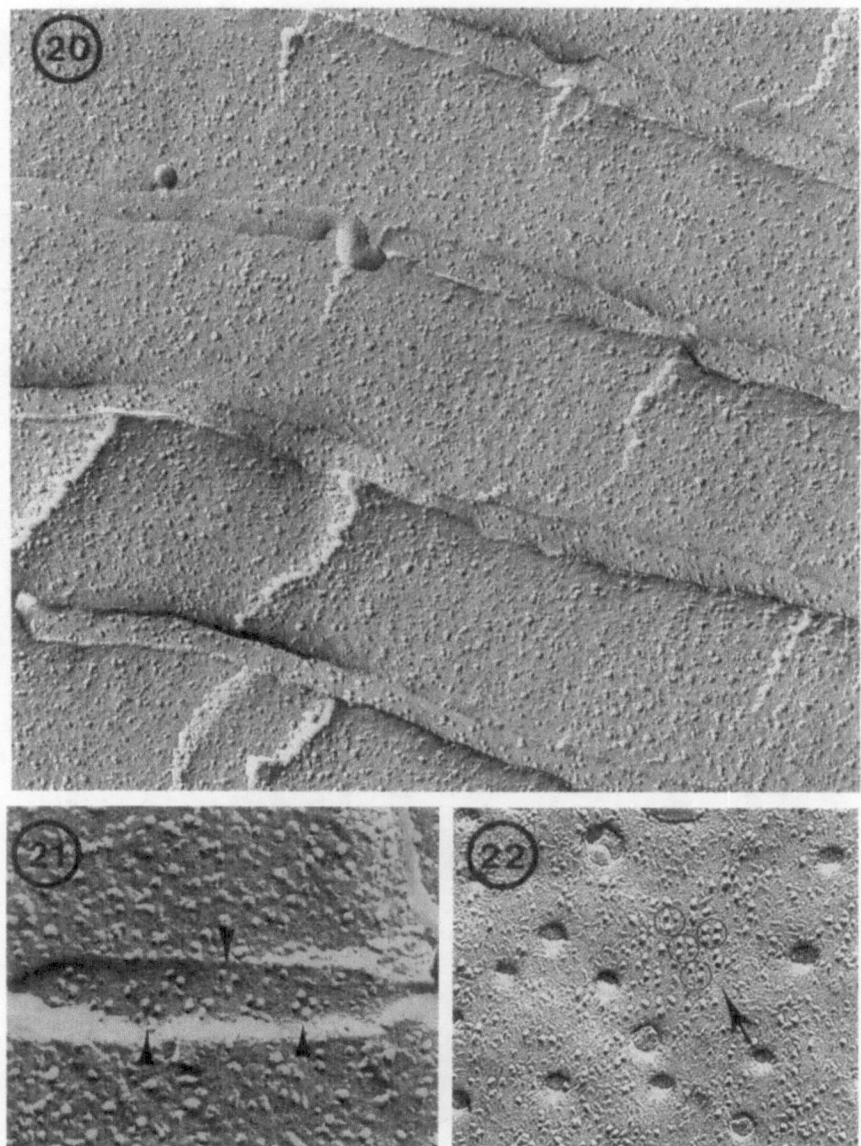

Figures 20 and 21. Cytoplasmic leaflet of junctional T-tubule membrane, immediately adjacent to junctional feet, is decorated by groups of four particles (JT tetrads). Tetrads (arrowheads, Fig. 21) form two rows overlying two rows of junctional feet, but their spacing along rows is twice that between feet, indicating that each tetrad interacts with alternate feet. (From Ref. 24.) ×66,000 and 133,000, respectively.

Figure 22. Freeze-fracture of plasmalemma of a frog slow fiber at a site of a peripheral coupling (junction between SR and surface membrane). Tetrads of particles (circled) very similar to those in JT membrane are disposed over alternate feet. Arrow indicates expected orientation of rows of junctional feet located below this membrane. (From Ref. 41.) ×106,000.

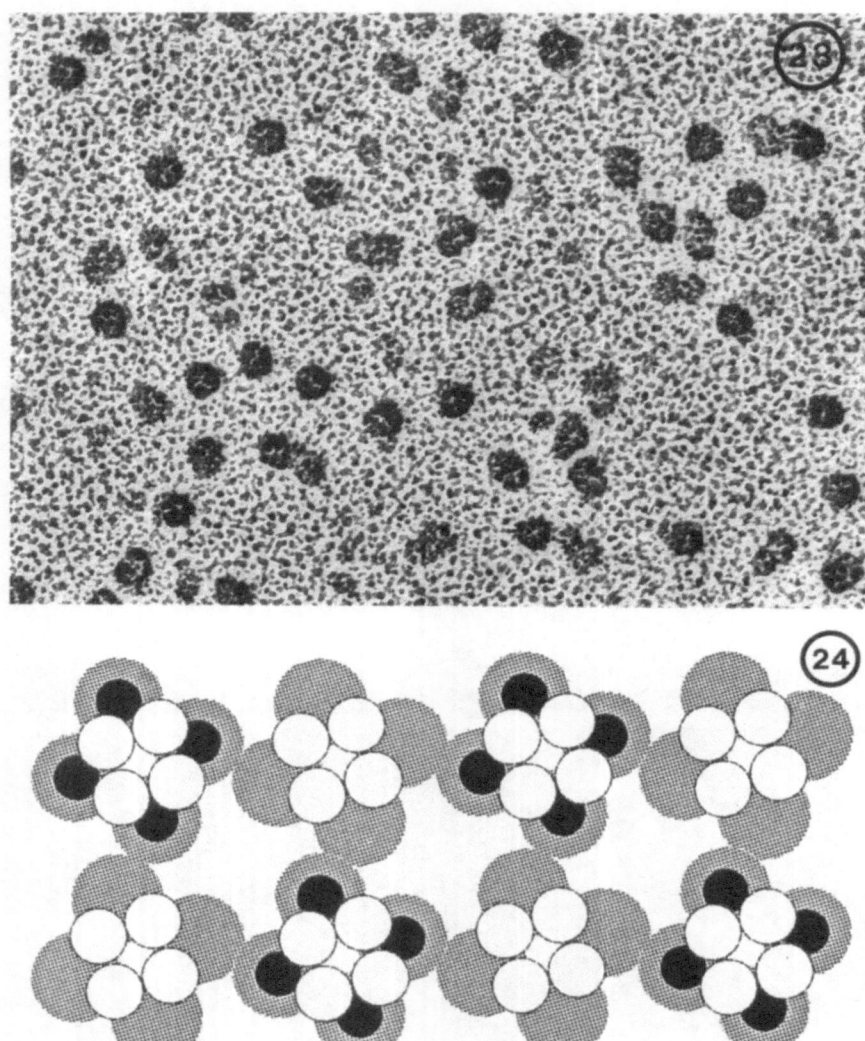

Figure 23. Freeze-dried, rotary-shadowed dihydropyridine receptors purified from rabbit muscle. Molecules are slightly elongated and show two subunits. (From Ref. 23.) ×376,000.

Figure 24. Diagrammatic representation of relationship between foot proteins and JT tetrads. Cytoplasmic components of foot protein (or junctional foot) is represented by four shaded balls. White balls represent four subunits of intramembranous component of foot proteins. JT tetrads are indicated by black balls located over subunits of alternate feet. (From Ref. 24.)

4. T-TUBULE JUNCTIONAL PARTICLES

Evidence for a direct interaction between feet and components either of the JT membrane or of the junctional domain of the exterior membrane is so far indirect, and based on the two following observations: (i) junction formation between JSR membranes, presumably with attached feet, and T-tubule membranes can be demonstrated *in vitro*[22]; (ii) following freeze-fracture, the JT membrane has a population of particles having the same diameter and similar fracturing properties, probably belonging to a single molecular species. As shown below, these particles have a location corresponding to that of the junctional feet. Under appropriate fixation conditions, the particles remain preferentially on the cytoplasmic leaflet of the membrane, where they are arranged in an orderly disposition (Fig. 20). The repeating component is a group of four particles (here called JT tetrads), which delineate a diamond shape and are arranged with a slight skew relative to the long axis of the T-tubules (Figs. 20 and 21). In triads that have two rows of junctional feet, the JT tetrads are arranged along two parallel rows, but their center-to-center spacing is exactly twice that of the junctional feet. If the outline of a tetrad is superimposed on that of a foot, the four components of the tetrad fall above the centers of the four subunits of the foot. However, JT tetrads form connections only with alternate feet. Tetrads of prominent particles are also located at areas of junctions between surface membrane and SR (Fig. 22).

Before constructing a complete image of the triad, we will mention a major component of the T-tubule membranes. This is a molecule that binds dihydropyridines, and has been called the dihydropyridine receptor (DHPR). DHPRs are the only known large-molecular-weight component of the T-tubules, and for that reason they are good candidates for the large particles of the JT membrane. DHPRs are discussed in detail in Chapter 20 by Talvenheimo et al. Figure 23 illustrates the appearance of the isolated dihydropyridine receptor. It has a slightly ovoidal shape, approximately 16×22 nm dimensions, and two subunits.[23]

5. THE STRUCTURE OF THE TRIAD AND MODELS FOR EXCITATION–CONTRACTION COUPLING

The relationship between the cytoplasmic and intramembranous components of the feet and the JT tetrads is schematically shown in Fig. 24 (see figure legend for details). The major remaining puzzle about this structure is that only every other foot is associated with the T-tubule components.

From the similarities and differences in the structure of the isolated and *in situ* feet, and from the coincidences in the spacings of JT and JSR components, we derive the composite image of the junction illustrated in Fig. 25.[24] The highlights of the structure are as follows: (i) the foot protein has a dual role: its cytoplasmic component spans the gap between JT and JSR membranes, making direct contact with JT components, and its central portion penetrates into the SR, forming an ionic channel; (ii) the components of JT tetrads make direct contact with the subunits of alternate feet; (iii) the JT domain contains only the elements that are related to the junction formation. JSR contains the roots of the junctional feet, mingled with anchoring sites for calsequestrin.[25] Consider-

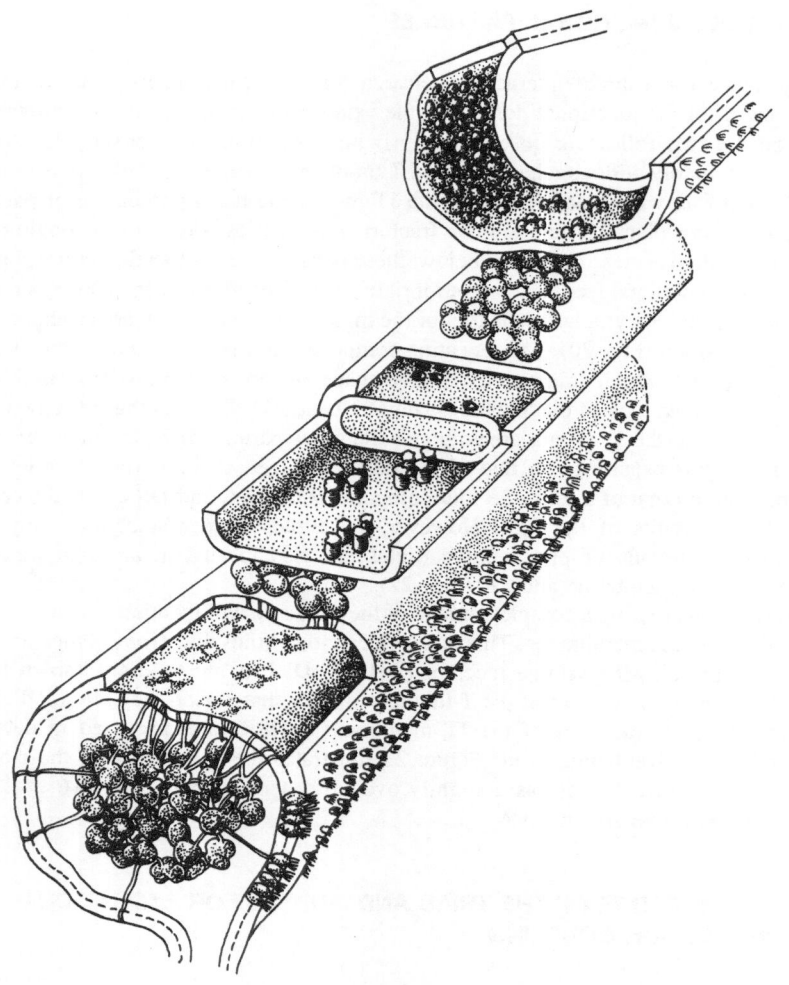

Figure 25. Three-dimensional reconstruction of a triad, showing relative positions of calcium ATPase, calsequestrin, foot proteins and JT tetrads. ATPase occupies nonjunctional domain of membrane, and is absent from junction. Calsequestrin occupies lumen of SR, and is associated with JSR and in part with free SR membranes by intermediate thin strands which anchor it to membrane.[25] Feet, represented by four balls, occupy T-SR junctional gap, and intramembranous portion of foot protein penetrates into JSR membrane, forming bumps with fourfold structure. JT tetrads are located in JT membrane, opposite alternate feet. (From Ref. 24.)

ing the complexity of protein composition of SR and T-tubules it is likely that this is an oversimplification (see Chapter 23, this volume).

The JT tetrad–foot association forms a large junctional complex which reaches from the lumen of the T-tubule, across the junctional gap, to the lumen of the SR. By means of this junctional complex a specific interaction may occur between T-tubule components and the calcium release channel from the SR. The molecular continuity provides a structural basis for a hypothesis of T-SR coupling which is appealing in its simplicity and represents a reasonable current explanation for the tight dependence of the SR calcium release on T-tubule membrane voltage. The hypothesis[26] is that the initial step in excitation–contraction coupling is a voltage-dependent rearrangement of molecules located in the T-tubule membrane, resulting in a detectable "charge movement." Direct interaction between the voltage sensor and the calcium release channel of the SR would result in opening of the latter. The T-tubule particles are most appropriate candidates as voltage sensors, and foot proteins in this scheme would perform the dual role of receiving the signal from the voltage sensor and being the release sites. This model is not as far-fetched as it may seem at first sight. Other examples of related long-range molecular interactions exist in muscle. The interactions between troponins and tropomyosin may be compared with the proposed interaction between JT tetrads and feet, while the opening of the calcium release channel as a result of charge movement in the JT components may be compared to the ligand gating of the ACh receptor.

The identity of the voltage sensors and of the JT particles has not yet been established. On the basis of physiological experiments Rios and Brum[27] have proposed that some of the dihydropyridine receptors in T-tubule membranes are silent as calcium channels, but maintain voltage sensitivity. These modified calcium channels would have the function of initiating excitation–contraction coupling. On the basis of various morphological considerations, we propose that the particles in the JT tetrads represent dihydropyridine receptors directly involved in excitation–contraction coupling.[24] The next step in the understanding of T-SR interactions will be the identification of the JT particles and the isolation of the intact junctional complex (see Chapter 23, this volume). Identification of the JT component of the large junctional complex with dihydropyridine receptors would greatly strengthen the direct molecular interaction hypothesis of excitation–contraction coupling.

Two other models of T-SR interaction that have gained and maintained attention involve an indirect step mediated by a "transmitter." The two transmitters that may have such a role are calcium ions and IP_3, both of which are capable of inducing release of calcium from the SR under appropriate conditions. These two components are discussed in Chapters 25 and 27, this volume. Calcium-activated calcium release may play a role in cardiac muscle, but it is less likely to be a first step in skeletal muscle. Although the metabolic machinery for the production of IP_3 is present in muscle (see Chapter 30, this volume), and the metabolite is produced in an active muscle (see Chapter 27, this volume), acceptance of a key role for IP_3 still requires demonstration of the following points: (i) Is a sufficient quantity produced?; (ii) is the effect sufficiently rapid?; (iii) can the rapidity of the turning off during T-tubule depolarization be accounted for?; (iv) is the calcium concentration in the resting muscle fiber at the level needed for IP_3 effect? In skinned muscle fiber (Chapter 29, this volume) and in reconstituted channels (Chapter 33, this volume) a minimum calcium concentration ($10^{-7}M$) is needed for IP_3 action; (v) is the need for a minimum free calcium concentration responsible for the failure of IP_3 injected into intact muscle fibers to elicit a response?[28]

Finally, in modelling transmission at the T-SR junction, we should consider one specific constraint imposed by the structure of the junction. The foot protein is an unusually large molecule and for that reason its packing density (number/μm^2) in the

JSR membrane is very low. This in turn limits the amount of amplification that can be expected to occur in T-SR coupling step during excitation–contraction coupling. To understand this problem it is very useful to consider what we know about chemical synapses and in particular the neuromuscular junction in vertebrates. The postsynaptic membrane has a very high density of ACh receptors (20,000–30,000/μm^2), located in membrane patches in proximity of the vesicle release sites. A large number of ACh molecules (about 10,000 per vesicle) are released and act on the channels, thus producing a large amplification of the presynaptic signal. The amplification in this case is a prerequisite of effective transmission, because the large difference in the size of pre- and postsynaptic elements results in an impedance mismatch such that direct coupling with no amplification would not be effective.[29] The molecular components for the amplification process are the large numbers of ACh molecules and of receptors. By contrast, the molecular components of the T-SR junctions are not designed for amplification. The density of junctional feet in the SR membrane (homologous to ACh receptors in this comparison) is 20- to 30-fold less than that of ACh receptors, and, more importantly, it is approximately equal to the density of voltage-sensing sites.[26] Thus, in transmission between T-tubule and SR membranes, there is likely to be a one-to-one correlation between pre- and postsynaptic events, and this makes a chemical hypothesis of transmission less attractive. The situation in excitation–contraction coupling is significantly different from that in visual transduction, where amplification occurs at several successive steps of the system (see Chapter 2, this volume).

6. DISTRIBUTION OF FEET AMONG DIFFERENT MUSCLE FIBERS

The recent identification of feet with calcium release channels gives significance to comparative observations on the abundance of feet in various fiber types. While feet are invariable components of the T-SR junctions, the overall size, shape, and disposition of the junctions between SR and T-tubules are variable in a manner that depends on the organism of origin and more significantly on the functional characteristics of the muscle fiber. Considerations of the variable characteristics of the junction are important in understanding the role played by the junctions in excitation–contraction coupling.

Variations in the junctions between various muscle fibers are both qualitative and quantitative. The qualitative differences, mostly catalogued during the Sixties and Seventies, may not have direct functional relevance, and have been reviewed.[30,31] Triads, dyads, and peripheral coupling are all variations of the junctions between JSR and either JT tubules or the surface membrane. No direct relationship between the shape of the junction and fiber properties seems to exist. The slow tonic fibers of the frog,[32,33] and the fast abdominal flexor fibers of the crayfish,[34] for example, both have round-shaped junctions, while slow tonic and fast twitch fibers in a fish both have elongated junctions.[35]

Quantitative differences are obviously more important than qualitative ones, particularly if one considers that the size and frequency of junctions in a muscle fiber directly determine the density of feet (i.e., calcium release sites). One would expect that the density of junctional feet is the parameter that most directly relates to the rapidity of onset of contraction following stimulation, and this turns out to be true. In the last few years we have used the Golgi infiltration technique for a quantitative analysis of the contributions of junctional and free segments of the T-tubules to the overall T-tubule network in fibers with known functional and metabolic properties. From these data we estimate the density of junctional feet per fiber volume. The approach is based on an old recipe for silver impregnation, which was shown to delineate the T-tubules and make them highly visible in the light microscope (Ref. 36, from 1902, translated in 1961).

A slight alteration was used to adapt the technique to electron microscopy, with better preservation of the structure of the tubules.[37] The infiltrated tubules are very electron dense, so that relatively thick sections, containing a lot of information, can be used, and more importantly the shape distinction between the segments of the T-tubule network that form junctions with the SR (JT segments) and those that do not (FT segments) are clearly maintained.

Figures 26 to 29 illustrate the power of the technique. The muscle fibers are all from the swimbladder of the toadfish, which is known to have a band of very rapidly acting muscle fibers. The very fast (300–400 Hz) contraction–relaxation cycles of the muscle fibers are responsible for the drumlike sound of the swimbladder. In examining these fibers we were surprised to find a clear distinction between fibers from females, in which T-tubules form long junctions (Fig. 26), and those from males in which the short junctional T-tubule segments alternate with free T-tubule segments (Fig. 27). Thus, the overall area of junction is more limited in muscle fibers from males. In addition to this sexual dimorphism, indicating some subtle differences in excitation–contraction coupling, we found totally unexpected bundles of fibers, which, judging from the extent of junction formation by T-tubules, should belong to a slow twitch (Fig. 28) and slow tonic category (Fig. 29). The latter identification was actually confirmed by the use of the old trick of producing a potassium contracture, which is of long duration in tonic and not in twitch-type fibers. Thus, T-tubule shapes and the extent of T-SR formation, which are readily observed by the Golgi infiltration, are effective indicators of fiber types.

Functional differences between fiber types in the hind legs of mammals are less dramatic than in the toadfish swimbladder, but are also correlated to extent of T-SR junction formation and density of junctional feet. Three types of fibers in the guinea pig were examined.[38] These are fast twitch glycolytic (fatigueable) fibers, fast twitch oxidative–glycolytic (fatigue-resistant), and slow twitch oxidative (fatigue-resistant) fibers. We measured the total area of T-tubules and the percentage of T-tubule lengths which form junctions with the SR (JT). From this data we calculated the density of feet per unit fiber volume. The three types of fiber have a very similar total surface area of T-tubular membrane, but the relative length of the network participating in junctions varies with the fiber type (Fig. 30). The average density of feet is 159, 151, and $64/\mu m^3$, respectively, for the three types of fibers. Comparison of these data with the twitch contraction times of motor units with similar properties in cat muscle[39] is revealing (Fig. 30). The distribution of time to peak tension values and of relative lengths of JT tubules (which is directly proportional to feet density) for the three types of fibers show a very similar pattern. This indicates that availability of calcium to the myofibrils may play a role in the rate of tension development in a twitch.

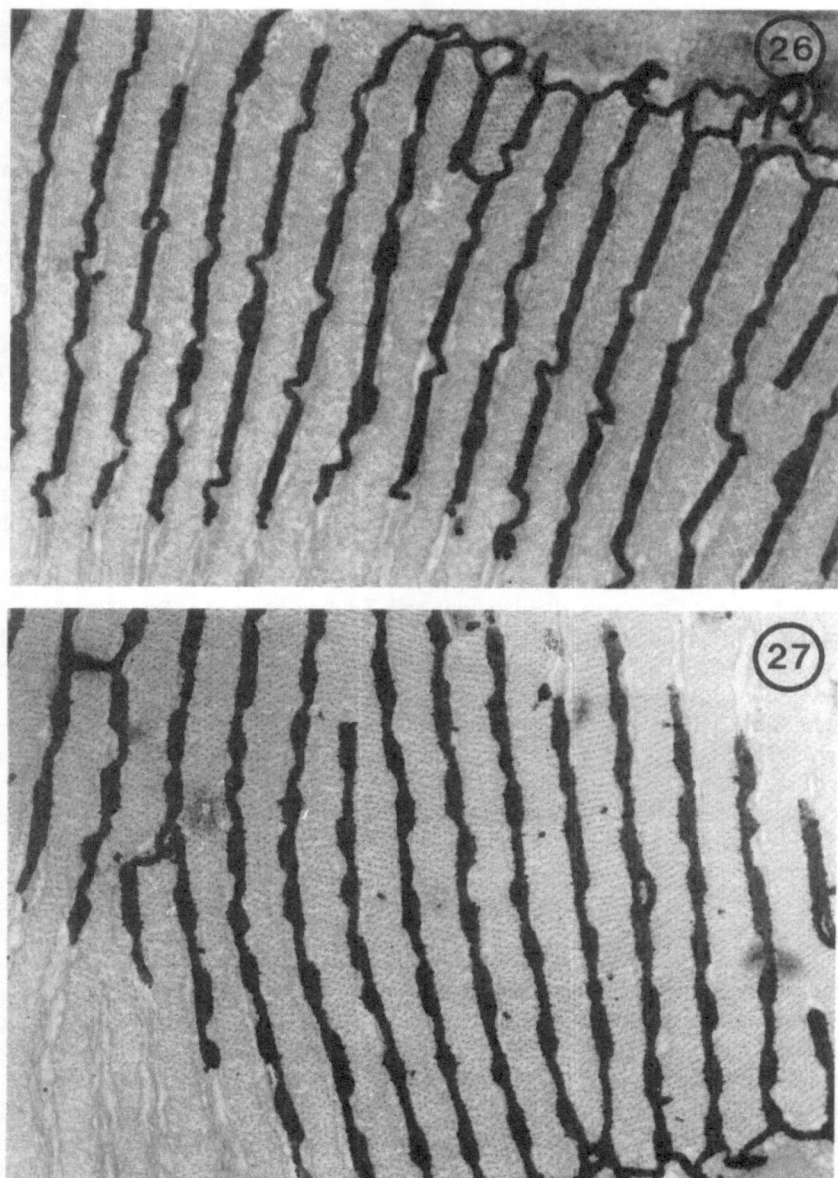

Figures 26–29. Cross sections through fast fibers from swimbladder of female (Fig. 26) and male (Fig. 27) toadfishes. The swimbladder also contains slow twitch (Fig. 28) and slow tonic (Fig. 29) fibers. T-tubules are infiltrated by black reaction of Golgi, and differences in fiber properties are reflected in variable amount of T-tubules forming junctions with SR. Junctional T-tubules, most frequent in Fig. 26, are wider and flat; nonjunctional T-tubule segments are smaller and have a round diameter. In slow tonic fibers most of the T-tubule network is nonjunctional. ×21,000.

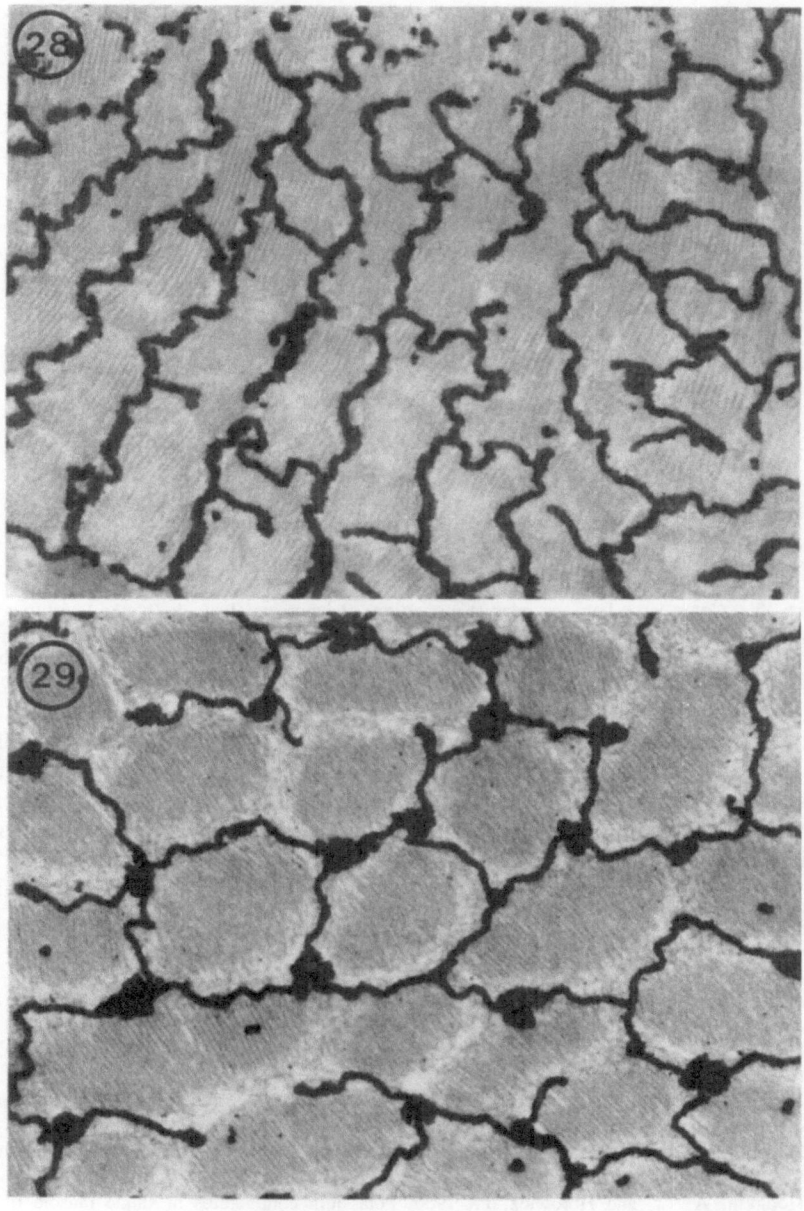

Figures 26–29 (*Continued*)

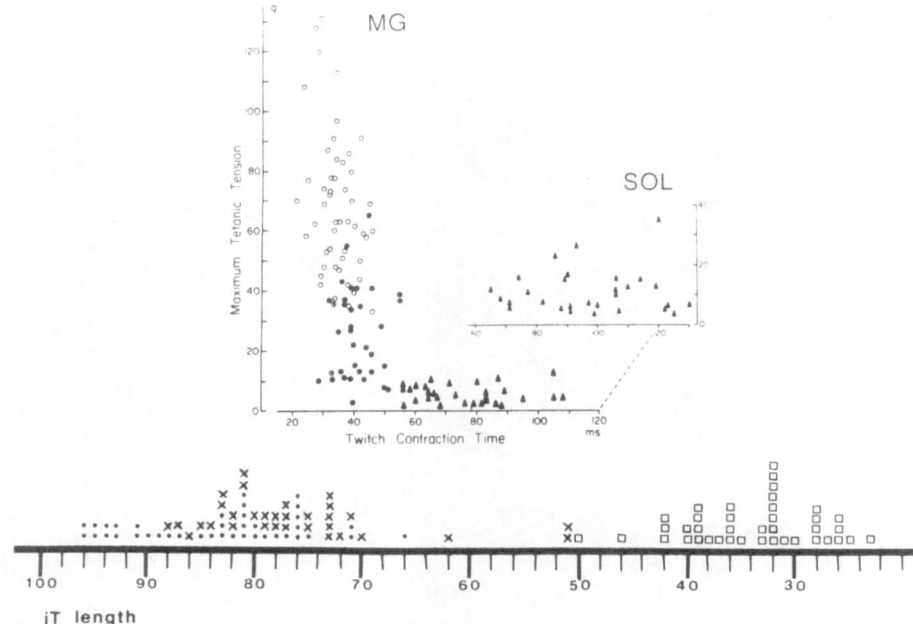

Figure 30. Comparison between relative proportions of junctional T-tubule (bottom) and twitch contraction time (center) for three types of twitch fibers. Each fiber is represented by a symbol. Fast twitch glycolytic and fast twitch oxidative glycolytic fibers form overlapping families, but slow twitch fibers are clearly distinct. (Bottom from Ref. 38; center from Ref. 39.)

ACKNOWLEDGMENTS.   We thank Drs. Kevin P. Campbell, Steve Kahl, C. Michael Knudson, Albert Leung, and Toshiaki Imagawa for giving us the purified protein fractions. We are grateful to Ms. Denah Appelt and Nosta Glaser for expert help. Some of the micrographs were obtained in collaboration with Dr. Grazia Nunzi. We thank Dr. Avril Somlyo for donating some of her blocks of toadfish muscle. Supported by MDA (H. M. Watts Center) and NIH 15735 to Pennsylvania Muscle Institute. Dr. B. Block is a fellow of MDA.

REFERENCES

1. Huxley, A. P., and Taylor, R. E., 1958, Local activation of striated muscle fibers, *J. Physiol. (Lond).* **144:** 426–441.
2. Porter, K. R., and Palade, G. E., 1957, Studies of the endoplasmic reticulum. III. Its form and distribution in striated muscle cells, *J. Biophys. Biochem. Cytol.* **3:** 269–300.
3. Andersson-Cedergren, E., 1959, Ultrastructure of motor end plate and sarcoplasmic components of mouse skeletal muscle fibers, *J. Ultrastructure Res. Suppl.* **1:** 1–191.
4. Hodgkin, A. L., and Horowicz, P., 1960, Potassium contractures in single muscle fibers, *J. Physiol. (Lond.)* **153:** 386–403.
5. Franzini-Armstrong, C., 1970, Studies of the triad. I. Structure of the junction in frog twitch fibers, *J. Cell. Biol.* **47:** 488–499.

6. Peachey, L. D., 1961, Structure of the longitudinal body muscles of amphioxus, *J. Biophys. Biochem. Cytol.* **10**(Suppl. 4): 159–178.

7. Sanger, J. W., 1971, Sarcoplasmic reticulum in the cross-striated adductor muscle of the bay scallop *Aquipecten iridians*, *Z. Zellforsch.* **118**: 156–161.

8. Nunzi, M. G., and Franzini-Armstrong, C., 1981, The structure of smooth and striated portions of the adductor muscle of the valves in a scallop, *J. Ultrastructure Res.* **76**: 134–148.

9. Grocki, K., 1981, Ultrastruktur der Pumpfmuskulatur von Branchiostoma Lanceolatum, dissertation thesis, Ruhr Universität Bochum.

10. Jorgensen, A. O., Shen, A. C. Y., MacLennan, D. H., and Tokuyasu, K. T., 1982, Ultrastructural localization of the Ca^{2+}, Mg^{2+}-dependent ATPase of sarcoplasmic reticulum in rat skeletal muscle by immunoferritin labelling of ultrathin frozen sections, *J. Cell. Biol.* **92**: 409–416.

11. Franzini-Armstrong, C., 1974, Freeze-fracture of striated muscle from a spider: Structural differentiations of sarcoplasmic reticulum and transverse tubular system membranes, *J. Cell. Biol.* **61**: 501–513.

12. Campbell, K. R., Franzini-Armstrong, C., and Shamoo, A. E., 1980, Further characterization of light and heavy sarcoplasmic reticulum vesicles: Identification of the sarcoplasmic reticulum feet associated with heavy sarcoplasmic reticulum vesicles, *Biochem. Biophys. Acta* **602**: 97–116.

13. Meissner, G., 1975, Isolation and characterization of two types of sarcoplasmic reticulum vesicles, *Biochim. Biophys. Acta* **389**: 51–68.

14. Ferguson, D. G., Schwartz, H., and Franzini-Armstrong, C., 1984, Subunit structure of junctional feet in triads of skeletal muscle: A freeze-drying, rotary-shadowing study, *J. Cell. Biol.* **99**: 1735–1742.

15. Kawamoto, R. M., Brunschwig, J. P., Kim, K. C., and Caswell, A. H., 1986, Isolation, localization, and characterization of the spanning protein of the skeletal muscle triad, *J. Cell. Biol.* **103**: 1405–1414.

16. Volpe, P., Gutweniger, H. E., Montecucco, C., 1987, Photolabelling of the integral proteins of skeletal muscle sarcoplasmic reticulum: Comparison of junctional and nonjunctional membrane fractions, *Arch. Biochim. Biophys.* **253**: 138–145.

17. Lai, F. A., Erickson, H. P., Block, B. A., and Meissner, G., 1986, Evidence for a junctional feet–ryanodine receptor complex from sarcoplasmic reticulum, *Biochem. Biophys. Res. Comm.* **143**: 704–709.

18. Inui, M., Saito, A., and Fleischer, S., 1987, Purification of the ryanodine receptor and identity with feet structures of junctional terminal cisternae of sarcoplasmic reticulum from fast skeletal muscle, *J. Biol. Chem.* **262**: 1740–1747.

19. Campbell, K. P., Knudson, C. M., Imagawa, T., Leung, A. T., Sutko, J. L., Kahl, S. D., Raab, C. R., and Madson, L., 1987, Identification and characterization of the high-affintiy (^{3}H)ryanodine receptor of the junctional sarcoplasmic reticulum Ca^{2+} release channel, *J. Biol. Chem.* **262**: 6460–6463.

20. Lai, F. A., Erickson, H. P., Rousseau, E., Liu, Q. Y., and Meissner, G., 1988, Purification and reconstitution of the calcium release channel from skeletal muscle, *Nature* **331**: 315–320.

21. Smith, J. S., Coronado, R., and Meissner, G., 1986, Single-channel measurements of the calcium release channel from skeletal muscle sarcoplasmic reticulum: Activation by Ca^{2+}, ATP, and modulation by Mg^{2+}, *J. Gen. Physiol.* **88**: 573–588.

22. Corbett, A. M., Caswell, A. R., Brandt, N. R., and Brunschwig, J-P., 1985, Determinants of triad junction reformation: Isolation and identification of an endogenous promoter of junction reformation in skeletal muscle, *J. Membrane Biol.* **86**: 267–276.

23. Leung, A., Imagawa, T., Block, B., Franzini-Armstrong, C., and Campbell, K. P., 1988, Biochemical and ultrastructual characterization of the dihydropyridine receptor from rabbit skeletal muscle: Evidence for a 52 kilodalton subunit, *J. Biol. Chem.* **263**: 994–1001.

24. Block, B., Imagawa, T., Campbell, K. P., and Franzini-Armstrong, C., 1988, Structural evidence for direct interaction between the molecular components of the T-SR junction in skeletal muscle, *J. Cell. Biol.* **107**: 2587–2600.

25. Franzini-Armstrong, C., Kenney, L., and Varriano-Marston, E., 1987, The structure of calsequestrin in triads of vertebrate muscle, *J. Cell. Biol.* **105:** 49–56.
26. Schneider, M. F., and Chandler, W. K., 1973, Voltage-dependent charge movement in skeletal muscle: A possible step in excitation–contraction coupling, *Nature* **242:** 747–751.
27. Rios, E., and Brum, G., 1987, Involvement of dihydropyridine receptors in excitation–contraction coupling in skeletal muscle, *Nature* **325:** 717–720.
28. Hannon, J. D., Lee, N. K. M., and Blinks, J. R., 1988, Calcium release by inositol triphosphate in amphibian and mammalian skeletal muscle is an artifact of cell disruption and probably results from depolarization of sealed-off T-tubules, *Biophys. J.* **53:** 607a.
29. Katz, B., 1966, *Nerve, Muscle, and Synapse*, McGraw-Hill, New York.
30. Peachey, L. D., and Franzini-Armstrong, C., 1983, Structure and function of membrane systems of skeletal muscle, in: *Handbook of Physiology, Section 10: Skeletal Muscle* (L. D. Peachey, R. H. Adrian, and S. R. Geiger, Eds.), Chapter 2, pp. 237–71, American Physiology Society, Bethesda, MD.
31. Franzini-Armstrong, C., 1986, The sarcoplasmic reticulum and the transverse tubules, in: *Myology* (A. G. Engel and B. Q. Banker, Eds.), pp. 125–154. McGraw-Hill, New York.
32. Page, S. G., 1965, A comparison of the fine structure of frog slow and twitch muscle fibers, *J. Cell. Biol.* **26:** 477–497.
33. Franzini-Armstrong, C., 1973, Studies of the triad. IV. Structure of the junction in frog slow fibers, *J. Cell. Biol.* **56:** 120–128.
34. Franzini-Armstrong, C., Eastwood, A. E., and Peachey, L. D., 1986, Shape and disposition of clefts, tubules, and sarcoplasmic reticulum in long- and short-sarcomere fibers of crab and crayfish, *Cell Tissue Res.* **244:** 9–19.
35. Franzini-Armstrong, C., Gilly, W. F., Aladjem, E., and Appelt, D., 1987, Golgi stain identifies three types of fibers in fish muscle, *J. Muscle. Res. Cell. Motility* **8:** 418–427.
36. Veratti, E., 1961, Investigations on the fine structure of the striated muscle fiber, *J. Biophys. Biochem. Cytol.* **10**(4): 3–59.
37. Franzini-Armstrong, C., and Peachey, L. D., 1982, A modified Golgi black reaction method for light and electron microscopy, *J. Histochem. Cytochem.* **30:** 99–105.
38. Franzini-Armstrong, C., Champ, C., and Ferguson, D. G., 1988, Discrimination between fast- and slow-twitch fibers of guinea pig skeletal muscle using the relative surface density of junctional transverse tubule membrane, *J. Muscle Res. Cell Motility* **9:** 403–414.
39. Burke, R. E., and Tsairis, P., 1974, The correlation of physiological properties with histochemical characteristics in single muscle units, *Ann. N.Y. Acad. Sci.* **228:** 145–159.
40. Franzini-Armstrong, C., and Nunzi, G., 1983, Junctional feet and membrane particles in the triads of a fast twitch muscle fiber, *J. Muscle Res. Cell Motility* **4:** 233–252.
41. Franzini-Armstrong, C., 1984, Freeze-fracture of frog slow tonic fibers: Structure of surface and internal membranes, *Tissue and Cell* **16**(3): 146–166.

Chapter 23

Proteins of the Triad Junction of Skeletal Muscle

Anthony H. Caswell, Neil R. Brandt, Shu-Rong Wen, and
Jane A. Talvenheimo

1. INTRODUCTION

The basic mechanism of excitation–contraction coupling in skeletal muscle is still not understood despite many years of active research on both the physiology and the biochemistry of the process. Several years ago we embarked on a project of identifying the constituents of the triad junction with the view that this junctional region represents not only the point of physical contact between the external and internal organelle but also the point of dynamic transmission in muscle excitation.[1,2] In 1986 we described the isolation of a protein of approximate subunit molecular weight 300,000 which spanned the gap between the transverse (T) tubule and the terminal cisternae, and therefore represented a portion of the junctional feet.[3] Later work by others has demonstrated that this protein is identical to the ryanodine receptor protein and contains Ca^{2+}-channel activity.[4–8] This then suggests that the process of Ca^{2+} release from the sarcoplasmic reticulum (SR) takes place at the level of the junctional foot. What is still not clear, however, is the relationship between the T-tubule and the junctional foot. The experiments described in this chapter represent an approach to the understanding of the

---

ANTHONY H. CASWELL, NEIL R. BRANDT, and JANE A. TALVENHEIMO • Department of Pharmacology, University of Miami School of Medicine, Miami, Florida 33101. SHU-RONG WEN • Department of Pharmacology, University of Miami School of Medicine, Miami, Florida 33101; and Department of Pharmacology, Beijing Medical University, Beijing, People's Republic of China.

constituents of the T-tubule and of the junction which forms the full triadic junctional structure.

2. RESULTS

2.1. Affinity Chromatography of Junctional Foot Protein

The isolated junctional foot protein was attached to CNBr-activated sepharose. T-tubular proteins dissolved in the detergent Zwittergent 3-14 in the presence of 30 mM K gluconate were passed through this column and specifically-bound protein was eluted by 1 M NaCl. The results in Fig. 1 show that predominantly two bands are specifically associated with the junctional foot protein. Lane A is molecular weight standards; lane B is T-tubular proteins that are not retarded by the column; lane C is the NaCl eluate of retarded proteins. These have molecular weights of 36,000 and 40,000 Da and correspond in position exactly to the proteins aldolase and glyceraldehyde phosphate dehydrogenase (GAPD), respectively. To test whether GAPD binds to the junctional foot protein, we ran a similar column in which the pure protein was passed through the affinity column and demonstrated the retention of enzymic activity of the protein on the

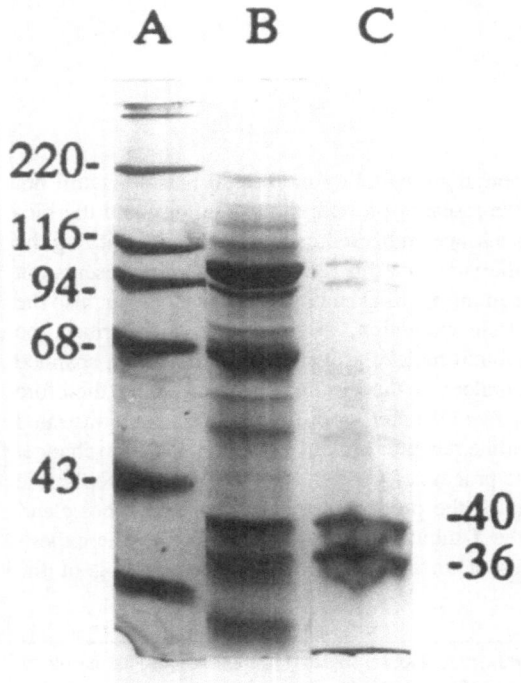

Figure 1. Association of T-tubular proteins with the junctional foot protein. (A) Molecular weight standards. (B) T-tubular proteins not retarded by a column of isolated junctional foot protein attached to CNBr-activated sepharose. (C) T-tubular proteins eluted with NaCl from the column described above.

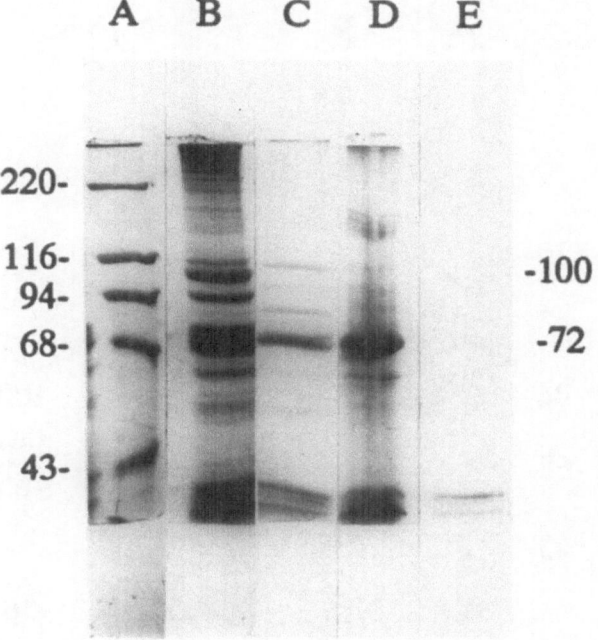

Figure 2. Association of T-tubular proteins with GAPD. (A) Molecular weight standards. (B) T-tubules. (C) T-tubular eluate not retained by a GADP affinity column. (D) T-tubular proteins retained by the GADP affinity column and subsequently eluted with 1*M* NaCl. (E) Same fraction as that shown in lane D but eluted from a control column lacking GADP.

column and its subsequent elution by 1 *M* NaCl. A control column lacking the junctional foot protein showed very little retention of GAPD. Because GAPD is an extrinsic protein of the T-tubule, we then evaluated which protein from the T-tubule associated with GAPD by making an affinity column of GAPD and passing dissolved T-tubular proteins through the column. The results are shown in Fig. 2. Lane A is standards; lane B is T-tubules; lane C is the T-tubular eluate that is not retained by the column, lane D is that which is retained and then eluted with 1 *M* NaCl. Lane D shows that although a number of proteins are associated with the GAPD, specifically retained proteins include one of molecular weight 72,000 and one in the range of 150,000 as well as GAPD itself. The control showed little binding.

2.2. Western Blot Overlay

A second approach to evaluate the T-tubular constituents that bind to the junctional foot protein is through overlay on electrophoretic blots. T-tubular proteins were subjected to SDS PAGE and then electrophoretically blotted onto nitrocellulose filters. The filters were blocked with bovine serum albumin and Tween 20. The filters were then

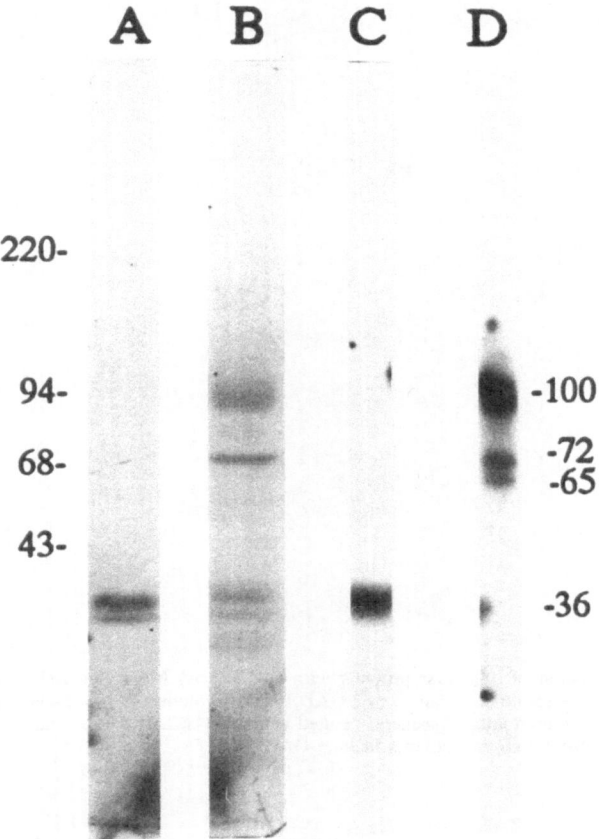

Figure 3. Overlay on electrophoretic blocks of T-tubular proteins plus junctional foot protein or GADP. (A) Amido Schwartz staining of electrophoretically blotted GADP. (B) Amido Schwartz staining of electrophoretically blotted T-tubular proteins. (C) Overlay of ^{125}I-labelled purified junctional foot protein and T-tubular proteins. (D) Overlay of ^{125}I-labelled GADP and T-tubular proteins.

incubated with ^{125}I-labelled protein followed by extensive washing and autoradiography. Figure 3 shows the results of such an experiment in which T-tubular proteins were overlaid with junctional foot protein and GAPD. Lanes A and B represent Amido Schwartz staining of electrophoretically blotted protein. Lane A is pure GAPD showing its position in the gel. Lane B is T-tubular proteins showing the two major proteins at molecular weight 100,000 and 72,000. Lane C is the overlay experiment using ^{125}I-labelled purified junctional foot protein. As can be seen in this figure, the label is almost exclusively associated with a protein of the same molecular weight as that of GAPD. Lane D represents a similar experiment employing ^{125}I-labelled GAPD for the overlay. In this case it can be seen that three major protein bands are labelled with molecular weights 100,000, 72,000, and 65,000, respectively. In this experiment we were unable to detect any association of GAPD with proteins in the higher molecular weight range that might be associated with the dihydropyridine receptor.

In experiments described in Fig. 4, partially purified dihydropyridine receptor was employed. This partially purified preparation was obtained from a crude T-tubular membrane preparation by dissolving it with digitonin, followed by passage through a wheat germ agglutinin column. The retained protein from that column was eluted with N-acetyl glucosamine and the output was run on SDS PAGE. Lane A represents the Amido Schwartz stain of the protein showing clearly the presence of the 170,000 and 150,000 subunits. Other protein constituents, which presumably represent contamination, are also present in this gel. Lane B represents an overlay using ^{125}I-labelled junctional foot protein. Little specific association is seen with any of the proteins of this partially purified dihydropyridine receptor preparation. Lane C represents an overlay in which the filter was first incubated with unlabelled GAPD, which was then washed from the filter, and ^{125}I-labelled junctional foot protein was incubated in a manner identical to lane B. As can be seen, considerable specific labelling now occurs. A strong labelling

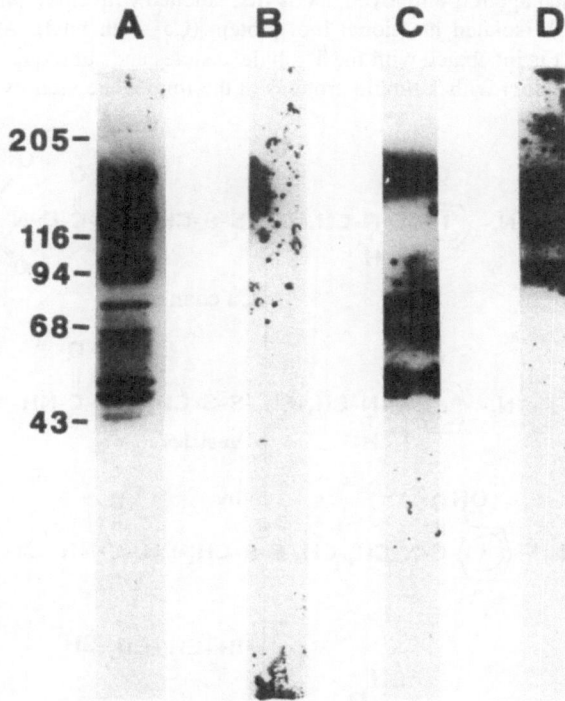

Figure 4. Overlay on electrophoretic blocks of partially purified dihydropyridine receptor and junctional foot protein or GADP. (A) Amido Schwartz staining of the partially purified dihydropyridine receptor. (B) Overlay of ^{125}I-labelled junctional foot protein onto partially purified dihydropyridine receptor. (C) Overlay in which the filter was first incubated with unlabelled GADP which was then washed from the filter and ^{125}I-labelled junctional foot protein was subsequently incubated in a manner identical to lane B. (D) Overlay of ^{125}I-labelled GADP and partially purified dihydropyridine receptor.

pattern is associated with the 170,000 Da subunit as well as two other proteins that band at molecular weight 80,000 and 50,000. It is possible that the 50,000 label corresponds to that of the 52,000 Da putative subunit of the dihydropyridine receptor. In lane D ^{125}I-labelled GAPD, was employed for the overlay. Again, considerable specific association with the 170,000 Da subunit is discernible as well as two other proteins of high molecular weight which may represent contamination. These observations combined with those of the affinity chromatography suggest that GAPD forms a direct association with the junctional foot protein and that it, in its turn, associates with T-tubular proteins including the dihydropyridine receptor. Whether the GAPD acts predominantly as a catalyst or whether there is a linear association of proteins in which the GAPD serves as the glue that joins the T-tubular proteins to the junctional foot is not as yet clear.

2.3. Crosslinking

A further mode for evaluating the interactions of the triad junction is through crosslinking agents. Figure 5 gives the basic protocol for the reaction. A photoreactive heterobifunctional agent is employed that is first labelled with chloramine T and ^{125}I and then reacted with isolated junctional foot protein (Ca^{2+} channel). After extensive dialysis the adduct is incubated with the T-tubule vesicles and subsequently illuminated to photoreact the adduct with T-tubule proteins in the immediate vicinity. The vesicles are

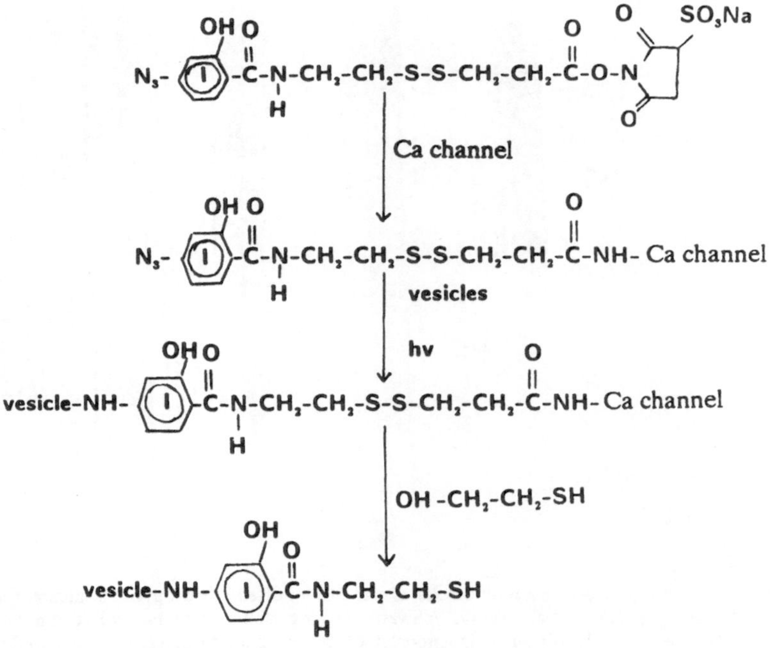

Figure 5. Basic protocol for the crosslinking reaction.

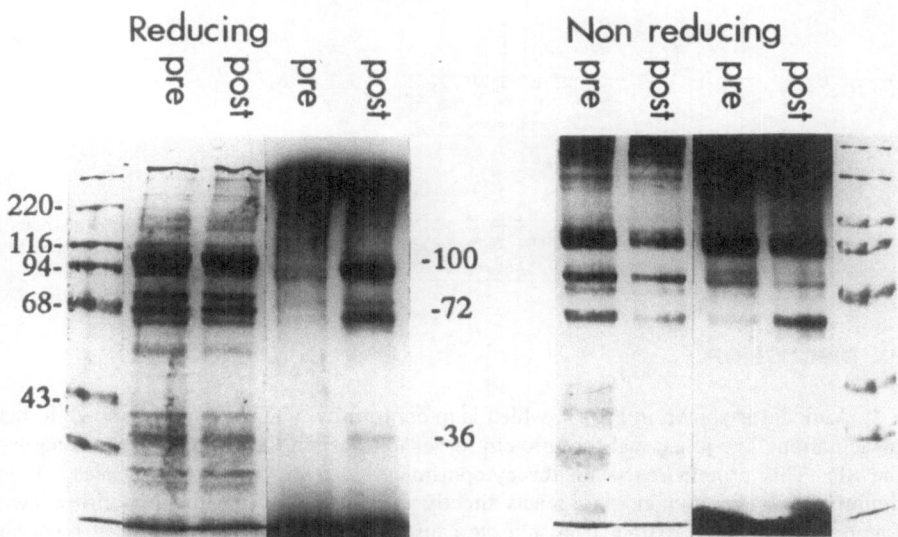

Figure 6. Crosslinking of T-tubular and isolated junctional foot protein. In each half of the figure there is one lane of standards, two lanes of Coomassie-blue stained gel, followed by two lanes of autoradiogram. The difference in autoradiographic intensity between pre- and postilluminated samples give the specific reactions.

then dissolved and electrophoresed. A reducing gel containing mercaptoethanol will cause breakage of the central disulfide of the crosslinking agent so that the radiolabel will be associated only with the photoreacted moiety and not with the junctional foot protein.

The experimental results are illustrated in Fig. 6. Two controls are employed. In the first, illumination precedes incubation of the adduct with the T-tubules so that the bifunctional agent is inactivated. In each half of the figure there is one lane of standards, two lanes of Coomassie-stained gel followed by two lanes of autoradiogram. On the left half of the figure the differences in autoradiographic intensity between pre- and postilluminated samples give the specific reactions. Major reacted proteins are observed at 100, 72, and 36 kDa. A second control (on right) is to employ a nonreducing gel. In this case the molecular weight of the labelled components should be the combined weight of each protein. Since the junctional foot protein has very high M_r, the expected label will be at the top of the gel. The data of Fig. 6 (right) show that the component at 100 kDa is present in both nonreducing and reducing gels and therefore does not conform to the crosslinking protocol of Fig. 5. In addition the component at 72 kDa, although it is not seen at that M_r in the nonreducing gel, may correspond to that observed at 65 kDa. Therefore, some ambiguity exists as to whether the labelling for this protein conforms to the protocol. On the other hand, the label at 36 kDa corresponding to GAPD is absent from the nonreducing gel.

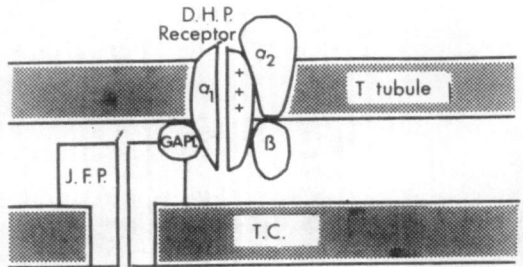

Figure 7. Proposed model of association of the junctional foot protein with the T-tubular membrane.

3. DISCUSSION

A model is shown in Fig. 7, which is in conformity with the data described in this presentation. The junctional foot protein serves to transport Ca^{2+} and other ions across the SR. This protein reacts on the cytoplasmic side with GAPD and aldolase. It is doubtful, however, that aldolase reacts directly with T-tubules. In earlier experiments we found that a soluble extract from muscle caused the reformation of the triad junction from isolate T-tubules and TC. This extract was purified and identified as GAPD.[9,10] Therefore, GAPD serves a specific role in triad formation even though other glycolytic enzymes may associate with junctional constituents.

In the model GAPD is shown associated with the DHP receptor. This is in accord with the hypothesis stated by Rios and Brum[11] that the DHP receptor is the voltage sensor that transmits the message directly to the SR Ca^{2+} release channel. Our experiments show, however, that this enzyme can associate with other T-tubular proteins. We had expected in the model of Rios and Brum that the specificity for association between the foot protein and the DHP receptor would have been more obvious experimentally. Alternate models of excitation–contraction coupling include transmitter–receptor interaction. In the currently favored model an extremely rapid machinery of phosphatidyl inositol diphosphate synthesis and breakdown at the junction is required for inositol trisphosphate formation. Therefore it is possible that the enzymes are associated with GAPD at the triad junction. The interaction of GAPD with membrane proteins appears to require the presence of extensive exposed acidic residues that bind to alkaline pockets in GAPD. This is fulfilled by the anion channel of erythrocytes and might be a prerequisite for all specific T-tubular proteins. In effect GAPD may serve as a glue not only to associate the junctional region of the triad but also to hold T-tubular-specific proteins preventing them from migrating to the surface membrane.

ACKNOWLEDGMENTS. This work was support by NIH grants AM21601 and HL36029 and by the American Heart Association (Florida affiliate).

REFERENCES

1. Cadwell, J. J. S., and Caswell, A. H., 1982, Identification of a constituent of the junctional feet linking terminal cisternae to transverse tubules in skeletal muscle, *J. Cell. Biol.* **93:** 543–550.

2. Caswell, A. H., and Brunschwig, J-P., 1984, Identification and extraction of proteins which compose the triad junction of skeletal muscle, *J. Cell. Biol.* **99:** 929–939.
3. Kawamoto, R. M., Brunschwig, J-P., Kim, K. C., and Caswell, A. H., 1986, Isolation, characterization, and localization of the spanning protein from skeletal muscle triads, *J. Cell. Biol.* **103:** 1405–1414.
4. Inui, M., Saito, A., and Fleischer, S., 1987, Purification of the ryanodine receptor and identity with feet structures of junctional terminal cisternae of sarcoplasmic reticulum from fast skeletal muscle, *J. Biol. Chem.* **262:** 1740–1747.
5. Lai, F. A., Erickson, H., Block, B. A., and Meissner, G., 1987, Evidence for a Ca^{2+}-channel within the ryanodine receptor complex from cardiac sarcoplasmic reticulum, *Biochem. Biophys. Res. Commun.* **143:** 704–709.
6. Campbell, K. P., Knudson, C. M., Imagawa, T., Leung, A. T., Sutko, J. L., Kahl, S. D., Raab, C. R., and Madson, L., 1987, Identification and characterization of the high-affinity [³H]-ryanodine receptor of the junctional sarcoplasmic reticulum calcium release channel, *J. Biol. Chem.* **262:** 6460–6463.
7. Imagawa, T., Smith, J. S., Coronado, R., and Campbell, K. P., 1987, Purified ryanodine receptor from skeletal muscle sarcoplasmic reticulum is the Ca^{2+}-permeable pore of the calcium release channel, *J. Biol. Chem.* **262:** 16636–16643.
8. Lai, F. A., Erickson, H. P., Rousseau, E., Liu, Q-Y., and Meissner, G., 1988, Purification and reconstitution of the calcium release channel from skeletal muscle, *Nature* **331:** 315–319.
9. Corbett, A. M., Caswell, A. H., Brandt, N. R., and Brunschwig, J-P., 1985, Determinants of triad junction reformation: Isolation and identification of an endogenous promoter of junction reformation in skeletal muscle, *J. Membrane Biol.* **86:** 267–276.
10. Caswell, A. H., and Corbett, A. M., 1985, Interaction of glyceraldehyde-3-phosphate dehydrogenase with isolated microsomal subfractions of skeletal muscle, *J. Biol. Chem.* **260:** 6892–6898.
11. Rios, E., and Brum, G., 1987, Involvement of dihydropyridine receptors in excitation–contraction coupling in skeletal muscle, *Nature* **325:** 717–720.

Chapter 24

Monoclonal Antibodies as Probes of Triad Structure and Excitation–Contraction Coupling in Skeletal Muscle

Mario S. Rosemblatt, Gonzalo Pérez, Bojena Antoniu,
Evelyn Reilley, and Noriaki Ikemoto

1. INTRODUCTION

It is generally accepted that the transverse tubules (T-tubules), a membrane system produced by invaginations of the plasma membrane in skeletal muscle, play a central role in the process of coupling the depolarization of the plasma membrane and muscle contraction (excitation–contraction coupling). However, the detailed molecular mechanism by which the T-tubules communicate with the intracellular membrane system of the sarcoplasmic reticulum (SR) where calcium is stored is largely unknown. Ultrastructural studies of skeletal muscle show that the T-tubules are connected to the SR at both sides through specific structures (the "feet"), forming a structural and functional unit known as a triad.[1]

Although recent evidence indicates that some T-tubule proteins and other proteins forming part of the feet may be involved in the excitation–contraction coupling pro-

MARIO S. ROSEMBLATT • Unidad de Immunología Celular, Instituto de Nutrición y Tecnología de los Alimentos, Universidad de Chile, Santiago, Chile; and Department of Muscle Research, Boston Biomedical Research Institute, Boston, Massachusetts 02114.   GONZALO PÉREZ • Unidad de Imunología Celular, Instituto de Nutrición y Tecnología de los Alimentos, Universidad de Chile, Santiago, Chile.   BOJENA ANTONIU and EVELYN REILLEY • Department of Muscle Research, Boston Biomedical Research Institute, Boston, Massachusetts 02114.   NORIAKI IKEMOTO • Department of Muscle Research, Boston Biomedical Research Institute, Boston, Massachusetts 02114; and Department of Neurology, Harvard Medical School, Boston, Massachusetts 02115.

cess[2] (see Chapter 22, this volume) many questions regarding T-tubule/SR communication remain unresolved. To elucidate the mechanism by which the T-tubules influence the permeability of the SR, it would be helpful to identify the components present either in the T-tubule, the SR, or the feet that may be responsible for the functional coupling between these two membrane systems. One of the most promising methods to identify the protein components involved in the coupling process is to produce monoclonal antibodies (mAb) that can affect either the triad structure or some step of the signal transmission pathway and thus can be used to identify the molecules (antigens) involved.

Pursuant to this goal we have performed several somatic fusions between mouse myeloma cells and splenocytes obtained from mice immunized with either T-tubules or SR membranes, and have obtained a number of mAbs directed against specific components of these membranes. These antibodies have been screened to determine their effect on (i) triad reassociation, (ii) specific T-tubule or SR functions, and (iii) T-tubule/SR communication (excitation–contraction coupling).

2. EFFECT OF mABs ON TRIAD REASSOCIATION

We have investigated the effect of some mAbs on the reassociation of dissociated triads in an attempt to identify the macromolecules involved in the structural linkage between the T-tubules and SR.[3] For these studies we have followed the protocols devised by Lau et al.[4] that have demonstrated that isolated, partially purified triads can be separated in their constituent membranes (T-tubule and SR) by treatment with a French press, and that these membranes can reassociate in the presence of 0.4 M

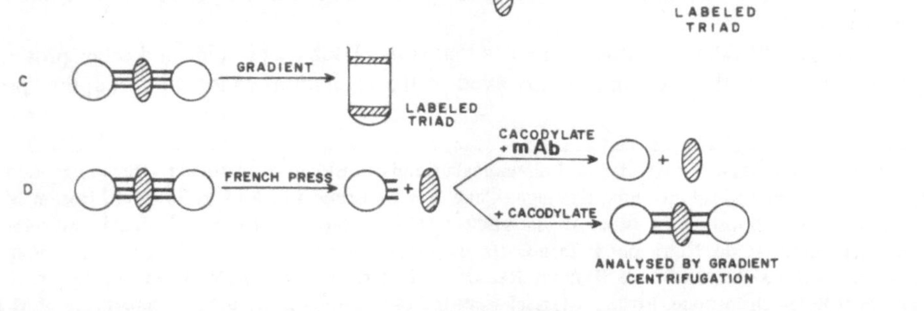

Figure 1. Protocol for the preparation of [3H]-triads and assay of the effect of anti-T-tubule mAbs on triad reassociation. For explanation see text.

Table I.   Effect of Various mAbs on Reassociation and Functional Integrity of Triads

mAb designation	Specificity[a]	Percent inhibition[b] of reassociation	Effect on depolarization-induced Ca^{2+} release	Effect on $(Ca^{2+}$ + caffeine)-induced release
2/1.1.1	SR	<25	None	None
TT2	T-T	~50	Inhibits ~ 50%	None
5/25.7	T-T	<20	None	None
5/49.9	T-T	0	None	None
5/131.7	T-T	<10	None	None
5/137.1	T-T	50	None	None
5/140.8	T-T	<20	None	None
6/261.3	SR	<10	None	None

[a]mAb specificity was determined by ELISA using purified T-tubules (T-T) and a light SR fraction purified by a Ca^{2+}-loading procedure.
[b]Inhibition of triad reassociation and mAb effect on Ca^{2+} release were measured as described in the text.

cacodylate. Figure 1 summarizes the protocol used in these experiments. In a first step, triad-enriched heavy microsomes obtained from rabbit skeletal muscle (HSR) were passed through a French press and the T-tubules recovered from a 35% sucrose cushion were labeled with [^{3}H]-succinimidyl propionate (Fig. 1A). These radiolabeled T-tubules were used to prepare labeled triads by reassociating them, in the presence of 0.4 M cacodylate, with HSR previously passed through the French press (Fig. 1B). These labeled triads were purified by sucrose gradient centrifugation (Fig. 1C) and used to study the effect of the mAbs on reassociation as follows. The labeled triads were dissociated with the French press (Fig. 1D) and incubated with cacodylate in the presence or absence of individual antibodies. The amount of T-tubules associated with the SR was determined after sucrose gradient fractionation of the reconstituted T-tubule/SR complexes. Figure 2A shows an example of one of the mAbs (5/137.1) that blocks the reassociation between the T-tubule and SR membranes, while Fig. 2B shows an example of an antibody (5/49.7) that has no effect on triad reassociation.

A summary of the effect of several mAbs tested for their effect on reassociation is shown in Table I. From these data one can infer that only some of the T-tubule molecules (as recognized by the apropriate mAb) participate in the interaction between T-tubules and SR. These proteins, which apparently do not participate in the attachment of the T-tubules to the SR, could belong to the nonjunctional portion of the T-tubules or they could belong to the junctional portion but have a function unrelated to actual membrane–membrane interaction. Furthermore, as we will describe below, not all of the proteins involved in T-tubule/SR binding are also involved in the actual functional communication between these two membranes.

3.   EFFECT OF mABs ON MEMBRANE FUNCTION

3.1.   Activation of the (Basal) Mg^{2+} ATPase

During the screening of our antibody library we found that preincubation of mAb 2/34.4 with T-tubule membranes obtained from frog skeletal muscle produces a marked activation of the enzyme Mg^{2+} ATPase (basal).[5] In frog T-tubules the Mg^{2+} ATPase

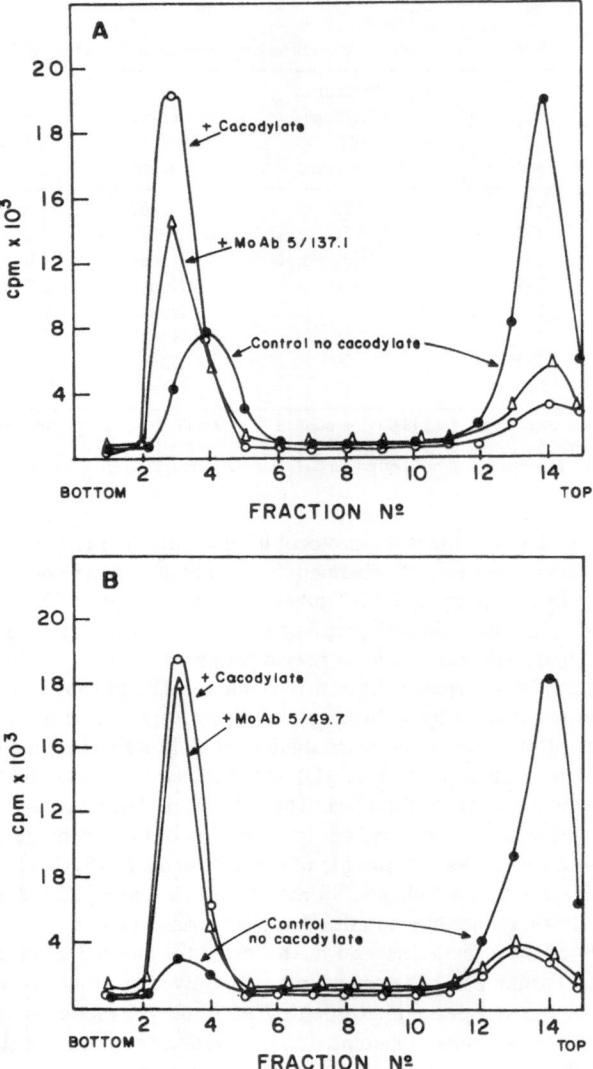

Figure 2. Sucrose gradient centrifugation diagram illustrating the effect of anti-T-tubule mAbs on T-tubule/SR reassembly. Tritium-labeled triads were prepared as indicated in Fig. 1 and incubated for 1 hr at 22°C and then overnight at 4°C with different T-tubule-specific mAbs (as established by ELISA). After this period K-cacodylate was added to a final concentration of 0.4 M and incubated for 1.5 hr to reconstitute the triads.[4] The incubation mixture was placed on top of a two-layer density gradient consisting, respectively, of 60% and 35% buffered sucrose and centrifuged at 35,000 rpm overnight. One-milliliter fractions were analyzed by scintillation counting. Panel A shows that in the control mixture containing cacodylate but no mAb (O—O), more than 90% of the [^{3}H]-T-tubules rejoined the SR to form triads (bottom of the gradient), while in an identical mixture preincubated with mAb 5/137.1 there is an inhibition of approximately 50% in the reassembly of the triads (△—△). Another control without mAb and cacodylate shows that most of the T-tubules remain uncoupled (top of the gradient) (●—●). Panel B shows an identical experiment as in A except that mAb 5/49.7 was used instead. As seen in the figure, this antibody has no effect on triad reassociation. A summary of identical experiments performed with other mAbs in shown in Table I.

has a two-phase kinetic of Pi production, an initial rapid phase followed by a slower rate of ATP hydrolysis.[6] It is this second, slower phase that is affected by the mAb, increasing the rate of Pi production, making it similar to that of the rapid phase. Figure 3 shows the effect of mAb 2/34.4 on the activity of this enzyme, and Fig. 4 shows the dependence of this effect on the antibody concentration. Immunoblots performed with frog T-tubules and mAb 2/34.4 show that this antibody recognizes a protein of 107,000 Da identified as the Mg^{2+} ATPase (data not shown). This antibody, which was prepared against T-tubules obtained from frog skeletal muscle, has no cross-reactivity with frog SR membranes either in ELISA or in immunoblots, but it does cross-react with rabbit T-tubules without affecting the Mg^{2+} ATPase of these membranes.[5]

The stimulation of the enzyme hydrolytic activity by this mAb is similar to the effect of lectins on the Mg^{2+} ATPase from frog (C. Hidalgo, personal communication), rat,[7] and chicken[8] T-tubules. In the case of the chicken enzyme, lectins prevent enzyme inactivation,[9] an effect we have also found for the mAb 2/34.2 on the frog enzyme. This effect could be the result of cross-linking of the enzyme in the plane of the membrane. Our results suggest, however, a different mechanism for this effect because mAb 2/34.4, which is an IgG, produces an activation of the enzyme higher than another mAb (mAb 2/25.2), which is of the IgM class and thus has a higher cross-linking capacity. These results are strengthened by the fact that polyclonal antibodies directed against the enzyme—and which are known to be good cross-linkers—produce an inhibi-

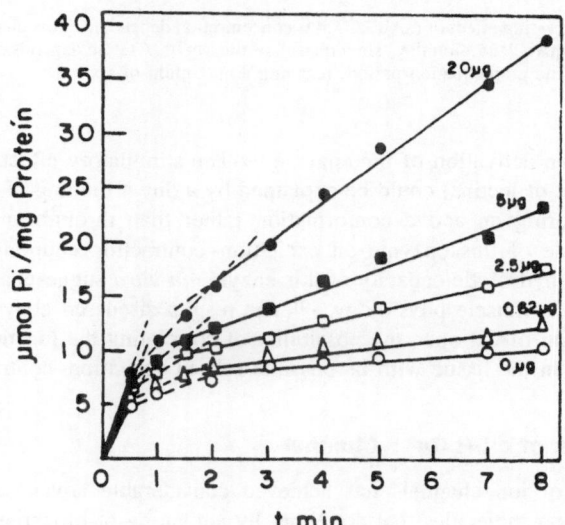

Figure 3. Effect of mAb 2/34.4 on the rate of ATP hydrolysis by the enzyme Mg^{2+}-ATPase from frog skeletal muscle T-tubules. T-tubule membranes prepared according to Hidalgo et al.[6] were preincubated with different amounts of purified mAb 2/34.4 for 30 min at 22°C, and then the rate of ATP hydrolysis was determined by measuring the amount of Pi liberated, as described.[6] As seen in the figure preincubation of the membranes with this antibody results in an increase in the rate of ATP hydrolysis during the "slow phase" of the enzyme. The different symbols represent increasing amounts of mAb in the incubation mixture as indicated in the figure. Control experiments done with an irrelevant immunoglobulin showed no effect.

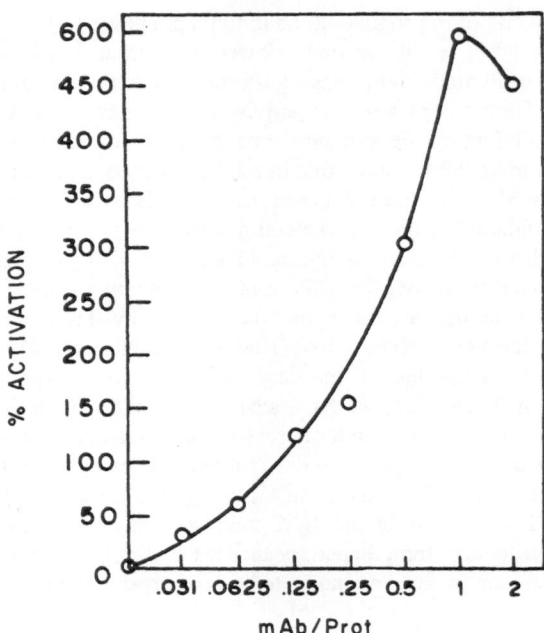

Figure 4. The activating effect of mAb 2/34.4 is concentration dependent. This diagram shows that the activating effect of mAb 2/34.4 on the "slow phase" of the Mg^{2+}-ATPase depends on the concentration of antibody during the preincubation period, reaching a maximum of sixfold.

tion rather that an activation of the enzyme.[8] The stimulatory effect of mAb 2/34.4 (and perhaps that of lectins) could be explained by a direct binding of this molecule to the enzyme favoring an active conformation rather than through an indirect effect. Although the role of this enzyme on excitation–contraction coupling remains unresolved, the high hydrolytic capacity of this enzyme *in vitro* suggests an important role for this enzyme in muscle physiology.[10] The results discussed above and the recent data obtained with mAbs open the possibility of correlating the location of the Mg^{2+} ATPase enzyme in the tissue with its possible role in excitation–contraction coupling.

3.2. Activation of a SR Ca^{2+} Channel

The study of ion channels has achieved considerable advances by the use of naturally occurring molecules that act either by activating or blocking ion movement. These channel modulators have been used not only to probe channel function, but in some cases they have served for the purification of the corresponding channel proteins, thus providing investigators with additional ways of studying the relationship between channel structure and its function. Another approach to the search for channel "modulators" for the purpose of studying structure–function relationships is to prepare specific probes such as mAbs. If the channel protein is known (i.e., the acetylcholine receptor protein), it is simpler to prepare such antibodies, but if the molecule carrying the channel activity is unknown, then one must use the "shotgun" approach—using whole membranes containing the channel to immunize the animals—in the hope of obtaining an antibody both capable of reacting with the channel protein and, most importantly, capable of *modifying* channel behavior.

It is generally accepted that release of Ca^{2+} from the SR occurs through a Ca^{2+}-channel that responds to a signal from the T-tubules.[11] More recently a number of investigators have described at least two different Ca^{2+}-channels for skeletal muscle SR capable of responding to different signals (messengers) (see Chapter 32, this volume). We have detected among our mAb library one antibody (mAb 6/29) that is capable of opening an SR Ca^{2+} channel.[12] Preincubation of Ca^{2+}-loaded HSR vesicles from rabbit muscle with this antibody shortens the time of spontaneous release of Ca^{2+} in a manner dependent on antibody concentration (Fig. 5). Furthermore, kinetic analysis of Ca^{2+} exit shows that the rate constant for Ca^{2+} release increases from a value of 0.156 sec^{-1} in the absence of antibody to 0.343 sec^{-1} in its presence. That these effects are due to binding of the antibody to the Ca^{2+} channel protein (or to a regulatory subunit) and not to some other indirect effect is supported by results that show that the antibody effect is blocked by ruthenium red, a known blocker of SR Ca^{2+} channels (Fig. 6).

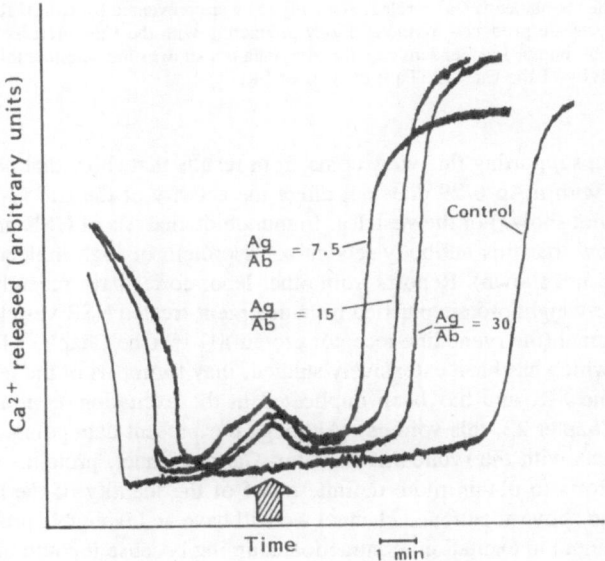

Figure 5. mAb 6/29 alters the Ca^{2+} uptake and release parameters of a triad-enriched SR fraction. Heavy SR (HSR) vesicles were incubated with the SR specific mAb 6/29 for 2.5 hr at 22°C, and the rates of Ca^{2+} uptake (in the presence of ATP) and release were measured in a dual-beam spectrophotometer using Arzenaso III as a Ca^{2+} indicator. In the figure a downward trace represents uptake, while an upward trace represents spontaneous Ca^{2+} release. Increasing the concentration of antibody in the mixture results in a reduction of the Ca^{2+} uptake rate and an acceleration of the reversible (arrow) and nonreversible Ca^{2+} release. Antibody concentrations used are indicated in the figure as the ratio of SR/mAb.

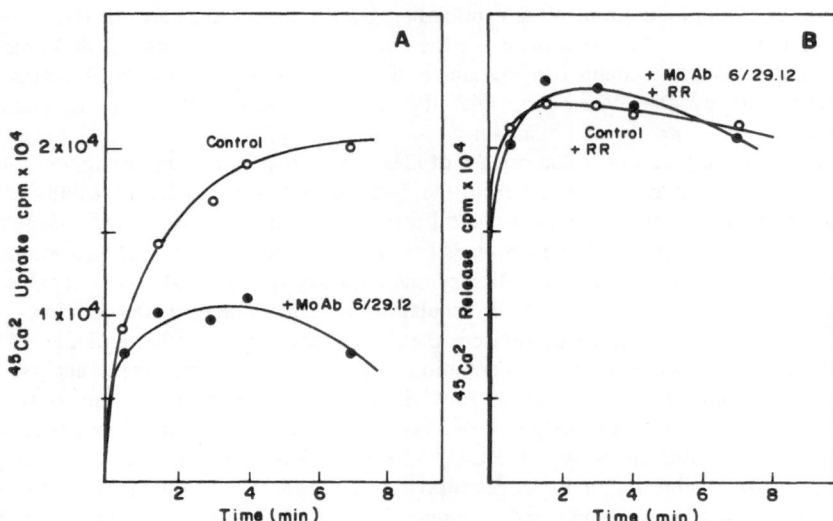

Figure 6. Influence of the Ca^{2+}-channel blocker ruthenium red (RR) on the effect produced by mAb 6/29 on Ca^{2+} uptake and release by HSR. The antibody-induced reduction on Ca^{2+} uptake rate and the acceleration of the spontaneous Ca^{2+} release (see Fig. 5) were prevented by $1.0\mu M$ RR, indicating that the observed effects are produced by the antibody interacting with the Ca^{2+} release machinery rather than with the Ca^{2+} pump. Independent experiments (data not shown) indicate that this antibody has no effect on the activity of the Ca^{2+}-ATPase enzyme of SR.

Other data supporting this view come from results that show that preincubation of HSR vesicles with mAb 6/29 does not affect the activity of the $Ca^{2+}-Mg^{2+}$ ATPase enzyme (data not shown) of the vesicles. Immunoblot analysis of HSR membranes with mAb 6/29 show that this antibody recognizes a protein of high molecular weight ($>$ 350,000) (data not shown). Reports from other laboratories have recently shown that a high-molecular-weight protein purified from detergent-treated HSR vesicles corresponds to a Ca^{2+}-channel (the ryanodine receptor protein)[13,14] (see Chapter 31, this volume). This protein, which has been extensively studied, may form part of the feet that connect T-tubules to the SR, and has been implicated in the excitation–contraction coupling process (see Chapter 22, this volume). Although our present data points to the fact that mAb 6/29 reacts with the ryonodine receptor–Ca^{2+}-channel, protein we are directing our present efforts to obtain more definite proof of the identity of the antigen. If this proves to be the above-mentioned channel we will have an invaluable probe to study the role of this channel in excitation–contraction coupling because it could be used to study the behavior of the isolated channel *in vitro* or to by-pass the T-tubule signal in a more physiological preparation using muscle fibers.

4. EFFECT OF mAb ON DEPOLARIZATION-INDUCED CALCIUM RELEASE

Monoclonal antibodies have been used with success in a number of systems to determine the subunit structure of channel proteins and to map the position of the subunits relative to one another and with respect to the lipid bilayer. Given that the SR Ca^{2+}-channel plays such an important role in the final steps of the excitation–contraction coupling process, it would be highly desirable to have a closer understanding, at the molecular level, of the workings of the SR Ca^{2+} channel. In this regard it would be important to understand not only the structural features of the channel vis-à-vis its ion transport properties but also the molecular mechanisms underlying its regulation by other molecules such as metabolites or by other feet proteins.

In an attempt to find mAbs capable of *modulating* T-tubule/SR communication we examined the capability of anti-T-tubule mAbs to block the T-tubule-mediated release of Ca^{2+} from SR in triads. HSR vesicles were incubated with the purified immunoglobulin fraction of various mAbs (obtained by ammonium sulfate fractionation of mouse ascites fluid) and then the time course of Ca^{2+} release induced either by chemical depolarization or by addition of Ca^{2+} plus caffeine, was monitored by a stopped-flow spectrophotometric method. Only one out of eight mAbs screened (mAb TT2) produced a significant inhibition of the depolarization-induced Ca^{2+} release without affecting the caffeine-induced release (see Fig. 7). This mAb reacts only with T-tubules membranes and with the triad containing HSR but does not react with purified longitudinal SR (Fig. 8).

Given the specificity of this antibody and that depolarization-induced Ca^{2+} release is triggered via the T-tubules, while the caffeine acts directly on the SR,[15] it is then reasonable to assume that mAb TT2 is actually binding to sites located in the T-tubule/SR communication pathway. Because this antibody is capable of inhibiting the reassociation of predisrupted triads (Fig. 9A and Table I) one could assume that the

Figure 7. mAb TT2 inhibits the depolarization-induced Ca^{2+} release by HSR but leaves unaffected the caffeine-induced release. Triad-enriched microsomes (see Ref. 15) were incubated overnight at 4°C with purified T-tubule mAb TT2 (HSR/mAb protein ratio 15). Aliquots of this incubation mixture were incubated at 27°C for 2.5 min with a loading solutions as described[15] and loaded in syringe A of a Durrum stopped-flow apparatus. Ca^{2+} release was triggered by mixing the content of syringe A with an equal volume from syringe B containing choline chloride instead of K-gluconate (depolarization-induced release) (panel A) or with an equal volume from syringe B containing 100 μM Ca^{2+} and 4.0 mM caffeine (caffeine-induced release) (panel B). As seen in the figure preincubation of the membranes with this mAb produces a decrease in the rate at the initial "rapid phase" of depolarization-induced Ca^{2+} release, while the caffeine-induced release is not affected.

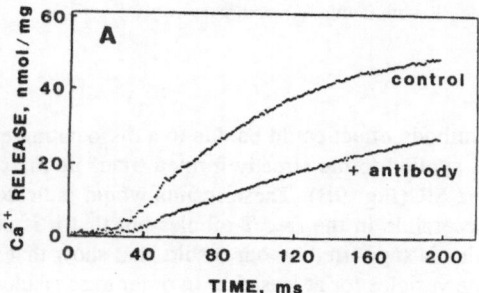

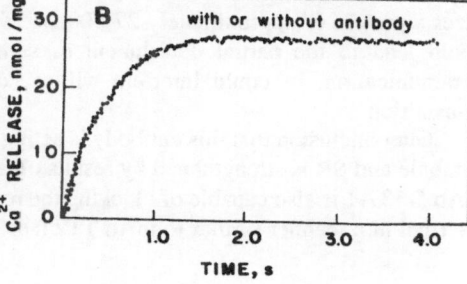

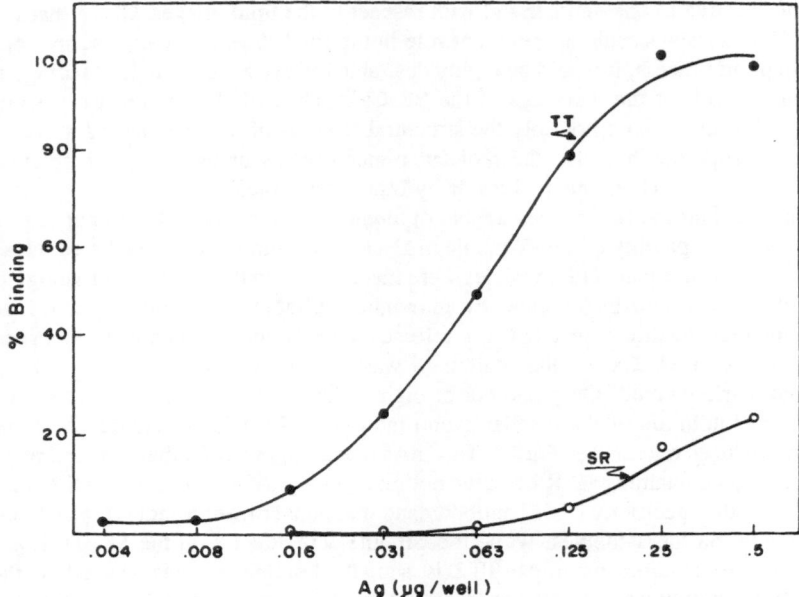

Figure 8. ELISA showing that mAb TT2 is T-tubule specific. ELISA plates were coated with decreasing amounts of T-tubule membranes (●—●) and SR membranes (○—○) diluted in PBS. Purified TT2 mAb diluted 1 : 1000 was added to the wells containing antigen and binding quantitated using horseradish peroxidase-labeled goat antimouse immunoglobulin. The results are expressed as percentage binding relative to the maximum value obtained with T-tubules. All experiments were done in duplicate. A small amount of binding of mAb TT2 to the SR membranes is probably due to a low level of contamination of the SR with T-tubule membranes.

antibody effect could be due to a dissociating effect on the triad. However, if mAb TT2 is applied to the already formed triads no dissociation occurs between the T-tubule and the SR (Fig. 9B). These results would indicate that the antigen to mAb TT2 is readily accessible in the free T-tubules but that it is actually occluded in the intact triad. This idea is supported by our results that show that the antibody has to be preincubated with the vesicles for at least 6 hr in order to see a clear inhibitory effect (Fig. 10) and could be explained by the fact that this antibody is of the IgM class. Immunoblot experiments with t-tubule or HSR membranes and mAb TT2 have shown that this antibody recognizes a protein of approximately 27,500 Da. Thus, binding of mAb TT2 to this protein could lead to the partial detachment of some feet protein blocking intermembrane communication, or could interfere with a conformational change needed for signal transmition.

Our conclusion that this antibody is acting on the signal transition pathway between T-tubule and SR is strengthened by results obtained with other antibodies. For example, mAb 5/137.1 is also capable of blocking the reassociation of T-tubules and SR to reform the triad in a manner similar to mAb TT2 (see Fig. 2 and Table I) but is unable to affect

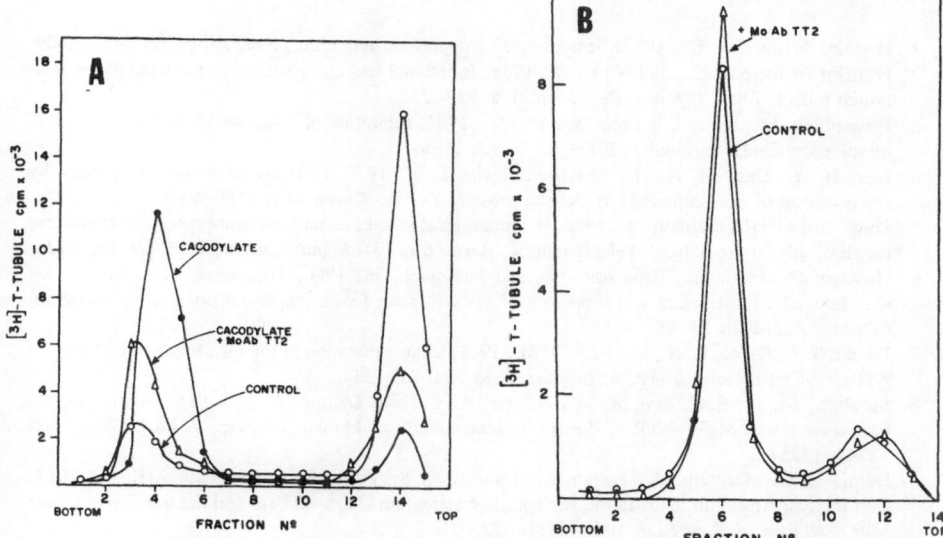

Figure 9.   Sucrose density fractionation diagram showing the effect of mAb TT2 on triad structure. Preincubation of radiolabeled triads with mAb TT2 under conditions similar to those shown in Fig. 2 demonstrate that binding of this antibody to the T-tubules interferes with the reassociation of the T-tubules and SR moieties of the triad (panel A). However, this blockage of the reassembly is not the cause of the inhibition produced by this antibody on the depolarization-induced Ca^{2+} release shown in Fig. 7. As seen in panel B incubation of the labeled triads with purified mAb TT2 under the same conditions that were used to determine depolarization-induced Ca^{2+} release produces no effect on the integrity of the preformed triads.

Figure 10.   The magnitude of the effect of mAb TT2 on the depolarization-induced Ca^{2+} release depends on the time of preincubation. Triad-enriched vesicles were preincubated at 4°C with purified mAb TT2. At the times indicated in the figure an aliquot was withdrawn and the amount of Ca^{2+} liberated during the initial rapid phase of the depolarization-induced release was analyzed as shown in Fig. 7. This amount is plotted as a function of the time of preincubation. Control experiments were done in the presence of an irrelevant antibody. The data show that the effect of this antibody is detectable only after 6 hr of preincubation.

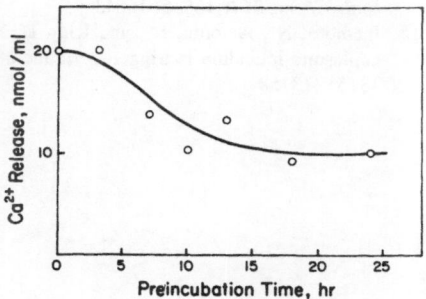

Ca^{2+} release regardless of the trigger. Table I also summarizes the effect of several other antibodies on the various activities assayed. Taken together these data would indicate that antibody binding to the T-tubules *per se* and inhibition of reassociation would not necessarily lead to functional blockage, and that the 27,500 Da molecule plays a crucial role both in the structural and functional coupling of the T-tubule and the SR.

REFERENCES

1. Franzini-Armstrong, C., 1977, Structure of sarcoplasmic reticulum, *Fed. Proc.* **39:** 2403–2409.
2. Franzini-Armstrong, C., and Ninzi, G., 1983, Junctional feet and particles in the triad of the fast-twitch muscle fiber, *J. Musc. Res. Motility* **4:** 233–252.
3. Rosemblatt, M., Nelson, R., and Ikemoto, N., 1984, Inhibition of T-tubule/SR linkage by anti-T-tubule monoclonal antibodies, *Biophys. J.* **455:** 398a.
4. Lau, H. Y., Caswell, A. H., and Brunschwig, J. P., 1977, Isolation of transverse tubules by fractionation of triad junctions of skeletal muscle, *J. Biol. Chem.* **252:** 5565–5574.
5. Perez, G., and Rosemblatt, M., 1986, Immunological studies on surface membranes and transverse tubules isolated from frog skeletal muscle, *Arch. Biol. Medicina Experiment. Chile* **19:** R239.
6. Hidalgo, C., Parra, C., Riquelme, G., and Jaimovich, E., 1986, Transverse tubules from frog skeletal uscle: Purification and properties of vesicles sealed with inside-out orientation, *Biochem. Biophys. Acta* **855:** 79–88.
7. Beeler, T. J., Gable, K. S., and Effer, J. M., 1983, Characterization of the membrane bound Mg^{2+}-ATPase of rat muscle. *Biochem. Biophys. Acta* **734:** 221–234.
8. Moulton, M. P., Sabbadini, R. A., Norton, K. C., and Dahms, A. S., 1986, Studies on the transverse tubule Mg^{2+}-ATPase: Lectin-induced alteration of kinetic behavior, *J. Biol. Chem.* **262:** 12244–12251.
9. Damiani, E., Margreth, A., Furlan, A., Dahms, A. S., Arnn, J., and Sabbadini, R. A., 1987, Common structural domains in the sarcoplasmic reticulum Ca^{2+}-ATPase and the transverse tubule Mg^{2+}-ATPase, *J. Cell Biol.* **104:** 46111–472.
10. Hidalgo, C., Gonzalez, E., and Lagos, R., 1983, Characterization of the Ca^{2+}- or Mg^{2+}-ATPase of transverse tubule membrane from rabbit skeletal muscle, *J. Biol. Chem.* **258:** 13937–13945.
11. Tada, M., Yamamoto, T., and Tonomura, Y., 1978, Molecular mechanism of active calcium transport by sarcoplasmic reticulum, *Physiol. Rev.* **58:** 1–79.
12. Rosemblatt, M., Cifuentes, M. E., and Ikemoto, N., 1988, Immunological studies of the Ca^{2+} release component of sarcoplasmic reticulum, *Biophys. J.* **61:** 32a.
13. Lai, F. A., Erickson, H. P., Rousseau, E., Liu, Q-Y., and Meissner, G., 1988, Purification and reconstitution of the calcium release channel from skeletal muscle, *Nature* **331:** 315–318.
14. Imagawa, T., Smith, J. S., Coronado, R., and Campbell K., 1987, Purified ryonodine receptor from skeletal muscle sarcoplasmic reticulum is the Ca-permeable pore of the calcium release channel, *J. Biol. Chem.* **262:** 16636–16643.
15. Ikemoto, N., Antoniu, B., and Kim, D. H., 1984, Rapid calcium release from the isolated sarcoplasmic reticulum is triggered via the attached transverse tubular system, *J. Biol. Chem.* **259:** 13151–13158.

D. Transduction at the Triad

Chapter 25

A Third Role for Calcium in Excitation–Contraction Coupling

Eduardo Ríos, Robert Fitts, Ismael Uribe, Gonzalo Pizarro, and
Gustavo Brum

1. INTRODUCTION

Two important roles for calcium ions in excitation–contraction coupling are traditionally described. First is the second-messenger role—Ca^{2+} undergoes a transient increase in the myoplasm and triggers the mechanochemical reaction of contraction.[1] This transient increase in myoplasmic free calcium concentration is usually called the *Ca transient*; it is the result of a flux of calcium release from the sarcoplasmic reticulum (SR). In this chapter we discuss the mechanisms by which the SR is made to open its release channels during normal muscle function. In regard to this question a second role of Ca^{2+} has been identified: Ca^{2+} is an effective agonist for opening the Ca channels of the SR. The increase in $[Ca^{2+}]_i$ in the vicinity of the SR release channels constitutes the normal mechanism of triggering release in the heart (Ca-induced Ca release).[2,3]

In regard to skeletal muscle, the functional significance of this second mechanism is still debated: Ca^{2+} has proved to be an effective trigger in both isolated channels reconstituted in bilayers[4] and in subcellular bundles of myofibrils.[5] Also, experiments with skinned skeletal fibers[6] demonstrate the possibility that Ca-induced Ca release is

EDUARDO RÍOS, ROBERT FITTS, ISMAEL URIBE, and GONZALO PIZARRO • Department of Physiology, Rush University, Chicago, Illinois 60612. GUSTAVO BRUM • Departamento de Biofísica, Facultad de Medicina, Universidad de la República, Montevideo, Uruguay. *Present address for I.U.:* Centro de Investigación del Instituto Politécnico Nacional, Mexico City, Mexico. *Permanent address for R. F.:* Department of Biology, Marquette University, Milwaukee, Wisconsin 53233.

playing a role. However, experiments of injection of calcium buffers in intact fibers have failed to stop Ca release and constitute evidence against a physiological role of the Ca-induced mechanism.[7] Ríos and Pizarro have recently proposed a way in which this mechanism could operate physiologically (see Ref. 36).

In this chapter we will review the work of our laboratory leading to the concept that there is another fundamental role of Ca^{2+} in excitation–contraction coupling; this role appears to take place at the "prejunctional" level, that is, at the T-tubular membrane of the T-SR junction.

There are different ideas of what exactly the prejunctional events are. In this book we can read arguments in favor of the possibility that a specific phospholipase, which cleaves inositol trisphosphate from PIP_2, is activated at a crucial prejunctional stage.[8] For some time it was thought that a Ca-channel opened, to let "trigger Ca^{2+}" enter the cell.[9] Others have proposed the existence of a charged molecule, connected directly, mechanically, to a gate in the Ca release SR channel.[10] Regardless of one's favorite coupling hypothesis, it is generally agreed that the process starts at a *voltage sensor*, connected in turn to a *mechanism of transmission*, the nature of which is not yet clear.

The fundamental third role appears to be played by Ca ions at the voltage sensor. Calcium ions are required to keep the voltage sensor in a working condition and prevent it from going into inactivation. This role is understood through the postulate of an essential cation-binding site on the voltage sensor to which Ca^{2+} must bind. As the process of recovery from inactivation is usually termed repriming, and the resting (not inactivated) state of the system could be considered "primed," we have named this hypothetical cation-binding site on the sensor the "priming" site. The evidence for this role will be presented after a brief description of our methods.

2. METHODS

The experiments to be described were performed on cut segments of singly dissected (fast twitch) fibers of the semitendinosus muscle of *Rana pipiens*. The segments are cut in a relaxing solution and mounted in a vaseline gap device where the middle portion ($\cong 500$ μm) of the fiber is voltage-clamped. The cut ends are in independent pools containing solutions that slowly equilibrate, by diffusion, with the intracellular space. This device thus has the advantage that both the intra- and extracellular media are relatively accessible and modifiable.

By extensively described optical techniques[11] we monitor the changes in time of $[Ca^{2+}]_i$, under voltage-clamp pulse stimulation. By less well known techniques[12–14] we can derive, from the Ca transients, the flux of Ca release that causes the increase in $[Ca^{2+}]_i$.

These methods thus lead to a measure of the main postjunctional event (the flux of Ca release). Independent electrical measurements are carried out simultaneously to determine a relevant prejunctional event, the intramembrane charge movement,[10] probably associated with voltage-driven conformational changes of the T-membrane voltage sensor. This measurement of charge movement and its problems have been discussed recently.[15,16] We use conventional pulse protocols (which in the lexicon of gating currents are termed P/n) in which the "control currents" for linear capacitance

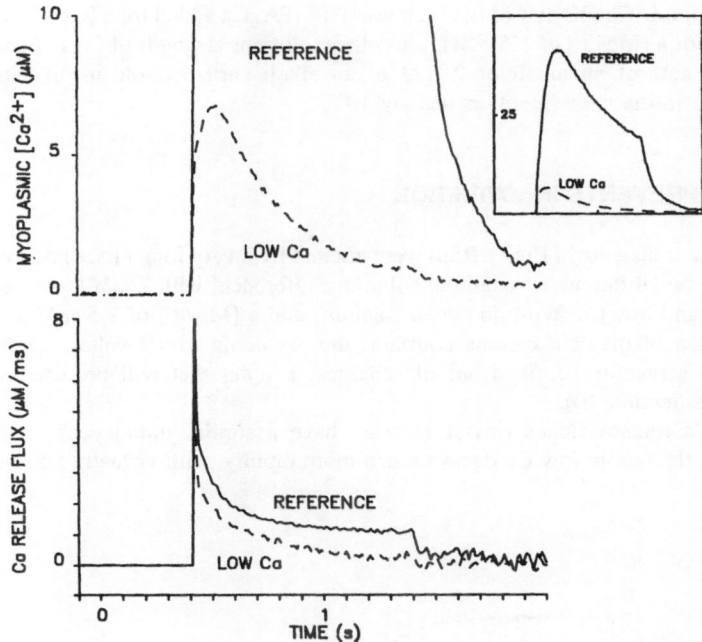

Figure 1. Effect of low calcium on Ca transients and calcium release. (Top) Ca transients (time course of change in free myoplasmic Ca^{2+} concentration) determined from changes in optical absorbance of a cut fiber equilibrated with an internal solution containing Antipyrylazo III. labeled reference, was obtained with the fiber immersed in an external solution containing 2 mM Ca^{2+}; the record labeled low Ca was obtained in a solution with no added calcium, 5 mM EGTA, and Mg added to obtain a $[Mg^{2+}]$ of 2.5 mM. Both records were obtained with pulse depolarization to 10 mV, for 1 sec, from a HP = -100 mV. The record in 2 mM Ca^{2+} is shown out of range to demonstrate the close superposition of both records at early times. The complete records are shown in the inset. (Bottom) Time course of flux of Ca release, derived from the Ca transients, following the procedure described in Ref. 14. Fiber 90, diameter 60 μm, dye concentration at the time of the measurements 320 to 350 μM, temperature 7°C. The internal solution contains 108 mM Cs glutamate, 5 mM Na_2ATP, 5 mM glucose, 5 mM $MgCl_2$, and is buffered to pH 7 with 15 mM Cs tris-maleate; the $[Ca^{2+}]_i$ is nominally set to 50 nM with 0.1 mM EGTA. The reference external solution is isotonic, with 127 mM tetraethylammonium (TEA) methanesulfonate (MES), 2 mM Ca-MES, 1 μM TTX, 1 μM 2,4-diaminopyridine, and is buffered to pH 7 with 5 mM TEA tris-maleate; the low-Ca solution is similar, except for the absence of Ca and the presence of 5 mM EGTA, and Mg, TEA, and MES are reduced to maintain osmolarity.

subtraction are obtained with negative-going pulses from the holding potential (V_h). We also use other protocols in which control currents are recorded in a very positive range of voltages[15] or in a very negative range.[17] Most experiments are carried out with Na, K, and sometimes Cl channels blocked pharmacologically.

The internal solution (occupying the end pools of the double-gap chamber) is as described in Ref. 14 (composition given in legend of Fig. 1); the external solutions are variants of a reference solution (legend of Fig. 1) which is essentially an isotonic TEA methanesulfonate with 2 mM Ca^{2+}; in the variants other cations replace Ca^{2+}. These external solutions are low Ca (no added Ca, 5 mM EGTA and Mg added for a $[Mg^{2+}]$ of

2.5 mM), 10 μM Ca (30 mM of the Ca buffer HEDTA, Ca added for a [Ca^{2+}] of 10 μM, Mg added for a [Mg^{2+}] of 2.5 mM), plus other solutions (legends of Figs. 7 and 8) with either 100 mM of an alkali or 2 mM of an alkali earth as sole metal cation. The temperature in the experiments is usually 10°C.

3. Ca^{2+} PREVENTS INACTIVATION

The Ca transients in Fig. 1 (top) were elicited by a very long (1 sec) depolarization to 0 mV in two different extracellular solutions: reference, with 2 mM Ca^{2+} as the sole metal ion, and low Ca (with no added calcium, and a [Mg^{2+}] of 2.5 mM to keep the concentration of divalent cations constant, thus avoiding trivial voltage shifts due to changes in screening of fixed anionic charges, a point that will be discussed more carefully in Section 10).

The Ca release fluxes (lower records) have a similar initial peak, but diverge afterwards: the one in low Ca decays much more rapidly, with virtually no release flux

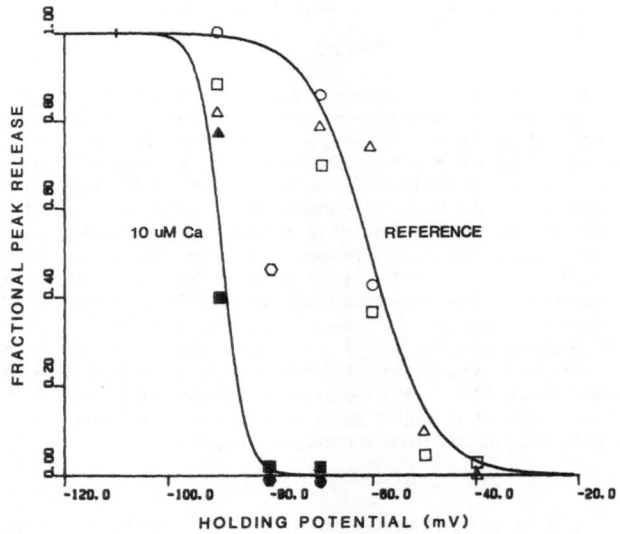

Figure 2. Low Ca^{2+} shifts the inactivation curve of Ca release. Peak release flux for pulses to 0 mV from a variable holding potential (maintained at least 5 min). The values of peak release were normalized to the value measured on the same fiber at -110 mV holding potential. Open symbols, reference solution (2 mM Ca^{2+}); filled symbols, 10 μM Ca solution ([Ca^{2+}]$_e$ = 10 μM, buffered with 30 mM HEDTA, 2.5 mM Mg^{2+}; other constituents as in reference). Continuous curves are best fits of the equation $R_p/R_p(-110) = 1/\{1+\exp[(V_h - \bar{V}_h)/k]\}$ where R_p is peak calcium release flux, V_h is holding potential, $\bar{V}_h$ is the half-inactivation holding potential, and k is a parameter that determines the steepness of the voltage dependence. $\bar{V}_h$ is -60 mV in 2 mM Ca^{2+} and -89.3 mV in 10 μM Ca. Different symbols represent different fibers. Temperature 7–9°C. (From Ref. 14.)

left by the end of the 1 sec pulse. Brum et al.[14] have demonstrated that this change in release occurs without change in the processes that remove Ca from the myoplasm.

The observation that the ability to release Ca is not very compromised initially, but that the difference becomes very marked late in the pulse, may be described as an increase in the inactivation rate; this kinetic change should have a translation in the steady-state inactivation curve, relating Ca release to the holding potential (first defined in Ref. 14). In other words, it should be possible to find large differences in peak release fluxes between the two conditions by holding at a more depolarized potential.

This expectation is confirmed by the experiment shown in Fig. 2. The steady-state inactivation curve is shifted to the left in 10 μM Ca by some 30 mV.[14]

4. Ca²⁺ ACTS AT THE VOLTAGE SENSOR

We have found evidence that this effect is due to a primary modification at the voltage sensor (a prejunctional effect of Ca²⁺ in the terminology used earlier). Figure 3 shows a set of charge movement records in reference (A), and in 10 μM Ca (B). It is clear that charge movement was reduced in this cell (as in every cell studied).[18]

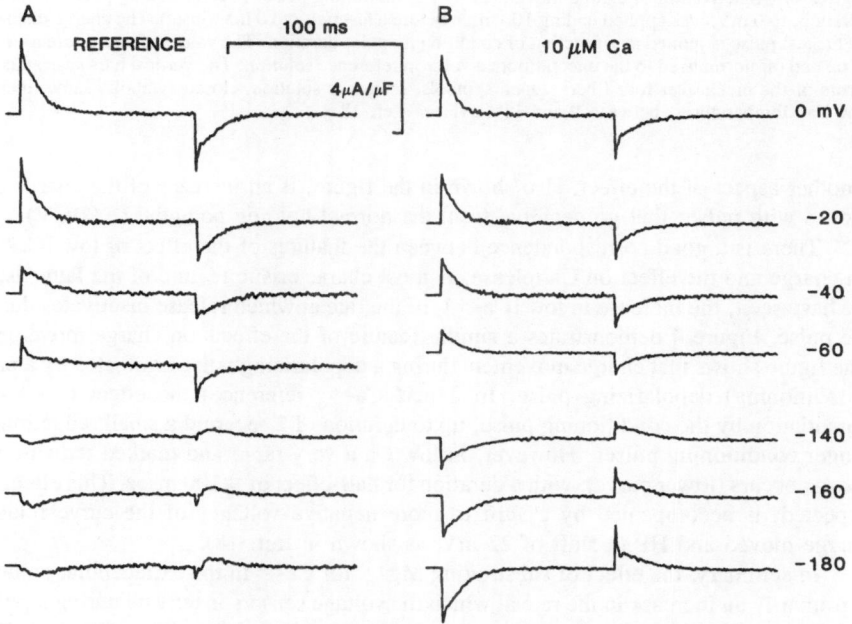

Figure 3. Intramembrane charge movements in normal and low [Ca²⁺]$_e$. Records obtained by subtraction of a scaled control current from test currents measured during 100 msec pulses from a holding potential of −100 mV to the potential indicated next to the records. The control currents were measured when the fiber was held at V_h = 0 mV with pulses to +80 mV. (A) Reference external solution; (B) 10 μM Ca. Fiber diameter 70 μm; temperature 7°C. (From Ref. 18.)

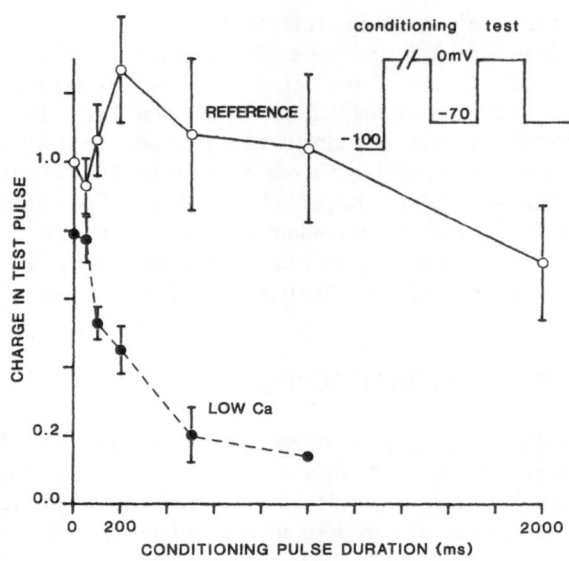

Figure 4. Inactivation of charge movement by a conditioning pulse. A conditioning pulse of variable duration, to 0 mV, was applied ending 100 msec before a test pulse to 0 mV (inset). The charge displaced by the test pulse is plotted as a function of conditioning pulse duration. The values of charge are averages of on and off normalized to the unconditioned value in reference solution. The vertical bars span standard errors of the mean over four fibers. Open symbols, reference solution; closed symbols, same fibers in low Ca. Temperatures, between 9 and 11°C. (From Ref. 18.)

Another aspect of the effect, also shown in the figure, is an increase of the charge that moves with pulses that go negative from the normal holding potential (-90mV).

There is a good correspondence between the features of the effect of low $[Ca^{2+}]_e$ on charge and the effect on Ca release. A most characteristic feature of the latter is, as we have seen, the increase in low $[Ca^{2+}]_e$ of the rate at which release inactivates during the pulse. Figure 4 demonstrates a similar feature of the effects on charge movement. The figure shows that charge movement during a depolarizing pulse is affected by a prior (conditioning) depolarizing pulse. In 2 mM Ca^{2+} (reference) the effect is a small potentiation by the conditioning pulse, up to duration of 2 sec, and a small reduction by longer conditioning pulses. However, in low Ca a very rapid and marked reduction in charge occurs (lower curve), with a duration for half-effect of $\cong 200$ msec. This effect, as expected, is accompanied by a shift to more negative voltages of the curve relating charge moved and HP (a shift of 22 mV, as shown in Ref. 18).

In summary, the effect of substituting Mg^{2+} for Ca^{2+} in the extracellular solution is primarily an increase in the rate at which the voltage sensors inactivate during a pulse. This has as a consequence an increased rate of inactivation of Ca release during a pulse, as well as parallel shifts of the steady-state inactivation curves, relating charge movement and calcium release to V_h. There is also an increase in the charge (charge 2) that moves in a very negative potential range.

5. THE VOLTAGE SENSOR REQUIRES METAL IONS

In the experiments described above, using both external solutions 10 μM Ca and low Ca we chose to substitute Mg^{2+} for Ca^{2+}. In a way it was unfortunate that we did not do until much later the simpler experiment of Fig. 5, in which we compared Ca transients in reference solution (2 mM Ca^{2+}) and another with no Ca^{2+} or any other metal ions. This metal-free solution had a radical effect. After 3–8 min the Ca transients and release fell to less than 5%, and the charge movement acquired all the characteristics of a fiber inactivated by prolonged depolarization. If a depolarized fiber is immersed in the metal-free solution and the V_h is then changed to a normal or hyperpolarized negative value, the fiber fails to show any evidence of repriming. The effect is rapidly reversible upon readmission of Ca^{2+} in the bath. The large Ca transient in the figure was actually obtained upon recovery, after 2 min in 2 mM Ca^{2+}. The speed of the recovery indicates that it does not involve replenishing of internal Ca stores; everything points at a superficial or T-membrane location of the site of Ca^{2+} action. These results are in remarkable agreement with the much earlier work of Caputo[19] measuring tension during K contractures.

The fact that the metal-free solution results in complete elimination of release, rather than the partial inhibition observed in 2.5 Mg^{2+} (solutions of low Ca and 10 μM

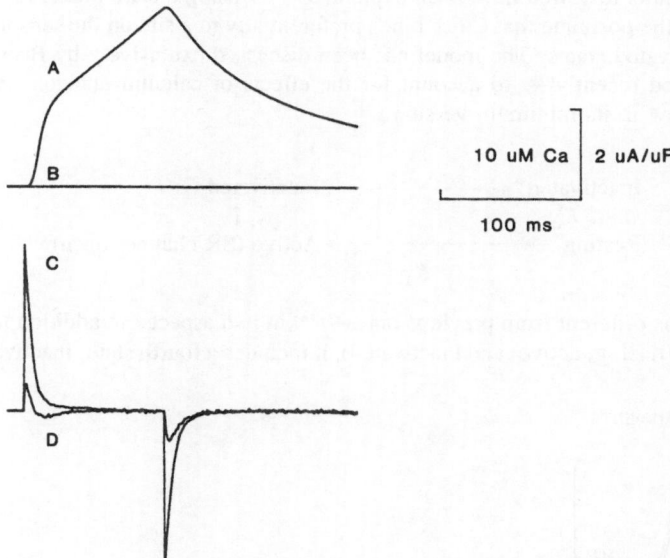

Figure 5. Ca transients and charge movements in a solution with no metal ions. The fiber was first exposed to the metal-free solution (essentially isotonic TEA methanesulfonate) and records B and D were obtained. Records A and C were obtained later, 3 min after changing to reference (2 mM Ca^{2+}) solution. Test pulses to 10 mV from a holding potential of −90 mV. Fiber 234; fiber diameter, 80 μm; dye concentration, between 284 and 314 μM; temperature, 12°C. Control currents obtained as in Fig. 3.

Ca) indicates that Mg^{2+} is able to support excitation–contraction coupling (or bind to the "priming site") to some extent.

6. Na⁺ MAY SUBSTITUTE FOR Ca²⁺

The result shown in Fig. 5 seems in contradiction with the classical result of Armstrong et al.[20] who showed that a muscle would keep contracting, for as long as 30 min, in a Ringer's saline with no Ca^{2+}. The Ringer's is different from the solution in Fig. 5 mainly in that it contains 125 mM Na^+. The experiment in Fig. 6, in which the Ca transient labeled "Na" was obtained in an external solution with Na^+ as sole metal ion, demonstrates that Na^+ can substitute for Ca^{2+} and support Ca transients. Records not shown demonstrate that solutions with Na^+ also support charge movements, although it is difficult to measure charge movements in solutions like the one used for the experiment of Fig. 6, given the loss of selectivity of the Ca-channels in the absence of extracellular Ca^{2+}.[21]

7. THE PRIMING SITE IS MODULATED

The results reviewed have been explained[15,18] using a state model of the voltage sensor plus the postulate that Ca^{2+} binds preferentially to a site on the sensor when it is in noninactivated states. The model has been discussed extensively by Brum et al.[18] and expanded recently[22] to account for the effects of calcium antagonists. It is described below in its minimum version.

$$\begin{array}{ccc} & K_2 & \\ \text{Inactivated*} \longleftarrow\!-\!-\!-\!-\!-\!-\!-\!-\!-\!-\!-\!-\!- & \text{Inactivated} \\ \downarrow K_c & & K_t \uparrow \\ \text{Resting} \;-\!-\!-\!-\!-\!-\!-\!-\!-\!-\!-\!-\!-\!-\!\longrightarrow & \text{Active (SR channel open)} \\ & K_1 & \end{array}$$

This model is different from previous ones[23,24] in two aspects: in addition to the usual three states (resting, active, and inactivated), it includes a fourth state, inactivated*, plus

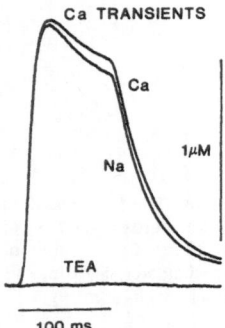

Figure 6. Ca transients in the absence of extracellular Ca^{2+}. The fiber, held at −90 mV, was first in reference solution, and record labeled Ca was obtained, then it was exposed to metal-free (record TEA). Na was obtained last, upon change to a solution with 160 mM Na methanesulfonate and no Ca^{2+}. Fiber 237; diameter, 130 μm; dye concentration between 321 and 390 μM; temperature 15°C.

the assumption that the horizontal transitions are fast and driven by voltage, therefore generating charge movement. The resting → active transition generates charge 1 (the charge that moves with depolarizing pulses in a normally polarized, primed fiber); the transition between inactivated states generates charge 2, the charge that moves in a fiber held inactivated at $V_h = 0$ mV.

The equilibrium properties of the model are specified by the four equilibrium ratios: K_t [= p(inactivated)/p(active)], K_2 (= p(inactivated*)/p(inactivated), K_c, and K_1. K_t and K_c can be specified through the known kinetic properties of inactivation and repriming. The sensors are known to inactivate spontaneously upon depolarization. Therefore, the active → inactivated transition must be spontaneous ($K_t \gg 1$). Because $K_t \gg 1$, the repriming transition cannot occur by simple reversal of the above. It must then occur through inactivated* → resting (implying $K_c > 1$). All four equilibrium ratios are linked by a microscopic reversibility condition ($K_t K_2 K_c K_1 = 1$), therefore $K_1 K_2$ must be much less than 1 for the model to work qualitatively. In the model, K_1 and K_2 are not arbitrary parameters, they are the equilibrium constants of the horizontal transitions, which, under specific postulates of the model, generate the measured distributions of charge 1 and charge 2. If the horizontal transitions generate the movement of ze charges through the transmembrane voltage difference V, where e is elemental electronic charge and z is the valence of the molecule in the transition, then the equilibrium constants must be related to voltage according to a canonical distribution function (Boltzmann).[25]

$$K_1 = \exp\left[(V - V_1)ze/kT\right] \quad \text{and} \quad K_2 = \exp\left[-(V - V_2)ze/kT\right]$$

These equations relate the equilibrium constants to the experimentally measurable parameters V_1 and V_2, the center voltages of the distributions of charge 1 and charge 2, respectively. More specifically, $V_1 = -20$ mV, $V_2 = -115$ mV, and kT/ze is close to 20 mV for both charge 1 and charge 2. Therefore,

$$K_1 K_2 = \exp[-(V_1 - V_2)/(kT/ze)] \cong \exp(-4.5) \ll 1$$

and the model performs as required, at least qualitatively.

We have recently extended the model to explain the effects of $[Ca^{2+}]_e$ and formalize the idea of a priming site, as follows[18]:

$$
\begin{array}{ccc}
 & K_2 & \\
\text{Inactivated*} \longleftarrow\!\!\!\!-\!\!\!\!-\!\!\!\!-\!\!\!\!-\!\!\!\!-\!\!\!\!-\!\!\!\!-\!\!\!\!-\!\!\!\!- & \text{Inactivated} \\
K_o \downarrow & & K_t \uparrow \\
 & K_1 & \\
Ca^{2+} + \text{resting} \longrightarrow\!\!\!\!-\!\!\!\!-\!\!\!\!-\!\!\!\!-\!\!\!\!-\!\!\!\!-\!\!\!\!-\!\!\!\!- & \text{Active} + Ca^{2+} \\
K_d \uparrow & & K_d \uparrow \\
 & K_1 & \\
\text{Resting: Ca} \longrightarrow\!\!\!\!-\!\!\!\!-\!\!\!\!-\!\!\!\!-\!\!\!\!-\!\!\!\!-\!\!\!\!- & \text{Active: Ca}
\end{array}
$$

In this version a specific site (the priming site) exists in the voltage sensor. This site is modulated: it has high affinity for Ca^{2+}, but only when the sensor is in its resting or active states. This assumption implies, from thermodynamic considerations, that the Ca^{2+}-bound form of the active sensor will not make the transition to "inactivated"; the normal path to inactivation will instead require detachment of the bound ion, followed by the active → inactivated transition, (which must strongly favor the inactivated state). The model explains all the observations described (Sections 3–7): the charge is reduced more rapidly by a conditioning pulse in low $[Ca^{2+}]_e$ because more active sensors are free of Ca^{2+} and can make the transition to inactivated. Also, decreases in charge moved by depolarizing pulses (charge 1) will be accompanied by increases in the charge that moves between inactivated states (charge 2). At intermediate holding potentials the relative occupancy of the primed (versus the inactivated) states will be directly favored by elevated $[Ca^{2+}]_e$, thus explaining the shift in the inactivation curve to more negative voltages in low $[Ca^{2+}]_e$. The effects on charge movement are directly extended to Ca release by the customary assumption that the charge is generated at voltage sensors that control SR channels with a fixed stoichiometry.

A detailed discussion of the quantitative properties of this model may be found in Refs. 18 and 22. In particular, it is shown there that the model accounts well for the main

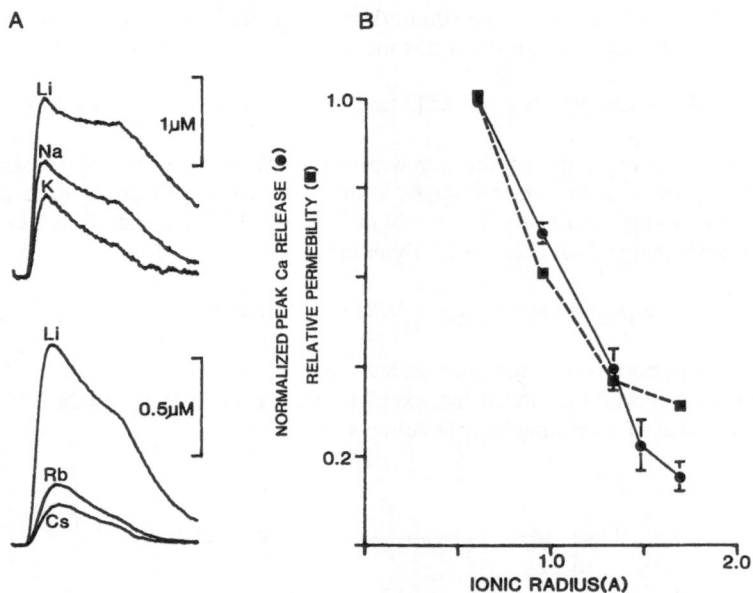

Figure 7. Ca transients recorded in the presence of 100 mM of the metal ion indicated next to each record. Additional components: TEA, 35 mM; EGTA, 0.1 mM; MES, TTX, tris buffer, and DAP as in reference solution. Fiber 296; fiber diameter, 79 μm; temperature 10°C. (B) Ca release flux records were derived from the Ca transients in (A); their peak value in the presence of each metal ion was normalized to the value of Li^+. The values in four experiments are plotted versus the ionic radius of the metal present (●); bars represent s.e.m. (■) relative permeabilities of cardiac Ca channels. From published values of reversal potentials E_i.[32,33] Relative permeabilities were derived as

$$P_i/P_{Li} = \exp[-zF/RT(E_{Li} - E_i)] \qquad \text{[equation (10,15) in Ref 33]}$$

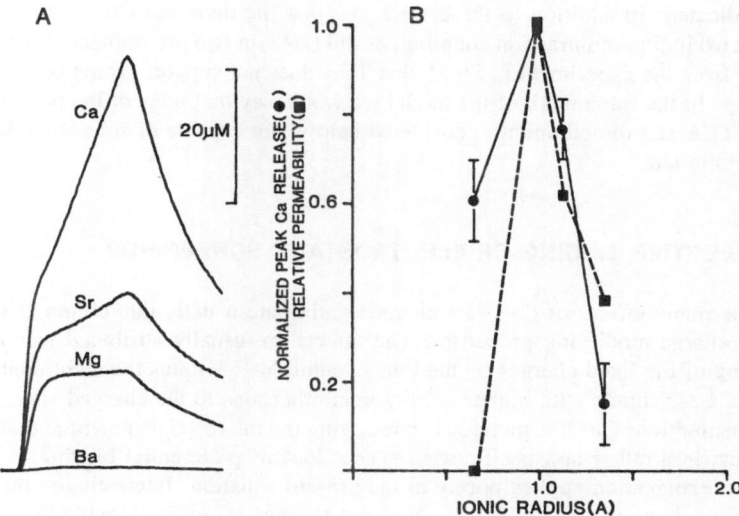

Figure 8. (A) Ca transients in a solution with 2 mM of the alkaline earth indicated in free ionic form; other components: TEA, 128 mM; EGTA added to a final free concentration of 0.1 mM, MES, TTX, tris buffer and DAP as in reference. Fiber 301; fiber diameter, 72 μm; temperature 9°C. (B) (●) peak values of Ca release flux, normalized to peak release in 2 mM Ca^{2+}; (■) relative permeabilities of the cardiac Ca channel, calculated as in Fig. 7.

kinetic aspect of the effect of Ca^{2+}, if the reasonable assumption is made that equilibration with Ca^{2+} is fast as compared with the inactivating—and repriming—transitions. As shown above, inactivation of Ca release and charge movement occurs more rapidly in low [Ca^{2+}]$_e$; it is proved in Ref. 18 that the model sensors inactivate at positive voltages following a first-order process of rate constant

$$k = k_{ia}/(1 + [Ca^{2+}]_e/K_d) + k_{ai}$$

where k_{ia} and k_{ai} are the rate constants of the active $\rightarrow$ inactivated transition and its reversal, and K_d is the dissociation constant of the Ca^{2+}-binding reaction. Therefore, at concentrations that do not saturate the site, inactivation is made slower at greater [Ca^{2+}]$_e$.

8. OTHER IONS SUPPORT EXCITATION–CONTRACTION COUPLING

According to the result of Fig. 6 the ion Na$^+$ may substitute for Ca^{2+}. Pizarro et al.[26] have recently explored the ability of other ions to substitute for Ca^{2+}. Figure 7 shows Ca transients obtained in the presence of 100 mM of the monovalent ions in group IA of the periodic table; all of them support excitation–contraction coupling to some extent. Figure 8 shows Ca transients obtained in the presence of 2 mM of the group IIA ions indicated. In addition to these we know that the divalents Co^{2+} and Cd^{2+} also support excitation–contraction coupling, as did La^{3+} in two preliminary experiments. It is clear from the experiment in Fig. 5 that TEA does not support excitation–contraction coupling. In the framework of the model we would say that none of the organic cations tested (TEA and dimethonium, considered below) are capable of substantial binding at the priming site.

9. SELECTIVE BINDING OR ELECTROSTATIC SCREENING?

The many effects of Ca^{2+} on channel gating are usually interpreted as due to its surface charge-modifying properties. The effects are usually attributed to nonspecific screening of the fixed charges by the ions in solution[27,34] plus the additional assumption that Ca^{2+} binds with higher affinity than other ions to the charged sites. Through these assumptions Ca^{2+} is pictured as modifying the microscopic potential that operates the gates, in a rather nonspecific way, except for this preferential binding.

The explanation applies poorly in the present situation. Interestingly, the divalent cation dimethonium, up to 50 mM, does not support excitation–contraction coupling. Dimethonium was introduced by McLaughlin et al.[28] as a tool to separate effects due to electrostatic screening, which dimethonium shares with all divalent metals, from those due to binding to fixed anionic sites (dimethonium binds very weakly to sites on phospholipids and proteins). The failure of this organic cation to support excitation–contraction coupling rules out completely electrostatic screening as a mechanism of these effects.

The effects described above could still be explained by preferential binding of Ca^{2+} to fixed charges near the gates. One problem with the explanation is that the effect is specific on the inactivation and not the activation phenomena.[29,30] To justify the specificity and remain in the framework of "fixed-charge" theory, it could be postulated that the fixed charges occur near the inactivation gates of the sensor. In this form, the fixed-charge theory becomes very close to our postulate of a priming site.

Finally, even in this local form, the fixed-charge formulation fails to account for the "modulated receptor" aspects of the effect. The fact that, for instance, a well-polarized fiber may have a normal initial phase of Ca release in solutions with low Ca^{2+} and 2.5 mM Mg^{2+} but late in the pulse will tend more rapidly to inactivation (Fig. 1). To justify all these aspects of the role of Ca^{2+} it would be necessary to postulate a "fixed anionic charge," located near the inactivation gates of the sensor, with an affinity profile that favors Ca^{2+} over Mg^{2+} over Na^+, and that is modulated by the state of the sensor. That is, in fact, equivalent to our definition of the priming site.

10. BINDING SELECTIVITY OF THE PRIMING SITE

In the framework of the model described, the occupancy by a cation of the hypothetical binding site is a requisite to prevent inactivation. It follows that the rate of Ca release obtained with a large depolarization is approximately proportional to the occupancy of the site. At a constant concentration of ions, and far from saturation, this occupancy should be determined solely by the affinity of the ion for the site. Therefore, it is possible to determine the relative affinities of various ions in experiments like the ones of Figs. 7 and 8, where the external solution contains only one metal ion species, and the ionic strength is maintained with TEA. In the experiment of Fig. 7 the solution contains 100 mM of a cation from group IA of the periodic table (listed next to the corresponding Ca transient). In Fig. 8 the cation is from group IIA, at a concentration of 2 mM. The system is indeed far from saturation (as revealed by the possibility of increasing release by going to greater $[Ca^{2+}]_e$). The circles on the graphs of Figs. 7 and 8 represent peak release flux as a function of ionic radius, normalized to the value of peak release flux in the presence of Li^+ in the first case, and Ca^{2+} in the second. It is therefore a measure of affinities of the corresponding ions, relative to Li^+ or Ca^{2+}.

In 1986 Hess et al.[31,32] carried out a determination of relative permeability of the L-type Ca-channel of ventricular myocytes, based on measurements of reversal potential and block. We have taken their measured reversal potentials and expressed them as relative permeabilities [using a generalized Goldman–Hodgkin–Katz expression; equation (10.15) of Ref. 33]. The results were plotted with squares and dotted lines in Figs. 7 and 8. The graphs of peak Ca release flux and relative permeabilities of the L-channel show a general similarity. For the monovalent ions they correspond to Eisenman's selectivity sequence XI,[34] characteristic of "strong-field-strength" sites; for the divalents the dependences are also similar, with a sharp maximum for Ca^{2+}.

The similarity is more remarkable if one considers that the permeability of the L-channel is probably a result of binding affinity for an intrapore anionic site.[32]

11. SIMILARITIES BETWEEN THE SENSOR AND THE Ca CHANNEL

This work has thus shown that there is a specific role of Ca^{2+} at the voltage sensor of excitation–contraction coupling, which can be understood as binding to a priming site. The affinity profile of this site is very similar to that of the intrapore binding site in the L-type Ca-channel. This is one of the many ways in which the voltage sensor of excitation–contraction coupling is similar to a Ca-channel; these similarities have been reviewed recently.[22] We consider this similarity at the level of the binding site to be a new argument in support of the hypothesis, proposed by Ríos and Brum in 1987,[35] that the high-affinity dihydropyridine receptors of skeletal muscle membranes are the voltage sensors of excitation–contraction coupling.

12. CONCLUSION

Calcium ions play an important role in excitation–contraction coupling as they stabilize the voltage sensors of the T-tubular membrane in noninactivated states. Apparently they do so by binding to a specific site, analogous to the intrapore site in Ca-channels. Given the similarity between the voltage sensor and the Ca-channel, Ca^{2+} might be playing a similar priming role in Ca-channels.

ACKNOWLEDGMENTS.   This work was supported by grants of the National Institutes of Health and the Muscular Dystrophy Association.

REFERENCES

1. Ebashi, S., Endo, M., and Ohtsuki, L., 1969, Control of muscle, *Quart. Rev. Biophys.* **2**: 351.
2. Endo, M., 1977, Calcium release from the sarcoplasmic reticulum, *Physiol. Rev.* **57**: 71–118.
3. Fabiato, A., 1985, Time and calcium dependence of activation and inactivation of calcium-induced release of calcium from the sarcoplasmic reticulum of a skinned canine cardiac Purkinje cell, *J. Gen. Physiol.* **85**: 247–289.
4. Smith, J. S., Coronado, R., and Meissner, G., 1986, Single-channel measurements of the calcium release channel from skeletal muscle sarcoplasmic reticulum: Activation by Ca^{2+} and ATP and modulation by Mg^{2+}, *J. Gen. Physiol.* **88**: 573–588.
5. Fabiato, A., 1984, Dependence of the calcium-induced release from the sarcoplasmic reticulum of skinned skeletal muscle fibers from the frog semitendinosus on the rate of change of free Ca^{2+} at the outer surface of the sarcoplasmic reticulum, *J. Physiol.* **353**: 56P.
6. Volpe, P., and Stephenson, E. W., 1986, Ca^{2+} dependence of transverse-tubule-mediated calcium release in skinned skeletal muscle fibers, *J. Gen. Physiol.* **87**: 261–288.
7. Baylor, S. M., and Hollingworth, S., 1988, Fura 2 Ca^{2+} transients in frog skeletal muscle fibers, *J. Physiol.* **403**: 151–192.
8. Vergara, J., and Asotra, K., 1987, The chemical transmission mechanism of excitation–contraction coupling in skeletal muscle, *NIPS* **2**: 182–185.
9. Frank, G. B., 1958, Inward movement of calcium as a link between electrical and mechanical events in contraction, *Nature* **182**: 1800–1801.
10. Schneider, M. F., and Chandler, W. K., 1973, Voltage-dependent charge movement in skeletal muscle: A possible step in excitation–contraction coupling, *Nature* **242**: 244–246.
11. Kovacs, L., Rios, E., and Schneider, M. F., 1983, Measurement and modification of free calcium transients in frog skeletal muscle fibers by a metallochromic indicator dye, *J. Physiol.* **343**: 161–196.
12. Melzer, W., Rios, E., and Schneider, M. F., 1984, Time course of calcium release and removal in skeletal muscle fibers, *Biophys. J.* **45**: 637–641.
13. Melzer, W., Rios, E., and Schneider, M. F., 1987, A general procedure for determining calcium release in skeletal muscle fibers, *Biophys J.* **51**: 849–864.
14. Brum, G., Rios, E., and Stefani, E., 1988, Effects of extracellular calcium on calcium movements of excitation–contraction coupling in frog skeletal muscle fibers, *J. Physiol.* **398**: 441–473.
15. Brum, G., and Rios, E., 1987, Intramembrane charge movement in frog skeletal muscle fibers: Properties of charge 2, *J. Physiol.* **387**: 489–517.
16. Lamb, G. D., 1987, Asymmetric charge movement in polarized and depolarized muscle fibers of the rabbit, *J. Physiol.* **376**: 85–100.
17. Caputo, C., and Bolanos, P., 1988, Effect of D600 and La^{3+} on charge movement in depolarized muscle fibers, *Biophys. J.* **53**: 604a.
18. Brum, G., Fitts, R., Pizarro, G., and Rios, E., 1988, Voltage sensors of the frog skeletal muscle membrane require calcium to function in excitation–contraction coupling, *J. Physiol.* **398**: 475–505.
19. Caputo, C., 1972, The time course of potassium contractures of single-muscle fibers, *J. Physiol.* **223**: 483–505.
20. Armstrong, C. M., Bezanilla, F. M., and Horowicz, P., 1972, Twitches in the presence of ethyleneglycol bis (β-aminoethyl ether)-N,N'-tetraacetic acid, *Biochim. Biophys. Acta* **257**: 605–608.
21. Almers, W., McCleskey, E. W., and Palade, P. T., 1984, A nonselective cation conductance in frog muscle membrane blocked by micromolar external calcium ions, *J. Physiol.* **353**: 565–583.
22. Pizarro, G., Brum, G., Fill, M., Fitts, R., Rodriguez, M., Uribe, I., and Rios, E., 1988, The

voltage sensor of skeletal muscle excitation–contraction coupling: A comparison with calcium channels, in: *The Calcium Channel: Structure, Function, and Implications*, (Morad, M., Nayler, W., Schramm, M., and Kazda, S., eds.) Springer-Verlag, Heidelberg, pp. 138–156.

23. Chandler, W. K., Rakowski, R. F., and Schneider, M. F., 1976. Effects of glycerol treatment and maintained depolarization on charge movement in skeletal muscle, *J. Physiol.* **254**: 285–316.

24. Rakowski, R. F., 1981, Immobilization of membrane charge in frog skeletal muscle by prolonged depolarization, *J. Physiol.* **317**: 129–148.

25. Reif, P., 1967, *Fundamentals of Statistical and Thermal Physics*, McGraw-Hill, New York.

26. Pizarro, G., Fitts, R., and Rios, E., 1988, Selectivity of a cation-binding membrane site essential for EC coupling in skeletal muscle, *Biophys. J.* **53**: 645a.

27. McLaughlin, S. G. A., 1977, Electrostatic potentials at membrane–solution interfaces, in: *Current Topics in Membrane Transport* (F. Bronner and A. Kleinzeller, Eds.) vol. 9, pp. 71–144.

28. McLaughlin, A., Eng, W. K., Vaio, G., Wilson, T., and McLaughlin, S. G. A., 1983, Dimethonium, a divalent cation that exerts only a screening effect on the electrostatic potential adjacent to negatively charged phospholipid bilayer membranes, *J. Membrane Biol.* **76**: 183–193.

29. Luttgau, H. C., and Spiecker, W., 1979, The effects of calcium deprivation upon mechanical and electrophysiological parameters in skeletal muscle fibers of the frog, *J. Physiol.* **296**: 411–429.

30. Stefani, E., and Chiarandini, D. J., 1973, Skeletal muscle: Dependence of potassium contractures on extracellular calcium, *Pflüg. Arch.* **343**: 143–150.

31. Hess, P., Lansman, J. B., and Tsien, R. W., 1986, Calcium channel selectivity for divalent and monovalent cations: Voltage and concentration dependence of single-channel current in ventricular heart cells, *J. Gen. Physiol.* **88**: 293–320.

32. Lansman, J. B., Hess, P., Tsien, R. W., 1986, Blockade of current through single calcium channels by Cd^{2+}, Mg^{2+}, and Ca^{2+}: Voltage and concentration dependence of calcium entry into the pore, *J. Gen. Physiol.* **88**: 321–348.

33. Hille, B., 1984, *Ionic Channels in Excitable Membranes*, Sinauer, Sunderland, Massachusetts.

34. Eisenman, G., 1962, Cation-selective glass electrodes and their mode of operation, *Biophys. J.* **2**(Suppl. 2): 259–323.

35. Rios, E., and Brum, G., 1987, Involvement of dihydropyridine receptors in excitation–contraction coupling in skeletal muscle, *Nature* **325**: 717–720.

36. Rios, E., and Pizarro, G., 1988, The voltage sensors and calcium channels of excitation–contraction coupling, *NIPS* **3**: 223–227.

Chapter 26

A Pharmacological Approach to the Physiological Mechanism of Excitation–Contraction Coupling

Philip Palade, Donald Brunder, Christine Dettbarn, and Philip Stein

1. INTRODUCTION

The use of pharmacologic agents as tools to aid in the characterization and separation of physiological processes is not a new idea. The availability of specific toxins like tetrodotoxin and saxitoxin made it easy to test for the presence of sodium channels in excitable cells and made possible the purification,[1] reconstitution,[2,3] and cloning[4] of such channels. We hope to utilize a pharmacologic approach here to determine whether a Ca^{2+}-channel isolated recently from skeletal muscle sarcoplasmic reticulum (SR) is the one of physiological importance. Is it the *right* channel? We know several methods exist to cause Ca^{2+} release from SR.[5,6] If we had specific inhibitors for each form of release, we could test whether the releases or channels they blocked were involved in excitation–contraction coupling. We will examine the role of Ca^{2+}-induced Ca^{2+} release channels first because these are the channels already isolated.

2. BACKGROUND

In the case of muscle SR most drug studies added little to our basic understanding of the process of muscle excitation–contraction coupling. Part of the problem stemmed

PHILIP PALADE, DONALD BRUNDER, CHRISTINE DETTBARN, and PHILIP STEIN • Department of Physiology and Biophysics, The University of Texas Medical Branch, Galveston, Texas 77550.

from different results from different investigators, and part involved the difficulty in ascertaining sites of action of compounds when only net ion movements (such as SR Ca^{2+} uptake, which represents influx minus efflux) were being measured. Research in muscle excitation–contraction coupling received a vital boost recently when the demonstration of Ca^{2+}-channels in the SR membrane[7] and suggestions of a role for inositol 1,4,5-trisphosphate in SR Ca^{2+} release[8,9] livened up the field.

More recently the alkaloid ryanodine has made possible startling breakthroughs in the field of SR research. Ryanodine was known for many years as an agent that produced contractures in skeletal muscle yet negative inotropic responses in cardiac muscle.[10] On the surface these results were generally interpreted as due to a ryanodine-induced Ca^{2+} release from skeletal SR[10] but a ryanodine-mediated inhibition of Ca^{2+} release from cardiac SR.[11] This difference between tissues no doubt discouraged investigators for many years until very recently when [3H] ryanodine was utilized to measure the density of SR Ca^{2+} release channels in terminal cisternae membranes.[12,13] Several laboratories then proceeded to isolate and purify the binding sites and finally incorporate them into planar lipid bilayers to demonstrate preservation of physiological function.[14,15] The apparent discrepancy between skeletal and cardiac effects of ryanodine has since been resolved by the suggestion that Ca^{2+} released slowly from cardiac SR would deplete its stores for subsequent physiological release.[16]

The SR Ca^{2+} release channels opened by ryanodine are ideally situated at the junctional face of the terminal cisternae where the SR abuts closest to the transverse tubules to form the triad.[15] Thus, whatever signal is required to elicit SR Ca^{2+} release has the shortest possible distance to traverse. The peculiar thing about these channels from a physiological point of view is that they are more closely associated with the Ca^{2+}-induced Ca^{2+} release mechanism[17] than with any other postulated messenger. The reason this is problematic is that vertebrate skeletal muscle has long been known not to require influx of extracellular Ca^{2+} for contraction.[18]

If these really are the SR Ca^{2+}-channels initiating Ca^{2+} release during excitation–contraction coupling, then either the source of "trigger" Ca^{2+} to open them would have to be voltage-dependent Ca^{2+}-binding sites on the cytoplasmic surface of the transverse tubule membrane, or the same channels would have to be gated open in some other fashion. Neither possibility should be dismissed, particularly the latter, since the channel has been shown by Palade[19] and by others (reviewed in Ref. 6) to be opened by so many drugs that it must have multiple different agonist-binding sites. On the other hand, other forms of SR Ca^{2+} release are known[5,6] and other SR Ca^{2+}-channels have been reported.[20,21]

What we propose to do is examine the question of which form of Ca^{2+} release (or which SR Ca^{2+}-channel) is involved in the physiological excitation–contraction coupling process. We hasten to note that involvement of one channel or mechanism does not rule out simultaneous involvement of another. Indeed, models of this sort have been formulated previously.[22] We would argue that the Ca^{2+}-induced Ca^{2+} release channel is ideally suited to serve as an amplifier of a smaller independent SR Ca^{2+} release that may have preceded it.

Our notion of how to proceed in this task is also not new. With respect to muscle physiology it owes its origins to pharmacologists like Paul Bianchi, a proponent of the hypothesis that there was a Ca^{2+} "trigger" for SR Ca^{2+} release,[23] but the true father of this particular approach is Makoto Endo, one of the actual discoverers of Ca^{2+}-induced Ca^{2+} release in skinned muscle fibers.[24] Endo reasoned that if Ca^{2+}-induced Ca^{2+} release was involved, then blockers of this release should inhibit tension elicited by depolarizing the fibers.[5] We wish to examine whether the biochemists have isolated the physiologically relevant SR Ca^{2+}-channel or an impostor.

The best choice for a blocker of this channel was ruthenium red, generally acknowledged to inhibit this form of Ca^{2+} release from isolated SR at micromolar[22] or even submicromolar concentrations.[19] Unfortunately, ruthenium red is neither very pure nor very specific,[25] being known among other things to inhibit mitochondrial Ca^{2+} uptake and to stain various membranes intensely. Accordingly, we began to hunt for a ruthenium red substitute.

3. EXPERIMENTS WITH ISOLATED SR

We started using isolated SR, in particular isolated triads,[26] and we then began trying to find experimental conditions that would allow us to conveniently measure caffeine-induced SR Ca^{2+} release because that was one of the best-known activators of the Ca^{2+}-induced Ca^{2+} release mechanism.[5,6] We utilized a spectrophotometric system with antipyrylazo III as a Ca^{2+}-sensing dye, using the Ca^{2+}-precipitating anion

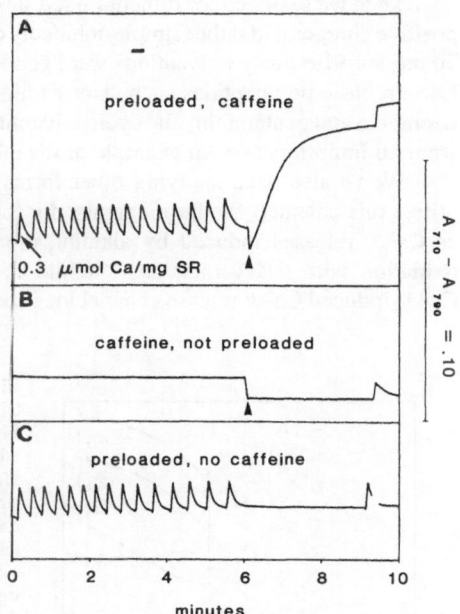

Figure 1. Caffeine-induced Ca^{2+} release from isolated triads. (A) Triads (42 μg SR protein) isolated from rabbit skeletal muscle[26] are incubated at 30°C in 1 ml medium consisting of 88 mM KCl, 16 mM KMOPS, 0.25 mM antipyrylazo III, 7.5 mM $Na_4P_2O_7$, 1 mM MgATP, 5 mM Na_2 phosphocreatine, and 20 μg/ml creatine phosphokinase. The trace is monitored spectrophotometrically as $A_{710} - A_{790}$. With each 12.5 nmol $CaCl_2$ addition the trace rises as the dye senses the added Ca^{2+} by increasing its absorbance at 710 nm. The trace declines between additions as the Ca^{2+} is sequestered inside the vesicles (away from the dye). Following a series of such Ca^{2+} additions, addition of 10 mM caffeine causes an immediate downward deflection in the trace followed by a larger upward rise. The final deflection at the end of the trace represents a procedure to recalibrate the Ca^{2+} indicator in the presence of the applied drugs. (B) The experiment was repeated without any Ca^{2+} additions to demonstrate that the downward deflection is an artifact of caffeine's interaction with antipyrylazo III but that the upward rise in the trace depends on the presence of accumulated Ca^{2+} inside the SR. (C) The experiment was repeated a third time but without any caffeine addition, demonstrating that the release observed in (A) truly was caffeine dependent. (Reproduced in modified form from Ref. 27 by permission of the American Society for Biochemistry and Molecular Biology, Inc.)

pyrophosphate in an attempt to both load more Ca^{2+} into less SR (thereby conserving sample; this prep takes ~12 continuous hours to prepare) and to slow down the rate of release so as to be able to monitor it on a time scale of seconds rather than milliseconds, which would have required more expensive and temperamental rapid mixing devices. We were able to fashion conditions[27] that allowed a net Ca^{2+} release to be seen at a threshold of approximately 2.5 mM caffeine, that being the lowest concentration that causes muscle fibers to undergo contracture in the absence of other stimuli. As seen in Fig. 1, calcium release can be induced in isolated triads by addition of 10 mM caffeine. One accompanying control experiment demonstrates that the immediate downward deflection in the trace is an artifact of caffeine addition that takes place even in the absence of any Ca^{2+} available to be released inside the SR. In contrast, the slower, larger rise in the trace does require Ca^{2+} inside the SR and consequently is indicative of Ca^{2+} release. The second control experiment demonstrates that no spontaneous release (e.g., Ref. 28) takes place in the absence of added caffeine under these conditions.

We found additionally that any number of other chemical agents or drugs caused Ca^{2+} release by opening this channel, a conclusion drawn from the result that all these releases were blocked by ruthenium red and three other unrelated caffeine-induced Ca^{2+} release blockers.[19] One other release-inducing agent was the calcium ion itself (Fig. 2). We reasoned that only a few of the more than a dozen release-activating substances tested could be acting at the same ligand-binding site and that since the blockers were unlikely to all interfere at several different binding sites, they were most likely to be blocking the channel itself.[19]

Then we searched for ruthenium red substitutes. We noted that ruthenium red has six positive charges and rather simple-mindedly decided to test other polycationic substances. To our surprise many polycations were good inhibitors of this form of SR Ca^{2+} release. Certain basic polypeptides such as protamine, many aminoglycoside antibiotics such as neomycin and gentamicin, and even polyamines endogenous to muscle, such as spermine, were all inhibitory.[29] An example of the effect of neomycin is shown in Fig. 3.

We've also been studying other forms of SR Ca^{2+} release that do not appear to utilize this channel. We have examined a form of spontaneous Ca^{2+} release[28] as well as Ca^{2+} releases induced by alkalinization,[30] by chloride addition,[31] and by SH oxidation with heavy metals.[32] While certain heavy metals do indeed open up the Ca^{2+}-induced Ca^{2+} release channel localized at the SR terminal cisternae, they can also

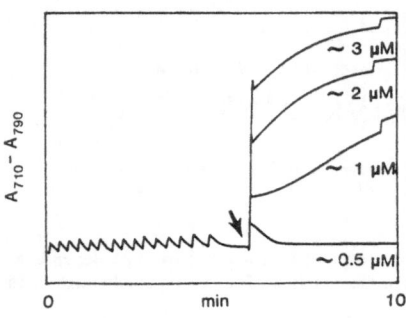

Figure 2. Ca^{2+}-induced Ca^{2+} release by isolated triads. Experiments were performed as described for Fig. 1A except that variable amounts of Ca^{2+} were added. The approximate free $[Ca^{2+}]$ immediately after the addition is given in micromolar units to the right of the each trace. Small Ca^{2+} additions elicit only net uptake; larger additions elicit net release. The presence of such Ca^{2+}-induced Ca^{2+} release provides the rationale for the incremental Ca^{2+} preloading procedure for these experiments. Larger Ca^{2+} additions would have necessitated longer experiments. (Reproduced in modified form from Ref. 19 by permission of the American Society for Biochemistry and Molecular Biology, Inc.)

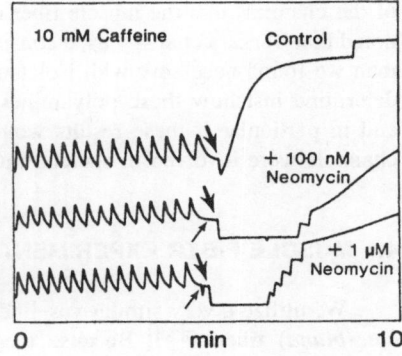

Figure 3. Inhibition of caffeine-induced Ca^{2+} release by neomycin. Experiments were performed as described for Fig. 1A except that neomycin was added at different concentrations at the small arrows, jut prior to caffeine addition (larger arrows). Note that in the presence of neomycin the caffeine addition was unable to directly elicit Ca^{2+} release but that subsequent Ca^{2+} additions were able to elicit such a release. We interpret this effect as a caffeine-induced increase in the sensitivity of the channel to Ca^{2+} itself. (Reproduced in modified form from Ref. 29 by permission of the American Society for Biochemistry and Molecular Biology, Inc.)

elicit ruthenium red-insensitive Ca^{2+} release from light SR subfractions derived from the longitudinal elements of the SR, as can alkalinization, chloride addition, or spontaneous Ca^{2+} release. Preliminary evidence from some of our initial drug screening suggests that these four additional forms of Ca^{2+} release exhibit quite different pharmacological profiles and most probably involve different Ca^{2+} efflux pathways, possibly channels. None of these forms of Ca^{2+} release from light SR were inhibited by the ruthenium red-like polyamines at concentrations that completely block caffeine-induced Ca^{2+} release. Thus we assumed we could use these polyamines as relatively specific blockers of the SR Ca^{2+}-induced Ca^{2+} release channel.

These molecules do have additional effects, however, including inhibition of IP_3 formation by complexation with its predecessor, the phospholipid phosphatidylinositol bisphosphate.[33] Indeed, Vergara et al.[8] introduced neomycin and polylysine into muscle fibers and suggested that their inhibition of SR Ca^{2+} release argued for a role for IP_3 in excitation–contraction coupling. In our hands inositol 1,4,5-trisphosphate does not cause Ca^{2+} release under the same experimental conditions that permit caffeine-induced Ca^{2+} release (Fig. 4). Our results with isolated SR which was unresponsive to IP_3 suggested another interpretation for the results of Vergara et al.,[33] namely, that these polyamines block the Ca^{2+}-induced Ca^{2+} release channel,[29] which might be one

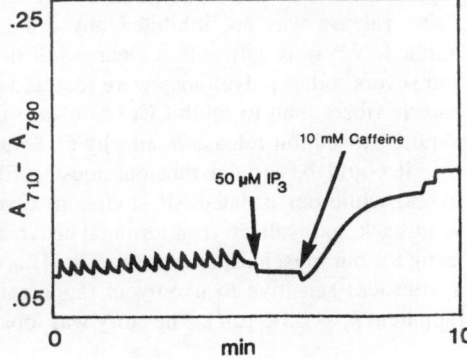

Figure 4. Inositol 1,4,5-trisphosphate (IP_3) fails to release Ca^{2+} from isolated triads under the same experimental conditions. The experiment was performed as for Fig. 1, except that 50 μM IP_3 was added following preloading with no effect. Subsequent caffeine-induced Ca^{2+} release was unaffected by the prior addition of IP_3. (Reproduced from Ref. 29 by permission of the American Society for Biochemistry and Molecular Biology, Inc.)

of the channels that the muscle fiber relies on for excitation–contraction coupling. We noted that Vergara et al.[33] used considerably higher concentrations in their experiments than we found necessary with isolated SR.[29] So we set out to reexamine this issue to determine just how these polyamines were affecting excitation–contraction coupling, and in particular if these results would indicate whether Ca^{2+}-induced Ca^{2+} release channels were involved in skeletal muscle excitation–contraction coupling or not.

4. MUSCLE FIBER EXPERIMENTS

We utilize a very similar vaseline gap voltage clamp preparation of bullfrog (*Rana catesbiana*) fibers.[8,34] Because these polyamines are membrane-impermeant, they have to be applied to the fiber's cut ends.[8] In most of our experiments we monitor directly neither SR Ca^{2+} release with Ca^{2+}-sensing dyes nor tension. We measure the shortest possible pulse (minimum stimulus duration) to $+100$ mV in TTX Ringer that elicits a just barely detectable fiber movement. The tetrodotoxin is added to the Ringer to prevent action potential propagation in the transverse tubules.[35] With such short pulses the contractile threshold is reached so rapidly that the Ca^{2+} release rate from the SR must vastly exceed the rate of Ca^{2+} reuptake. Consequently this minimum stimulus duration (or MSD) is inversely related to the rate of Ca^{2+} release.[36] Substances that slow down the rate of Ca^{2+} release would prolong the MSD.[36] We prefer these MSD experiments to Ca^{2+} transient measurements with dyes because they are easier and shorter, but we do cross-check with Ca^{2+} transients recorded with antipyrylazo III[37,38] to verify all major conclusions because drug effects on the contractile proteins could also affect the MSD.

Taking the example of neomycin, we find that concentrations that obliterated caffeine-induced Ca^{2+} release from isolated SR did very little to Ca^{2+} release in the muscle fiber. We had to go to much higher concentrations to achieve any appreciable inhibition of Ca^{2+} release (Fig. 5). Computer modeling of the time course of drug effects enabled us to estimate both the apparent diffusion coefficient and the ratio of the concentration applied to the fiber end to the apparent inhibitory constant for the drug effect in the fibers. Our MSD results were in good agreement with the results of Ca^{2+} transient experiments performed by others with neomycin.[8] Our own Ca^{2+} transient experiments with other polyamines (not shown) similarly suggested that intracellular Ca^{2+} release was not inhibited any more strongly than our MSD experiments had implied. We were left with a clear result that much higher concentration of neomycin and several other polyamines were required to block excitation–contraction coupling in muscle fibers than to inhibit Ca^{2+} release from isolated SR. With neomycin the dose required to inhibit release *in situ* by 50% (apparent k_i) was 3.7 ± 1.7 μM.

It could be argued that our muscle fiber experiments were carried out on frog fibers, while our isolated SR studies involved SR isolated from rabbit muscle. So we went back and isolated frog terminal cisternae[39] from the same bullfrogs we had been using for our muscle fiber experiments. Treated the same way, isolated frog SR was only a little less sensitive to neomycin (apparent $k_i = 0.11$ μM) than rabbit SR had been (apparent $k_i = 0.06$ μM). The story was looking more and more as if blockers of Ca^{2+}-

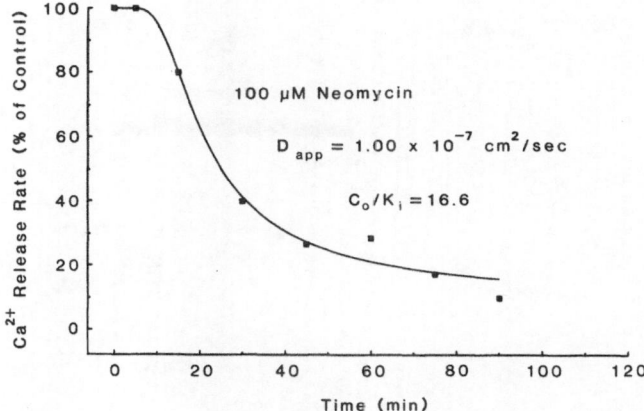

Figure 5. Effects of neomycin on a measure of excitation–contraction in cut frog skeletal muscle fibers. Cut segments of bullfrog semitendinosus muscle were mounted in a Hille–Campbell triple vaseline gap voltage-clamp chamber and maintained at a holding potential of −90 mV with Ringer solution (115 mM NaCl, 2.5 mM KCl, 1.8 mM CaCl$_2$, 3 mM MOPS, pH 7.2) containing 1μM tetrodotoxin (TTX) in the A pool and an "internal" solution containing 120 mM K aspartate, 3 mM tris maleate, 0.1 mM tris EGTA buffered to 0.1 μM free Ca^{2+}, 2 mM MgSO$_4$, 3 mM Na$_2$ATP, 5 mM Na$_2$ phosphocreatine, pH 7.1 in the B, C, and E pools. Neomycin (100 μM) was applied at time zero to the E pool only. Pulses of increasing duration were applied to +100 mV until the contractile threshold was reached (MSD). Ca^{2+} release rates were calculated as a normalized function of the inverse of the MSD. Fiber #031787E2. The data were fit to a diffusion equation to yield a D_{app} of 10^{-7} cm^2/sec. The term C_0/K_i refers to the ratio of the concentration applied to the end pool divided by the apparent concentration required to inhibit Ca^{2+} release by 50%.

induced Ca^{2+} release fail to inhibit excitation–contraction coupling at appropriate concentrations, and that would indeed suggest that Ca^{2+}-induced Ca^{2+} release or the channel that mediates it are not involved in physiological excitation–contraction coupling.

But we were reminded that the experimental conditions of our isolated SR studies might have differed substantially from those present in the myoplasm of our cut fibers. We needed to determine the effectiveness of neomycin at blocking the Ca^{2+}-induced Ca^{2+} release channels *in the muscle fiber*. Accordingly we applied a subcontracture concentration of caffeine (0.5 mM) and examined its effects on the MSD in the absence and then in the subsequent presence of neomycin. As shown in Fig. 6, 0.5 mM caffeine clearly does potentiate Ca^{2+} release reversibly (see also Refs. 40 and 41), and this potentiation is only partially inhibited by the same high concentration (50 μM) of neomycin that completely inhibited caffeine-induced Ca^{2+} release from isolated SR.[29] This result suggests that the Ca^{2+}-induced Ca^{2+} release channels in the muscle fiber are no more inhibited by these concentrations of neomycin than are the channels utilized by the muscle fiber during excitation–contraction coupling. We have obtained similar results with gentamicin and ruthenium red. This is quite different from the tentative conclusion we had been tempted to draw from simply comparing dose–response relations between muscle fiber and isolated SR experiments.

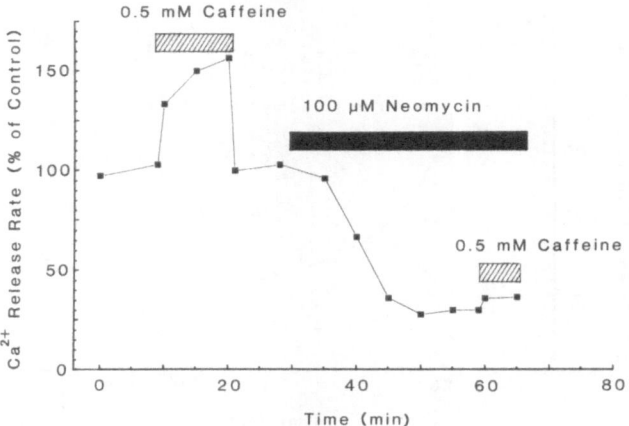

Figure 6. Effects of neomycin on *in situ* Ca^{2+} release rates in the presence and absence of caffeine. The experiment was performed as described for Fig. 5, but 0.5 mM caffeine was reversibly introduced into the A pool solution at two different times in the experiment, before and after introduction of 100 μM neomycin via the cut fiber end in the E pool. Fiber #040588A1.

5. MUSCLE FIBERS AND ISOLATED SR: COMPARISONS AND TENTATIVE CONCLUSIONS

There may be an object lesson here. We doubt that it is to avoid working with isolated SR. After all, it was the isolated SR experiments that suggested a way to test the hypothesis that the Ca^{2+}-induced Ca^{2+} release channel was involved in physiological excitation–contraction coupling. More likely the correct lesson is that comparisons should be made with caution. *In vivo* and *in vitro* conditions may differ sufficiently to greatly affect quantitative results, perhaps due to the artificially low free [Mg^{2+}] of our isolated SR experiments, $\sim 16\mu M$ due to the presence of pyrophosphate,[27] compared to ~ 1 mM estimated for myoplasm.[42] We are still in the process of trying to determine the source of the differences in polyamine sensitivity between isolated SR and fiber experiments. Thus, one conclusion from our studies is that particular caution needs to be exercised when trying to make inferences about the physiological state of affairs from studies on isolated organelles alone.

What other conclusions can be drawn? It is possible, even likely, that the same SR Ca^{2+} release channels that mediate Ca^{2+}-induced Ca^{2+} release may mediate physiological excitation–contraction coupling, but it is still too early to be certain. We won't know for sure until we determine whether neomycin produces its effects instead by inhibiting the charge movements believed to gate open the SR Ca^{2+}-channels or by interfering with the production of IP_3. The latter possibility is less probable given that interference with IP_3 production would not have been expected to inhibit caffeine-sensitive Ca^{2+} release (Fig. 6) since caffeine's effects do not depend on IP_3 (Fig. 4). Unfortunately, our experiments thus far do not enable us to conclude definitively which effect is relevant here. If we find a number of additional drugs that inhibit excitation–

contraction coupling at the same concentrations that they affect caffeine potentiation, alternative modes of action via charge movement or inhibition of IP_3 production will be rendered less likely. Even if this proves to be the case, it won't rule out a role for IP_3 in excitation–contraction coupling, particularly if IP_3 opens the same Ca^{2+} release channel that is inhibited by polyamine-like substances.[43]

In the future we intend to investigate the effects of blockers of SR Ca^{2+} release induced by Cl^- addition, by alkalinization, by SH oxidation, and, if possible, by IP_3 addition. Even if Ca^{2+}-induced Ca^{2+} release channels should be involved in excitation–contraction coupling, it doesn't mean that we can exclude participation of other possible SR Ca^{2+} efflux pathways or channels, particularly since the Ca^{2+}-induced Ca^{2+} release channel is so well-suited to amplify any smaller preceding Ca^{2+} release. In the meantime we are studying the Ca^{2+}-induced Ca^{2+} release channel further.

6. ELECTROPHYSIOLOGICAL STUDY OF Ca^{2+}-INDUCED Ca^{2+} RELEASE CHANNELS IN NATIVE SR MEMBRANE

We have developed what we feel is a relatively novel approach to the study of this channel. Rather than inserting purified channels or SR vesicles into planar lipid bilayers or liposomes, we have found a way to study SR channels in what is most probably their native lipid environment. We don't even have to prepare isolated SR, and we can begin recording within minutes of dissecting a fiber. Phil Stein discovered that skinning muscle fibers in the presence of Ca^{2+} or applying Ca^{2+} to skinned fibers causes them to contract vigorously and start exuding blebs of membrane on their skinned surfaces. These blebs coalesce into larger hemispheres that may approach a 100 μm in diameter (Fig. 7).

We call these protrusions sarcoballs.[44] They form within minutes, and excised patches from these membranes may be readily patch-clamped, most easily in the inside-out patch recording mode. In contrast to reports of SR Ca^{2+} release channels inserted into planer lipid bilayers[7,17,20,43] the channels we see exhibit a very prominent subconductance state are shown most clearly in Fig. 8. In this patch the full open level of the channel in symmetrical Ca^{2+} was 105 pS, and the substrate was 73 pS. Most patches contain one or more SR Ca^{2+} release channels. Once again we rely on pharmacologic tools for this identification. The channels we see are activated by Ca^{2+}, by thymol, by ryanodine, and in Fig. 9 by caffeine, and they may be blocked by ruthenium red. We have also determined that this channel is very permeable to K^+ as well as Ca^{2+}, and that it exhibits a very pronounced voltage dependence under nearly all conditions we have examined.[44]

This naturally raises the question of what might be the physiological activator of such a channel. It still could be Ca^{2+}, but it may be something else, another intracellular messenger, a conformational change in some associated T-tubule protein like the dihydropyridine receptor,[45] or it may even involve one of these agents altering SR permeability to some other ion in such a way as to produce a transient voltage change across the SR membrane that might trigger the Ca^{2+} release channels open. Perhaps a pharmacologic approach could help here as well if specific inhibitors are found that selectively interfere with gating of the channel by each of these processes.

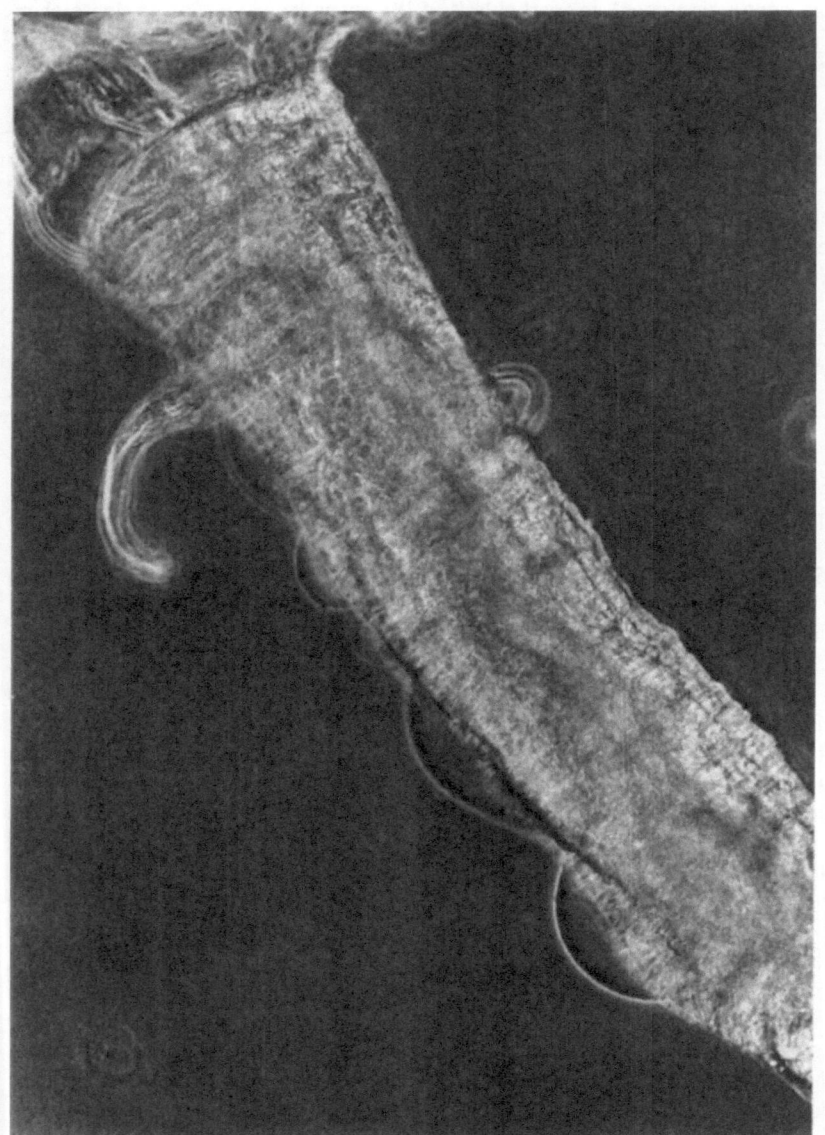

Figure 7. Phase-contrast light micrograph of sarcoball membrane blebs on a frog skeletal muscle fiber mechanically skinned in Ringer's solution and allowed to contract.

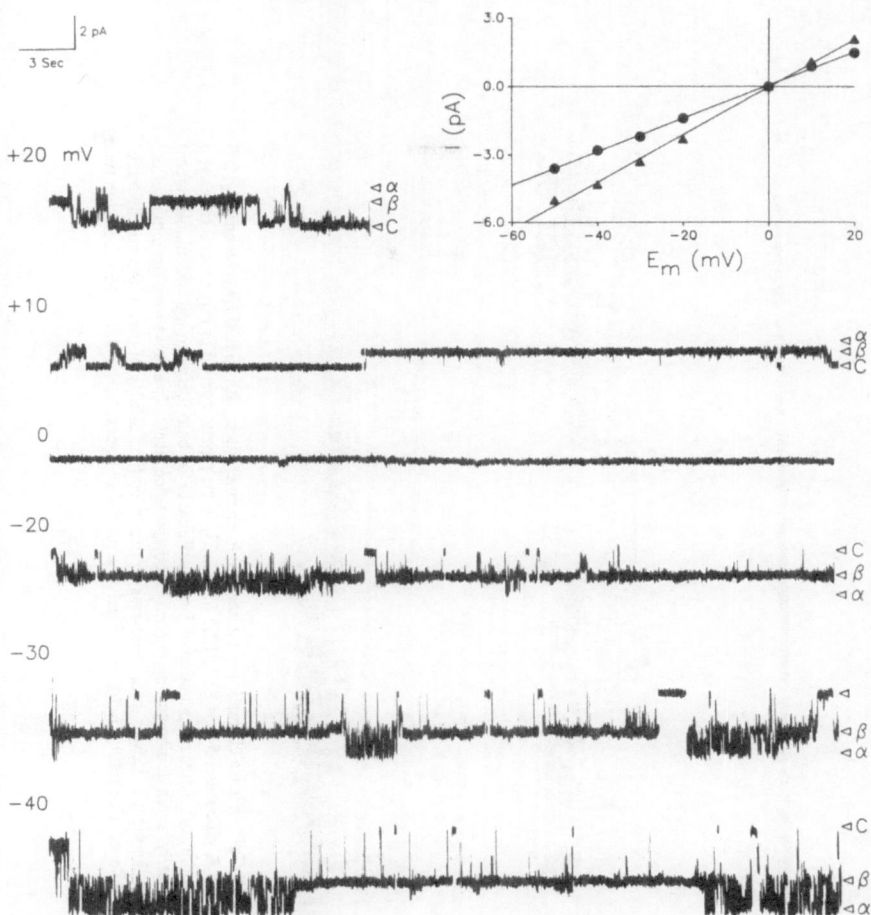

Figure 8. Single-channel recording of SR Ca^{2+} release channel from an excised inside-out sarcoball patch at the membrane potentials indicated. The solution in the bath and in the patch pipette was 50 mM Ca (gluconate)$_2$ plus 2.5 mM $CaCl_2$, pH 7.0. Note that the channel exhibits two conductance levels, labeled α and β, and that full closures can occur from either open state.

We're not satisfied that these are the only Ca^{2+}-channels in SR. We've seen evidence in isolated SR of Ca^{2+} releases that are not restricted to the terminal cisternae and not inhibited by polyamines. We hope to assess their role in excitation–contraction coupling, if any, following the same pharmacologic approach outlined earlier. We will then study any that appear physiologically relevant in sarcoballs. While much has been revealed about excitation–contraction coupling in recent years, much still remains to be worked out.

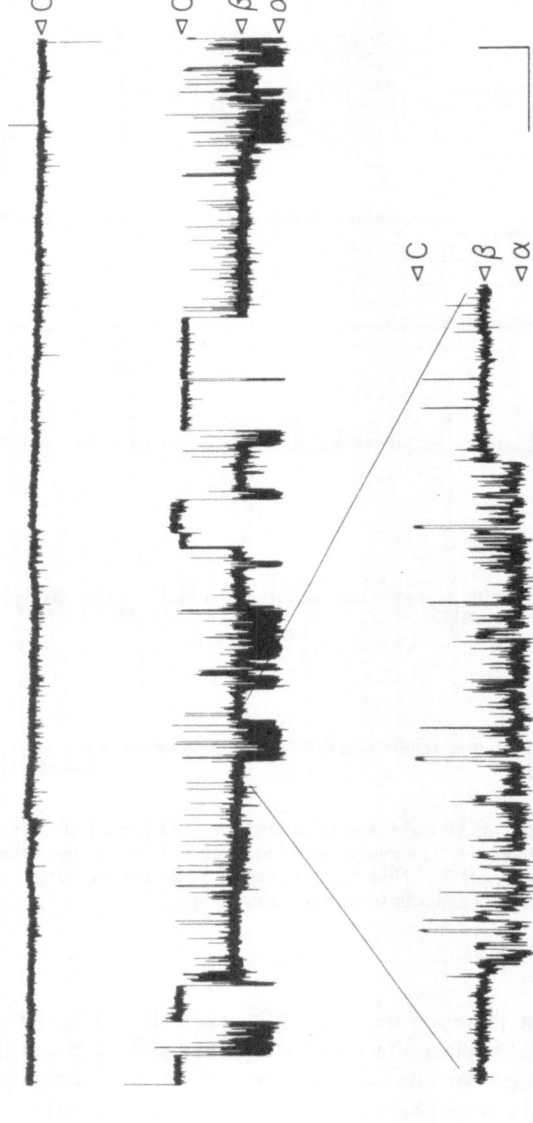

Figure 9. Caffeine activation of sarcoball Ca^{2+} release channel. An excised single-channel patch with 50 mM Ca (gluconate)$_2$, 2.5 mM CaCl$_2$, pH 7.0 in the bath and 100 mM K gluconate, 5 mM KCl, 12.5 μM CaCl$_2$ in the pipette opened briefly several times at 0 mV and then remained closed for 5 min (top trace). After introduction of 5 mM caffeine to the bath (between top and middle traces) the channel activity was increased dramatically, again demonstrating two major conductance levels. The downward openings reflect Ca^{2+} conduction through the channel into the pipette. Lower trace represents an expanded time display of part of the middle trace. Note full closures from either conductance level. Calibration marker: vertical = 2 pA for all traces, horizontal = 15 sec for top two traces, 1.2 sec for bottom trace.

REFERENCES

1. Agnew, W. S., Levinson, S. R., Brabson, J. S., and Raftery, M. A., 1978, Purification of the tetrodotoxin-binding component associated with the voltage-sensitive sodium channel from *Electrophorus electricus* electroplax membranes, *Proc. Natl. Acad. Sci. USA* **75:** 2606–2610.
2. Talvenheimo, J. A., Tamkun, M. M., and Catterall, W. A., 1982, Reconstitution of neurotoxin-stimulated sodium transport by the voltage-sensitive sodium channel purified from rat brain, *J. Biol. Chem.* **257:** 11868–11871.
3. Weigele, J. B., and Barchi, R. L., 1982, Functional reconstitution of the purified sodium channel protein from rat sarcolemma, *Proc. Natl. Acad. Sci. USA* **79:** 3651–3655.
4. Noda, M., Shimizu, S., Tanabe, T., Takai, T., Kayano, T., Ikeda, T., Takahashi, H., Nakayama, H., Kanaoka, Y., Minamino, N., Kangawa, K., Matsuo, H., Raftery, M. A., Hirose, T., Inayama, S., Hayashida, H., Miyata, T., and Numa, S. 1984, Primary structure of *Electrophorus electricus* sodium channel deduced from cDNA sequence, *Nature* **312:** 121–127.
5. Endo, M., 1977, Calcium release from the sarcoplasmic reticulum, *Physiol. Rev.* **57:** 71–108.
6. Martonosi, A. N., 1984, Mechanisms of Ca^{2+} release from sarcoplasmic reticulum of skeletal muscle, *Physiol. Rev.* **64:** 1240–1320.
7. Smith, J. S., Coronado, R., and Meissner, G., 1985, Sarcoplasmic reticulum contains adenine nucleotide-activated calcium channels, *Nature* **316:** 446–449.
8. Vergara, J., Tsien, R. Y., and Delay, M., 1985, Inositol 1,4,5-trisphosphate: A possible chemical link in excitation–contraction coupling in muscle, *Proc. Natl. Acad. Sci. USA* **82:** 6352–6356.
9. Volpe, P., Salviati, G., Di Virgilio, F., and Pozzan, T., 1985, Inositol 1,4,5-trisphosphate induces calcium release from sarcoplasmic reticulum of skeletal muscle, *Nature* **316:** 347–349.
10. Jenden, D. J., and Fairhurst, A. S., 1969, The pharmacology of ryanodine, *Pharm. Rev.* **21:** 1–25.
11. Jones, L. R., Besch, H. R., Jr., Sutko, J. L., and Willerson, J. T., 1979, Ryanodine-induced stimulation of net Ca^{2+} uptake by cardiac sarcoplasmic reticulum vesicles, *J. Pharmacol. Exp. Ther.* **209:** 48–55.
12. Fleischer, S., Ogunbunmi, E. M., Dixon, M. D., and Fleer, E. A. M., 1985, Localization of Ca^{2+} release channels with ryanodine in junctional terminal cisternae of sarcoplasmic reticulum of fast skeletal muscle, *Proc. Natl. Acad. Sci. USA* **82:** 7256–7259.
13. Pessah, I. N., Waterhouse, A. L., and Casida, J. E., 1985, Solubilization and separation of Ca^{2+}-ATPase from the Ca^{2+}-ryanodine receptor complex, *Biochem. Biophys. Res. Commun.* **128:** 449–456.
14. Imagawa, T., Smith, J. S., Coronado, R., and Campbell, K. P., 1987, Purified ryanodine receptor from skeletal muscle sarcoplasmic reticulum is the Ca^{2+}-permeable pore of the calcium release channel, *J. Biol. Chem.* **262:** 16636–16643.
15. Lai, F. A., Erickson, H. P., Rousseau, E., Liu, Q.-Y., and Meissner, G., 1988, Purification and reconstitution of the calcium release channel from skeletal muscle, *Nature* **331:** 315–319.
16. Hansford, R. G., and Lakatta, E. G., 1987, Ryanodine releases calcium from sarcoplasmic reticulum in calcium-tolerant rat cardiac myocytes, *J. Physiol. (London)* **390:** 453–467.
17. Smith, J. S., Coronado, R., and Meissner, G., 1986, Single-channel measurements of the calcium release channel from skeletal muscle sarcoplasmic reticulum: Activation by Ca^{2+} and ATP and modulation by Mg^{2+}, *J. Gen. Physiol.* **88:** 573–588.
18. Armstrong, C. M., Bezanilla, F. M., and Horowicz, P., 1972, Twitches in the presence of ethylene glycol bis(β-aminoethyl ether)-N,N'-tetraacetic acid, *Biochim. Biophys. Acta* **267:** 605–608.
19. Palade, P., 1987, Drug-induced Ca^{2+} release from isolated sarcoplasmic reticulum. II. Releases involving a Ca^{2+}-induced Ca^{2+} release channel, *J. Biol. Chem.* **262:** 6142–6148.
20. Smith, J. S., Coronado, R., and Meissner, G., 1986, Single-channel calcium and barium currents of large and small conductance from sarcoplasmic reticulum, *Biophys. J.* **50:** 921–928.
21. Suarez-Isla, B., Orozco, C., Heller, P. F., and Froehlich, J. P., 1986, Single calcium channels in native sarcoplasmic reticulum membranes from skeletal muscle, *Proc. Natl. Acad. Sci. USA* **83:** 7741–7745.

22. Miyamoto, H., and Racker, E., 1982, Mechanism of calcium release from skeletal sarcoplasmic reticulum, *J. Membrane Biol.* **66**: 193–201.

23. Bianchi, C. P., and Bolton, T. C., 1967, Action of local anesthetics on coupling systems in muscle, *J. Pharmacol. Exp. Ther.* **157**: 388–405.

24. Endo, M., Tanaka, M., and Ogawa, Y., 1970, Calcium-induced release of calcium from the sarcoplasmic reticulum of skinned skeletal muscle fibers, *Nature* **228**: 34–36.

25. Volpe, P., Salviati, G., and Chu, A., 1986, Calcium-gated calcium channels in sarcoplasmic reticulum of rabbit skinned skeletal muscle fibers, *J. Gen. Physiol.* **87**: 289–303.

26. Mitchell, R. D., Palade, P., and Fleischer, S., 1983, Purification of morphologically intact triad structures from skeletal muscle, *J. Cell. Biol.* **96**: 1008–1016.

27. Palade, P., 1987, Drug-induced Ca^{2+} release from isolated sarcoplasmic reticulum. I. Use of pyrophosphate to study caffeine-induced Ca^{2+} release, *J. Biol. Chem.* **262**: 6135–6141.

28. Palade, P., Mitchell, R. D., and Fleischer, S., 1983, Spontaneous calcium release from sarcoplasmic reticulum: General description and effects of calcium, *J. Biol. Chem.* **258**: 8098–8107.

29. Palade, P., 1987, Drug-induced Ca^{2+} release from isolated sarcoplasmic reticulum. III. Block of Ca^{2+}-induced Ca^{2+} release by organic polyamines, *J. Biol. Chem.* **262**: 6149–6154.

30. Nakamura, Y., and Schwartz, A., 1972, The influence of hydrogen ion concentration on calcium binding and release by skeletal muscle sarcoplasmic reticulum, *J. Gen. Physiol.* **59**: 22–32.

31. Kasai, M., and Miyamoto, H., 1973, Depolarization-induced calcium release from sarcoplasmic reticulum membrane fragments by changing ionic environments, *FEBS Lett.* **34**: 299–301.

32. Abramson, J. J., Trimm, J. L., Weden, L., and Salama, G., 1983, Heavy metals induce rapid calcium release from sarcoplasmic reticulum vesicles isolated from skeletal muscle, *Proc. Natl. Acad. Sci. USA* **80**: 1526–1530.

33. Lodhi, S., Weiner, N. D., and Schacht, J., 1976, Interactions of neomycin and calcium in synaptosomal membranes and polyphosphoinositide monolayers, *Biochim. Biophys. Acta* **426**: 781–785.

34. Hille, B., and Campbell, D. T., 1976, An improved vaseline gap voltage clamp for skeletal muscle fibers, *J. Gen. Physiol.* **67**: 265–293.

35. Heiny, J. A., and Vergara, J., 1982, Optical signals from surface and T-system membranes in skeletal muscle fibers. Experiments with the potentiometric dye NK2367, *J. Gen. Physiol.* **80**: 203–230.

36. Almers, W., 1977, Local anesthetics and excitation–contraction coupling in skeletal muscle: Effects on a Ca^{2+} channel, *Biophys. J.* **18**: 355–357.

37. Kovacs, L., Rios, E., and Schneider, M. F., 1979, Calcium transients and intramembrane charge movement in skeletal muscle fibers, *Nature* **279**: 391–396.

38. Palade, P., and Vergara, J., 1982, Arsenazo III and antipyrylazo III calcium transients in single skeletal muscle fibers, *J. Gen. Physiol.* **79**: 679–707.

39. Volpe, P., Bravin, M., Zorzato, F., and Margreth, A., 1988, Isolation of terminal cisternae of frog skeletal muscle: Calcium storage and release properties, *J. Biol. Chem.* **263**: 9901–9908.

40. Kovacs, L., and Szucs, G., 1983, Effect of caffeine on intramembrane charge movement and calcium transients in cut skeletal muscle fibers of the frog, *J. Physiol. (London)* **341**: 559–578.

41. Delay, M., Ribalet, B., and Vergara, J., 1986, Caffeine potentiation of calcium release in frog skeletal muscle fibers, *J. Physiol. (London)* **375**: 535–559.

42. Alvarez-Leefmans, F. J., Gamino, S. M., Giraldez, F., and Gonzales-Serratos, H., 1986, Intracellular free magnesium in frog skeletal muscle fibers measured with ion-selective microelectrodes, *J. Physiol. (London)* **378**: 461–483.

43. Suarez-Isla, B., Irribarra, V., Bull, R., Oberhauser, A., Larralde, L., Jaimovich, E., and Hidalgo, C., 1988, Inositol 1,4,5-trisphosphate activates a calcium channel in isolated sarcoplasmic reticulum (SR) membranes, *Biophys. J.* **53**: 467a.

44. Stein, P., and Palade, P., 1988, Sarcoballs: Direct access to sarcoplasmic reticulum Ca^{2+}-channels in skinned frog muscle fibers, *Biophys. J.* **53**: 455a; and **54**: 357–363.

45. Rios, E., and Brum, G., 1987, Involvement of dihydropyridine receptors in excitation–contraction coupling in skeletal muscle, *Nature* **325**: 717–720.

Chapter 27

A Chemical Mechanism for Excitation–Contraction Coupling in Skeletal Muscle

Julio Vergara, Nestor Lagos, and Deida Compagnon

1. INTRODUCTION

The early process of excitation–contraction coupling in skeletal muscle primarily involves the participation of a network of transverse tubules (T-tubules) and a coupling structure at which the T-tubular membrane encounters the sarcoplasmic reticulum (SR) membrane (T-SR junction). The transverse tubular system (T-system) is open to the extracellular space and connects with the surface membrane where the action potential propagates along the fiber. Currently it is known that the action potential also propagates radially into the deepest regions of the T-system,[1–4] and that somehow the electrical activation of the T-system membrane promotes the release of Ca^{2+} from the SR membrane,[1,5–8] which is about 20 nm apart.[9] There are three models proposing mechanisms to explain the coupling at the T-SR junction in skeletal muscle that are currently considered: (i) The Ca^{2+}-induced Ca^{2+} release hypothesis [10,11] suggests that an entry of Ca^{2+} ions from the extracellular space results in an amplified release from the SR. (ii) The direct coupling hypothesis implies that the depolarization of the T-system membranes initiates a perturbation that mechanically propagates across the 20 nm gap to induce the release of Ca^{2+} from the SR.[12] (iii) The chemical coupling hypothesis[13] states that the depolarization of the T-system membrane modulates the synthesis of a

JULIO VERGARA, NESTOR LAGOS, and DEIDA COMPAGNON • Department of Physiology, University of California–Los Angeles, Los Angeles, California 90024.

chemical transmitter (IP$_3$), which diffuses across the gap, binds to a postjunctional receptor (Ca^{2+}-channel), and induces the release of Ca^{2+} from the SR.[13,14]

2. RESULTS AND DISCUSSION

2.1. Ca^{2+}-Induced Ca^{2+} Release Hypothesis

2.1.1. Ca^{2+} Transients in Skeletal Muscle

Physiologically, the early events of excitation–contraction coupling are approximately bracketed between an early electrical signal, the action potential which propagates longitudinally along the muscle fiber (Fig. 1B), and the release of calcium from the SR (trace a–c, Fig. 1A). To obtain the latter signal, a single fiber from semitendinosus muscle of the frog *Rana catesbeiana* was stained intracellularly with the Ca^{2+}- indicating dye arsenazo III. The signals shown in Fig. 1A represent changes in absorbance proportional to the number of Ca^{2+} ions bound to dye molecules in the muscle fiber. To a first approximation [under the experimental conditions of Fig. 1,[15,16]], these absorbance signals reflect the total surplus amount of Ca^{2+} ions made available to the dye in response to the electrical stimulation.

2.1.2. Effect of External Ca^{2+}

Figure 1A shows Ca^{2+} transients under two experimental conditions: with normal (1.8 mM) and very low (0.2 μM; pCa 6.7) external Ca^{2+} concentrations. It can be observed that the Ca^{2+} transient recorded in standard Ringer's solution (trace a, Fig. 1A) is almost identical to the one obtained in low-Ca^{2+} solution (trace b, Fig. 1A). Because the [Ca^{2+}] inside the muscle fiber at rest is approximately 0.1 μM, at the peak of the action potential, Ca^{2+} ions would be driven out of the fiber instead of inwardly in low-Ca solution. Moreover, in the presence of 0.5 mM caffeine, a well-known poten-

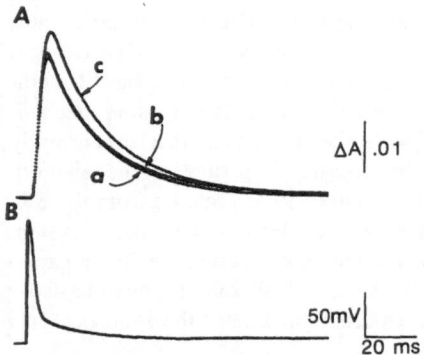

Figure 1. Arsenazo III Ca^{2+} transients and action potential in frog skeletal muscle fiber. The absorbance records (traces a–c, panel A) were recorded following action potential (panel B) stimulation from single muscle fiber using the triple vaseline gap technique[15] modified for optical recording.[4,16,17] The internal solution applied at the cut ends contained 100 mM K aspartate, 2 mM MgCl$_2$, 3 mM Na$_2$ ATP, 5 mM Na creatine phosphate, 20 mM K MOPS, 1.5 mM K EGTA, and 0.4 mM arsenazo III. The pH was adjusted at 7.1. The absorbance signals were obtained at a wavelength of 660 nm and the dye concentration inside the muscle fiber was calculated from the absorbance at 530 nm. Trace a was obtained with the fiber bathed in standard Ringer's solution containing 1.8 mM Ca^{2+}. Trace b was obtained with the external solution changed to a modified Ringer's solution containing 1 mM EGTA and trace amounts of Ca^{2+} to give a measured pCa of 6.7. Trace c was obtained in the same solution of trace b but with 0.5 mM caffeine added. The experiments were performed at 12°C.

Figure 2. The dependence of peak absorbance changes on membrane potential in frog skeletal muscle fibers, measured using arsenazo III. A single skeletal muscle fiber was voltage clamped, using the vaseline gap technique,[17] at a holding potential of -100 mV. The duration of the voltage clamp pulses was 20 msec. The internal solution had the same composition as that in Fig. 1, except that K was replaced by Cs. The arsenazo III concentration in the pools where the ends of the fiber were cut was 1.5 mM. The external solution contained 115 mM tetramethylammonium-Cl replacing sodium with all the other unchanged components of the normal Ringer's solution. The peak absorbance change (ΔA, ordinate) is plotted versus the absolute membrane potential (mV, abscissa). The temperature of the experiment was 12°C.

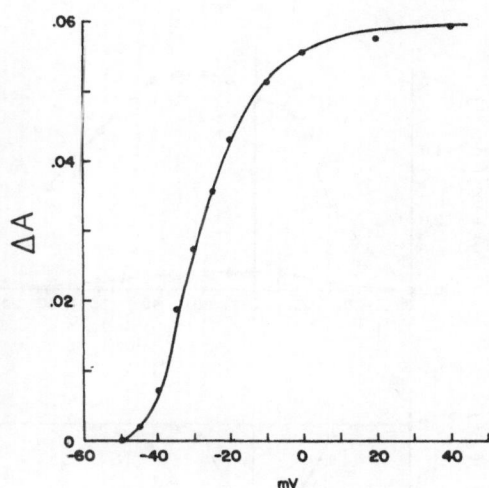

tiator of excitation–contraction coupling in skeletal muscle, the calcium release from the SR is still significantly potentiated in the low external Ca^{2+} solution (trace c, Fig. 1A). It can be concluded from this experiment that external Ca^{2+} is not required for the normal excitation–contraction coupling in skeletal muscle, and in addition that the caffeine potentiation of the Ca^{2+} release process is not significantly affected by the presence or absence of this ion in the external solution either. This view has been well substantiated in the past[17–19] and puts serious constraints on the plausibility that a Ca^{2+}-induced Ca^{2+} release mechanism operates in the excitation–contraction coupling mechanism in skeletal muscle.

2.1.3. Voltage Dependence of Ca^{2+} Release

Another property of the excitation–contraction coupling in skeletal muscle that argues against any mechanism invoking a major role to the external $[Ca^{2+}]$ is its voltage dependence. Figure 2 is a plot of the amplitude of calcium absorbance transients detected with the dye arsenazo III, as a function of the membrane potential from a fiber, under voltage-clamp condition, bathed in an external solution containing 1.8 mM Ca^{2+} and no major permeable ion. It can be observed that amplitude of the Ca^{2+} transients increases progressively with the applied voltage until it reaches a saturating amplitude, following a sigmoidal behavior characteristic of the excitation–contraction coupling process. The shape and amplitude of this curve is not significantly affected by the removal of external Ca^{2+} ions.

In contrast to the data shown in Fig. 2 for skeletal muscle, Fig. 3 shows the results of a similar experiment performed in giant muscle fibers from the barnacle *Balanus nubilus,* in which the muscle contractility depends on external calcium.[20,21] It can be observed in this case that the amplitude of Ca^{2+} transients (Fig. 3A), here detected with the Ca^{2+}-indicating dye azo 1, follow a bell-shaped voltage dependence. This pattern is well correlated with the voltage dependence of the Ca^{2+} influx into the muscle fiber,

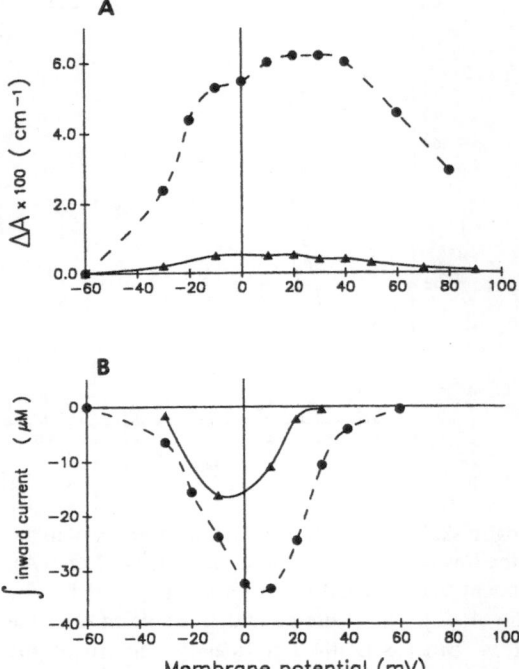

Figure 3. The dependence of peak absorbance changes on membrane potential in single barnacle muscle fibers, measured with azo 1. A single giant barnacle muscle fiber was voltage clamped and internally perfused, using an axial wire perfusion technique.[21] The holding potential was −60 mV and the duration of the voltage clamp pulses was 80 msec. The internal perfusion solution contained 180 mM Cs aspartate, 10 mM K HEPES, 5 mM Na_2 ATP, 2 mM $MgCl_2$, and 500 mM sucrose, pH 7.15. The Ca^{2+}-indicating dye azo 1 was added to this solution at a final concentration of 80 μM. The external solutions were Millipore filtered natural sea water (NSW) for the records plotted in the filled circles, and an artificial sea water with only 0.5 mM $[Ca^{2+}]$ instead of the 5 mM $[Ca^{2+}]$ of the NSW, but with all the other components unchanged. In panel A, the peak absorbance change (ΔA, ordinate) is plotted versus the absolute membrane potential (mV, abscissa). In panel B, the Ca^{2+} influx, calculated a the time integral of the current records (expressed in μM) are plotted against the membrane potential. The temperature of the experiment was 14°C.

carried by the prominent Ca^{2+} current, which decreases as the membrane potential approaches the Ca^{2+} reversal potential. This is shown in Fig. 3B. In addition, reduction of the external Ca^{2+} concentration to 10% of the normal concentration in the natural sea water (NSW), leads to a significant reduction of the Ca^{2+} transients concomitant with the decrease in the Ca^{2+} influx (see Figs. 3A and 3B, continuous traces).

We have obtained additional evidence suggesting that the amplitude of the Ca^{2+} transients recorded in barnacle muscle fibers cannot be explained entirely in terms of the actual influx of Ca^{2+} ions from the extracellular solution, but that there may be a component of intracellular release.[22] It may be concluded that the excitation–contraction coupling in the barnacle muscle preparation may follow a Ca^{2+}-induced Ca^{2+} release mechanism. In contrast, the sigmoidal voltage dependence shown in Fig. 2 for skeletal muscle seems incompatible with this hypothesis.

2.2. Direct Coupling Hypothesis

The early observation of the existence of nonlinear dielectric currents (charge movements) in skeletal muscle fibers,[23] was followed by the proposal that these electrical phenomenon may be linked by a direct mechanical coupling to the release of Ca^{2+} from the SR.[12] It seems important at this moment to distinguish between the experimental observation of these nonlinear charge movements and the hypothesis of the direct mechanical link. The former are undeniable experimental occurrences, the latter a proposal. There are no proven data supporting such a remote control mechanism. Moreover, the experimental evidence suggests that the charge movements are complex entities with several contributing components and there are various reports in the literature with different claims about the relevance of each particular component.[24–30] Of these diverse components, the one that follows the closer relationship (in terms of voltage dependence, effect by pharmacological agents, and time course), in regard to the Ca^{2+} release from the SR, is the Q_γ component.[26–28] The other components are more remotely correlated with the Ca^{2+} release from the SR, although it has been claimed recently that the subtraction of subthreshold charge movements may result in a closer association.[31] In any event, charge movements are quite likely to be an antecedent of the coupling process at the T-SR junction, but their direct causality on the Ca^{2+} release from the SR is disputable. We will discuss this in more detail later in this chapter, specifically in relation to the absence of a one-to-one correlation between pre- and postjunctional events.

2.3. Chemical Coupling Hypothesis

Our laboratory[13,32–34] and Volpe et al.[14] proposed that a chemical transmitter, inositol 1,4,5-trisphosphate (IP_3), may participate in the excitation–contraction coupling process at the level of the T-SR junction in skeletal muscle. Following the work of Berridge and Irvine[35] it has been demonstrated that IP_3 acts as a link between the receptor agonist-binding reaction at the external membrane and the internal release of Ca^{2+} from the endoplasmic reticulum in many other cells. We suggested that, in skeletal muscle, instead of a receptor mediated effect, the T-tubule depolarization would activate a local phospholipase-C, leading to hydrolysis of PtdI (4,5)bisphosphate (PIP_2) to release IP_3 and diacylglycerol intracellularly. The former would diffuse across the T-SR junction to open an IP_3-sensitive Ca^{2+}-channel at the SR, which is nothing but the endoplasmic reticulum of skeletal muscle fibers. Many metabolic processes required in this cycle were proposed be present in skeletal muscle. It should be mentioned that there was virtually no information about the phosphoinositide–inositol phosphate cycle in this preparation.

2.3.1. Triadic Delay

Before dealing with the metabolic aspects of this hypothesis, one of the first dilemmas that needed to be addressed was the fact that, since a chemical transmission process was proposed, it had to follow the paradigm of such a mechanism in the coupling at the T-SR junction. A classical, chemically mediated coupling hypothesis is the neuromuscular junction.[36] The evidence for chemical transmission in that preparation was strongly reinforced by experiments showing a strong temperature-dependent latency between pre- and postsynaptic currents (the synaptic delay).[37,38] In particular, we investigated whether the coupling process at the T-SR junctions in skeletal muscle fibers showed a junctional delay (the "triadic delay"), analogous to the synaptic delay of the neuromuscular junction, and if the temperature dependence of such a delay is compatible with the involvement of a chemical transmitter.[39]

The experiments in this case were complicated by the inaccessibility of both pre- and postjunctional membrane systems to direct study of their electrophysiological characteristics. We overcame this difficulty using optical techniques to record two signals involved in the triadic coupling: the T-tubule membrane depolarization, a key prejunctional event; and the Ca^{2+} release from the SR, an early postjunctional event. These signals were monitored respectively using the voltage-sensing dye NK2367,[4] and the Ca^{2+}-indicating dye azo 1.[39,40]

Trace *b* of Fig. 4 displays the time course of the depolarization of the T-system membrane in response to activation of the muscle fiber by an action potential at the surface membrane (trace *c*, Fig. 4). The Ca^{2+} release from the SR is shown in trace *a* of Fig. 4. This figure shows that it is possible to detect, in two successive records, these three fundamental physiological events. Comparing the time course of these records, it can be observed that the T-system depolarization clearly precedes the Ca^{2+} release from the SR. Moreover, the T-system membrane potential transient has passed its peak value before the Ca^{2+} signal takes off. Nevertheless, the "total lag" of about 2.5 msec that separates both signals cannot be accounted for exclusively by the T-SR junction coupling because the radial action potential propagation in the T-tubule is responsible for part of this delay. We modelled this[39] and concluded that the "propagational lag" accounts for only a small fraction (0.7 msec in the experiment shown in Fig. 4) of this total period of latency. The remaining 1.8 msec can be defined as a genuine junctional lag or "triadic delay," which is not only sufficient but also indicative of an underlying chemical coupling mechanism. A sequence of more than 12 stages of exponential delay are necessary to simulate such a prominent delay. Other physiological coupling processes, also reportedly requiring large number of sequential steps, which are worth mentioning here, are the phototransduction processes in *Limulus* ommatidia[41] and turtle cones.[42] These coupling mechanisms are now widely accepted to be mediated by second messengers such as IP_3[43,44] and cGMP.[45]

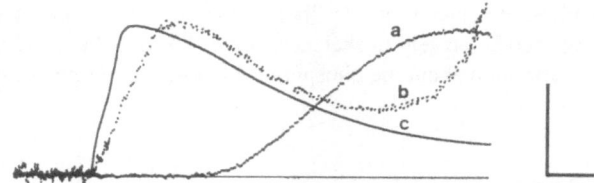

Figure 4. Transmission delay of coupling at the T-SR junction in a frog single muscle fiber, measured with the Ca^{2+} dye azo 1 and the potentiometric dye NK2367. A single muscle fiber was mounted in a vaseline gap chamber using a similar methodology to that described in Fig. 1, except that instead of arsenazo III, 200 μM of the Ca^{2+} dye azo 1 was used in this experiment. In addition, the fiber segment in pool A was stained with a Ringer's solution containing 0.4 mg/ml of the dye NK2367 for 15 min. This figure shows the surface membrane action potential (trace *c*), tubular potential signal (trace *b*), and Ca^{2+} signal (trace *a*). The optical traces were obtained in successive recordings at illuminating wavelengths of 670 nm (16 sweeps averaged, trace *b*) and 480 nm (2 sweeps averaged, trace *a*). The vertical calibrating bar represents an increase in absorbance of 0.24×10^{-3} for the 670 nm T-tubular signal, a decrease in absorbance of 5×10^{-3} for the 480 nm Ca^{2+} signal, and a depolarization of 84 mM for the action potential. The horizontal calibration bar represents 2 msec. The late falling phase of the T-system signal shows interference (seen in ths figure as a late upswing) from fiber movement subsequent to the stimulated Ca release. The experiment was performed at 13°C.

If the published values for the kinetic rate constants of charge movements[46,47] are used in a direct coupling model to simulate the Ca^{2+} release on the time scale of our experiments, we find that (i) the predicted triadic delay is too short, and (ii) the rate of rise of the Ca^{2+} transients is too slow compared to our records. This is not surprising because the charge movement provides a single-time-constant rate-limiting step; as discussed above, the Ca^{2+} release process requires many more such steps to predict the observed triadic delay.

2.3.2. Temperature Dependence of Excitation–Contraction Coupling

It is well accepted that a prominent temperature dependence characterizes chemical reaction schemes. It is therefore important to study the T-SR coupling process in skeletal muscle fibers at different temperatures. This is shown in Fig. 5. It should be observed that the rate of rise of the Ca^{2+} transients becomes slower at lower temperature, reflecting a prominent temperature dependence. Nevertheless, the most dramatic effect of lowering the temperature is actually in the time taken by the Ca^{2+} signal to take off (total lag). If the propagational lag is taken into consideration as described above, the triadic delay, computed at several temperatures ranges, from 0.5 msec at 26°C to 4 msec at 4°C, and predicts a temperature dependence compatible with a Q_{10} of about 2.7.[39] Even the small triadic delay (near 0.5 msec) at high temperatures allows ample time for occurrence of a chemical process, considering that delays across the neuromuscular junction fall within the same range.[38] Consequently, the triadic delay not only is a relatively long quiescent period but is extremely temperature dependent as well. A further difficulty with the direct coupling hypothesis is the significant discrepancy between the values of Q_{10} obtained for the kinetic rate constants of the charge movements and the Q_{10} reported here for the triadic delay; the former are in the range of 1.5 to 2.0[46,47] and the latter near 2.7. Just about the simplest way to reconcile these measurements would be to propose that charge movements may indeed represent a dielectric effect associated with the conformational change of the voltage sensor in the T-SR coupling process, followed by subsequent chemical steps yielding a higher Q_{10}.

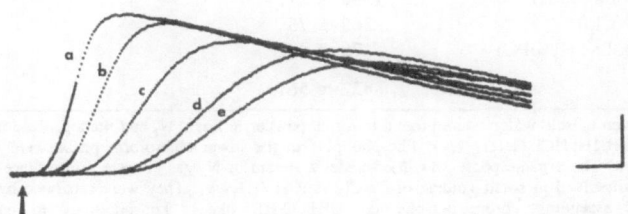

Figure 5. Temperature dependence of the early time course of Ca^{2+} transients in skeletal muscle fibers. A series of Ca^{2+} transients were recorded from a single muscle fiber stimulated to elicit action potentials at several transients: 26°C, trace *a*; 17°C, trace *b*; 12°C, trace *c*; 7°C, trace *d*; 2°C, trace *e*. The vertical calibration bar represents a decrease in absorbance of 0.01 at 480 nm; the horizontal calibration bar represents 2 msec.

2.3.3. Phosphoinositide–Inositol Phosphate Metabolism in Skeletal Muscle

Further support for the involvement of IP$_3$ in the excitation–contraction coupling process in skeletal muscle comes from more biochemical research studying the enzymatic machinery of synthesis an metabolization of inositol phosphates in muscle preparations.

2.3.3.1. Quantitation of Phospholipids in Skeletal Muscle. When we proposed the possibility of a chemical coupling process in skeletal muscle, there was little work on the lipid composition in muscle that would have addressed the issue of the importance of phosphoinositides in this biological preparation.

Total Phospholipids. The most important steps in these experiments are briefly as follows: Excised muscles are ground and lipids extracted; they are separated in two-dimensional and one-dimensional high-performance thin-layer chromatography (HPTLC); and all the major phospholipids were identified and quantitated. Table I describes the total lipid composition of frog skeletal muscle.

Phosphoinositides. Table I shows that the PIP$_2$, PIP, and PI contents of skeletal muscle is about 28, 180, and 400 nmol per gram of wet weight, respectively. Although they constitute a minor proportion of the total phospholipids, their presence indicates that the synthetic enzymatic machinery exists in skeletal muscle.

The next question to be answered is, What do these PIP$_2$ and PIP concentrations mean in terms of their possible participation in a chemical coupling process at the T-SR junction? Hidalgo et al.[48] first demonstrated that skeletal muscle T-tubule membrane

Table I. Phospholipid Composition of Frog Skeletal Muscle[a]

Phospholipid	Mean ± S.D. (nmol/g w.w.)	Number of determinations	Contribution (%)
PtdI(4,5)bisP (PIP$_2$)	28 ± 11	7	0.32
Lyso-PtdI(4,5)bisP (lysoPIP$_2$)	8 ± 2	3	0.09
PtdI4P (PIP)	180 ± 31	7	2.04
PtdI (PI)	406 ± 55	9	4.60
PtdSerine (PS)	126 ± 26	4	1.42
Sphingomyelin	952 ± 172	4	10.79
PtdCholine (PC)	5286 ± 413	4	59.91
PtdEthanolamine (PE)	1390 ± 277	4	15.75
Cardiolipin (Clp)	265 ± 75	4	3.00
Lyso-PtdCholine (lysoPC)	170 ± 40	4	1.93
TOTAL	8822 ± 561		99.85

[a]The fast-frozen muscle wafers were ground to a fine powder in liquid N$_2$ and homogenized in a mixture of CHCl$_3$:MetOH:1N HCl (1:1:1, v/v). Phospholipids in the lower chloroform phase were removed after centrifugation. The organic phase was dried under a stream of N$_2$ gas at room temperature and the phospholipids redissolved in small volume of CHCl$_3$:MetOH (1:1,v/v). They were separated by one- or two-dimensional ascending chromatography on HETLC-HL plates. Phospholipids in one-dimensional chromatography were developed in a mobile phase comprising CHCl$_3$:MetOH:H$_2$O:conc. NH$_4$OH (40:48:10.5, v/v). For two-dimensional chromatography, the phospholipids were developed in the first direction n CHCl$_3$:MetOH:conc. NH$_4$OH (65:25:5, v/v) and in the second direction in CHCl$_3$:MetOH:glacial acetic acid (65:50:5, v/v). Phospholipid spots on the HETLC plates were visualized with I$_2$ vapors, scraped, and extracted for quantitative analysis. Total phosphorus content for every spot was determined by a Malachite green method with a lower detection limit of 0.1 nmol of phosphorous.[54]

Figure 6. Specific activity of ^{32}P-labeling in skeletal muscle at rest. The individual spots for each phospholipid were separated using the chromatographic methodology described in Table 1. The radiolabeled phospholipid spots were scraped, extracted, and counted. Specific activities were computed from the ^{32}P radioactivity (efficiency 85%) and from the total phosphorus contents for each phospholipid. The ATP specific activity was obtained from the separation of high-pressure liquid chromatography (HPLC) of the solutes in aqueous phase extracts from the same muscles used for

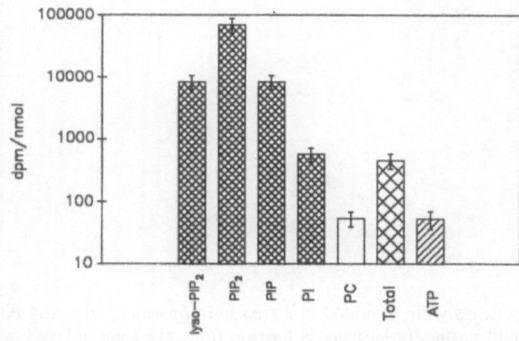

phospholipid analysis. Specific activities of phosphoinositides are represented in narrow crosshatched bars, PC in empty bar, total phospholipids in wide crosshatched bar, and ATP in right diagonal hatch.

vesicles have very active PI kinase and PIP kinase enzymatic activities. The latter is exclusively present in this membrane fraction, being absent in SR membrane vesicles.[48] These results have been confirmed by other laboratories and by us[49,50] (see later, and Fig. 8). An important implication of these *in vitro* results on the phosphoinositide content of muscle reported in Table I is that the PIP_2 content of resting muscles is contributed exclusively from the T-tubule membrane. Moreover, we have shown[51,33] that tetanic electrical stimulation of muscle leads to release of IP_3 at concentrations of units of nmoles per gram wet weight. Thus, tens of nmoles of PIP_2 per gram of wet weight, even if considered a static quantity of precursor availability at the T-tubules prior to electrical stimulation, is sufficient to support the release of IP_3 at the measured amount.

2.3.3.2.   Specific Activity of ^{32}P Labeling of Muscle Phospholipids.   The incubation of excised muscles with $^{32}PH_3PO_4$ carrier free in the extracellular bath, leads to ^{32}P labeling of these phospholipids. Figure 6 shows that ^{32}P labels PIP_2 attaining specific activities 10-fold that of the second highest (lysoPIP$_2$ and PIP), 100-fold higher than PI, and 1000-fold higher than PC, which is the major phospholipid in skeletal muscle. It should be mentioned that (a) PC is the only nonphosphoinositide phospholipid that gets any significant ^{32}P label, and (b) the specific activity of ^{32}P labeling of PC (expressed in dpm/nmol) is identical to that measured for the total intracellular ATP. This results suggest strongly a highly compartmentalized use of the ATP in the preferential labeling of PIP_2 and PIP above the average level of other phospholipids. This is in definite agreement with the model[13] in which the enzymatic machinery responsible for the phosphorylation of phosphoinositides (PI kinase and PIP kinase) is localized in the T-tubule membrane (T-SR junction), close to the extracellular medium.

2.3.3.3.   Mobilization of Phosphoinositides by Electrical Stimulation of Skeletal Muscle.   We developed a liquid nitrogen hammer apparatus to rapidly freeze and smash a stimulated muscle at the point of maximum stimulation and tension development. This process has been measured to occur within about 15–20 msec. It allows us to perform the above-mentioned lipid analysis in relaxed (control) and stimulated (experimental)

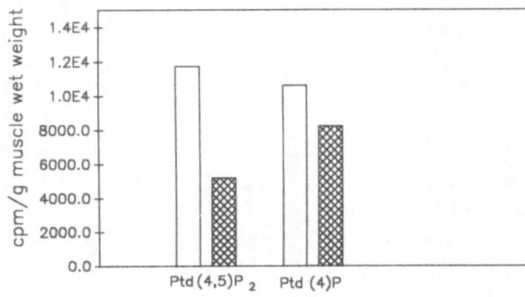

Figure 7. Effect of electrical stimulation on ^{32}P labeling of skeletal muscle phosphoinositides. An intact pair of semitendinosus muscles of the frog *Rana catesbeiana* were incubated for 1.5 hr at room temperature in Ringer's solution containing ^{32}P at (333 uCi carrier-free/ml of solution). The muscles were washed extensively at 4°C in isotope-free Ringer's solution containing 10^{-4} *M* curare. The control muscle (resting) was treated, in addition, with 10^{-7} *M* tetrodotoxin (TTX). Both control and experimental muscles were mounted in a smasher apparatus, specially built for these experiments. The muscles were held vertically between a tension transducer and a fixed holder and were equilibrated for 15 min with Ringer's solution in a temperature-controlled bath at 12°C. The experimental muscle was directly stimulated tetanically at 50 Hz for 2 sec with supramaximal 0.5 msec current pulses. At the end of the tetanic stimulation period, the muscles were smashed between two copper hammers at −200°C. The thickness of the frozen "muscle wafer" after smashing was 0.5 mm. The procedure followed with the control muscles was identical, except that it was not electrically stimulated before smashing. The extraction procedure followed thereafter is described in Table 1. Control muscle: empty bars. Experimental muscle (2 sec stimulation): crosshatched bars.

muscles. The muscles have been incubated previously for 1-hr with (0.5 mCi carrier free) ^{32}P-H_3PO_4 in frog Ringer's solution. After incubation, the muscles were washed thoroughly with cold Ringer's. Figure 7 reveals that electrical stimulation of skeletal muscle leads a rapid decrease in the ^{32}P labeling of phosphoinositides, but preferentially in PIP$_2$ labeling. Comparable experiments with ^3H-myoinositol-labeled phosphoinositides show a similar effect. Our results show, for the first time, a rapid mobilization of phosphoinositides in response to electrical stimulation in skeletal muscle. The most straightforward interpretation of these data is that electrical stimulation activates an endogenous phospholipase-C as originally proposed.[13] They are also in agreement with the finding of IP$_3$ release by electrical stimulation.[13,33,51]

2.3.4. Enzymatic Studies in Isolated T-Tubule Vesicles

We have studied kinase activities in the T-tubule membrane vesicles *in vitro*. Figure 8 illustrates a typical time course of synthesis of PIP$_2$ and PIP in this preparation. The profile of incorporation of ^{32}P into these phospholipids shows transient rising phase followed, after 1 min, by a fast decline. In an attempt to understand the cause of the transient nature of this labeling process, we measured the quantity of nucleotides present at each incubation time interval in the above preparation. This is shown in Fig. 9. It can be observed that there is a very rapid consumption of ATP, a consequently fast rate of formation of ADP, and a delayed synthesis of AMP. A comparative analysis of the profile of incorporation of ^{32}P into PIP$_2$ and the profile of ATP consumption, shows that, sharply at the point in time when ATP virtually disappears, PIP$_2$ labeling starts rapidly falling down. The rapid ATP consumption by these membranes is accounted for by the activity of the prominent Mg^{2+}-ATPase present in the T-tubule.[52,53] The results shown

Figure 8. Time course of phosphoinositide synthesis in T-tubule membrane vesicles. T-tubule vesicles were obtained by the method of Hidalgo et al.[48] Vesicles at a concentration of 35 μg of membrane protein diluted final volume of 0.1 ml were incubated at 25°C with a solution of the following composition: 0.1 M K-Aspartate, 20 mM HEPES-K, 1 mM Mg-Aspartate, 12 mM LiCl, 5 mM EGTA-K, pH 7.0, pCa 7. The reaction was started by addition of gamma-^{32}P ATP diluted K-ATP to a final concentration of 1.5 mM and a specific activity of 0.32 Ci/mmol. The reaction was stopped

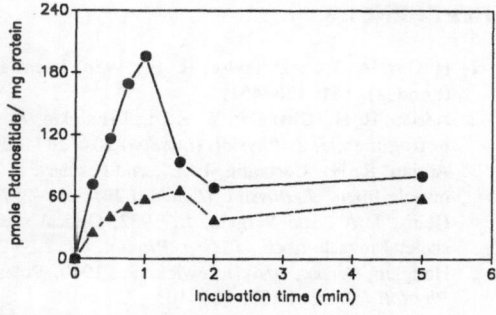

by addition of 1 ml of 1 N HCl, followed by addition of 2 ml CHCl$_3$:MetOH (1:1, v/v). The organic phase was collected and evaporated under a stream of N$_2$ and the component phospholipids were resolved by HPTLC. Filled circles: PIP$_2$; filled triangles: PIP.

in Fig. 9 suggests the possibility that these T-tubule membrane vesicles also contain an endogenous phospholipase-C whose activity becomes evident when the availability of ATP becomes the limiting factor in the synthesis of PIP$_2$. Recent control experiments support this possibility.[54] In conclusion, we are beginning to obtain experimental evidence suggesting that the T-tubule membrane not only manifests phosphoinositide kinase activities but also displays phospholipase-C activity plausibly responsible for the voltage-triggered release of IP$_3$ in skeletal muscle.

ACKNOWLEDGMENTS. We thank Dr. R. Y. Tsien for the gift of azo 1, and Mr. Dennis Gorospe for his skillful technical assistance. This work was supported by grants from NIH (AR-25201), MDA (project 6JLNRC), and NSF (INT 86-13052). N. L. was supported by a postdoctoral fellowship from the AHA-GLAA. The experiments in barnacle muscle fibers were performed at the Friday Harbor Laboratories, University of Washington.

Figure 9. Time course of nucleotide metabolization by T-tubule membrane vesicles. The aqueous phase of the experiment shown in Fig. 7 was injected to through a HPLC Whatman SAX column. The nucleotides were identified by comparative retention times of elution of calibrating standards. The quantitation was done by integrating the absorbance profiles for the elution peak of each nucleotide at 254 nm. Open circles: ATP; filled circles: ADP; open triangle: AMP.

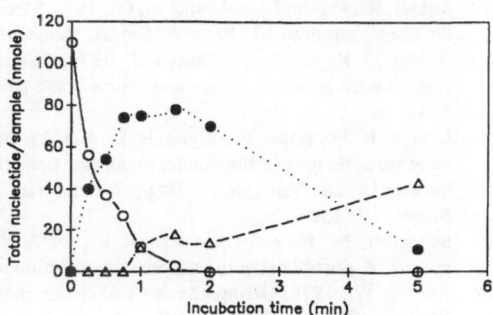

REFERENCES

1. Huxley, A. F., and Taylor, R. E., 1958, Local activation of striated muscle fibers. J. Physiol. (London). **144:** 426–451.
2. Adrian, R. H., Chandler, W. K., and Hodgkin, A. L., 1969, The kinetics of mechanical activation in frog muscle, *J. Physiol. (London)* **204:** 207–230.
3. Adrian, R. H., Costantin, L. L., and Peachey, L. D , 1969, Radial spread of contraction in frog muscle fibers, *J. Physiol. (London)* **204:** 231–257.
4. Heiny, J. A., and Vergara, J., 1982, Optical signals from surface and T-system membranes in skeletal muscle fibers, *J. Gen. Physiol.* **80:** 203–230
5. Hodgkin, A. L., and Horowicz, P., 1960, Potassium contractures in single muscle fibers, *J. Physiol. (London)* **153:** 386–403.
6. Kovacs, L., Rios, E., and Schneider, M. F., 1979, Calcium transients and intramembrane charge movement in skeletal muscle fibers, *Nature (London)* **279:** 391–396.
7. Baylor, S. M., Chandler, W. K., and Marshall, M. W., 1983, Sarcoplasmic reticulum calcium release in frog skeletal muscle fibers estimated from arsenazo III calcium transients, *J. Physiol. (London)* **344:** 625–666.
8. Palade, P., and Vergara, J., 1982, Arsenazo III and antipyrylazo III calcium transients in single skeletal muscle fibers, *J. Gen. Physiol.* **79:** 679–707.
9. Franzini-Armstrong, C., 1970, Studies of the triad. I. Structure of the junction in frog twitch fibers, *J. Cell Biol.* **47:** 488–499.
10. Endo, M., 1977, Calcium release from the sarcoplasmic reticulum, *Physiol. Rev.* **57:** 71–108.
11. Fabiato, A., 1985, Simulated calcium current can both cause calcium loading in and trigger calcium release from the sarcoplasmic reticulum of a skinned canine cardiac cell, *J. Gen. Physiol.* **85:** 291–320.
12. Chandler, W. K., Rakowski, R. F., and Schneider, M. F., 1976, A nonlinear voltage-dependent charge movement in frog skeletal muscle, *J. Physiol. (London)* **254:** 245–283.
13. Vergara, J., Tsien, R. Y., and Delay, M., 1985, Inositol 1,4,5-trisphosphate: A possible chemical link in excitation–contraction coupling in muscle, *Proc. Natl. Acad. Sci. USA* **82:** 6352–6356.
14. Volpe, P., Salviati, G., Di Virgilio, F., and Pozzan, T., 1985, Inositol 1,4,5-trisphosphate induces calcium release from sarcoplasmic reticulum of skeletal muscle, *Nature (London)* **316:** 347–349.
15. Hille, B., and Campbell, D. T., 1976, An improved vaseline gap voltage clamp for skeletal muscle fibers, *J. Gen. Physiol.* **67:** 265–293.
16. Vergara, J., Bezanilla, F., and Salzberg, B. M., 1978, Nile blue fluorescence signals from cut single muscle fibers under voltage or current clamp conditions, *J. Gen. Physiol.* **72:** 775–800.
17. Delay, M., Ribalet, B., and Vergara, J., 1986, Caffeine potentiation of calcium release in frog skeletal muscle fibers, *J. Physiol. (London)* **375:** 535–559.
18. Armstrong, C. M., Bezanilla, F., and Horowicz, P., 1972, Twitches in the presence of ethylene glycol bis(-aminoethylether)-*N*-tetracetic acid, *Biochem. Biophys. Acta* **267:** 605–608.
19. Miledi, R., Parker, I., and Schalow, G., 1977, Measurement of calcium transients in frog muscle by the use of arsenazo III, *Proc. R. Soc. B. (London)* **198:** 201–210.
20. Atwater, I. Rojas, E., and Vergara, J., 1974, Calcium influxes and tension development in perfused single barnacle muscle fibers under membrane potential control, *J. Physiol. (London)* **243:** 523–555.
21. Keynes, R. D., Rojas, E., Taylor, R. E., and Vergara, J., 1973, Calcium and potassium systems of a giant barnacle muscle fiber under membrane potential control, *J. Physiol. (London)* **229:** 409–455.
22. Vergara, J., and Verdugo, P., 1988, Calcium transients in voltage-clamped barnacle muscle fibers, *Biophys. J.* **53:** 647a.
23. Schneider, M. F., and Chandler, W. K., 1973, Voltage-dependent charge movement in skeletal muscle: A possible step in excitation–contraction coupling. *Nature (London)* **242:** 244–246.
24. Almers, W., 1978, Gating currents and charge movements in excitable membranes, *Rev. Physiol. Biochem. Pharmacol.* **82:** 96–190.

25. Adrian, R. H., and Peres, A., 1979, Charge movement and membrane capacity in frog muscle, *J. Physiol. (London)* **289:** 83–97.

26. Huang, C. I.-H., 1982, Pharmacological separation of charge movement components in frog skeletal muscle, *J. Physiol. (London)* **324:** 375–387.

27. Hui, C. S., 1983, Pharmacological studies of charge movement in frog skeletal muscle, *J. Physiol. (London)*. **337:** 509–529.

28. Vergara, J., and Caputo, C., 1983, Effects of tetracaine on charge movements and calcium signals in frog skeletal muscle fibers, *Proc. Natl. Acad. Sci. USA* **80:** 1477–1481.

29. Brum, G., and Rios, E., 1987, Intramembrane charge movement in frog skeletal muscle fibers, Properties of charge 2, *J. Physiol. (London)* **387:** 489–517.

30. Lamb, G. D., 1986, Components of charge movement in rabbit skeletal muscle: The effect of tetracaine and nifedipine, *J. Physiol. (London)* **376:** 85–100.

31. Simon, B. J., and Schneider, M. F., 1987, A comparison of the kinetics of charge movement and activation of SR calcium release during excitation in frog skeletal muscle, *Biophys. J.* **51:** 550a.

32. Vergara, J., and Tsien, R. Y., 1985, Inositol-triphosphate-induced contractures in frog skeletal muscle fibers, *Biophys. J.* **47:** 351a.

33. Vergara, J., Asotra, K., and Delay, M., 1987, A chemical link in excitation–contraction coupling in skeletal muscle, in: *Cell Calcium and the Control of Membrane Transport* (L. J. Mandel and D. C. Eaton, Eds.), pp. 133–151, Rockefeller Press, New York.

34. Vergara, J., and Asotra, K., 1987, The chemical transmission mechanism of excitation–contraction coupling in skeletal muscle, *News Physiol. Sci.* **2:** 182–186.

35. Berridge, M. J., and Irvine, R. F., 1984, Inositol triphosphate, a novel second messenger in cellular signal transduction, *Nature (London)* **312:** 315–321.

36. Katz, B., 1969, *The Release of Neural Transmitter Substances* (C. C. Thomas, Ed.), pp. 1–60, Charles C. Thomas, Springfield, Illinois.

37. Katz, B., and Miledi, R., 1965, The measurement of synaptic delay, and the time course of acetylcholine release at the neuromuscular junction, *Proc. R. Soc. B.* **161:** 483–495.

38. Katz, B., and Miledi, R., 1965, The effect of temperature on the synaptic delay at the neuromuscular junction, *J. Physiol. (London)* **181:** 656–670.

39. Vergara, J., and Delay, M., 1986, A transmission delay and the effect of temperature at the triadic junction of skeletal muscle, *Proc. R. Soc. Lond. Ser. B.* **229:** 97–110.

40. Vergara, J., and Delay, M., 1985, The use of metallochromic Ca indicators in skeletal muscle, *Cell Calcium* **6:** 119–132.

41. Fuortes, M. G. F., and Hodgkin, A. L., 1964, Changes in time scale and sensitivity in the ommatidia of *Limulus*, *J. Physiol. (London)* **172:** 239–263.

42. Baylor, D. A., Hodgkin, A. L., and Lamb, T. D., 1974, The electrical response of turtle cones to flashes and steps of light, *J. Physiol. (London)* **242:** 685–727.

43. Fein, A., Payne, R., Corson, D. W., Berridge, M. J., and Irvine, R. F., 1984, Photoreceptor excitation and adaptation by inositol 1,4,5-trisphosphate, *Nature* **311:** 157–160.

44. Brown, J. E., Rubin, I. J., Ghalayani, A. J., Tarver, A. P., Irvine, R. F., Berridge, M. J., and Anderson, R. E., 1984, Myo-inositol polyphosphate may be a messenger for visual excitation in *Limulus* photoreceptor, *Nature* **311:** 160–163.

45. Yau, K-W., and Nakatani, K., 1985, Light-induced reduction of cytoplasmic tree calcium in retinal rod outer segment, *Nature* **313:** 579–582.

46. Hollingworth, S., and Marshall, M. W., 1981, A comparative study of charge movements in rat and frog skeletal muscle fibers, *J. Physiol. (London)* **321:** 583–602.

47. Simon, B. J., and Beam, K. G., 1985, The influence of transverse tubular delays on the kinetics of charge movement in mammalian skeletal muscle, *J. Gen. Physiol.* **85:** 21–42.

48. Hidalgo, C., Carrasco, M. A., Magendzo, K., and Jaimovich, E., 1986, Phosphorylation of phosphatidylinositol by transverse tubule vesicles and its possible role in excitation–contraction coupling, *FEBS Lett.* **202:** 69–73.

49. Varsanyi, M., Messer, M., Brandt, N., and Heilmeyer, L. M. G., 1986, Phosphatidylinositol 4,5-

bisphosphate formation in rabbit skeletal and heart muscle membranes, *Biochem. Biophys. Res. Comm.* **138:** 1395–1404.

50. Lagos, N., and Vergara, J., 1989, Phosphoinositide kinase and phospholipase-C activities in T-tubule membrane vesicles of frog skeletal muscle, *Biophys. Soc. Abstracts* **55:** 236a.

51. Asotra, K., and Vergara, J., 1986, Levels of inositol phosphates in stimulated and relaxed muscles, *Biophys. J.* **49:** 190a.

52. Hidalgo, C., Parra, C., Riquelme, G., Jaimovich, E., 1986, Transverse tubules from frog skeletal muscle: Purification and properties of vesicles sealed with the inside-out orientation, *Biochim. Biophys. Acta* **855:** 79–88.

53. Kirley, T. L., 1988, Purification and characterization of the Mg^{2+}-ATPase from rabbit skeletal muscle transverse tubule, *J. Biol. Chem.* **263:** 12682–12689.

54. Iless, H. H., and Derr, J. F., 1975, Assay of inorganic and organic phosphorus in the 0.1–5 nanomole range, *Anal. Biochem.* **63:** 607–613.

Chapter 28

What We Know and What We Would Like to Know About the Role of Inositol 1,4,5-Trisphosphate in Skeletal Muscle

Pompeo Volpe, Francesco Di Virgilio, and Tullio Pozzan

1. INTRODUCTION

In skeletal muscle, the depolarization of the transverse tubule (TT) membrane system evokes Ca^{2+} release from the terminal cisternae (TC) of the sarcoplasmic reticulum (SR), a specialized endomembrane network. The coupling between excitation and contraction occurs at the triad where TT and TC are associated junctionally via bridging structures, also referred to as "feet."

The TT–TC coupling has been explained on the basis of either a mechanical,[1] electrical,[2] or chemical hypothesis. The chemical hypothesis states that a specific chemical transmitter is released within the triadic junction in response to an action potential. Simple diffusion across the 120–150 Å junctional space requires less than 1 μsec, whereas the latency between the upswing of the TT action potential and the rise of myoplasmic free Ca^{2+} is about 2.5 msec,[3] that is, TT–TC coupling is not too fast to be mediated chemically. The mechanical and chemical hypotheses have attracted considerable interest in recent years, and both are supported by several lines of experimental evidence. We emphasize, however, that decisive proof in favor or against any proposed

POMPEO VOLPE • Department of Physiology and Biophysics, The University of Texas Medical Branch, Galveston, Texas 77550. FRANCESCO DI VIRGILIO and TULLIO POZZAN • Centro di Studio per la Fisiologia dei Mitocondri del CNR, Istituto di Patologia Generale dell' Università di Padova, 35131 Padua, Italy.

models is lacking—i.e., the mechanism of excitation–contraction coupling is not known.

Inositol 1,4,5-trisphosphate (IP_3) has been proposed as the messenger coupling extracellular stimuli to Ca^{2+} release from intracellular stores in a variety of cell types.[4] The general scheme outlined by Berridge and Irvine (see Fig. 1 in Ref. 4) dictates that the appropriate extracellular stimulus triggers the hydrolysis of phosphatidylinositol 4,5-bisphosphate (PIP_2), located in the inner leaflet of the plasma membrane, into diacylglycerol and IP_3, the latter compound being a water-soluble second messenger. IP_3 is hydrolyzed by specific phosphatases to inositol 1,4-bisphosphate (IP_2) and inositol 1-phosphate (IP_1). A simplified *model* involving IP_3 in excitation–contraction coupling is as follows (see Ref. 5): TT depolarization evokes IP_3 production at the level of TT membranes via a PIP_2 phosphodiesterase. IP_3 released within the triadic junction opens IP_3-sensitive Ca^{2+}-channels localized in TC, and myoplasmic free Ca^{2+} rises. IP_3 is then hydrolyzed by a specific inositol 1,4,5-trisphosphate phosphatase (IP_3ase) to IP_2.

The aim of this chapter is to summarize briefly what we *really* know about the role of IP_3 in excitation–contraction coupling and to address the major, unanswered questions. A few pertinent overviews have been published recently.[6–8]

2. WHAT WE KNOW ABOUT IP_3 IN SKELETAL MUSCLE

Biochemical pathways involved in IP_3 generation and degradation have been described. Vergara et al.[8] have estimated the amount of PIP_2 in resting muscle to be in the range of 4–8 nmol/g wet weight. PIP_2 has been reported to be selectively localized in TT membranes purified from either rabbit[9] or frog[10] skeletal muscle. Isolated TT membranes appear to retain the kinases which phosphorylate phosphatidylinositol to phosphatidylinositol 4-phosphate (PIP_1) and PIP_1 to PIP_2: The first phosphorylation reaction is Ca^{2+} independent below 1 μM and inhibited by Ca^{2+} above 100 μM, whereas phosphorylation of PIP_1 to PIP_2 is optimal above 1 μM. These findings have been partially confirmed by Jay and Campbell[11] in TT membranes obtained by French press treatment of isolated triads of rabbit skeletal muscle: PIP_2 production was reported to be specific for TT membranes, "rapid" ($t_{1/2} < 30$ sec) and independent of calmodulin- and cAMP-dependent protein kinases. Micromolar Ca^{2+} appeared to have an inhibitory effect on the phosphorylation.

Circumstantial evidence has been published suggesting the existence of PIP_2 phosphodiesterase (or phospholipase-C) in isolated triads.[12] More direct evidence has been provided by Salviati et al.[13] who measured PIP_2 hydrolysis on partially purified TT membranes and homogenates of rabbit skeletal muscle. Phospholipase-C activity of TT membranes was found to be stimulated by 100 μM Ca^{2+} and 10 mM NaF.

Direct electrical stimulation of intact frog muscle (e.g., a tetanus lasting more than 3 sec) increased the levels of IP_3, IP_2, and IP_1 two- to fourfold above control.[14] By ion-exchange high-pressure liquid chromatography, the levels of IP_3 in muscles tetanized for 5, 10, or 20 sec at 20°C were found to be high (8 nmol/g wet weight) compared to negligible amounts in control muscle (<0.3 nmol/g wet weight). These experiments show that IP_3 can be produced in skeletal muscle. Given the selective localization of

PIP$_2$ in the TT membranes,[9-12] the reports by Vergara et al.[8,14] indirectly indicate that IP$_3$ can be released within the triad junction.

If IP$_3$ plays some role as second messenger within the muscle fiber, it must be hydrolyzed by a specific IP$_3$ase, as happens in other tissues.[15,16] The existence of a powerful IP$_3$ase in skeletal muscle has been postulated previously on the basis of experiments where the Ca^{2+}-releasing action of IP$_3$ was enhanced by manipulations capable of inhibiting a putative IP$_3$ase. Vergara et al.[14] showed that low Mg^{2+} or Cd^{2+}, Ni^{2+}, and 2,3-diphosphoglycerate all potentiate the action of IP$_3$ on frog skinned fibers. A similar observation was made by Rojas et al.,[17] but not by Lea et al.[18] Moreover, Donaldson et al.[19] observed that microinjected IP$_3$ (1 μM) in rabbit skinned fibers was as effective as 100–300 μM IP$_3$ added to the bathing solution. The explanation for this result was that "endogenous phosphatases create a radially decreasing concentration gradient" of exogenously added IP$_3$.[19] IP$_3$ase was also implicated in the transient nature of IP$_3$-induced Ca^{2+} release because "microinjection of more IP$_3$ during the decline of an IP$_3$-induced tension transient elicits an abrupt increase in fiber tension."[19] Rapid IP$_3$ degradation has been indirectly shown by Vergara et al.[8] in intact frog skeletal muscles.

Milani et al. have recently provided direct evidence for the occurrence of a specific powerful IP$_3$ase in rabbit fast twitch muscles,[20] which shares several biochemical properties with the homologous enzyme from many different tissues (Ref. 20 and references therein): It is present in membrane-bound (60–65%) and soluble form (35–40%), has an absolute requirement for Mg^{2+}, has a K_m for IP$_3$ around 20 μM, and is inhibited by compounds such as 2,3-diphosphoglycerate and Cd^{2+}. The IP$_3$ase of a partially purified TT fraction displays the highest V_{max}, but the enzyme is present in SR fractions as well. Thus, the skeletal muscle IP$_3$ase seems to have a strategic localization at the triad level, both on TT and TC membranes, where signal transduction for muscle activation occurs.

Walker et al.[21] have instead reported that: (i) IP$_3$ase activity of frog skinned fibers is 35 times slower than that of skinned strips of rabbit main pulmonary artery (smooth muscle); (ii) there is no diffusible ("soluble"?) IP$_3$ase either in skeletal or smooth muscle; and (iii) IP$_3$ase activity was negligible in both types of muscle. They concluded that rapid IP$_3$ degradation cannot be accounted for by an active IP$_3$ase in skeletal muscle. The experimental models and protocols of our study[20] are different from those of Walker et al.,[21] and many ad hoc explanations for the discrepancy could be put forward. With respect to the comparison between smooth and skeletal muscle of the rabbit, however, we note that the total IP$_3$ase activity of skeletal muscle homogenates seems to be comparable to that of smooth muscle homogenates (see Table I in Ref. 20).

3. IP$_3$-INDUCED Ca^{2+} RELEASE FROM THE SR OF SKELETAL MUSCLE

The basic question of whether IP$_3$ induces Ca^{2+} release from the SR still awaits a definitive answer and many conflicting reports have appeared since 1985 (Table I). It is difficult to identify clear-cut reasons that may account for the discrepancy in experimental results.

Table I.   Does IP_3 Induce Ca^{2+} Release from the SR of Skeletal Muscle?

Reference		Species
Volpe et al.[22]	Yes[a,b]	Rabbit
Vergara et al.[14]	Yes[b]	Frog
Sherer and Ferguson[25]	No[a]	Rabbit, lobster
Adunjah and Dean[26]	No[a]	Rabbit
Nosek et al.[23]	Yes[b]	Frog
Lea et al.[18]	No[b,c]	Frog, barnacle
Smith et al.[29]	No[d]	Rabbit
Rojas et al.[17]	Yes[b]	Frog, barnacle
Donaldson et al.[19]	Yes[b]	Rabbit
Donaldson[24]	Yes[b]	Rabbit (slow twitch fibers)
Palade[27]	No[a]	Rabbit
Walker et al.[21]	Yes[b]	Frog
Mikos and Snow[28]	No[a]	Rabbit
Hannon et al.[31]	No[e], yes[f]	Frog, mouse, guinea pig
Suarez-Isla et al.[30]	Yes[d]	Frog

[a]Isolated membrane preparations.
[b]Skinned fiber preparations.
[c]A few barnacle fibers did respond to IP_3.
[d]SR vesicles incorporated into planar lipid bilayer.
[e]Intact fibers microinjected with IP_3.
[f]Detubulated fibers—fibers in which electrical discontinuity and morphological disruption of the TT system is accomplished by glycerol shock.

There is general agreement that *skinned muscle fibers* are responsive to IP_3.[14,17,19,21–24] Only one negative report has been published describing the inability of IP_3 to induce Ca^{2+} release from the SR of skinned fibers of frog and barnacle muscles.[18] Effective concentrations of *exogenous* IP_3 are very variable (from 1 μM to 450 μM) and have been ascribed both to the presence of a powerful IP_3ase[14,19] and to the mode of IP_3 addition.[19]

As to the effect of IP_3 in *isolated SR fractions* there are five published reports. Two of them describe lack of effect of IP_3 on crude SR fractions[25,26] and are misleading because experiments were not carried out on a purified terminal cisternae fraction (i.e., the SR subfraction responsible for Ca^{2+} release). However, Palade,[27] Mikos and Snow,[28] and Volpe et al.[22] reported contradictory effects using enriched TC preparations from rabbit skeletal muscle. Only Volpe et al.[22] have been able so far to obtain IP_3-induced Ca^{2+} release from isolated TC. Reasons for the discrepancy are not readily apparent. Comparison of results obtained in skinned fibers and isolated TC seems to indicate that the IP_3 sensitivity is a labile SR function.

Single-channel recordings of SR vesicles incorporated into lipid bilayers also provide conflicting results (Table I). Smith et al.[29] could not detect any effect of IP_3 on single-channel properties of rabbit SR vesicles. On the other hand, Suarez-Isla et al.[30] reported that IP_3 activates a high-conductance calcium channel of frog TC vesicles by increasing channel fractional open time. IP_3 effects are observed in the presence of high-Mg^{2+} concentrations (5 mM) and appear to be Ca^{2+}-dependent.

4. WHAT WE WOULD LIKE TO KNOW ABOUT IP₃ IN SKELETAL MUSCLE

Several questions, which await clarification, are listed below.

4.1. How Fast is IP₃-Induced Ca²⁺ Release from SR?

It is estimated that the latency of Ca^{2+} release from SR after an action potential is 2–3 msec.[5] This quite obviously sets an upper limit to IP_3-induced Ca^{2+} release, if IP_3 is considered the messenger for excitation–contraction coupling.

Walker et al.[21] have reported experiments in which 1-(2-nitrophenyl)ethylinositol (1,4,5)trisphosphate, or "caged" IP_3, was used. Caged IP_3 was added to skinned fibers of the frog, a few minutes were allowed for equilibration, and then IP_3 was regenerated to a concentration of about 80 μM by laser photolysis. Under these experimental conditions, tension developed without appreciable lag, yet the t 1/2 to peak tension was three orders of magnitude greater than $t1/2$ associated with a propagated TT action potential. The interpretation of Walker et al.[21] was that IP_3 cannot be the primary messenger for excitation–contraction coupling. However, Walker et al.[21] did not provide an internal control for their experiments (e.g., the effect of photolysis breakdown products on the properties of the SR membrane) (see also Ref. 6). Moreover, a long $t_{1/2}$ to peak tension might be due to a small extent of Ca^{2+} release rather than to a slow Ca^{2+} release rate (see below). In contrast to Walker et al.,[21] Vergara et al.[8] using a similar preparation (i.e., frog skinned muscle fibers), claimed that the lag between iontophoretic application of IP_3 and tension development was below 33 msec.

It is also worth pointing out that, due to the anatomical organization of the TC–TT junction, neither addition of IP_3 to the bathing medium,[22] microinjection of IP_3,[13] nor possibly equilibration of skinned fibers with caged IP_3[21] mimic faithfully the pattern of IP_3 generation that might occur within the triad junction following TT depolarization. Thus, it is fair to conclude that the time course of IP_3-induced Ca^{2+} release has not been assessed fully.

4.2. Is IP₃-Induced Ca²⁺ Release Ca²⁺-Dependent or Ca²⁺-Independent?

Among the several studies reporting that IP_3 does induce Ca^{2+} release from the SR, there is controversy as to the Ca^{2+} dependence of IP_3 action. Reports by Volpe et al.[22] and Suarez-Isla et al.[30] seem to support a role for Ca^{2+}, whereas Donaldson et al.[19,24] deny it. Thus, a related question is whether IP_3 induces Ca^{2+} release via Ca^{2+}-, caffeine-, and ruthenium red-sensitive Ca^{2+} release channels[27,29] or via distinct yet to be identified efflux pathways.

4.3. Signal Transduction: How Can TT Depolarization Be Linked to PIP₂ Hydrolysis and Ca²⁺ Release?

The first observations suggesting that depolarization activated phosphoinositide turnover were reported by Novotny et al.[32] These authors showed that K^+ contractures of frog sartorius muscles increased incorporation of ^{32}P into the phosphoinositide pool.

Direct electrical stimulation (tetanus) of intact frog muscles[8,14] elicted IP_3 production. These reports, however, *do not* address the key issue, that is, whether IP_3 is causally related to the twitch. In Vergara et al.,[8,14] IP_3 might very well be the consequence rather than the cause of prolonged contractures. To show unambiguously that IP_3 has a primary role in excitation–contraction coupling, IP_3 production must be measured after a single twitch given via the motor nerve to muscles that are then completely frozen within 3–4 msec. This experiment should also clarify whether or not the rate of phosphoinositide breakdown is compatible with the excitation–contraction coupling time scale.

IP_3 accumulation in chick embryo myotubes following nicotinic cholinergic stimulation has also been reported.[33] This is a potentially interesting observation because the phosphoinositide response has always been linked to muscarinic cholinergic stimulation. It is possible that the linkage of phosphoinositide turnover to nicotinic stimulation in myotubes is indirect and due to membrane depolarization. The relationship between activation of nicotinic receptor, IP_3 production, and membrane potential in both myotubes and mature muscle fibers remains to be further investigated.

GTP-binding proteins in other cell types have been involved in coupling extracellular stimuli to PIP_2 phosphodiesterase activation.[34] Di Virgilio et al.[35] obtained circumstantial evidence that GTP-binding protein(s) may be involved in excitation–contraction coupling in skeletal muscle: (i) GTP-$_\gamma$S, a nonhydrolyzable analogue of GTP, caused tension development in skinned fibers; (ii) GTP$_\gamma$S did not act *directly* on the SR, as indicated by lack of effect on Ca^{2+} fluxes in isolated SR fractions; (iii) GTP$_\gamma$S, most likely, evoked Ca^{2+} release from the SR by activating PIP_2 phosphodiesterase because its effect was blocked partially by pertussis toxin, which is believed to inactivate stimulatory GTP-binding protein(s). Di Virgilio et al.[35] speculated that skeletal muscle could be equipped with a family of G-proteins sensitive to TT membrane potential. According to this hypothesis,[35] TT depolarization triggers the exchange of bound GDP for GTP at the level of a G-protein. The activated G-protein can thus interact with phospholipase-C, which finally splits PIP_2 into IP_3 and diacylglycerol. This possibility is further strengthened by a recent report[36] on the presence of G-proteins in the sarcolemma and TT of rabbit skeletal muscle. Some G-proteins, however, might have a role in the regulation of PIP_2 hydrolysis, in particular those sensitive to pertussis toxin. Interestingly, GTP$_\gamma$S stimulated the formation of IP_3 in chick myotubes,[37] and distribution of G-proteins coincided with the distribution of PIP_2 and of PIP_1 kinase in skeletal muscle TT membranes.[9,12]

The relationship between TT membrane potential and IP_3-induced Ca^{2+} release was investigated by Donaldson et al.,[38] who claim to have demonstrated that TT voltage controls SR sensitivity to IP_3 using mechanically skinned rabbit skeletal muscle fibers. In this preparation,[35] the TT–TC junction is reported to be functionally intact and the TT seal-off and form closed compartments after skinning. TT potential could be "set" by bathing solution composition. Microinjected IP_3 (0.5 μM) elicited maximum tension transients when TT were "depolarized" (see below). The same fibers failed to respond to IP_3 when TT were "polarized." The authors' conclusion was that the TT membrane potential may control the SR sensitivity to basal IP_3 concentrations in the sarcoplasm rather than IP_3 production. However, Donaldson et al.[38] did not provide estimates about the permissive TT membrane potential and did not explain what turns on (and off) PIP_2 hydrolysis, and whether IP_3ase is also controlled by TT voltage or whether IP_3 is bound to escape degradation by the IP_3ase.

On the other hand, Hannon et al.[31] reported that microinjection of IP_3 (1 μM–1 mM) in intact muscle fibers was without effect. However, the same fibers responded to IP_3 (1 μM) after a glycerol shock. (In detubulated fibers, the electrical activity of TT is greatly reduced; TT are believed to seal off; TT are irregularly swollen, and the integrity of the TT–TC junction is compromised; see Refs. 39, 40–42). In detubulated fibers, IP_3-induced contractions were abolished by pretreatment with digitoxin, a membrane-permeant inhibitor of the (Na^+-K^+)-ATPase. The interpretation of Hannon et al.[31] is that IP_3 is effective only when TT are *polarized*, and is ineffective when TT are depolarized by digitoxin-dependent inhibition of the (Na^+-K^+)-ATPase (see Ref. 43). These authors[31] also argued that "IP_3 triggers Ca^{2+} release by depolarizing sealed-off T-tubules." How IP_3 may bring about TT depolarization is far from clear. We feel the conclusions drawn by Hannon et al.[31] are unwarranted since the object of their speculations is somewhat elusive. In fact, not all TT and TT–TC junctions are involved during the glycerol shock as indicated by capacitance measurements[43] and EM studies.[44] The scheme proposed by Hannon et al.[31] would apply only in those unpredictable situations where TT have been severed from the sarcolemma, yet the TT–TC junction is still intact.

4.4. Additional/Alternative Roles for IP_3 in Skeletal Muscle

The role of IP_3 as the *messenger* of excitation–contraction coupling in skeletal muscle is far from proven. The causal relationship between TT depolarization, Ca^{2+} release from SR, and ensuing twitch is one of the crucial, yet unanswered, questions.

If IP_3 is not the primary messenger (and this remains to be ascertained), it might still be involved in excitation–contraction coupling as a *modulator*: (i) IP_3 might be produced at times coincident or subsequent to the deployment of the physiological messenger within the triad. For instance, ATP, which is stored in the nerve terminal of the motor endplate and coreleased with acetylcholine, has been reported to activate PIP_2 turnover, generate IP_3, and increase myoplasmic free Ca^{2+}.[37] ATP, at the neuromuscular junction, might prolong the acetylcholine effects via IP_3 production. (ii) IP_3 might be generated as a result of Ca^{2+} release by a Ca^{2+}-dependent PIP_2 phosphodiesterase and, thus, cause additional Ca^{2+} release. A secondary rise in IP_3 may be important in tuning the amount of Ca^{2+} released from TC or in amplifying the response to Ca^{2+}. (iii) IP_3 might play a role in contractile events other than the twitch (e.g., facilitation, contracture, fatigue).

An alternative role for IP_3 can be postulated. A few receptors coupled to PIP_2 turnover have been recently identified in skeletal muscle (e.g., muscarinic,[37,44] P_2-purinergic,[37,45] and V_1-vasopresin[46]). IP_3 might thus function as a *second messenger for metabolic pathways*. This hypothesis would also suggest the occurrence of two distinct, parallel intracellular Ca^{2+} pools: one sensitive to IP_3, the other to the "physiological" excitation–contraction coupling messenger.

5. CONCLUDING REMARKS

Since 1985 the hypothesis involving the role of IP_3 in excitation–contraction coupling has caused a steady flow of publications and abstracts, which, unfortunately, do not fit into a single homogeneous picture. We think that more experimental work is needed to address a number of important, unanswered questions and that time has not come yet to dismiss or accept IP_3 as the messenger for excitation–contraction coupling.

REFERENCES

1. Schneider, M. F., and Chandler, W. K., 1973, Voltage-dependent charge movement in skeletal muscle: A possible step in excitation–contraction coupling, *Nature* **242**: 244–246.
2. Mathias, R. T., Levis, R. A., and Eisenberg, R. S., 1980, Electrical models of excitation–contraction coupling and charge movement in skeletal muscle, *J. Gen. Physiol.* **76**: 1–31.
3. Vergara, J., and Delay, M., 1986, A transmission delay and the effect of temperature at the triadic junction of skeletal muscle, *Proc. R. Soc. (London)* **229**: 97–110.
4. Berridge, M. J., and Irvine, R. F., 1984, Inositol trisphosphate, a novel second messenger in cellular signal transduction, *Nature* **312**: 315–321.
5. Volpe, P., Di Virgilio, F., Pozzan, T., and Salviati, G., 1986, Role of inositol 1,4,5-trisphosphate in excitation–contraction coupling in skeletal muscle, *FEBS Lett.* **197**: 1–4.
6. Volpe, P., Di Virgilio, F., Bruschi, G., Regolisti, G., and Pozzan, T., 1989, Phosphoinositide metabolism and excitation–contraction coupling in smooth, cardiac, and skeletal muscles, in: *Inositol Lipids and Cell Signaling* (R. H. Mitchell, A. R. Drummond, and C. P. Downes, Eds.), Academic, New York. pp. 377–404.
7. Volpe, P., Pozzan, T., and Di Virgilio, F., 1989, Is inositol 1,4,5-trisphosphate the chemical transmitter for excitation–contraction coupling in skeletal muscle? in: *Neuromuscular Junction* (Sellig, L. C., Sibelius, R., and S. Thesleff, Eds.) Elsevier, Amsterdam. pp. 381–393.
8. Vergara, J., Asotra, K., and Delay, M., 1987, A chemical link in excitation–contraction coupling in skeletal muscle, in: *Cell Calcium and the Control of Membrane Transport* (L. J. Mandel and D. G. Eaton, Eds.), pp. 133–151, The Rockefeller University Press, New York.
9. Hidalgo, C., Carrasco, M. A., Magendzo, K., and Jaimovich, E., 1986, Phosphorylation of phosphatidylinositol by transverse tubule vesicles and its possible role in excitation–contraction coupling, *FEBS Lett.* **202**: 69–73.
10. Carrasco, M. A., Magendzo, K., Jaimovich, E., and Hidalgo, C., 1988, Calcium modulation of phosphoinositide kinases in transverse tubule vesicles from frog skeletal muscle, *Arch. Biochem. Biophys.* **262**: 306–366.
11. Jay, S. D., and Campbell, K. P., 1988, Characterization of phosphatidylinositol 4,5-bisphosphate production in skeletal muscle triads, *Biophys. J.* **53**: 467a.
12. Varsanyi, M., Messer, M., Brandt, N. R., and Heilmeyer, L. M. G., 1986, Phosphatidylinositol 4,5-bisphosphate formation in rabbit skeletal and heart muscle membranes, *Biochem. Biophys. Res. Commun.* **138**: 1395–1401.
13. Salviati, G., Betto, R., Tegazzin, V., and Della Puppa, A., 1988, Subcellular localization of G-protein and phospholipase-C activity in rabbit skeletal muscle, *Biophys. J.* **53**: 332.
14. Vergara, J., Tsien, R. Y., and Delay, M., 1985, Inositol 1,4,5-trisphosphate: Possible chemical link in excitation–contraction coupling in muscle, *Proc. Natl. Acad. Sci. USA* **82**: 6352–6356.
15. Downes, C. P., and Mitchell, R. H., 1981, The polyphosphoinositide phosphodiesterase of erythrocyte membranes, *Biochem. J.* **198**: 133–140.
16. Storey, D. J., Shears, S. B., Kirk, C. J., and Michell, R. H., 1984, Stepwise enzymatic dephosphorylation of inositol 1,4,5-trisphosphate to inositol in liver, *Nature* **312**: 374–376.

17. Rojas, E., Nassar-Gentina, V., Luxoro, M., Pollard, M. E., and Carrasco, M. A., 1987, Inositol 1,4,5-trisphosphate-induced Ca²⁺ release from the sarcoplasmic reticulum and contraction in crustacean muscle, *Can. J. Physiol. Pharacol.* **65:** 672–680.
18. Lea, T. J., Griffiths, D. J., Treagar, R. T., and Ashley, C. C., 1986, An examination of the ability of inositol 1,4,5-trisphosphate to induce calcium release and tension development in skinned skeletal muscle fibers of frog and crustacean, *FEBS Lett.* **207:** 153–161.
19. Donaldson, S. K., Goldberg, N. D., Walseth, T. F., and Huettemann, D. A., 1987, Inositol trisphosphate stimulates Ca²⁺ release from peeled skeletal muscle fibers, *Biochem. Biophys. Acta* **927:** 92–99.
20. Milani, D., Volpe, P., and Pozzan, T., 1989, D-myo-inositol 1,4,5-trisphosphate phosphatase in skeletal muscle, *Biochem. J.* **254:** 525–529.
21. Walker, J. W., Somlyo, A. V., Goldman, Y. E., Somlyo, A. P., and Trentham, D. A., 1987, Kinetics of smooth and skeletal muscle activation by laser pulse photolysis of caged inositol 1,4,5-trisphosphate, *Nature* **327:** 249–252.
22. Volpe, P., Salviati, G., Di Virgilio, F., and Pozzan, T., 1985, Inositol 1,4,5-trisphosphate induces Ca²⁺ release from the sarcoplasmic reticulum of skeletal muscle, *Nature* **316:** 347–349.
23. Nosek, T. M., Williams, M. F., Zeigler, J. T., and Godt, R. E., 1986, Inositol trisphosphate enhances calcium release in skinned cardiac and skeletal muscle, *Am. J. Physiol.* **250:** C807–C810.
24. Donaldson, S. K., 1986, Mammalian muscle fiber types: Comparison of excitation–contraction coupling mechanisms, *Acta Physiol. Scand.* **128:** 157–166.
25. Sherer, N. M., and Ferguson, J. E., 1985, Inositol 1,4,5-trisphosphate is not effective in releasing calcium from skeletal muscle sarcoplasmic reticulum membranes, *Biochem. Biophys. Res. Commun.* **316:** 347–349.
26. Adunjah, S. A., and Dean, M., 1986, Ca²⁺ transport in human platelet membranes: Kinetics of active transport and passive release, *J. Biol. Chem.* **261:** 3122–3127.
27. Palade, P., 1987, Drug-induced Ca²⁺ release from isolated sarcoplasmic reticulum. III. Block of Ca²⁺-induced Ca²⁺ release by organic polyamines, *J. Biol. Chem.* **262:** 6149–6157.
28. Mikos, G. J., and Snow, T. R., 1987, Failure of inositol 1,4,5-trisphosphate to elicit or potentiate Ca²⁺ release from isolated skeletal muscle sarcoplasmic reticulum, *Biochim. Biophys. Acta* **927:** 256–260.
29. Smith, J. S., Coronado, R., and Meissner, G., 1986, Single-channel measurements of the calcium release channel from skeletal muscle sarcoplasmic reticulum, *J. Gen. Physiol.* **88:** 573–588.
30. Suarez-Isla, B. A., Irribara, V., Bull, R., Oberhauser, A., Larrale, L., Jaimovich, E., and Hidalgo, C., 1988, Inositol 1,4,5-trisphosphate activates a calcium channel in isolated sarcoplasmic reticulum (SR) membranes, *Biophys. J.* **53:** 467a.
31. Hannon, J. D., Lee, N. K. M., and Blinks, J. R., 1988, Calcium release by inositol trisphosphate in amphibian and mammalian skeletal muscle is an artifact of cell disruption, and probably results from depolarization of sealed-off T-tubules, *Biophys. J.* **53:** 607a.
32. Novotny, I., Saleh, F., and Novotna, R., 1983, K⁺ depolarization and phospholipid metabolism in frog sartorius muscle, *Gen. Physiol. Biophys.* **2:** 329–337.
33. Adamo, S., Zani, B. M., Nervi, C., Senni, M. I., Molinaro, M., and Eusebi, F., 1985, Acetylcholine stimulates phosphatidyl inositol turnover at nicotinic receptors of cultured myotubes, *FEBS Lett.* **190:** 161–164.
34. Cockroft, S., and Gomperts, B. D., 1985, Role of guanine nucleotide-binding proteins in the activation of polyphosphoinositide phosphodiesterase, *Nature* **314:** 534–536.
35. Di Virgilio, F., Salviati, G., Pozzan, T., and Volpe, P., 1986, Is a guanine nucleotide-binding protein involved in excitation–contraction coupling in skeletal muscle? *EMBO J* **5:** 259–262.
36. Sherer, N. M., Toro, M-J., Entman, M. L., and Birnbaumer, L., 1987, G-protein distribution in canine cardiac sarcoplasmic reticulum and sarcolemma: Comparison to rabbit skeletal muscle membranes and to brain and erythrocyte G-proteins, *Arch. Biochem. Biophys.* **259:** 431–440.
37. Heilbronn, E., and Haggblad, J., 1987, Transmitter/modulator-induced events related to excitation–contraction coupling in skeletal muscle of vertebrates, Action of ATP. N.A.T.O. ASI Series.

38. Donaldson, S. K., Goldberg, N. D., Walseth, T. F., and Huetteman, D. A., 1988, Transverse tubule voltage control of inositol-trisphosphate-induced Ca^{2+} release in peeled skeletal muscle fibers, *Biophys. J.* **53:** 468a.

39. Donaldson, S. K. B., 1985, Peeled mammalian skeletal muscle fibers: Possible stimulation of Ca^{2+} release via a transverse tubule–sarcoplasmic reticulum mechanism, *J. Gen. Physiol.* **86:** 501–523.

40. Eisenberg, R. S., and Gage, P. W., 1967, Frog skeletal muscle fibers: Changes in electrical properties after disruption of transverse tubular system, *Science* **158:** 1700–1703.

41. Eisenberg, B., and Eisenberg, R. S., 1968, Selective disruption of the sarcotubular system in frog sartorius muscle, *J. Cell. Biol.* **39:** 451–467.

42. Franzini-Armstrong, C., Venosa, R. A., and Horowicz, P., 1973, Morphology and accessibility of the "transverse" tubular system in frog sartorius muscled after glycerol treatment, *J. Membr. Biol.* **14:** 197–212.

43. Volpe, P., and Stephenson, E. W., 1986, Ca^{2+} dependence of transverse-tubule-mediated calcium release in skinned skeletal muscle fibers, *J. Gen. Physiol.* **87:** 271–288.

44. Eusebi, F., Grassi, F., Nervi, C., Caporale, C., Addrens, S., Zani, B. M., and Molinaro, M., 1987, Acetylcholine may regulate its own nicotinic receptor-channel through the C-kinase system, *Proc. R. Soc. Lond. B* **230:** 355–365.

45. Heggblad, J., and Heilbronn, E., 1987, Externally applied adenosine-5'-triphosphate causes inositol triphosphate accumulation in cultured chick myotubes, *Neurosci. Lett.* **74:** 199–204.

46. Wakelam, M. J. O., Patterson, S., and Hanley, M. R., 1987, L6 skeletal muscle cells have functional V_1-vasopressin receptors coupled to stimulated inositol phospholipid metabolism, *FEBS Lett.* **210:** 181–184.

Chapter 29

Calcium Release in Skinned Muscle Fibers: Effect of Inositol 1,4,5-Trisphosphate

Enrique Jaimovich, Cecilia Rojas, and Eduardo Rojas

1. INTRODUCTION

A key issue when postulating a role for inositol 1,4,5-trisphosphate (IP_3) as a messenger in excitation–contraction coupling in skeletal muscle is whether the effect of exogenous IP_3 on skeletal muscle fibers is compatible with the effect postulated for a physiological intracellular transmitter. It is important to establish then whether IP_3 acts at concentrations that are low enough, whether its action is fast enough, and whether it is specific and compatible with the known physiological properties of the process.

2. CALCIUM RELEASE IN INTACT AND SKINNED FIBERS

As discussed by Volpe et al. (Chapter 28, this volume) the effect of IP_3 in calcium release from either intact skeletal muscle fibers or from permeabilized fibers has been the subject of controversy during the past few years. Vergara et al.[1] reported that microinjection of IP_3 induced contraction in frog skeletal muscle fibers and that this event was seen before 33 msec, the time elapsed between two frames of the video

ENRIQUE JAIMOVICH AND CECILIA ROJAS • Departamento de Fisiología y Biofísica, Facultad de Medicina, Universidad de Chile, Santiago, Chile.   EDUARDO ROJAS • Laboratory of Cell Biology and Genetics, National Institute of Diabetes and Digestive and Kidney Diseases, National Institutes of Health, Bethesda, Maryland 20892.

recorder.[2] Contraction in intact or skinned fibers in response to IP_3 was also reported by Volpe et al.,[3,4] Nosek et al.,[5] and Donaldson et al.[6] in mammalian muscle, and by Rojas et al.[7] in crustacean muscle; all of them used IP_3 concentrations on the order of 10 μM. The hypothesis of IP_3 being the chemical messenger in excitation–contraction coupling has been favored by a number of nondirect evidences (see Chapters 27, 28, 30, and 33, this volume); but the direct effect of IP_3 at low concentrations on muscle fibers was challenged by several negative results obtained in isolated SR vesicles[8–10] and in isolated and skinned fibers as well. Lea et al.[11] used the aequorin light signal as a measure of calcium release from skinned fibers and found no effect of high concentrations of IP_3 in fibers that were sensitive to caffeine; on the other hand, Blinks et al.[12] did find an effect of IP_3 on contraction but only in fibers that were detubulated after an osmotic shock. Walker et al.[13] used IP_3 released after flash photolysis of a photosensitive derivative and obtained a slow contractile response, too slow if compared to the physiological twitch.

These controversial results induced us to try to answer some basic questions on the effect of IP_3 on calcium release from permeabilized skeletal muscle fibers. The first question we wanted to address is whether IP_3, at low concentrations, releases calcium in muscle fibers. The second question is whether this release is fast enough to be of physiological meaning. Further questions on the specificity of the IP_3 response and on possible explanations for the negative results obtained by other laboratories can also be approached. Part of this work has been published elsewhere.[14,15]

3. A METHOD TO MEASURE A LARGE CALCIUM RELEASE SIGNAL

The apparatus described in Fig. 1A consists of a light-tight chamber in which a plastic tube is in close proximity to a photomultipler window. A pipette allows additions and mixing of a small volume of aqueous media where the muscle fiber bundles were incubated in the presence of aequorin. The muscle represents about 50% of the volume, and the permeabilized fibers remain loose within the chamber. As shown in Fig. 1B, in the absence of muscle, with calcium concentrations between pCa 7 and pCa 6, measurable light signals were obtained and our system saturates at pCa 6. In the presence of skinned muscle fibers (Fig. 1C), the addition of IP_3 produced a large light signal, and we were able to clearly distinguish these signals from mixing artifacts and scattering. The usual protocol in these experiments is as follows.

All experiments were performed at room temperature on sartorious muscles isolated from the frog *Caudiverbera caudiverbera*. Muscles were dissected and placed for 30 min in solution A (118 mM NaCl, 2 mM MgCl2, and 5 mM HEPES sodium, pH 7.4, containing 100 μM EGTA in most cases). Permeabilization of the fibers was carried out mechanically, scratching with a sharp needle the muscle immersed in solution B (130 mM sodium glutamate or other ions when indicated, 2.5 mM ATP-sodium, 3 mM MgCl$_2$, and 5 mM HEPES sodium, pH 7.2).

Sarcoplasmic reticulum was loaded by incubation of fiber bundles with 1–10 μM CaCl$_2$ in solution B for 10–60 min. Before recording cytosolic Ca^{2+} changes, fiber bundles were washed twice in solution B with no added Ca^{2+}.

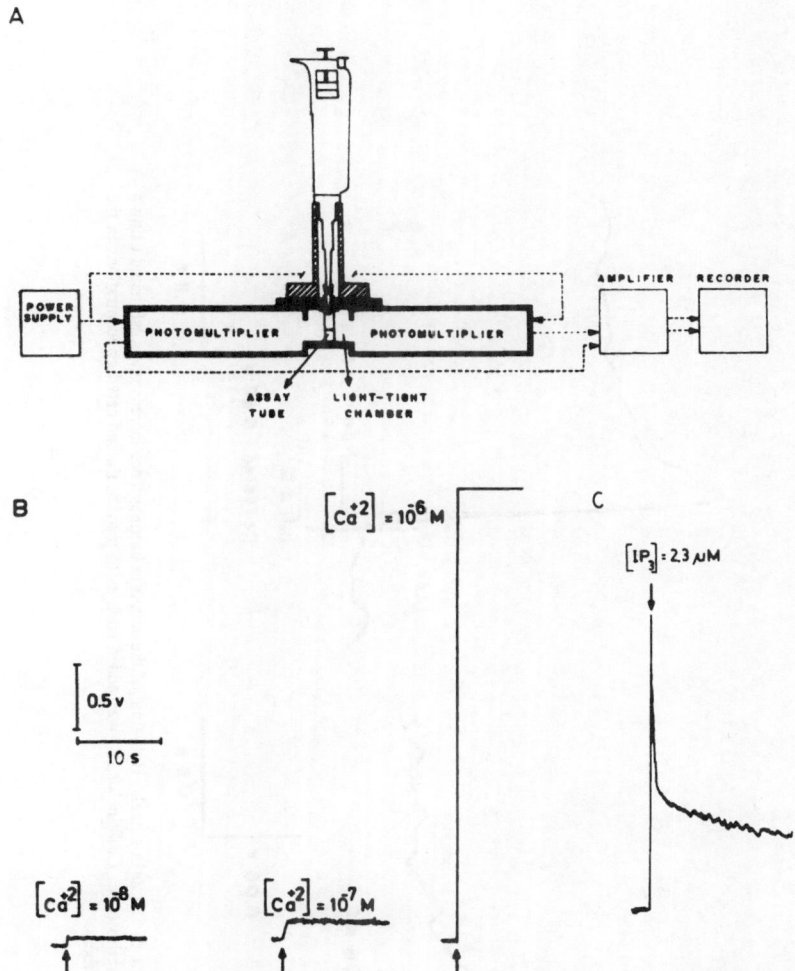

Figure 1. (A) Experimental set-up for incubation of skinned fibers and recording of the light signals induced by calcium. (B) Calibration of the aequorin light signal with calcium. The luminescence records obtained when aequorin was added to calcium buffers of indicated concentrations are shown in the time and voltage scale generally used afterward with muscle fibers. (C) Aequorin light signal following addition of IP₃ to a bundle of skinned fibers.

The "skinned" fiber bundles (20–50 mg) were transferred into 1 ml plastic test tubes, containing 0.1 ml of solution B, and were then placed in the light-tight chamber. Aequorin was diluted in solution B, such as to provide 0.1 μg/ml protein and 50 μM EDTA in the assay medium. Drugs used to induce Ca^{2+} liberation were diluted to the required concentration in solution B and were added in a constant fixed volume of 20 μl.

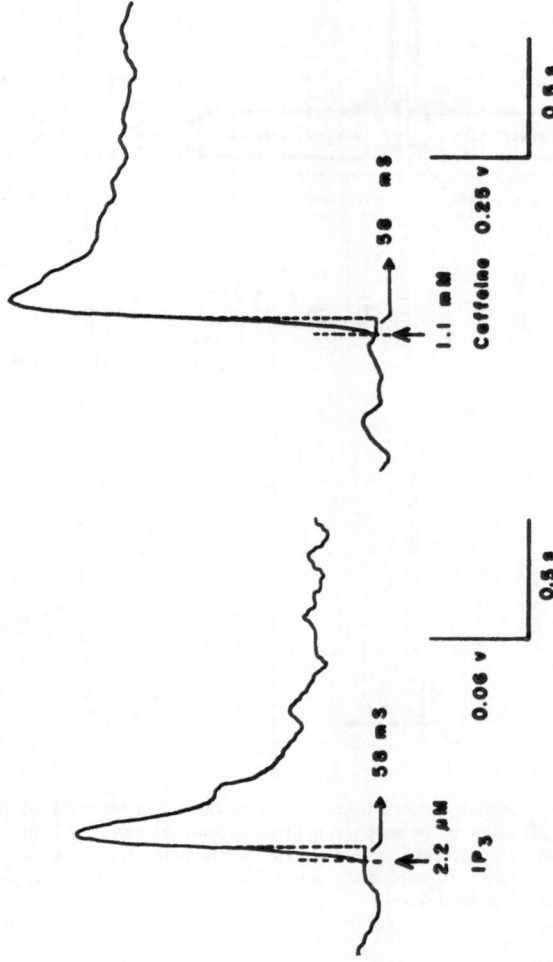

Figure 2. Time course of aequorin signals. Half-time of the aequorin luminescence increase was estimated using a Nicolet 4094 digital oscilloscope. Caffeine or IP$_3$ was added such as to provide the indicated concentrations in the solution, bathing the fibers.

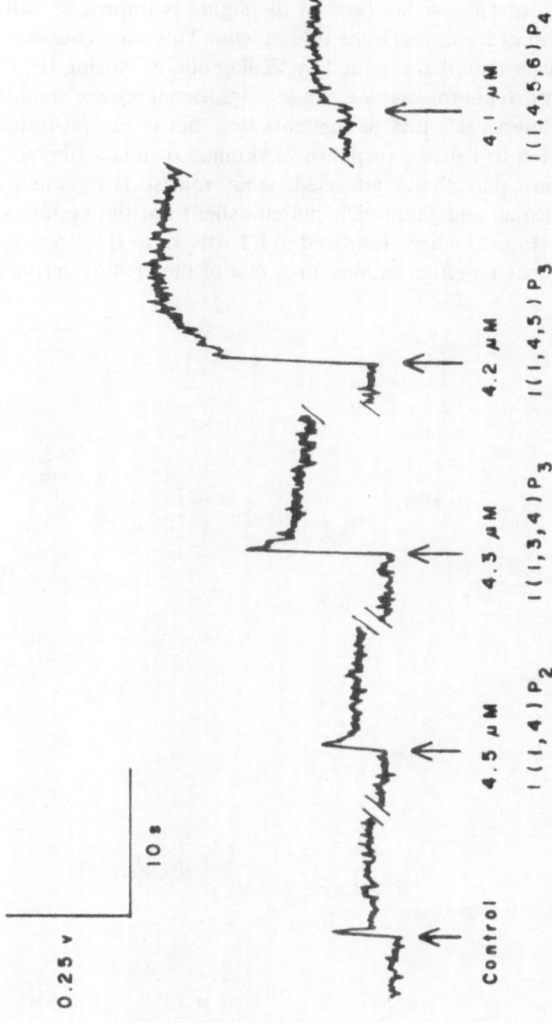

Figure 3. Effect of IP₃ isomeric compounds and metabolites on myoplasmic calcium concentrations. The effect of consecutive additions of I(1,3,4)P₃, I(1,4)P₂, I(1,4,5,6)P₄ and I(1,4,5)P₃ to the same fiber bundle in similar concentrations can be compared to the control signal with buffer saline alone.

4. IP₃ INDUCES A FAST RELEASE SIGNAL AT LOW CONCENTRATIONS

As shown above, a significant light signal was observed when micromolar concentrations of IP₃ were added to the incubation medium containing the skinned fibers in the presence of aequorin. The signal is fast (Fig. 2), and the half rising time is between 55 and 65 msec. The same time course is obtained with caffeine (Fig. 2) or with calcium (not shown) suggesting that the rising time of the signal is limited by diffusion of the agonist into the fibers in order to reach the release site. This time course is at least two orders of magnitude faster than that reported by Walker et al.[13] using IP₃ released from a caged compound by flash photolysis. We can see significant release signals induced by IP₃ concentrations as low as 0.2 μM, a concentration that is several times lower than those previously reported to cause a response in skinned or intact fibers.[1–6]

When other inositol phosphates are used, some release is obtained at relatively large concentrations (Rojas and Jaimovich, unpublished) but the system seems rather specific for I(1,4,5)P₃ (Fig. 3) when compared to I(1,4)P₂ or to I(1,4,5,6)P₄. I(1,3,4)P₃ is also active but produces an effect smaller than that of the (1,4,5) derivative (Fig. 3).

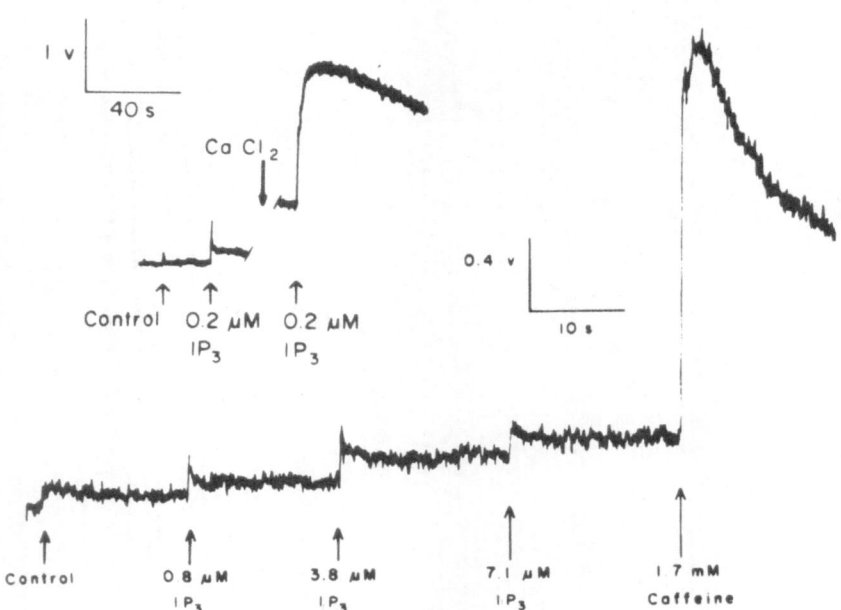

Figure 4. Calcium requirement for IP₃-induced response. Response to consecutive additions of IP₃ and caffeine to fiber bundles incubated in low calcium (lower record) solution. When calcium is added to the medium (upper record) the fiber becomes sensitive to IP₃.

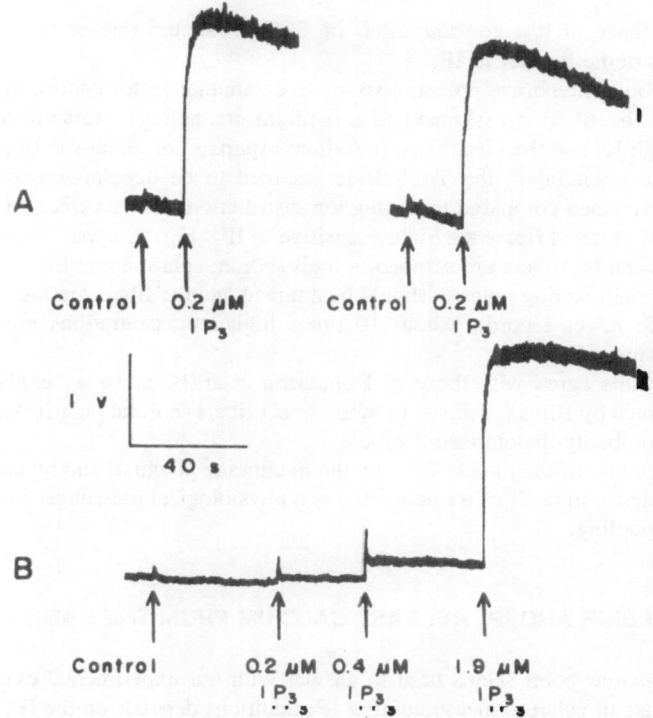

Figure 5. Influence of the ionic composition of the medium on the effect of IP₃ on calcium release. (A) Light signals evoked by low concentrations of IP₃ in fibers skinned in high K and incubated in high Na solutions. (B) Much larger concentrations of IP₃ are needed to induce calcium release in fibers skinned in high Na and incubated in high K solutions

5. THE EFFECT OF IP₃ IS MODULATED BY CALCIUM AND BY THE T-TUBULE MEMBRANE POTENTIAL

As discussed above, concentrations of IP₃ as low as 0.2 μM are able to induce calcium release under certain conditions, but this is not always the case. Lea et al.[11] reported lack of effect of IP₃ in a similar preparation treated with EGTA, and Blinks et al.[12] upon microinjection of IP₃ saw effects on detubulated but not on intact fibers. On the other hand, Donaldson et al. measuring tension in skinned fibers reported that the effect of IP₃ was dependent on the membrane potential of the sealed transverse tubules.[16] In some of our experiments (Fig. 4) the skinned fibers did not respond to relatively large concentrations of IP₃, but they did respond to caffeine as reported by Lea et al.[11] Nevertheless, when calcium was added to the incubation medium and the pCa value was increased to a value of approximately 7,[15] a large signal was obtained in response to fairly low concentrations of IP₃ (Fig. 4, upper record). The calcium effect was also seen when EGTA was added or removed from the incubation medium;[15] that is, in the presence of low concentrations of EGTA, calcium release was obtained in response to caffeine but not to IP_3.

The T-tubule membrane potential seems to be another factor controlling the effect of IP_3; when the fibers are skinned in a medium containing potassium aspartate or glutamate (high K) and then incubated in sodium aspartate, or glutamate (high sodium), the membrane potential in the T-tubule is assumed to be depolarized or presenting reverse polarity when compared to resting ion distribution in intact fibers. Under these conditions, the skinned fibers are highly sensitive to IP_3 (Fig. 5, upper records). On the other hand, when the fibers are skinned in high-sodium solution and incubated in high potassium, normal resting polarity should be attained and the fibers are less sensitive to IP_3 (Figure 5, lower record). About 10 times higher concentrations are needed to produce the same effect.

These results agree with those of Donaldson et al.[16] and may explain the differences obtained by Blinks et al.[12] between intact fibers (normal polarity) and detubulated fibers (probably depolarized T-tubules).

The regulation of the effect of IP_3 by the membrane potential and by calcium is an important evidence in favor of a role for IP_3 as a physiological messenger in excitation–contraction coupling.

6. DO CAFFEINE AND IP_3 RELEASE CALCIUM FROM THE SAME POOL?

This particular point seems hard to answer with our experimental evidence. The maximal release of calcium measured after IP_3 additions depends on the IP_3 concentration used (Fig. 6) as it does when caffeine is used. Nevertheless, the maximal signals induced by caffeine are always larger than the IP_3-induced signals. This may be due to the fact that IP_3 and caffeine may be releasing calcium from somehow different pools or it may be explained by the presence of inositol trisphosphatase (see Chapter 30, this volume) that would never allow IP_3 to act on the receptor for a time long enough to release all the available calcium. There may be as well a buffer capacity for IP_3 in the fiber, represented by relatively low-affinity binding sites; these combined effects would never allow IP_3 concentrations to increase over certain values near its receptor site.

In summary, the results discussed here agree with those of Vergara et al.,[1,2] Volpe et al.,[3,4] Nosek et al.,[5] and Donaldson et al.[6] showing that IP_3 releases calcium from skeletal muscle fibers. This release is fast (the actual kinetics remain to be determined) and can occur at fairly low IP_3 concentrations; the effect is rather specific for $I(1,4,5)P_3$ as compared to other inositol phosphates, and the amount of calcium released by IP_3 is generally smaller but of the same order than that released by caffeine. Interestingly, the effect of IP_3 requires a minimal calcium concentration $(10^{-7} M)$ and is modulated by the membrane potential from the transverse tubules.

All of these evidences have to be taken together with those discussed in other chapters of this volume (Hidalgo et al., Vergara et al., Suarez-Isla et al.), namely, that the enzymes that metabolize the phosphoinositides and the inositol phosphates are present in the muscle fibers, most of them localized at the T-tubule membrane; that IP_3 is actually produced upon muscle stimulation; that the time course of excitation–contraction coupling is compatible with a chemical process; and that IP_3 activates the calcium

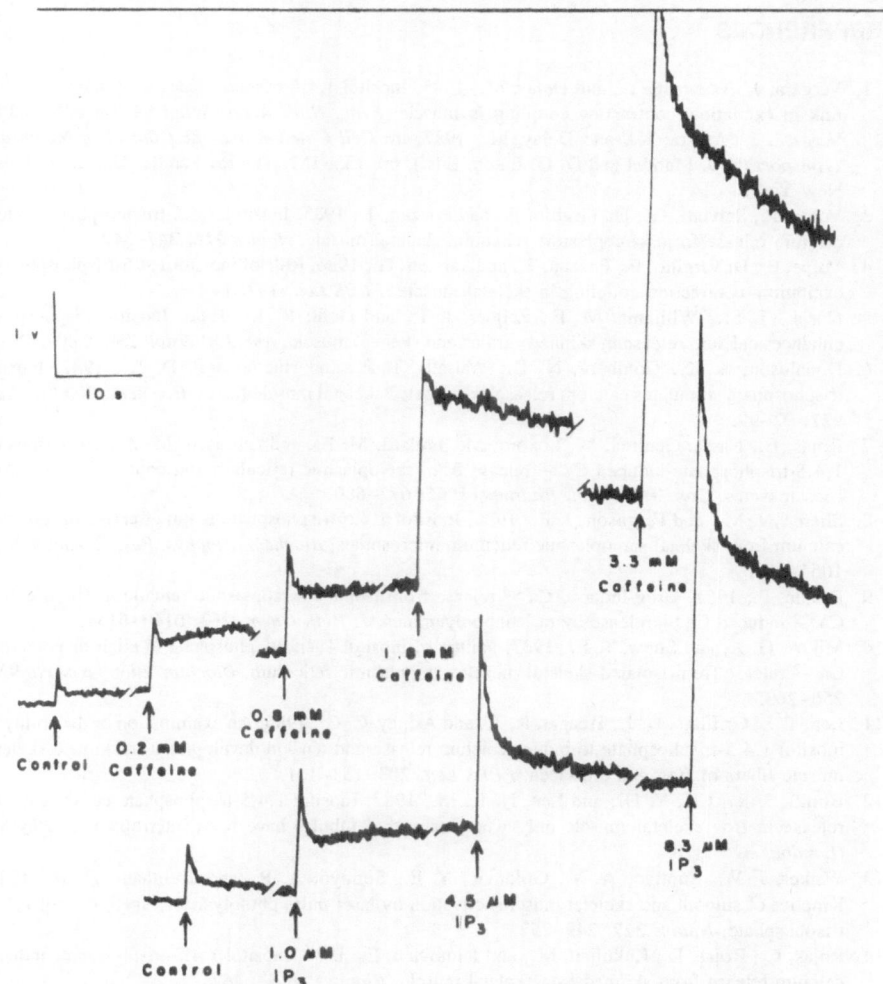

Figure 6. Effect of various concentrations of caffeine (upper record) and IP₃ (lower record) on calcium release. Maximal doses of IP₃ produce smaller signals than caffeine.

release channel from sarcoplasmic reticulum. It has to be concluded that a role for IP₃ in the regulation of calcium release in skeletal muscle is highly likely. Whether IP₃ is the prime messenger in excitation–contraction coupling during a single twitch still awaits to be demonstrated using techniques with higher sensitivity and time resolution than those described here.

ACKNOWLEDGMENTS. This work was financed by the Muscular Dystrophy Association of America, NIH GM 35981, FONDECYT 972, and Universidad de Chile DTI 2123.

REFERENCES

1. Vergara, J., Tsien, R. Y., and Delay, M., 1985, Inositol 1,4,5-trisphosphate: A possible chemical link in excitation–contraction coupling in muscle, *Proc. Natl. Acad. Sci. USA* **82:** 6352–6356.
2. Vergara, J., Asotra, K., and Delay, M., 1987, in: *Cell Calcium and the Control of Membrane Transport* (L. J. Mandel and D. G. Eaton, Eds.), pp 133–151, The Rockefeller University Press, New York.
3. Volpe, P., Salviati, G., Di Virgilio, F., and Pozzan, T., 1985, Inositol 1,4,5-trisphosphate induces calcium release from sarcoplasmic reticulum skeletal muscle, *Nature* **316:** 347–349.
4. Volpe, P., Di Virgilio, F., Pozzan, T., and Salviati, G., 1986, Role of inositol 1,4,5-trisphosphate in excitation–contraction coupling in skeletal muscle, *FEBS Lett.* **197:** 1–4.
5. Nosek, T. M., Williams, M. F., Zeigler, J. T., and Godt, R. E., 1986, Inositol trisphosphate enhances calcium release in skinned cardiac and skeletal muscle, *Am. J. Physiol.* **250:** C807–C810.
6. Donaldson, S. K., Goldberg, N. D., Walseth, T. F., and Huettemann, D. A., 1987, Inositol trisphosphate stimulates calcium release from peeled skeletal muscle fibers, *Biochem. Biophys. Acta* **927:** 92–99.
7. Rojas, E., Nassar-Gentina, V., Luxoro, M., Pollard, M. E., and Carrasco, M. A., 1987, Inositol 1,4,5-trisphosphate-induced Ca^{+2} release from sarcoplasmic reticulum and contraction in crustacean muscles, *Can. J. Physiol. Pharmacol.* **65:** 672–680.
8. Sherer, N. M., and Ferguson, J. E., 1985, Inositol 1,4-5-trisphosphate is not effective in releasing calcium from skeletal sarcoplasmic reticulum microsomes, *Biochem. Biophys. Res. Commun.* **128:** 1064–1070.
9. Palade, P., 1987, Drug-induced Ca^{+2} release from isolated sarcoplasmic reticulum. III. Block of Ca^{+2}-induced Ca^{+2} release by organic polyamines, *J. Biol. Chem.* **262:** 6149–6154.
10. Mikos, G. J., and Snow, T. R., 1987, Failure of inositol 1,4,5-trisphosphate to elicit or potentiate Ca^{+2} release from isolated skeletal muscle sarcoplasmic reticulum, *Biochim. Biophys. Acta* **927:** 256–260.
11. Lea, T. J., Griffiths, D. J., Treagar, R. T., and Ashley, C C., 1986, An examination of the ability of inositol 1,4,5-trisphosphate to induce calcium release and tension development in skinned skeletal muscle fibers of frog and crustacea, *FEBS Lett.* **203:** 153–161.
12. Blinks, J. R., Cai, Y. D., and Lee, N. K. M., 1987, Inositol 1,4,5-trisphosphate causes calcium release in frog skeletal muscle only when transverse tubules have been interrupted, *J. Physiol. (London)* **394:** 39P.
13. Walker, J. W., Somlyo, A. V., Goldman, Y. E., Somlyo, A. P., and Trentham, D. R., 1987, Kinetics of smooth and skeletal muscle activation by laser pulse photolysis of caged inositol 1,4,5-trisphosphate, *Nature* **327:** 249–252.
14. Rojas, C., Rojas, E., Kukuljan, M., and Jaimovich, E., 1988, Inositol 1,4,5-trisphosphate induces calcium release from skinned frog skeletal muscle, *Biophys. J.* **53:** 467a.
15. Rojas, C., and Jaimovich, E., 1990, Calcium release modulated by inositol trisphosphate in skinned fibers from frog skeletal muscle, *Eur. J. Physiol.* (in press).
16. Donaldson, S. K., 1988, Transverse tubule voltage control of inositol trisphosphate-induced Ca^{2+} release in peeled skinned muscle fibers, *Biophys. J.* **53:** 468a.

Chapter 30

Metabolism of Phosphoinositides in Skeletal Muscle Membranes

Cecilia Hidalgo, Ximena Sánchez, and M. Angélica Carrasco

1. INTRODUCTION

The activation of a number of cell surface receptors results in the formation of inositol phosphates and diacylglycerol (DAG) through the cleavage of membrane phosphoinositides by phospholipase-C.

The role of inositol 1,4,5-trisphosphate (IP_3) and DAG as second messengers has been well established and is described in detail in several recent reviews.[1-5] IP_3 mobilizes calcium from intracellular nonmitochondrial stores[6] (Chapter 13, this volume), and DAG is an essential cofactor for protein kinase-C.[7,8]

The metabolic pathways of phosphoinositides and inositol phosphates are shown in Fig. 1. The numerous studies in the field have led to an increasingly complex scheme. Phosphatidylinositol (PI), which accounts for around 90% of the total phosphoinositides present in the cell membranes, is sequentially phosphorylated to phosphatidylinositol 4-phosphate (PIP) and to phosphatidylinositol (4,5-bisphosphate (PIP_2). Phosphatases degrade them back to PI. Recently, a third phosphatidyl-inositol, phosphatidylinositol 3-phosphate, has been described.[9]

The primary event after receptor stimulation is the hydrolysis of PIP_2 to DAG and

CECILIA HIDALGO • Departamento de Fisiología y Biofísica, Facultad de Medicina, Universidad de Chile, and Centro de Estudios Científicos de Santiago, Santiago, Chile. XIMENA SÁNCHEZ and M. ANGÉLICA CARRASCO • Departamento de Fisiología y Biofísica, Facultad de Medicina, Universidad de Chile, Santiago, Chile.

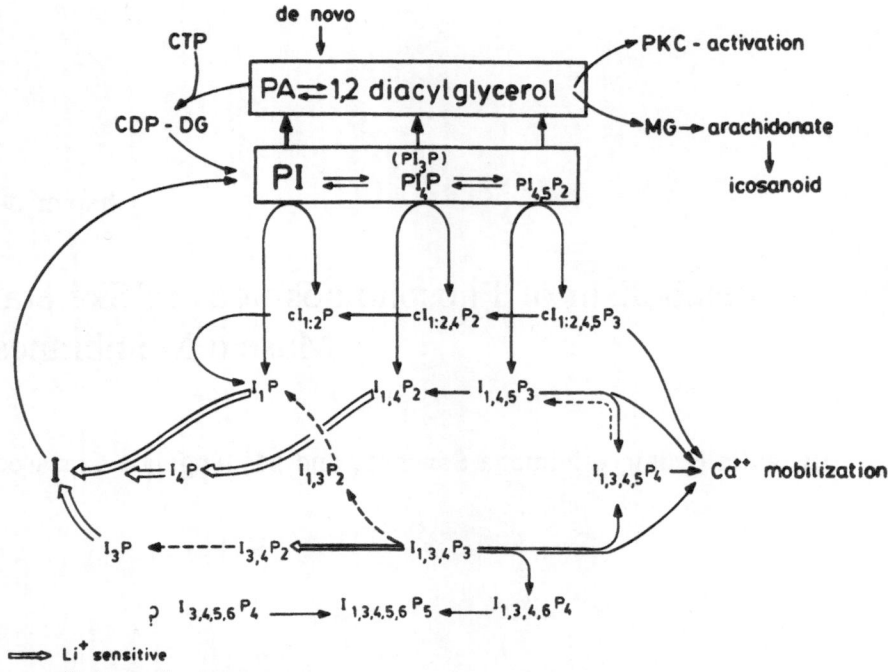

Figure 1. Pathways of phosphoinositide and inositol phosphate metalbolism. The membrane-bound lipid components are shown in boxes at the top. PI: phosphatidylinositol; PI_3PL phosphatidylinositol 3-phosphate; PI_4: phosphatidylinositol 4-phosphate; $PI_{4,5}P_2$: phosphatidylinositol (4,5)-bisphosphate; PA: phosphatidic acid, and CDP-DG: cytidine diphosphate diacylglycerol. On the lower levels are shown the inositol phosphates: I, inositol; P, phosphate. The numbers preceding P represent the positions of the phosphate groups, and those following P represent the number of phosphate groups. PKC, protein kinase-C and MG, 2-monoacylglycerol. (Adapted from Ref. 5.)

IP_3. This process requires calcium in the nanomolar range and involves GTP-binding proteins.[10] There are at least three different GTP-binding proteins responsible for regulating receptor–phospholipase-C coupling, as evidenced by studies carried out with bacterial toxins.[11]

Inositol monophosphate (IP) and inositol (1,4)-bisphosphate (IP_2) are produced either by degradation of IP_3 by phosphomonoesterases, or by the cleavage of PI and PIP, respectively. The 1:2 cyclic counterparts of these inositol phosphates have been studied less, mainly due to technical difficulties.[5]

In addition to being degraded to IP_2,[12,13] IP_3 is phosphorylated to inositol (1,3,4,5)-tetraphosphate (IP_4), followed by the cleavage of the phosphate in the 5 position to yield inositol (1,3,4)-phosphate.[14–17] This metabolic route has been studied extensively, and it has been shown in some cells that calcium modulates this 3-kinase activity. This finding implies that following stimulation, the rise in internal calcium concentration favors the phosphorylation of IP_3 over its degradation to IP_2. Actually, the levels of IP_3 and of inositol (1,3,4)-trisphosphate increase with cell stimulation.[18,19] It has been proposed that IP_4 acts together with IP_3 as a second messenger for promoting extracellular calcium entry,[20,21] and by itself as a promoter of intracellular calcium sequestration.[22] Inositol (1,3,4)-trisphosphate has not yet been ascribed a physiological function. IP_4 can also be dephosphorylated to IP_3, as described recently.[23]

Two other isomers of IP_4 also increase following cell stimulation: inositol (3,4,5,6)-tetraphosphate, of unknown metabolic origin, and inositol (1,3,4,6)-tetraphosphate, derived from phosphorylation of inositol (1,3,4)-trisphosphate.[24-26] These two isomers of IP_4 are precursors of inositol (1,3,4,5,6)-pentaphosphate (IP_5).[26] The presence of both IP_5 and inositol hexaphosphate has been reported in several cell systems.[27,28]

2. METABOLISM OF PHOSPHOINOSITIDES IN SKELETAL MUSCLE

In analogy with other cell systems, IP_3 was proposed in 1985 as a chemical messenger responsible for excitation–contraction coupling in skeletal muscle (see Chapters 27 and 28, this volume). It remained to be established, however, whether the metabolic machinery responsible for the production and hydrolysis of IP_3 was present in skeletal muscle, as postulated in the model proposed by Vergara et al.[29] and Volpe et al.[30]

Biochemical studies have been carried out in muscle cells and in isolated muscle membranes, with the general aim of investigating the metabolism of phosphoinositides in these cells. However, contrary to the considerable information regarding phosphoinositide metabolism in other cell systems, much less is known in skeletal muscle, and more experiments are needed to clarify in detail the different aspects of the cycle in this cell system.

In this chapter, we review published findings regarding phosphoinositide metabolism in membranes isolated from skeletal muscle, and we present a description of some new findings obtained in our laboratory. For a discussion of cellular aspects, see Chapter 27, this volume.

2.1. Phosphoinositide Kinases

2.1.1. Localization

T–T membranes isolated from frog skeletal muscle phosphorylate endogenous PI to PIP and PIP_2, whereas sarcoplasmic reticulum (SR) membranes only form PIP but not PIP_2 (Table I).[31,32] Similar observations have been reported using rabbit muscle membranes.[32-35] Thus, the kinase responsible for the formation of PIP_2, the direct precursor of IP_3, seems to be restricted to the T–T membranes.

2.1.2. Calcium Modulation

Both PI kinase and PIP kinase, the enzymes responsible for the formation of PIP and PIP_2, respectively, are regulated by calcium in frog skeletal muscle.[32] In T–T and SR membranes, PI kinase activity is maximal at calcium concentrations higher than 1

Table I. Phosphorylation of Phosphatidylinositol by Transverse
Tubule and Sarcoplasmic Reticulum Membranes[a]

	PIP	PIP$_2$
	(pmol per μmol lipid Pi)	
Transverse tubules	57.7 $\pm$ 7.1 (3)	23.4 $\pm$ 7.2 (3)
Sarcoplasmic reticulum	7.1 $\pm$ 1.8 (2)	0.4 $\pm$ 0.1 (2)

[a]Transverse tubule and sarcoplasmic reticulum vesicles were obtained as described by Hidalgo et al.[36] Membranes were incubated for 1 min with 0.5 mM
[γ-^{32}P]-ATP (0.35 Ci/mmol) at 25°C in a solution containing 0.1 M KCl, 5.0
mM MgCl$_2$, 20 mM tris maleate, pH 7.0, pCa 6.0. For further experimental
details, see Ref. 31. Numbers represent mean $\pm$ S.D. The number of preparations studied is given in parentheses.

μM, decreasing to 20–10% at mM calcium concentrations ($K_{0.5} = 5\ \mu M$). The PIP
kinase activity of T–T membranes has a different calcium dependence, with maximal
activity at calcium concentrations higher than 2 μM and decreasing to 30% at calcium
concentration lower than 0.2 μM ($K_{0.5} = 1\ \mu M$). As mentioned above, SR membranes
do not exhibit PIP kinase activity in the entire range of calcium concentrations tested.[32]

If the behavior of the *in vitro* system reflects the situation *in vivo*, the increase in
PIP$_2$ formation observed following an increase in calcium concentration from 0.1 μM to
2 μM might have physiological implications. The PIP$_2$ present in resting conditions
might suffice to form enough IP$_3$ for one or a few twitches: the ensuing increase in
intracellular calcium following stimulation would increase PIP$_2$ formation, ensuring an
adequate supply for repetitive stimulation.

2.1.3. K$_m$ for ATP

There are no published results describing values for the K$_m$ for ATP of either
skeletal muscle kinase. We have found that the time course of ^{32}P incorporation from the
gamma phosphate of [γ-^{32}P]-ATP into PIP and PIP$_2$ shows rapid labeling of both
phosphoinositides (Fig. 2). Frog skeletal muscle T–T membranes have a high Mg-
ATPase activity[36] that hydrolyzes most of the ATP present in the solution in a few
seconds, concomitant with the leveling off of the phosphorylation reactions (Fig. 2).
Thus, to maintain a constant substrate concentration to measure the K$_m$ for ATP the Mg-
ATPase activity had to be inhibited. We used Triton X-100 for this purpose, since this
detergent inhibits the Mg-ATPase[37] but not the kinases.[38–40]

Higher amounts of PIP and PIP$_2$ were formed in the presence of Triton X-100 (Fig.
3) than in the controls (Table I). A study of the rate of PIP formation as a function of
[ATP] in the presence of Triton X-100 allowed us to determine an apparent K$_m$ for the PI
kinase of 0.1 mM (Fig. 4). This value is similar to other K$_m$ values reported in the
literature for this enzyme.[40,41] We also observed that the formation of PIP$_2$ increased
with increasing [ATP] (Fig. 4) but in this case it was not possible to estimate a value for
a K$_m$ since the amounts of PIP formed by the membrane are low and could be limiting
the rate of PIP$_2$ formation.

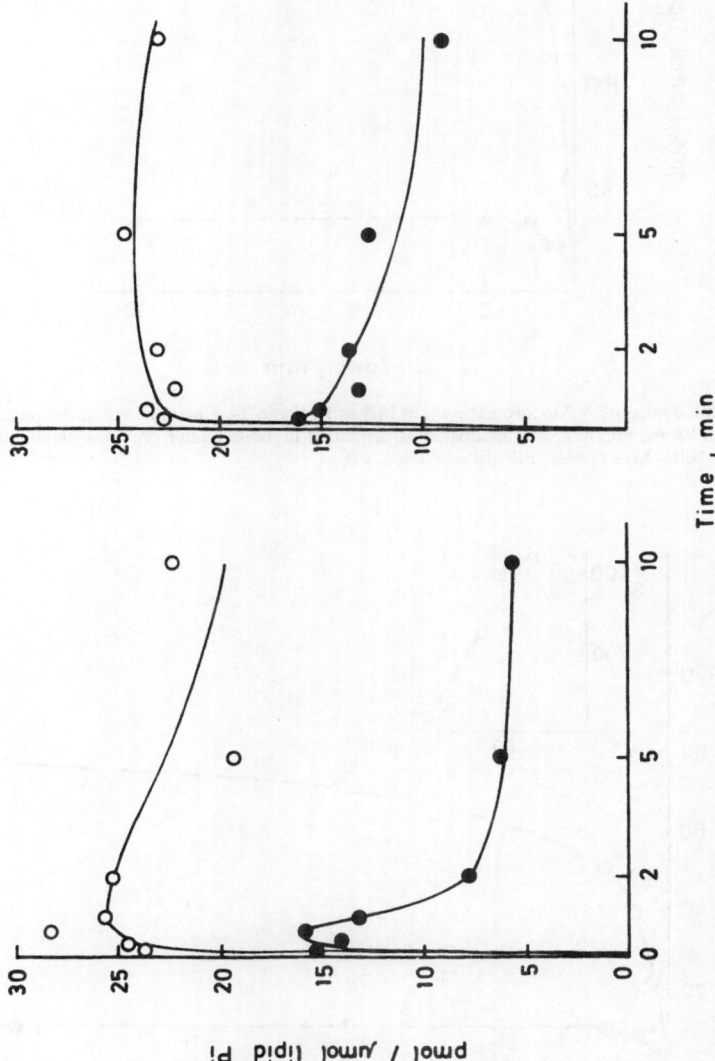

Figure 2. Time course of ^{32}P incorporation into PIP and PIP_2 in two different T–T membrane preparations. Experimental conditions were as described in Table I. Open circles: PIP; filled circles: PIP_2.

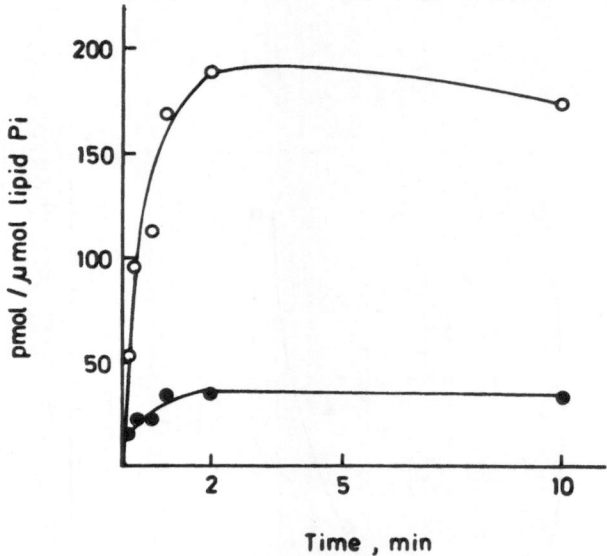

Figure 3. Time course of ^{32}P incorporation into PIP and PIP_2 in T–T membranes in the presence of Triton X-100. T–T membranes were incubated as described in Table I after previous incubation with 0.1% Triton X-100. Open circles: PIP; filled circles: PIP_2.

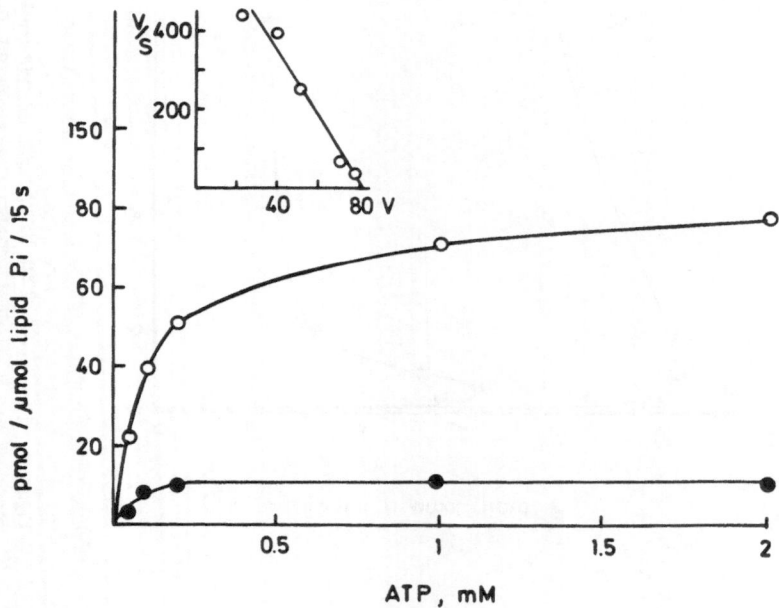

Figure 4. Incorporation of ^{32}P into PIP and PIP_2 in T–T membranes at different ATP concentrations. T–T vesicles were incubated with varying [ATP] in the presence of 0.1% Triton X-100. Open circles: PIP; filled circles: PIP_2. Inset figure: Eadie–Scatchard plot for the PIP formation.

2.1.4. Effect of Excitation–Contraction Coupling Modulators

Scant information is available in regard to the effect of excitation–contraction coupling modulators on the activity of both kinases. We have carried out experiments to investigate the effect of blockers of voltage-dependent calcium channels, because the dihydropyridine receptors of skeletal muscle T-tubules have been proposed as the voltage sensors involved in excitation–contraction coupling (addition of dihydropyridines to muscle fibers inhibits charge movements and calcium release from SR; Ref. 42 and Chapter 25, this volume). We tested the effect of different types of calcium-channel blockers on the activity of the kinases and we found that neither nifedipine nor verapamil affected the activity of these enzymes (Table II).

Of the other excitation–contraction coupling modulators tested, we found that dantrolene had no effect on the kinases, while tetracaine stimulated the production of PIP (Table II and Ref. 31). The stimulatory effect of tetracaine on PIP formation could reflect inhibition of PIP kinase activity. If this were the case, one could explain the inhibition of excitation–contraction coupling produced by tetracaine as inhibition of phosphoinositide metabolism.[31] This hypothesis should be tested experimentally.

2.1.5. Effect of GTP Analogs

A role for GTP-binding proteins on excitation–contraction coupling has been proposed.[43] In agreement with recent findings reported by Varsanyi et al.,[34] we have found that nonhydrolyzable analogs of GTP had no effect on the activity of both kinases (Table II). Furthermore, addition of GTP analogs to T–T membranes previously phosphorylated with $[\gamma\text{-}^{32}\text{P}]$-ATP had no effect on the levels of PIP or PIP_2 (not shown), indicating that in these membranes phospholipase-C activity is not activated by GTP

Table II. Phosphorylation of Phosphatidylinositol by Transverse Tubule Membranes: Effect of Excitation–Contraction Coupling Modulators and GTP Analogs[a]

	PIP	PIP_2
	(% of control)	
Nifedipine, 1 μM	117	113
Nifedipine, 5 μM	92	125
Verapamil, 10 μM	103	124
Dantrolene, 50 μM	107	108
Dantrolene, 500 μM	110	100
Tetracaine, 0.1 mM	83	76
Tetracaine, 0.5 mM	187	105
Tetracaine, 2.0 mM	330	96
Gpp(NH)p, 1 μM	115	114
Gpp(NH)p, 100 μM	105	98
GTP-γ-S, 100 μM	138	95

[a]Experimental conditions were as described in Table I.

analogs. This finding suggests that G-proteins are not involved in the metabolism of phosphoinositides in these membranes.

2.2. Phospholipase-C Activity

2.2.1. Studies Using Exogenous Substrate

Partially purified T–T membranes from rabbit skeletal muscle display phospholipase-C activity toward exogenous $[^3H]$-PIP_2[44]; this reaction is stimulated by 0.1 mM calcium and by 10 mM NaF. We have also found that T–T membranes from frog skeletal muscle hydrolyze exogenous $[^3H]$-PIP_2. The decrease of $[^3H]$-PIP_2 is not accompanied by appearance of $[^3H]$-PIP, indicating that we are measuring phospholipase-C activity and not PIP_2 phosphatases.

Using exogenous $[^3H]$-PIP_2 as substrate, Varsanyi et al.[34] found values for phospholipase-C activity in isolated T–T membranes of the order of 10 pmol per min per mg of protein at 25°C, when measured in the presence of μM $CaCl_2$. This rate is at least three orders of magnitude lower than that needed to raise the concentration of IP_3 in the entire muscle fiber to 1 μM.[34] However, if we consider only the volume comprised in the triadic space (one thousand times smaller than that of the fiber), the amounts of IP_3 produced in one twitch would be sufficient to raise the concentration of IP_3 in the triad to 1 μM, a concentration high enough to activate calcium release in skinned muscle fibers (see Ref. 45 and Chapter 29, this volume) and to activate SR calcium channels.[46]

2.2.2. Studies Using Endogenous Substrate

There are no published studies of phospholipase-C activity measured with endogenous substrate, except for a short communication (see below). We have found that in the absence of detergent the incorporation of ^{32}P into PIP and PIP_2 levels off in a few seconds due to substrate depletion. Thereafter, the levels of ^{32}P–PIP remain relatively constant with time, while the levels of ^{32}P–PIP_2 decrease to variable extents (Fig. 2), most likely by hydrolysis by phospholipase-C. The variability of the rates of PIP_2 decay observed in different membrane preparations might reflect the lability of this enzyme. The highest rate of decline of endogenous ^{32}P-labeled PIP_2 was 10 pmol per mg of protein per min, orders of magnitude lower that the rates of 10 nmol per mg per min reported by Varsanyi et al.[34] with exogenous substrate. One possibility to explain why we measured such low activity would be the fact that mM Mg drastically inhibits the phospholipase-C activity of isolated T–T membranes.[34] Because Mg is needed to form PIP and PIP_2,[31] we used 5 mM $MgCl_2$ in all our assays with endogenous substrate.

A direct demonstration of phospholipase-C activity requires measurement of ^{32}P–IP_3 production concomitant to the decrease of ^{32}P–PIP_2 levels. Although it is difficult to separate the residual levels of $[\gamma\text{-}^{32}P]$-ATP used as substrate from the ^{32}P–IP_3 produced during the reaction, the HPLC separation of nucleotides and IP_3 has been accomplished, demonstrating significant phospholipase-C activity in a similar T–T membrane preparation by direct measurement of the IP_3 produced.[47]

Neomycin binds to polyphosphoinositides,[48,49] rendering them unavailable to enzyme action, a property that has been widely used in the studies related to phos-

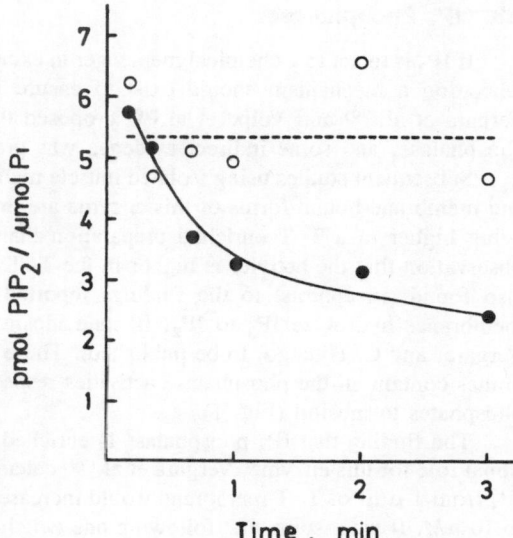

Figure 5. Effect of neomycin on ^{32}P incorporation into PIP_2 in T–T membranes. Membranes were incubated as described in Table I after previous incubation with 1 mM neomycin. Open circles: 1 mM neomycin; filled circles: control.

phoinositide metabolism. The decay of ^{32}P–PIP_2 with time, a reaction most likely due to phospholipase-C action, was blocked by 1 mM neomycin (Fig. 5), while PIP was not affected. Incorporation of ^{32}P from $[\gamma\text{-}^{32}P]$-ATP into both PIP and PIP_2 was considerably increased by concentrations of neomycin of 5–10 mM. This stimulation is not due to an increase in the effective concentration of $[\gamma\text{-}^{32}P]$-ATP, because the Mg-ATPase activity of T–T was not affected by neomycin. Neomycin concentrations in the millimolar range are required for the interaction with PIP_2 in biological membranes.[50] The increase in PIP and PIP_2 formation observed after addition of 5–10 mM neomycin might represent inhibition of hydrolysis of both phosphoinositides, whereas the specific inhibition of PIP_2 hydrolysis caused by 1 mM neomycin (Fig. 5) might simply reflect the higher affinity of neomycin for PIP_2.[51]

It has been reported that neomycin blocks excitation–contraction coupling.[29] It is not known whether this effect is due to inhibition of IP_3 formation by inhibition of PIP_2 phospholipase-C activity or is caused by direct inhibition of calcium release from SR (see Chapter 26, this volume).

2.2.3. Role in Excitation–Contraction Coupling

It remains to be established whether the phospholipase-C activity is directly controlled by the T–T membrane potential, as originally proposed,[29,30] or if other mechanisms exist to release IP_3 from the T-tubules in response to depolarization (see Ref. 52 for a discussion of this point). Nevertheless, the rates measured by Varsanyi et al.[34] are certainly high enough to account for sufficient IP_3 production in the triad to activate calcium release from SR.

2.3. IP$_3$ Phosphatase

If IP$_3$ is to act as a chemical messenger in excitation–contraction coupling, to allow relaxation a mechanism should exist to ensure its removal from the triadic space. Vergara et al.[29] and Volpe et al.[30] proposed that IP$_3$ would be hydrolyzed by IP$_3$ phosphatase, and some indirect evidence was presented supporting this proposal.[29]

Subsequent studies using isolated muscle membranes have shown that both soluble and membrane-bound forms of this enzyme are present in skeletal muscle, the activity being higher in a T–T enriched preparation than in SR.[53] We have confirmed the observation that the activity is higher in the T–T membranes (Table III), and we have also found, in contrast to the findings reported by Milani et al.,[53] that the T–T membranes hydrolyze IP$_3$ to IP$_2$, IP, and inositol (X. Sánchez, M. A. Carrasco, J. Vergara, and C. Hidalgo, to be published). These results indicated that the T–T membranes contain all the phosphatase activities responsible for the degradation of inositol phosphates to inositol (Fig. 1).

The finding that IP$_3$ phosphatase is enriched in the T–T system suggests a functional role for this enzyme. Vergara et al.[29] calculated that release of 100 molecules of IP$_3$ from 1 μm^2 of T–T membrane would increase its concentration in the triadic space to 10 μM. If we assume that following one twitch 10 molecules of IP$_3$ are released per μm^2 to reach 1 μM in the triad, an IP$_3$ase rate of 2 molecules per μm^2 per msec would be needed to remove all the IP$_3$ released in 5 msec. This calculated rate corresponds to 3 $\times$ 10^{-12} nmol of IP$_3$ hydrolyzed per μm^2 per sec. Assuming that there are 10^6 molecules of phospholipid per μm^2 of T–T surface,[54] we can further convert this value to 2 nmol of IP$_3$ hydrolyzed per mg of protein per sec, since the T–T membranes contain 2 μmol of phospholipid per mg of protein.[36]

The IP$_3$ase rates measured by us, 40 nmol per mg per min, or 0.7 nmol per mg per sec, are in this range (Table III). However, the Km for IP$_3$ is relatively high, 20 μM (Ref. 53, and our own observations). If this enzyme is responsible for the removal of IP$_3$ required for relaxation, it should hydrolyze IP$_3$ at rates compatible with the physiological relaxation rates and at the low substrate concentrations (0.5–1 μM) that activate the SR calcium channel. Accordingly, we have measured IP$_3$ase activity in T–T membranes

Table III. Hydrolysis of [^{3}H]-Inositol 1,4,5-Trisphosphate by Isolated Transverse Tubule and Sarcoplasmic Reticulum Membranes[a]

	Hydrolysis rate (nmol per min per mg of protein)
Transverse tubules	41.0 ± 13.4 (3)
Sarcoplasmic reticulum	7.4 ± 0.6 (3)

[a]Membranes were incubated at 25°C in a solution containing 0.1 M KCl, 3 mM MgCl$_2$, 0.2 mM EGTA, 50 mM HEPES-tris, pH 7.2, and 0.1 mg of protein per ml. The reaction was started by addition of 40 μM [^{3}H]-myoinositol 1,4,5-trisphosphate (final concentration), and was stopped by adding ice-cold 20% trichloroacetic acid. After brief centrifugation, the supernatant was extracted with 4 × 3.5 ml of diethyl ether, neutralized with NH$_4$OH, and loaded on Dowex-1 columns (formate form). Elution was performed as described in Ref. 12. Data are given as mean ± S.D.

using low concentrations of IP_3 (0.5–1.0 μM), and although we found significant activity even in in these conditions, with values ranging from 30–50 pmol per mg per sec, the rates are lower than those needed to account for relaxation. It is possible that (i) the *in vivo* rates are higher, since we might be missing physiological cofactors in the *in vitro* assays that might produce lower *in vitro* activities, (ii) the K_m is regulated so that lower K_m values might be obtained in physiological conditions, or (iii) these results imply that the IP_3 phosphatase is not involved in the removal of IP_3 and that other mechanisms exist to decrease the IP_3 levels to the resting values.

2.4. Incorporation of [^{3}H]-Inositol into PI

As a further step in investigating phosphoinositide metabolism in muscle membranes, we incubated SR and T–T membranes with [^{3}H]-inositol under different experimental conditions. We found that SR membranes incorporated [^{3}H]-inositol into PI more efficiently than the T–T membranes, and that this incorporation required addition of either exogenous CDP–DG or of phosphatidic acid plus ATP to the reaction solution. Negligible [^{3}H]-inositol incorporation took place into PIP or PIP_2, most likely due to the very small fraction that these lipids represent relative to PI. PI synthesis was stimulated by ATP and by $MnCl_2$ (not shown). Concomitant with PI synthesis release of IP_1 to the aqueous phase took place, indicating that the isolated T–T membranes contain the phospholipase-C that hydrolyzes PI to DAG and IP_1 (to be published).

To our knowledge, there are no other studies regarding the incorporation of [^{3}H]-inositol in isolated muscle membranes, although some information exists in whole muscle cells (Chapter 27, this volume).

ACKNOWLEDGMENTS.   We wish to thank Drs. Enrique Jaimovich and Julio Vergara for their help in setting up experimental procedures and for many helpful discussions during the course of this work. This work was supported by National Institutes of Health grant GM35981, by a grant from the Muscular Dystrophy Association of America, by a grant from the Tinker Foundation, Inc., to the Centro de Estudios Cientificos de Santiago, by Universidad de Chile grant DIB-2149, and by FONDECYT grant 972.

REFERENCES

1. Majerus, P. W., Connolly, T. M., Deckmyn, H., Ross, T. S., Bross, T. E., Ishii, H., Bansal, V. S., and Wilson, D. B., 1986, The metabolism of phosphoinositide-derived messenger molecules, *Science* **234:** 1519–1526.
2. Sekar, M. C., and Hokin, L. E., 1986, The role of phosphoinositides in signal transduction, *J. Membrane Biol.* **89:** 193–210.
3. Berridge, M. J., 1987, Inositol trisphosphate and diacylglycerol: Two interacting second messengers, *Ann. Rev. Biochem.* **56:** 159–193.
4. Berridge, M. J., 1988, Inositol lipids and calcium signalling, *Proc. Royal Soc. London* **234:** 359–378.
5. Majerus, P. W., Connolly, T. M., Bansal, V. S., Inhorn, R. C., Ross, T. S., and Lips, D. L., 1988, Inositol phosphates: Synthesis and degradation, *J. Biol. Chem.* **263:** 3051–3054.
6. Guillemette, G., Balla, T., Baukal, A. J., and Catt, K. J., 1988, Characterization of inositol 1,4,5-

trisphosphate receptors and calcium mobilization in a hepatic plasma membrane fraction, *J. Biol. Chem.* **263**: 4541–4548.

7. Nishizuka, Y., 1984, The role of protein kinase-C in cell surface signal transduction and tumor promotion, *Nature* **308**: 693–697.

8. Nishizuka, Y., 1986, Studies and perspectives of protein kinase-C, *Science* **233**: 305–312.

9. Whitman, M., Downes, C. P., Keeler, M., Keller, T., and Cantley, L., 1988, Type I phosphatidylinositol kinase makes a novel inositol phospholipid, phosphatidylinositol-3-phosphate, *Nature* **332**: 644–646.

10. Cockcroft, S., 1986, The dependence on Ca^{2+} of the guanine-nucleotide-activated polyphosphoinositide phosphodiesterase in neutrophil plasma membranes, *Biochem. J.* **240**: 503–507.

11. Lo, W. W. Y., and Hughes, J., 1987, Receptor–phosphoinositidase-C coupling: Multiple G-proteins? *FEBS Lett.* **224**: 1–3.

12. Downes, C. P., Mussat, M. C., and Michell, R. H., 1982, The inositol trisphosphate phosphomonoesterase of the human erythrocyte membrane, *Biochem. J.* **203**: 169–177.

13. Storey, D. J., Shears, S. B., Kirk, C. J., and Michell, R. H., 1984, Stepwise enzymatic dephosphorylation of inositol 1,4,5-trisphosphate to inositol in liver, *Nature* **312**: 374–376.

14. Batty, I. R., Nahorski, S. R., and Irvine, R. F., 1985, Rapid formation of inositol 1,3,4,5-tetrakisphosphate following muscarinic receptor stimulation of rat cerebral cortical slices, *Biochem. J.* **232**: 211–215.

15. Burguess, G. M., McKinney, J. S., Irvine, R. F., and Putney, Jr., J. W., 1985, Inositol 1,4,5-trisphosphate and inositol 1,3,4-trishosphate formation in Ca^{2+}-mobilizing hormone-activated cells, *Biochem. J.* **232**: 237–243.

16. Irvine, R. F., Anggard, E. E., Letcher, A. J., and Downes, C. P., 1985, Metabolism of inositol 1,4,5-trisphosphate and inositol 1,3,4-trisphosphate in rat parotid glands, *Biochem. J.* **229**: 505–511.

17. Hansen, C. A., Mak, S., and Williamson, J. R., 1986, Formation and metabolism of inositol 1,3,4,5-tetrakisphosphate in liver, *J. Biol. Chem.* **261**: 8100–8103.

18. Biden, T. J., and Wollheim, C. B., 1986, Ca^{2+} regulates the inositol tris/tetrakisphosphate pathway in intact and broken preparations of insulin-secreting RIN mSF cells, *J. Biol. Chem.* **261**: 11931–11934.

19. Daniel, J. J., Dangelmaier, C. A., and Smith, J. B., 1988, Calcium modulates the generation of inositol 1,3,4-trisphosphate in human platelets by the activation of inositol 1,4,5-trisphosphate 3-kinase, *Biochem. J.* **253**: 789–794.

20. Irvine, R. F., and Moor, R. M., 1986, Microinjection of inositol 1,3,4,5-tetrakisphosphate activates sea urchin eggs by a mechanism dependent on external Ca^{2+}, *Biochem. J.* **240**: 917–920.

21. Morris, A. P., Gallacher, D. V., Irvine, R. F., and Petersen, O. H., 1987, Synergism of inositol trisphosphate and tetrakisphosphate in activating Ca^{2+}-dependent K^+ channels, *Nature* **330**: 653–655.

22. Hill, T. D., Dean, N. M., and Boynton, A. L., 1988, Inositol 1,3,4,5-tetrakisphosphate induces Ca^{2+} sequestration in rat liver cells, *Science* **242**: 1176–1178.

23. Doughney, C., McPherson, M. A., and Dormer, R. L., 1988, Metabolism of inositol 1,3,4,5-tetrakisphosphate by human erythrocyte membranes: A new mechanism for the formation of inositol 1,4,5-trisphosphate, *Biochem. J.* **251**: 927–929.

24. Balla, T., Guillemette, G., Bambol, A. J., and Catt, K. J., 1987, Metabolism of inositol 1,3,4-trisphosphate to a new tetrakisphosphate isomer in angiotensin-stimulated adrenal glomerulosa cells, *J. Biol. Chem.* **262**: 9952–9955.

25. Shears, S. B., Parry, J. B., Tang, E. K. Y., Irvine, R. F., Michell, R. H., and Kirk, C. J., 1987, Metabolism of D-myo-inositol 1,3,4,5-tetrakisphosphate by rat liver, including the synthesis of a novel isomer of myo-inositol tetrakisphosphate, *Biochem. J.* **246**: 139–147.

26. Stephens, L. R., Hawkins, P. T., Barber, C. J., and Downes, C. P., 1988, Synthesis of myo-inositol 1,3,4,5,6-pentakisphosphate from inositol phosphates generated by receptor activation, *Biochem. J.* **253**: 721–733.

27. Heslop, J. P., Irvine, R. F., Tashjian, A. H., Jr., and Berridge, M. J., 1985, Inositol tetrakis- and pentakisphosphates in GH_4 cells, *J. Exp. Biol.* **119**: 395–401.
28. Vallejo, M., Jackson, T., Lightman, S., and Hanley, M. R., 1987, Occurrence and extracellular actions of inositol pentakis- and hexakisphosphate in mammalian brain, *Nature* **330**: 656–658.
29. Vergara, J., Tsien, R. Y., and Delay, M., 1985, Inositol 1,4,5-trisphosphate: Possible chemical link in excitation–contraction coupling in muscle, *Proc. Natl. Acad. Sci. USA* **82**: 6352–6356.
30. Volpe, P., Salviati, G., Di Virgilio, F., and Pozzan, T., 1985, Inositol trisphosphate induces Ca^{2+} release from the sarcoplasmic reticulum of skeletal muscle, *Nature* **316**: 347–349.
31. Hidalgo, C., Carrasco, M. A., Magendzo, K., and Jaimovich, E., 1986, Phosphorylation of phosphatidylinositol by transverse tubule vesicles and its possible role in excitation–contraction coupling, *FEBS Lett.* **202**: 69–73.
32. Carrasco, M. A., Magendzo, K., Jaimovich, E., and Hidalgo, C., 1988, Calcium modulation of phosphoinositide kinases in transverse tubule vesicles from frog skeletal muscle, *Arch. Biochem. Biophys.* **262**: 306–366.
33. Varsanyi, M., Messer, M., Brandt, N. R., and Heilmeyer, L. M. G., 1986, Phosphatidylinositol 4,5-bisphosphate formation in rabbit skeletal and heart muscle membranes, *Biochem. Biophys. Res. Commun.* **138**: 1395–1404.
34. Varsanyi, M., Messer, M., and Brandt, N. R., 1989, Intracellular localization of inositol-phospholipid-metabolizing enzymes in rabbit fast twitch skeletal muscle: Can D-myo-inositol 1,4,5-trisphosphate play a role in excitation–contraction coupling? *Eur. J. Biochem.* **179**: 473–479.
35. Jay, S. D., and Campbell, K. P., 1988, Characterization of phosphatidylinositol 4,5-bisphosphate production in skeletal muscle triads, *Biophys. J.* **53**: 467a.
36. Hidalgo, C., Parra, C., Riquelme, G., and Jaimovich, E., 1986, Transverse tubules from frog skeletal muscle: Purification and properties of vesicles sealed with the inside-out orientation, *Biochem. Biophys. Acta* **855**: 79–88.
37. Hidalgo, C., Gonzalez, M. E., and Lagos, R., 1983, Characterization of the Ca^{2+}- or Mg^{2+}-ATPase of transverse tubule membranes isolated from rabbit skeletal muscle, *J. Biol. Chem.* **258**: 13937–13945.
38. Collins, C. A., and Wells, W. W., 1983, Identification of phosphatidylinositol kinase in rat liver lysosomal membranes, *J. Biol. Chem.* **258**: 2130–2134.
39. Guisto, N. M., and Ilincheta de Boschero, M. G., 1986, Synthesis of polyphosphoinositids in vertebrate photoreceptor membranes, *Biochim. Biophys. Acta* **877**: 440–446.
40. Yamakawa, A., and Takenawa, T., 1988, Purification and characterization of membrane-bound phosphatidylinositol kinase from rat brain, *J. Biol. Chem.* **263**: 17555–17560.
41. Tooke, N. E., Hales, C. N., and Hatton, J. C., 1984, Ca^{2+}-sensitive phosphatidylinositol 4-phosphate metabolism in a rat β-cell tumor, *Biochem. J.* **219**: 471–480.
42. Rios, E., and Brum, G., 1987, Involvement of dihydropyridine receptors in excitation–contraction coupling in skeletal muscle, *Nature* **325**: 717–720.
43. Di Virgilio, F., Salviati, G., Pozzan, T., and Volpe, P., 1986, Is a guanine nucleotide-binding protein involved in excitation–contraction coupling in skeletal muscle? *EMBO J.* **5**: 259–262.
44. Salviati, G., Betto, R., Tegazzin, V., and Della Puppa, A., 1988, Subcellular localization of G-protein and phospholipase-C activity in rabbit skeletal muscle, *Biophys. J.* **53**: 332a.
45. Donaldson, S. K., Goldking, N. D., Walseth, T. F., and Huettemann, D. A., 1987, Inositol trisphosphate stimulates Ca^{2+} release from peeled skeletal muscle fibers, *Biochem. Biophys. Acta* **927**: 92–99.
46. Suarez-Isla, B., Irribarra, V., Bull, R., Oberhauser, A., Larralde, L., Hidalgo, C., and Jaimovich, E., 1988, Inositol 1,4,5-trisphosphate activates a calcium channel in isolated sarcoplasmic reticulum membranes, *Biophys. J.* **54**: 737–741.
47. Lagos, N., and Vergara, J., 1989, Phosphoinositide kinase and phospholipase-C activities in T-tubule membrane vesicles of frog skeletal muscle, *Biophys. J.* **55**: 236a.
48. Schacht, J., 1976, Inhibition by neomycin of polyphosphoinositide turnover in subcellular fractions of guinea-pig cerebral cortex *in vitro*, *J. Neurochem.* **27**: 1119–1124.

49. Reid, D. G., and Gajjar, K., 1987, A proton and carbon-13 nuclear magnetic resonance study of neomycin B and its interactions with phosphatidylinositol 4,5-bisphosphate, *J. Biol. Chem.* **262:** 7967–7972.

50. Kasianowicz, J., Gabev, E., and McLaughlin, S., 1988, The binding of neomycin to phosphatidylinositol 4,5-bisphosphate (PIP_2), *Biophys. J.* **53:** 517a.

51. Tysnes, O-B., Verhoeven, A. J. M., and Holmsen, H., 1987, Neomycin inhibits agonist-stimulated polyphosphoinositide metabolism and responses in human platelets, *Biochem. Biophys. Res. Commun.* **144:** 454–462.

52. Hidalgo, C., and Jaimovich, E., 1989, Inositol trisphosphate and excitation–contraction coupling in skeletal muscle, *J. Bioenerg. Biomemb.* **21:** 267–281.

53. Milani, D., Volpe, P., and Pozzan, T., 1988, D-myo-inositol 1,4,5-trisphosphate phosphatase in skeletal muscle, *Biochem. J.* **254:** 525–529.

54. Janiak, M. J., Small, D. M., and Shipley, G. G., 1979, Temperature and compositional dependence of the structure of hydrated dimyristoyl lecithin, *J. Biol. Chem.* **254:** 6068–6078.

E. Calcium Release

Pharmacology of the *Ryania* Alkaloids: The Ester A, a Ryanodine Analog That Only Increases Sarcoplasmic Reticulum Calcium Permeability

John L. Sutko, Esther Robinson, Frank A. Lattanzio, Jr., Robert G. Schlatterer, Pierre Deslongchamps, and Luc Ruest

1. INTRODUCTION

Ryanodine, an alkaloid extracted from the wood of *Ryania speciosa*, is a specific ligand of the "foot" protein of the triad junction[1] and a modulator of sarcoplasmic reticulum (SR) calcium permeability in both striated and smooth muscles.[2-5] Ryanodine modulates the calcium permeability of sarcoplasmic reticulum terminal cisternae membranes in a complex manner. Depending on its concentration and the experimental conditions employed, ryanodine can either increase or decrease this variable.[2-4] It is important to note that either effect will reduce activation-dependent release of SR calcium in intact muscle cells.[6]

Ryanodine and the 1,4-dihydropyridines are similar in that both can increase and decrease calcium fluxes through the channels they affect.[3,4,7] Different dihydropyridines vary in their relative abilities to cause these two effects.[7] For example, Bay K 8644 and nitrendipine are more potent as an activator and inhibitor, respectively, of plasma membrane calcium currents. The availability of analogs of ryanodine[8] has

JOHN L. SUTKO, ESTHER ROBINSON, FRANK A. LATTANZIO, JR., and ROBERT G. SCHLATTERER • Department of Pharmacology, University of Nevada School of Medicine, Reno, Nevada 89557. PIERRE DESLONGCHAMPS and LUC RUEST • Laboratory of Organic Synthesis, Department of Chemistry, Faculty of Science, University of Sherbrooke, Sherbrooke, Quebec J1K 2R1, Canada.

made possible similar determinations of the structure–activity relationships underlying the actions of the *Ryania* alkaloids on calcium fluxes through SR channels.

In the present study we have compared the actions of the diterpene ester A,[8] a structural analog of ryanodine, with those of ryanodine on both the passive influx and efflux of calcium across, and the ATP-dependent accumulation of calcium by rabbit fast twitch skeletal muscle SR terminal cisternae membranes.[9] At low concentrations, ryanodine increased the passive fluxes of calcium, an action consistent with an enhancement of native SR membrane calcium permeability. At high ryanodine concentrations, calcium influx and efflux were diminished and ATP-dependent calcium accumulation was markedly enhanced, effects consistent with a decreased calcium permeability. In contrast, at all concentrations tested, the ester A only increased the passive fluxes of calcium and did not alter ATP-dependent calcium accumulation. Coadministration of high concentrations of the ester A and ryanodine reduced the effects of the latter compound on both the passive calcium fluxes and ATP-dependent calcium accumulation. These results suggest, first, that ryanodine is an agonist at either a single site capable of assuming different conformations or at two distinct sites that increase (enhancing site) and decrease (inhibitory site) the flux of calcium through a SR channel. Second, the ester A is an agonist at the enhancing site, and either a direct or indirect antagonist of the inhibitory actions of ryanodine.

2. RESULTS AND DISCUSSION

The ryanodine used in these studies (S. B. Penick & Co., Lot #704 RWP-1) was a mixture primarily of ryanodine and dehydroryanodine.[8,10,11] These two compounds were purified using HPLC[11] and found to be equipotent in the paradigms used for these studies (data not shown). Therefore, the use of the mixture did not complicate the present experiments. The ester A was isolated from powdered *Ryania* wood (Agri-Systems International) as previously described,[8] with the exception that a final purification was achieved using a HPLC procedure.[12] The structures of ryanodine, dehydroryanodine and the ester A are shown in Figure 1.

Ryanodine and the ester A increased calcium efflux from skeletal terminal cisternae membranes (Fig. 2). The effects of these agents were clearly observed with 1.0 μM, but not with 0.03 μM, concentrations. At a concentration of 300 μM, ryanodine reduced calcium efflux to a level less than that observed in its absence. In contrast, calcium efflux was only further increased by 300 μM ester A to an extent comparable to the maximal effects of ryanodine.

The changes in the influx of calcium into terminal cisternae vesicles caused by ryanodine and the ester A were similar to those in calcium efflux. As shown in Fig. 3, 0.03 and 1.0 μM concentrations of ryanodine increased calcium influx during a 1-sec period over that observed in its absence by 12.6 and 43.6%, respectively; whereas, the values obtained in the presence of 300 μM ryanodine were similar to those obtained under control conditions. At concentrations of 0.03, 1.0, and 300 μM, the ester A increased calcium influx over the values obtained in the absence of drug by 4.6, 28.8, and 43.2%, respectively.

At the concentrations tested, ryanodine and the ester A exhibited differences in

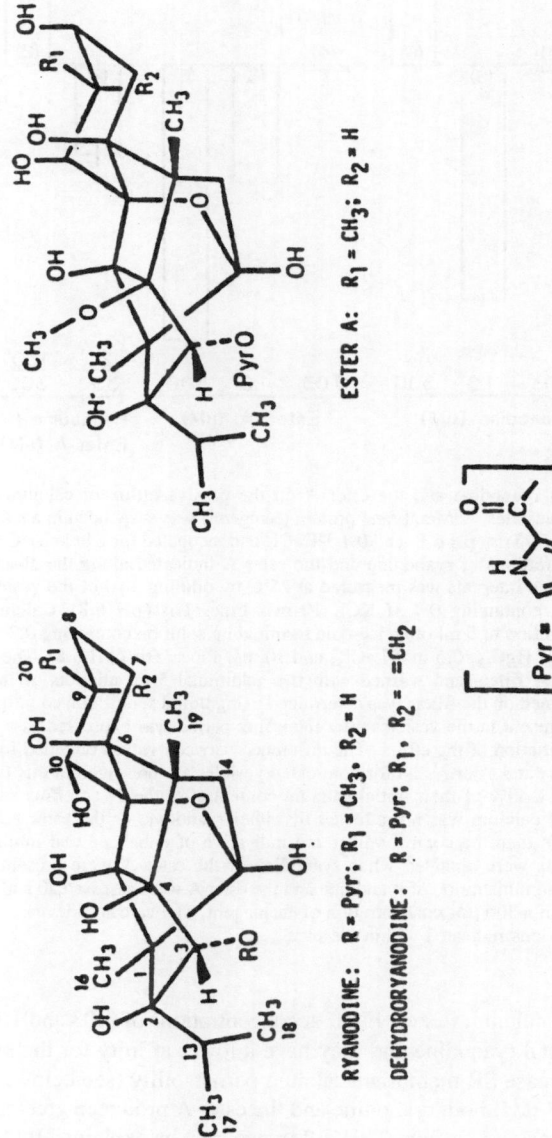

RYANODINE: R = Pyr; R₁ = CH₃, R₂ = H

DEHYDRORYANODINE: R = Pyr; R₁, R₂ = =CH₂

ESTER A: R₁ = CH₃; R₂ = H

Figure 1. Structures of ryanodine, dehydroryanodine, and the ester A.

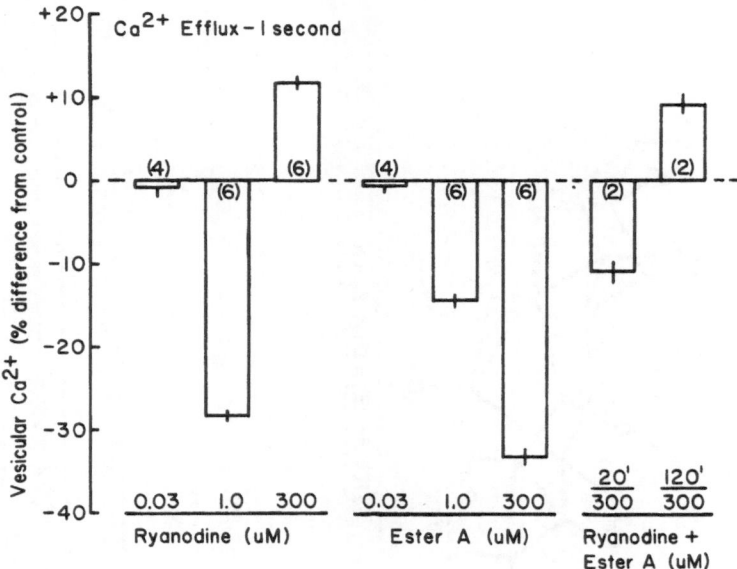

Figure 2. The effects of ryanodine and the ester A on the passive efflux of calcium from skeletal muscle terminal cisternae vesicles.[9] Membrane protein (5 mg/ml) was suspended in a solution containing 0.1 M KCl, 10 mM Pipes/Tris (pH 6.8) and 0.1 $^{45}CaCl_2$ and incubated for 2 hr at 37°C in the absence or presence of the concentrations of ryanodine and the ester A indicated along the abscissa. Calcium efflux occurring during 1-sec intervals was measured at 25°C by diluting 5 μl of the vesicle suspension into 500 μl of a solution containing 0.1 M KCl, 10 mM Pipes/Tris (pH 6.8). Calcium efflux was terminated by the rapid addition of 5 ml of an ice-cold terminating solution containing 0.3 M sucrose, 10 μM ruthenium red, 0.5 mM HgCl$_2$, 0.5 mM LaCl$_3$, and 10 mM Pipes/Tris (pH 6.8). The vesicles were trapped on Whatman GF/A filters and washed with two additional 5-ml aliquots of the terminating solution. The calcium retained on the filters was determined using liquid scintillatin counting techniques. The quantity of calcium present in the vesicles after the efflux period was expressed as a percentage of that present prior to the initiation of the efflux. The difference between values obtained for the preparations treated with ryanodine and ester A and for nontreated vesicles are presented in this figure. Control vesicles retained 61.45 ± 1.48% of their initial calcium content after the 1-sec efflux period, and the initial vesicular content of calcium was not affected by either ryanodine or the ester A. Values more negative and more positive than the control values are indicative of enhanced and inhibited effluxes, respectively. Similar results were obtained when ryanodine or the ester A were present in the efflux medium. The effects of coadministration of ryanodine and the ester A were assessed 20 and 120 min after the simultaneous addition of a 300 μM concentration of each agent. The values shown are means ± S.E. for the number of preparations indicated within the bars.

their effects on passive calcium fluxes. First, at concentrations of 0.03 and 1.0 μM, ester A was less effective than ryanodine and may have a lower affinity for the site at which these agents act to increase SR membrane calcium permeability (see below). Second, at a concentration of 0.03 μM, both ryanodine and the ester A produced greater changes in calcium influx than in calcium efflux. This difference may be explained by the presence of calcium concentrations that decrease ryanodine binding to the enhancing site or site conformation[4,13] during exposure of the membranes to these agents. Third, the inhibition of the passive calcium fluxes produced by 300 μM ryanodine was larger and observed more consistently during measurements of calcium efflux than for those of calcium influx. Our data do not address the basis of this last difference.

Under the appropriate experimental conditions, high concentrations (0.1–1.0 mM) of ryanodine increase ATP-dependent calcium accumulation by cardiac SR and skeletal terminal cisternae vesicles by decreasing SR membrane calcium permeability.[4,14–16] Therefore, we investigated whether inhibitory effects by the ester A might depend on the assay conditions used. As shown in Fig. 4, 300 μM ryanodine produced the expected increase in calcium accumulation, while the same concentration of the ester A did not significantly affect this variable. These and the preceding results suggest that the ester A is not effective as an agonist at the inhibitory site or site conformation. Under the experimental conditions used for measurement of active calcium transport, the SR calcium channels appear to be open[4,14–16] and this accounts for the low levels of calcium accumulated under control conditions and explains why neither ryanodine nor the ester A caused a further reduction in calcium accumulation.

Ryanodine and the ester A compete for binding at the same high-affinity site, which has approximately a 20-fold higher affinity for ryanodine (Fig. 5). This site appears to be

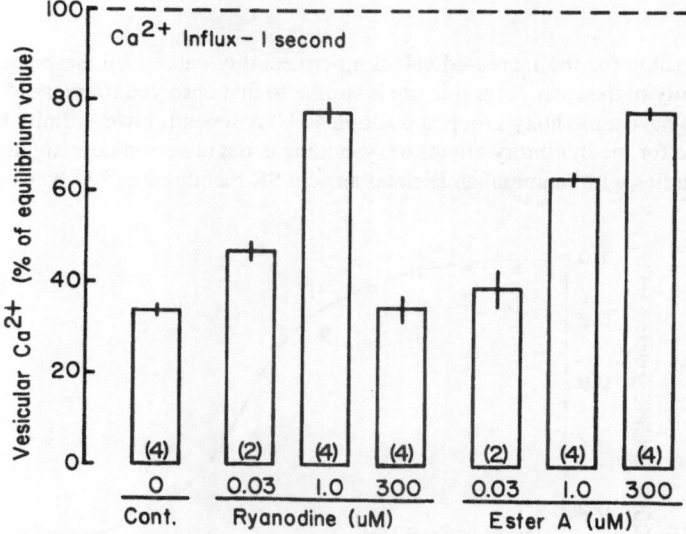

Figure 3. The effects of ryanodine and the ester A on the passive influx of calcium into skeletal muscle terminal cisternae vesicles. Membrane protein (5 mg/ml) was suspended in a solution containing 0.1 M KCl, 10 mM Pipes/Tris (pH 6.8) and incubated for 2 hr at 37°C in the absence and presence of ryanodine and the ester A at the concentrations shown along the abscissa. The quantity of calcium entering the vesicles during a 1-sec period was determined at 25°C by mixing 46.4 μl of the protein suspension with 3.6 μl of $^{45}CaCl_2$. The final concentration of calcium was either 0.1 or 1.0 mM. The results obtained with either concentration were similar once normalized, and have been combined. Calcium influx was terminated and vesicular calcium measured as described for Fig. 2. The quantity of calcium entering the vesicles during the 1-sec uptake interval is expressed as a percentage of the equilibrium vesicular calcium content obtained after 20 min in the presence of the calcium ionophore ionomycin (1 μM). Neither ryanodine nor the ester A effected the latter quantity. The values presented are means ± S.E. for the number of preparations shown in each bar.

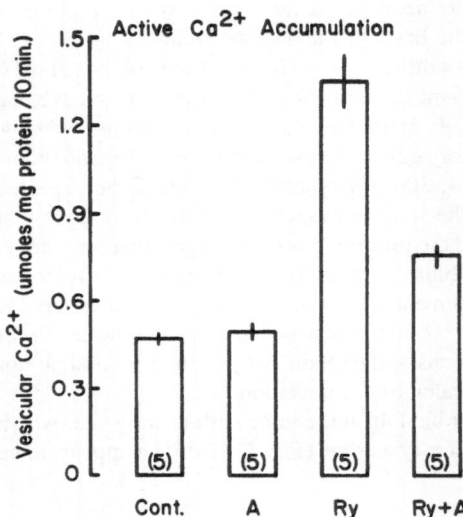

Figure 4. The effects of ryanodine (Ry) and the ester A (A) on ATP-dependent calcium accumulation by skeletal muscle terminal cisternae membranes. Membrane protein (0.2–0.8 mg/ml) was added to a medium containing 50 mM histidine (pH 7.0), 3 mM MgCl$_2$, 3 mM K oxalate, 100 mM KCl, and 38 μM ^{45}CaCl$_2$ and incubated for 10 min at 37°C in the absence and presence of 300 μM concentrations of ryanodine or the ester A. Calcium uptake was initiated by the addition of 3 mM Na$_2$ATP and allowed to proceed for 10 min. Calcium uptake was terminated and vesicular calcium measured as described for Fig. 2. The values shown have been corrected for calcium binding occurring in the absence of ATP (which was unaffected by either ryanodine or the ester A) and represent means ± S.E. for the number of preparations indicated in the bars.

that responsible for the increased calcium permeability caused by these agents.[4] The lower affinity of the ester A for this site is similar to that observed for ryanoid derivatives which also have a methoxy group at position 4.[17] A second, lower-affinity binding site responsible for the inhibitory effects of ryanodine is not observed consistently in ligand-binding studies with mammalian skeletal muscle SR membranes.[4,13,16] Consequently,

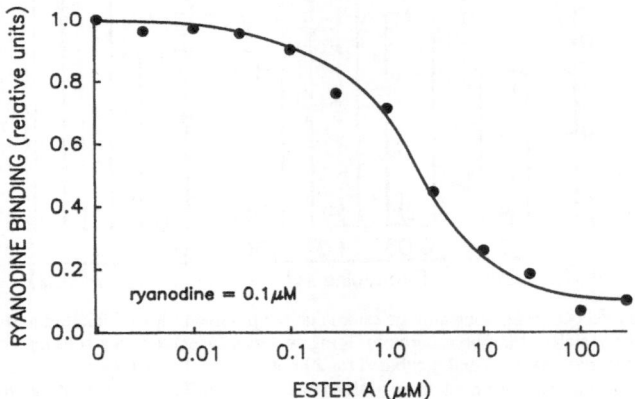

Figure 5. Competitive binding to skeletal muscle terminal cisternae membranes between [^{3}H]ryanodine and the ester A. Membrane protein (0.1 mg/ml) was incubated in a medium containing 20 mM Tris, pH 7.4, 0.5 M KCl, 0.020 mM free calcium at 37°C for 2 hr in the presence of [^{3}H]ryanodine (0.1 μM) and the concentrations of the ester A indicated on the abscissa. Nonspecific binding was measured in the presence of 10 μM unlabeled ryanodine. The assay was terminated by dilution with ice-cold incubation buffer and filtration through Whatman GF/A filters. The filters were subsequently washed three times with 4-ml aliquots of this buffer.

the ability of the ester A to interact with this latter site could not be tested directly in this model. Therefore, to determine whether the ester A could antagonize the actions of ryanodine at the inhibitory site or site conformation, we investigated the consequences of coadministration of 300 μM concentrations of ryanodine and the ester A on both passive calcium fluxes and the ATP-dependent accumulation of this cation. Consistent with this possibility, the ester A reversed the inhibition of the passive calcium fluxes caused by ryanodine, as shown for calcium efflux in Fig. 3, and antagonized the enhancement of active calcium accumulation by ryanodine (Fig. 4). Under the conditions used for the passive flux measurements, the interaction between the ester A and ryanodine was observed when the membranes were exposed to the ester A before ryanodine, or when both agents were added simultaneously and relatively short incubation periods of 30 min or less were used. This interaction was not seen in the latter case when the membranes were simultaneously exposed to both agents for longer times (e.g., 2 hr, Fig. 3).

The extent of the antagonism exerted by the ester A on the inhibitory actions of ryanodine was dependent on their relative concentrations (data not shown) and could result from direct competitive interactions at the inhibitory site or site conformation. The temporal and addition sequence dependencies of this antagonism suggest that with sufficient time, ryanodine can successfully compete with the ester A for the inhibitory site. We have found that the effects of 1 μM ryanodine on passive calcium fluxes increase progressively over a period of 1–2 hr. It is possible that a single binding site capable of assuming different functionally significant conformations is responsible for both the enhancing and inhibitory effects of these agents. In this case, only the ryanodine-liganded site would be capable of assuming the inhibitory site conformation. In contrast to the results obtained for the passive calcium fluxes, the antagonism by the ester A of the enhancement of ATP-dependent calcium accumulation by ryanodine was observed after both 20-min and 2-hr periods of simultaneous drug exposure. The effects of ryanodine on active calcium transport were developed fully after an exposure of 20–30 min (data not shown). An additional component(s) of the active transport assay system, such as ATP, may affect the characteristics of the binding of ryanodine and the ester A to the inhibitory site or site conformation.

Alternatively, the ester A could indirectly antagonize the inhibitory effects of ryanodine by acting at the enhancing site to increase SR membrane permeability. Due to the large conductance of the ryanodine-sensitive SR calcium channel,[18] the opening of a single channel in an isolated membrane vesicle by the ester A could effectively negate inhibition of other channels by ryanodine. This limitation of the vesicle system precludes determining the exact nature of the effects by ester A on the inhibitory actions of ryanodine in this system. Measurement of SR membrane calcium currents in planar lipid bilayers should provide a useful approach to this issue.

In conclusion, our results indicate that, as shown previously,[3,4] ryanodine is an agonist at sites that both increase and decrease calcium fluxes through a SR channel. Like ryanodine, the ester A is an agonist at the site or site conformation involved in increasing calcium permeability, but can antagonize the effects of ryanodine at the site or site conformation that decreases this variable. The high ryanodine concentrations necessary for inhibition of the calcium permeability of isolated SR membranes[4,14,15] suggest that this effect could be due to actions by this agent at a nonspecific (i.e., nonsaturable) site or sites. However, the ability of relatively minor, structural changes to preclude expression of the inhibitory effects by the ester A, and of the ester A to antagonize the inhibitory effects of ryanodine may indicate the involvement of a specific (saturable) binding site or site conformation in this effect.

The structural diversity existing among the *Ryania* alkaloids[8,17] may permit an elucidation of the structural requirements for specific effectors of the regulatory sites associated with the SR calcium channel. In this regard, a potentially interesting parallel may exist between the mechanism(s) used to regulate the plasma and SR membrane calcium channels. In both cases there is evidence for the existence of a channel component(s) that can both increase and decrease calcium fluxes through the channel. The identification of exogenous substances that affect this component(s) when present in nanomolar concentrations raises the possibility that endogenous channel effectors may exist. In addition, the ability of compounds, such as ryanodine and the ester A, with minor structural differences to produce different functional effects suggests that caution must be exercised when using either crude preparations of *Ryania* alkaloids, or conditions or systems that could result in structural modifications of the ryanodine molecule.

ACKNOWLEDGMENTS.   This work was supported by NSF grant PCM 8402100 and NIH grants HL 27470 and HL 17669. F.A.L. received support as a postdoctoral fellow from NIH Training Grant HL 07360 during part of this project. J.L.S. is an Established Investigator of the American Heart Association. We thank Drs. J. L. Kenyon, G. Meissner, and J. T. Willerson and Ms. J. Nichol for comments during the course of this work and on this report.

REFERENCES

1. Franzini-Armstrong, C., 1970, Studies of the triad. I. Structure of the junction in frog twitch fibers, *J. Cell. Biol.* **47:** 488–499.
2. Fairhurst, A. S., and Hasselbach, W., 1970, Calcium efflux from a heavy sarcoplasmic reticulum fraction: Effects of ryanodine, caffeine, and magnesium, *Eur. J. Biochem.* **13:** 504–509.
3. Meissner, G., 1986, Ryanodine activation and inhibition of the Ca^{2+} release channel of sarcoplasmic reticulum, *J. Biol. Chem.* **261:** 6300–6306.
4. Lattanzio, F. A., Jr., Schlatterer, R. G., Nicar, M., Campbell, K. P., and Sutko, J. L., 1986, The effects of ryanodine on passive calcium fluxes across sarcoplasmic reticulum membranes, *J. Biol. Chem.* **262:** 2711–2718.
5. Ito, K., Takakura, S., Sato, K., and Sutko, J. L., 1986, Ryanodine inhibits the release of calcium from intracellular stores in guinea pig aortic smooth muscle, *Circ. Res.* **58:** 730–734.
6. Sutko, J. L., Ito, K., and Kenyon, J. L., 1985, Ryanodine: A modifier of sarcoplasmic reticulum calcium release in striated muscle, *Fed. Proc.* **44:** 2984–2988.
7. Thomas, G., Gross, R., and Schramm, M., 1984, Calcium channel modulation: Ability to inhibit or promote calcium influx resides in the same dihydropyridine molecule, *J. Cardiovasc. Pharmacol.* **6:** 1170–1176.
8. Ruest, L., Taylor, D. R., and Deslongchamps, P., 1985, Investigation of the constituents of *Ryania speciosa, Can. J. Chem.* **63:** 2840–2843.
9. Saito, A., Seiler, S., Chu, A., and Fleischer, S., 1984, Preparation and morphology of SR terminal cisternae from rabbit skeletal muscle, *J. Cell. Biol.* **99:** 875–885.
10. Waterhouse, A. L., Holden, I., and Casida, J. E., 1984, 9,21-Didehydroryanodine: A new principal toxic constituent of the botanical insecticide *Ryania, J. Chem. Soc. Commun.* 1265–1266.

11. Sutko, J. L., Thompson, L. J., Schlatterer, R. G., Lattanzio, F. A., Fairhurst, A. S., Campbell, C., Martin, S. F., Deslongchamps, P., Ruest, L., and Taylor, D. R., 1985, Separation and formation of ryanodine from dehydroryanodine: Preparation of [³H]ryanodine, *J. Label. Compd. Radiopharmaceut.* **23**: 215–222.
12. Robinson, E., and Sutko, J. L., unpublished observations.
13. Pessah, I. N., Waterhouse, A. L., and Casida, J. E., 1985, The calcium–ryanodine receptor complex of skeletal and cardiac muscle, *Biochem. Biophys. Res. Comm.* **128**: 449–456.
14. Jones, L. R., Besch, H. R., Jr., Sutko, J. L., and Willerson, J. T., 1979, Ryanodine-induced stimulation of net Ca^{2+} uptake by cardiac sarcoplasmic reticulum vesicles, *J. Pharmacol. Exp. Ther.* **209**: 48–55.
15. Seiler, S., Wegener, A. D., Whang, D. D., Hathaway, D. R., and Jones, L. R., 1984, High molecular weight proteins in cardiac and skeletal muscle junctional sarcoplasmic reticulum vesicles bind calmodulin, are phosphorylated, and are degraded by Ca^{2+}-activated protease, *J. Biol. Chem.* **259**: 8550–8557.
16. Fleischer, S., Ogunburmi, E. M., Dixon, M. C., and Fleer, E. A. M., 1985, Localization of Ca^{2+} release channels with ryanodine in junctional terminal cisternae of sarcoplasmic reticulum of fast skeletal muscle, *Proc. Natl. Acad. Sci. USA* **82**: 7256–7259.
17. Waterhouse, A. L., Pessah, I. N., Francini, A. O., and Casida, J. E., 1987, Structural aspects of ryanodine action and selectivity, *J. Med. Chem.* **30**: 710–716.
18. Smith, J. S., Coronado, R., and Meissner, G., 1985, Sarcoplasmic reticulum contains adenine-nucleotide-activated calcium channels, *Nature (London)* **316**: 446–449.

Chapter 32

Ca^{2+} Release Channel of Sarcoplasmic Reticulum: Characterization of the Regulatory Sites

Gerhard Meissner

1. INTRODUCTION

This chapter focuses on an unresolved aspect of muscle function, namely, how does sarcoplasmic reticulum (SR) release calcium ions, thereby causing muscle to contract? A schematic drawing of a small segment of a muscle cell is depicted in the upper part of Fig. 1. Two key structures in excitation–contraction coupling—the surface membrane, composed of the sarcolemma and transverse (T-)system, and the SR—are shown. SR is an intracellular membrane system whose main function is to regulate muscle contraction and relaxation by releasing and taking up again the released Ca^{2+}. Rapid release from skeletal muscle SR is triggered by a surface membrane potential that is thought to be communicated to SR at specialized areas where the SR comes in close contact with the T-tubule, and protein bridges ("feet") are present which span the gap between the two membrane systems. However, how calcium ions are released from SR has remained an enigma[1,2] (also see Chapter 22, this book). One hypothesis is that T-tubule depolarization increases in the junctional gap the concentration of a chemical messenger such as Ca^{2+}, which then opens a Ca^{2+}-channel in the SR membrane. Another popular hypothesis suggests that the feet are in direct contact with a voltage-sensing molecule in the T-tubule membrane. During T-tubule depolarization the feet undergo a conformational change which regulates the opening of the SR Ca^{2+}-channels.

GERHARD MEISSNER • Departments of Biochemistry and Physiology, School of Medicine, University of North Carolina, Chapel Hill, North Carolina 27599-7260.

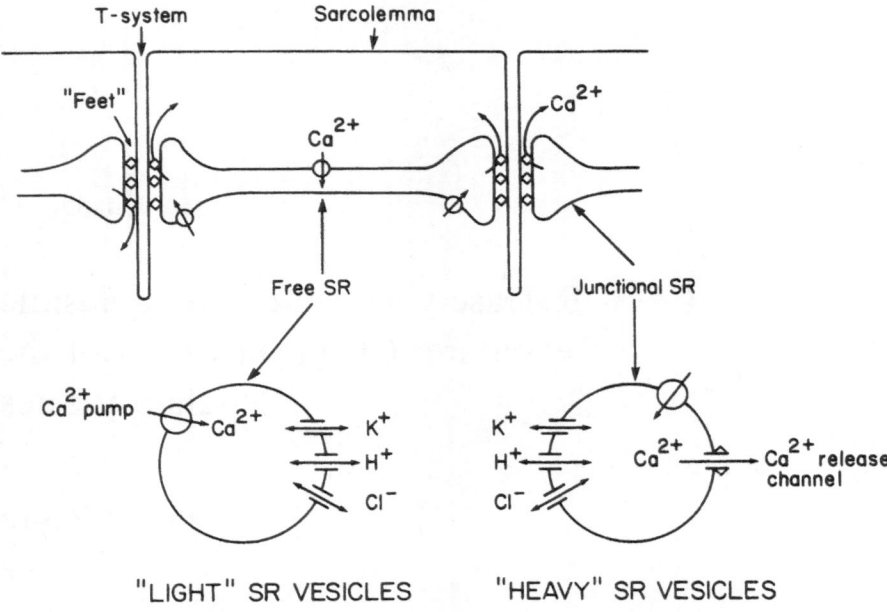

Figure 1. Schematic representation of a segment of a muscle cell and derived light and heavy SR vesicle fractions.

Although the mechanism of signal transmission across the junctional gap remains to be determined, recent single-channel recording measurements using isolated membrane fractions have suggested that SR Ca^{2+} release is mediated by a high-conductance, Ca^{2+}- and ATP-activated "Ca^{2+} release" channel.[3–5] Using the Ca^{2+} release channel-specific probe [^{3}H]ryanodine, a 30 S protein complex has been isolated from skeletal and cardiac SR vesicles,[6,7] which is composed of polypeptides of apparent molecular mass ~400,000.[6–9] Upon reconstitution of the 30 S complex into planar lipid bilayers, Ca^{2+}- and Na^+-conducting channels were evident with pharmacological properties similar to those observed when SR Ca^{2+} release vesicles were fused with the bilayers.[6] Negative-stain electron microscopy further revealed the four-leaf clover structure described for the feet (Chapter 22, this volume) that span the T-tubule–SR gap. These findings have suggested that the "feet" are synonymous with the high-conductance, ligand-gated Ca^{2+} release channel of sarcoplasmic reticulum[6] (Table I). Further, they support the idea that Ca^{2+} release from the SR in response to a surface membrane action potential may be effected through a direct interaction of T-tubule proteins with SR Ca^{2+} release channels.

In this chapter, we describe radioisotope flux–Millipore filtration measurements that we have carried out to determine the Ca^{2+} release behavior of a "heavy" skeletal muscle junctional SR-derived vesicle fraction. The heavy skeletal membrane fraction is largely free of T-tubule, which suggests that the feet or Ca^{2+} release channels, which remain with the SR membrane, have been dissociated from the T-tubule membrane

Table I. Properties of the Ca²⁺ Release Channel of Skeletal Muscle Sarcoplasmic Reticulum

---

1. 30 S complex composed of four polypeptides of $M_r \sim 400{,}000$

2. Single-channel conductance: 100 pS in 50 mM Ca²⁺
   600 pS in 500 mM Na⁺

3. Regulation: Activation by μM Ca²⁺ and mM ATP
   Inhibition by mM Mg²⁺ and μM calmodulin

4. Identical with protein bridges ("feet") that span the gap between T-tubule and junctional SR

---

during isolation.[10,11] Our studies suggest that the "uncoupled" SR Ca²⁺ release channel is regulated in a complex manner by Ca²⁺, Mg²⁺, adenine nucleotide, and calmodulin.

2. ISOLATION OF SR Ca²⁺ RELEASE VESICLES

During homogenization, the sarcoplasmic reticulum structure is disrupted into various kinds of sealed membranous vesicles that can be partially resolved in a centrifuge according to their size and buoyant density.[10] There are two major types that have been referred to as control ("light") and Ca²⁺ release ("heavy") vesicles. Light vesicles are thought to be primarily derived from the free or nonjunctional region of SR, and heavy SR vesicles from the terminal cisternae, junctional region of SR (Fig. 1). Both types contain the Ca²⁺, Mg²⁺-ATPase or Ca²⁺ pump and are therefore capable of active Ca²⁺ transport. Both are also readily permeable to monovalent ions like K⁺, H⁺, and Cl⁻.[12] What sets the two types of vesicles apart is that Ca²⁺ release vesicles contain a Ca²⁺-conducting channel or Ca²⁺ release channel which can mediate the rapid release of calcium ions from skeletal vesicles[13–15] and cardiac vesicles.[16]

3. MEASUREMENT OF RAPID ⁴⁵Ca²⁺ EFFLUX RATES FROM HEAVY SR Ca²⁺ RELEASE VESICLES

We are using two kinds of measurements to study the kinetic behavior of the SR Ca²⁺ release channel. In one, we incorporate Ca²⁺ release vesicles or the purified Ca²⁺ release channel complex into planar lipid bilayers, which allows us to look at single channels.[4,6] In the other, we measure rapid radioactive calcium ion fluxes across vesicle membranes using a rapid quench apparatus and filtration.[15] Although the information provided by the single-channel measurements is more direct, an advantage of the vesicle flux technique is that it averages the kinetic behavior of a large number of channels.

Figure 2 depicts a diagrammatic representation of the rapid quench set-up. A major part of the rapid quench apparatus (Update System 1000 Chemical Quench Apparatus, Madison, WI) is a ram that can push simultaneously up to four syringes filled with SR vesicles and three mixing solutions. Typically, vesicles are passively loaded with millimolar concentrations of radioactive Ca²⁺. In a first mixing chamber, extravesicular

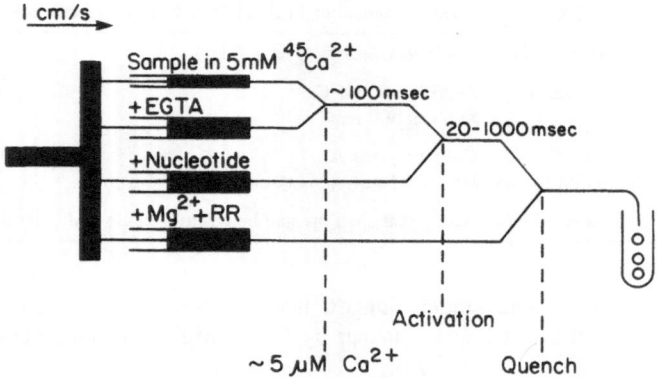

Figure 2. Diagrammatic representation of rapid $^{45}Ca^{2+}$ efflux quench experiments with heavy SR vesicles. (From Ref. 22.)

Ca^{2+} is reduced to a more physiological concentration of 5 μM or less by dilution into a solution containing EGTA, a Ca^{2+}-complexing agent. A nucleotide such as ATP or the nonhydrolyzable ATP analog AMP-PCP can then be added to induce very rapid Ca^{2+} release. In a third mixing chamber, $^{45}Ca^{2+}$ release is stopped by the addition of the two Ca^{2+} release channel inhibitors Mg^{2+} and ruthenium red. After the final mixing step, vesicles are collected, placed on a filter, and rapidly rinsed, and the radioactivity remaining with the vesicles is determined by liquid scintillation counting.

A typical Ca^{2+} release experiment with a heavy skeletal muscle SR Ca^{2+} release vesicle fraction is shown in Fig. 3. $^{45}Ca^{2+}$ efflux is slow when the vesicles are diluted into a medium containing the two Ca^{2+} release channel inhibitors Mg^{2+} and ruthenium red. This allows one to determine the amounts of $^{45}Ca^{2+}$ trapped by all vesicles. In 5 μM Ca^{2+}, a majority of the vesicles release their $^{45}Ca^{2+}$ in just a few seconds, resulting in nearly complete release within 30 sec. Some radioactivity remains with the vesicles for longer times because not all of them contain the Ca^{2+} release channel.[10] In the inset, we have used the rapid quench apparatus to stop $^{45}Ca^{2+}$ efflux at varying time intervals ranging from 25 to 1000 msec. In release media containing 5 μM free Ca^{2+}, the vesicles released half their $^{45}Ca^{2+}$ stores within 0.7 sec, corresponding to a first-order rate constant of about 1 sec^{-1}. Addition of 5 mM AMP-PCP, a nonhydrolyzable ATP analog, to the 5 μM Ca^{2+} release medium increased the initial release rate by a factor of about 50, resulting in nearly complete release within 50 msec.

4. pH DEPENDENCE OF $^{45}Ca^{2+}$ RELEASE

The rate of Ca^{2+} release is strongly dependent on pH (Fig. 4). Ca^{2+} release from skeletal muscle SR vesicles was measured in a medium containing close to 5 μM free Ca^{2+}, that is, in the presence of a maximally activating concentration of Ca^{2+}.[15] At pH 7.8, vesicles released about half their contents within 110 msec, which translates into a first-order rate constant of about 6 sec^{-1}. Most of our vesicle experiments are

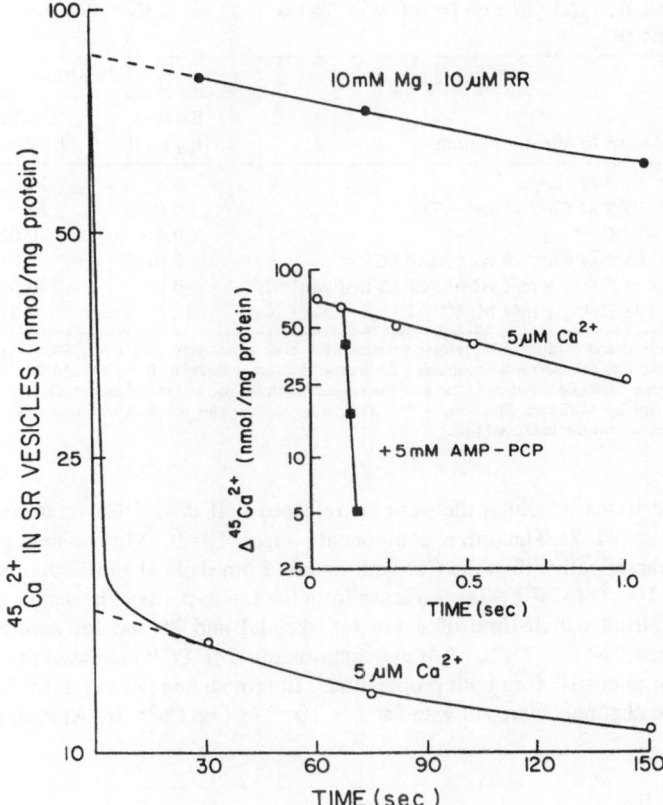

Figure 3. Measurement of $^{45}Ca^{2+}$ efflux rates. A heavy rabbit skeletal muscle SR vesicle fraction was loaded passively with 5 mM $^{45}Ca^{2+}$ and diluted into isoosmolal, unlabeled release media. $^{45}Ca^{2+}$ remaining with the vesicles was determined by Millipore filtration. Rapid $^{45}Ca^{2+}$ efflux was inhibited at time intervals ranging from 25 to 1000 msec by the addition of 10 mM Mg^{2+} and 10 μM ruthenium red (RR). The amount of $^{45}Ca^{2+}$ initially trapped by all vesicles (87 nmol/protein) as well as the amount not readily released by a subpopulation of vesicles (15 nmol/mg protein) were obtained by back extrapolation to the time of vesicle dilution. In the inset, the time course of $^{45}Ca^{2+}$ efflux from the vesicle population containing the Ca^{2+} release channel was obtained by subtracting the amount not readily released (15 nmol/mg protein). (Adapted from Ref. 22.)

done at pH 7, that is, at a pH where Ca^{2+}-induced Ca^{2+} release from skeletal SR has a first-order rate constant of about 1 sec⁻¹.

5. EFFECTS OF Ca²⁺, Mg²⁺, AND ADENINE NUCLEOTIDE ON ⁴⁵Ca RELEASE FROM SKELETAL AND CARDIAC VESICLES

We have studied the Ca^{2+} release behavior of rabbit skeletal vesicles in the presence of Ca^{2+}, Mg^{2+}, and adenine nucleotide, and have compared it with the Ca^{2+} release behavior of canine cardiac vesicles (Table II). In the last two columns are

Table II. Ca^{2+} Release Properties of Skeletal and Cardiac Ca^{2+} Release Vesicles[a]

	$^{45}Ca^{2+}$ efflux	
Additions to release medium	Skeletal $[t_{1/2}(sec)]$	Cardiac $[t_{1/2}(sec)]$
$2 \times 10^{-9}\ M\ Ca^{2+}$	8	25
$2 \times 10^{-9}\ M\ Ca^{2+}$, 5 m$M$ ATP	0.06	12
$10^{-5}\ M\ Ca^{2+}$	0.6	0.02
$2 \times 10^{-6}\ M\ Ca^{2+}$, 5 m$M$ AMP-PCP	0.01	0.01
$10^{-5}\ M\ Ca^{2+}$, 5 mM AMP-PCP, 5 mM Mg^{2+}	0.09	0.03
$10^{-5}\ M\ Ca^{2+}$, 1 mM Mg^{2+}	15	0.25

[a]Skeletal and cardiac Ca^{2+} release vesicles were loaded passively with 1 mM $^{45}Ca^{2+}$ and diluted into release media containing the indicated concentrations of free Ca^{2+}, Mg^{2+}, and adenine nucleotide. Release rates were determined with the use of a chemical quench apparatus and by Millipore filtration.[15,16] Ca^{2+}-permeable vesicles released half their $^{45}Ca^{2+}$ stores within the indicated times.

tabulated the times in which the vesicles released half their $^{45}Ca^{2+}$ stores. All were determined at pH 7. The three components—free Ca^{2+}, Mg^{2+}, and adenine nucleotide—dramatically affect the Ca^{2+} efflux rate from skeletal and cardiac SR vesicles.

At $2 \times 10^{-9}\ M\ Ca^{2+}$, Ca^{2+} release from the Ca^{2+}-permeable vesicle populations is slow, requiring a half-time of 8 sec for skeletal and 25 sec for cardiac vesicles. Increase in free Ca^{2+} to $2 \times 10^{-6}\ M$ and addition of AMP-PCP increased the release rate by a factor of about 1000 in both preparations. Intermediate release rates were observed when the two channels were activated at $2 \times 10^{-9}\ M$ free Ca^{2+} by ATP, or by $10^{-5}\ M$

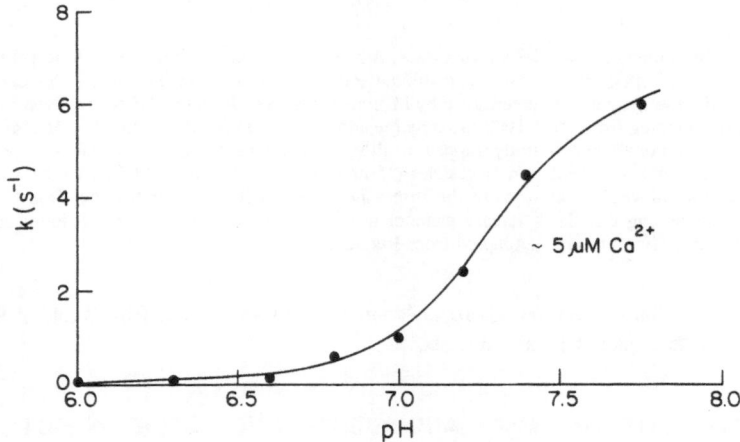

Figure 4.   Dependence of $^{45}Ca^{2+}$ efflux rate on pH. Rabbit skeletal SR Ca^{2+} release vesicles were loaded passively with 1 mM $^{45}Ca^{2+}$ and diluted into media containing close to 5 μM Ca^{2+} after the addition of the vesicles. $^{45}Ca^{2+}$ efflux rates from the Ca^{2+}-permeable vesicle fraction were determined as indicated in Fig. 3.

free Ca^{2+} alone. The last two rows show that Mg^{2+} has an inhibitory effect when added to a 10^{-5} M Ca^{2+} medium containing or lacking nucleotide.

Several important differences were noted to exist between cardiac and skeletal SR. At 10^{-5} M external Ca^{2+}, the half-time of $^{45}Ca^{2+}$ release was 20 msec for cardiac vesicles, as compared to 600 msec for skeletal vesicles. In contrast, adenine nucleotides were more effective in stimulating $^{45}Ca^{2+}$ release from skeletal than cardiac vesicles. Another significant difference was that $^{45}Ca^{2+}$ efflux from cardiac vesicles in the absence of nucleotide was only partially inhibited by 1 mM Mg^{2+}, whereas the Ca^{2+}-channel of skeletal SR was essentially fully inhibited by 1 mM Mg^{2+}.

We have formulated a minimum model to explain the effects of Ca^{2+}, Mg^{2+}, and adenine nucleotides on the Ca^{2+} release behavior of the vesicles (Fig. 5). The model assumes that the SR Ca^{2+} release channel is an allosteric enzyme with interacting regulatory ligand-binding sites. The channel is present in its ligand-free form designated E_0, in the absence of Ca^{2+}, Mg^{2+}, and adenine nucleotide. Distinct channel intermediates, e.g., E_{ca} or E_A, are formed by diluting the vesicles into media containing optimally activating concentrations of Ca^{2+} or adenine nucleotide, respectively.

We have used this model to obtain dose response curves that describe 8 of the 12 reactions leading to the formation of 6 of the 7 hypothetical ligand-occupied states of the skeletal channel.[15] What our measurements have suggested is that Ca^{2+}, ATP, and Mg^{2+} affect the channel in a saturable and cooperative or noncooperative manner, depending on the experimental conditions (Table III).

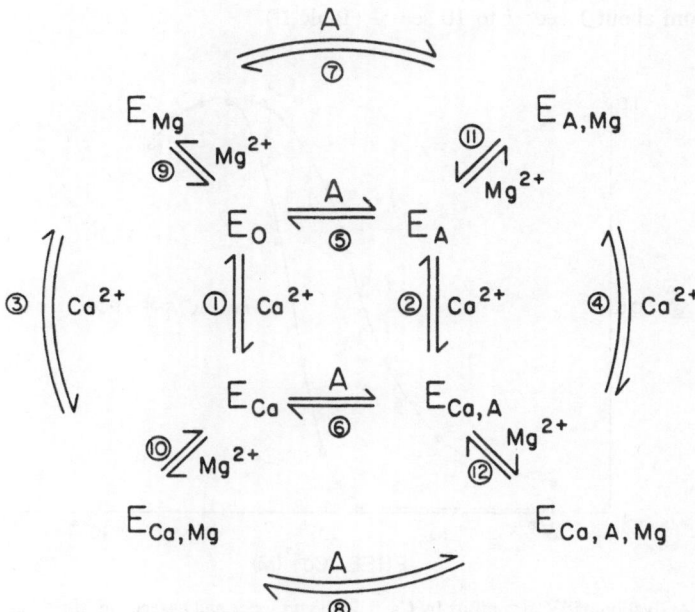

Figure 5. Model of regulation of the SR Ca^{2+} release channel by Ca^{2+}, Mg^{2+}, and adenine nucleotide. Abbreviations: E, SR Ca^{2+} release channel; A, adenine nucleotide. (From Ref. 15.)

Table III. Regulation of Ca^{2+} Release by Ca^{2+}, ATP (AMP-PCP), and Mg^{2+a}

Reaction	N_{app}	L_{50} (M)
$E_0 \leftrightarrow E_{CA}$	1.0	5×10^{-7}
$E_A \leftrightarrow E_{A,CA}$	0.8	9×10^{-7}
$E_{Mg} \leftrightarrow E_{Mg,Ca}$	1.3	$\sim 2 \times 10^{-6}$
$E_{A,Mg} \leftrightarrow E_{A,Mg,Ca}$	2.1	$\sim 2 \times 10^{-6}$
$E_0 \leftrightarrow E_A$	1.9	2×10^{-3}
$E_{Ca} \leftrightarrow E_{Ca,A}$	1.6	1.6×10^{-3}
$E_0 \leftrightarrow E_{Mg}$	1.1	2×10^{-5}
$E_{Ca} \leftrightarrow E_{Ca,Mg}$	1.5	1×10^{-4}

$^a L_{50}$ indicates the concentration of the variable ligand that caused a half-maximal increase or reduction in the rate constants of Ca^{2+} release. L_{50} with free Ca^{2+} as the variable ligand depends on the free Mg^{2+} concentration in the release medium.[15]

Two specific examples describing reactions 1 and 4 of the model are shown in Fig. 6. Dependence of the rate constant of Ca^{2+} release on free Ca^{2+} concentration in the presence and absence of 5 mM Mg^{2+} plus 5 mM AMP-PCP is compared. Flux data were normalized because of the large differences seen at 5 μM free Ca^{2+}. Addition of the Mg–nucleotide complex increased the rate constant of Ca^{2+} release from skeletal vesicles from about 1 sec^{-1} to 10 sec^{-1} (Table II).

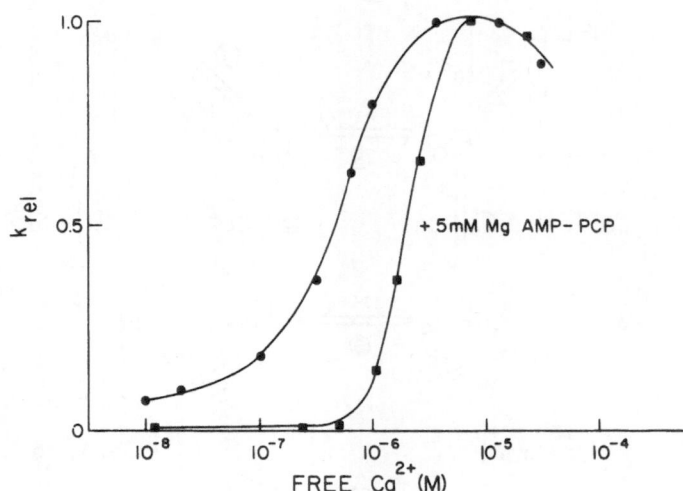

Figure 6. Activation of $^{45}Ca^{2+}$ efflux by Ca^{2+} in the presence and absence of Mg^{2+} and AMP-PCP. Skeletal SR Ca^{2+} release vesicles were passively loaded with 5 mM $^{45}Ca^{2+}$ and diluted into media containing the indicated concentrations of free Ca^{2+} in the absence or presence of 5 mM AMP-PCP plus 5 mM Mg^{2+}.

We were surprised to find that Mg^{2+} and nucleotide render the channel sensitive to external Ca^{2+} in a quite narrow concentration range (Fig. 6). This increase in Ca^{2+} sensitivity is probably of biological importance and provides a clue for a possible role of ATP and Mg^{2+} in the process of Ca^{2+} activation of Ca^{2+} release by SR. In contrast to Ca^{2+}, Mg^{2+} and ATP concentrations are thought to remain fairly constant during all phases of muscle activity. It is therefore unlikely that Mg^{2+} and ATP act like second messengers, as has been proposed for Ca^{2+}. Rather, a major function of these two ligands appears to be to increase the cooperativity of Ca^{2+} activitation of Ca^{2+} release, thereby rendering the channel sensitive to Ca^{2+} in a narrow concentration range.

We have attempted to mimic the ionic conditions in relaxed and contracted muscle. Mammalian muscle contains about 5 mM ATP.[17] The free Mg^{2+} concentration in skeletal muscle has been estimated to range from 0.2 to 4 mM.[18,19] As shown in Table IV, the rate constant of Ca^{2+} release from skeletal SR vesicles is affected greatly by the free Mg^{2+} concentration in the release medium. Vesicles were diluted into media containing a low (10^{-8} M) or optimally activating concentration (2–4×10^{-6} M) of Ca^{2+}, and 0, 5, or 9 mM Mg^{2+} plus 5 mM nucleotide. In the presence of 5 mM AMP-PCP and 5 mM Mg^{2+} (0.7 mM free Mg^{2+}), Ca^{2+} release was nearly fully inhibited at 10^{-8} M free Ca^{2+}. In contrast, at 4×10^{-6} M free Ca^{2+} there occurred only partial inhibition. A more dramatic decrease of the first-order rate constant at 4×10^{-6} M Ca^{2+} was observed when the free Mg^{2+} concentration was increased to 4 mM by the addition of 9 mM Mg^{2+}. One important conclusion that can be drawn from the data of Table IV is that heavy SR vesicles demonstrate Ca^{2+}- and nucleotide-induced Ca^{2+} release rates which, at least in the presence of low free Mg^{2+} concentrations, appear to approach those in muscle.

Data of Table IV however also point out one major difficulty we are currently facing, namely, that we are not able in our vesicle studies to fully close the Ca^{2+}- and nucleotide-activated channel using physiological concentrations of free Mg^{2+}. An unexpected finding was that in the presence of 4 mM free Mg^{2+}, which is the upper estimated level of free Mg^{2+} in muscle, the vesicles still released half their Ca^{2+} stores

Table IV. Inhibition of Ca^{2+} Release by Mg^{2+}[a]

Additions to release medium			
Free Ca²⁺ (M)	Mg²⁺ (M)	Nucleotide (M)	⁴⁵Ca²⁺ efflux [k_1 (sec⁻¹)]
10^{-8}		5×10^{-3}	16
10^{-8}	5×10^{-3b}	5×10^{-3}	0.06
10^{-8}	9×10^{-3c}	5×10^{-3}	0.03
2×10^{-6}		5×10^{-3}	56
4×10^{-6}	5×10^{-3b}	5×10^{-3}	14
4×10^{-6}	9×10^{-3c}	5×10^{-3}	1.3

[a]Rabbit skeletal SR Ca^{2+} release vesicles were loaded passively with 5 mM ⁴⁵Ca^{2+} and diluted into release media containing the indicated concentrations of free Ca^{2+}, Mg^{2+}, and adenine nucleotide. The first-order rate constants of Ca^{2+} release from the Ca^{2+}-permeable vesicle population were determined as indicated in Fig. 3.
[b]0.7 mM free Mg^{2+}.
[c]4 mM free Mg^{2+}.

in less than 1 sec. Two explanations for this apparently inconsistent behavior are possible. First, the channel may have altered its Ca^{2+} dependence on dissociation from the T-tubule membrane. Second, there are other factors besides Ca^{2+}, Mg^{2+}, and adenine nucleotide that influence Ca^{2+} release from SR.

6. INHIBITION OF $^{45}Ca^{2+}$ RELEASE BY CALMODULIN

One of the additional regulatory factors we have considered is calmodulin (Fig. 7). Calmodulin inhibited $^{45}Ca^{2+}$ release at Ca^{2+} concentrations in excess of 10^{-7} M. In release media containing 10^{-6}–10^{-3} M free Ca^{2+}, calmodulin reduced Ca^{2+} efflux rates by a factor of 2–3. At 10^{-7} M Ca^{2+}, calmodulin was less effective in slowing down Ca^{2+} release, whereas at 10^{-9} M Ca^{2+} it was without effect. Ca^{2+} dependence of calmodulin inhibition of Ca^{2+} release is in agreement with what has been observed for other calmodulin-regulated proteins. Ca^{2+} release was half-maximally inhibited by about 2×10^{-7} M calmodulin. This inhibition by calmodulin was reversible, occurring on a time scale of 0.1 to about 10 sec.[20] In other studies, we found that the rate constant of $^{45}Ca^{2+}$ efflux from the vesicles was reduced by calmodulin by a factor of about 2–3 in release media containing 10 μM free Ca^{2+} and varying concentrations of Mg^{2+} and nucleotide. As shown in Table II, the addition of Mg^{2+} or nucleotide to the 10 μM Ca^{2+} release medium varied the release rates by a factor of 1000 or more. Thus, under a wide range of release conditions, calmodulin reduced the $^{45}Ca^{2+}$ release rate by approximately the same percentage.

Our results with calmodulin are quite unusual in two respects. First, calmodulin inhibition occurs in the absence of ATP and therefore does not appear to involve a

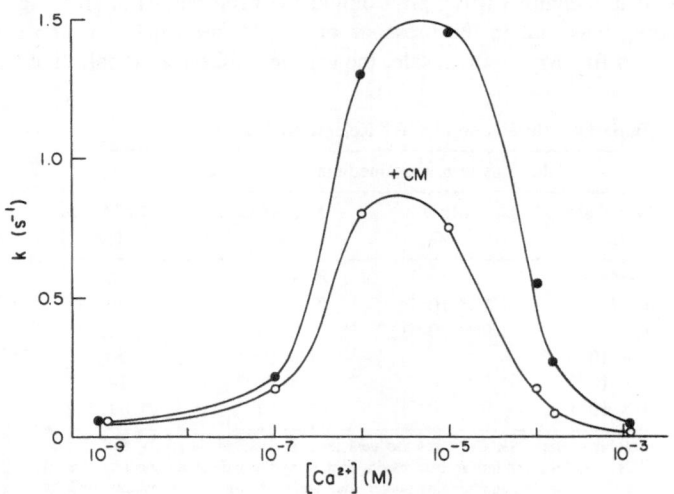

Figure 7. Ca^{2+} concentration dependence of calmodulin inhibition of $^{45}Ca^{2+}$ efflux. Skeletal SR Ca^{2+} release vesicles were loaded passively with 1 mM $^{45}Ca^{2+}$ in the absence or presence of 6 μM calmodulin and diluted 5-fold into media containing the indicated concentrations of free Ca^{2+}. (From Ref. 15.)

protein-kinase-mediated phosphorylation reaction, as suggested by Kim and Ikemoto.[21] Second, calmodulin does not completely stop the release of Ca^{2+}; rather, in all of our experiments we noted only a partial reduction of the release rate by calmodulin. The precise physiological function of calmodulin in excitation–contraction coupling is difficult to define at present. One attractive possibility is that calmodulin, by binding to the channel and thereby slowing down but not stopping the release of Ca^{2+} from SR, compensates for the incomplete removal of Ca^{2+} from the myoplasmic space during increased muscle activity.

7. CONCLUSION

The equation below summarizes in a simple fashion what we have learned so far about the regulation of the "uncoupled" SR Ca^{2+} release channel by Ca^{2+}, Mg^{2+}, and adenine nucleotide.

$$\text{"Closed"} \xrightleftharpoons[H^+, Mg^{2+}]{Ca^{2+} + ATP} \text{"Open"}$$

$$(Mg \cdot ATP, \text{calmodulin})$$

Our vesicle flux measurements suggest that the channel opens in the presence of micromolar calcium and millimolar nucleotide. The presence of both ligands is required to maximally open the channel. Mg^{2+} and H^+ are inhibitors of channel opening.

In muscle, regulation of the channel can be expected to be more complex than indicated by the above equation. One reason for this is that Ca^{2+} release from the vesicles is modulated by $Mg \cdot ATP$ and calmodulin. A second reason is that dissociation from the T-tubule membrane may have resulted in an alteration of channel regulation. While the vesicle studies strongly suggest a role for Ca^{2+} in the release of Ca^{2+}, our results are apparently inconsistent with the observation of a SR Ca^{2+} release process that is tightly coupled to T-tubule charge movement and voltage in skeletal muscle (Chapter 25, this volume). An additional complexity arises from the recent discovery that only every other foot appears to be linked to a group of four particles located in the T-tubule membrane (Chapter 22, this volume). It is therefore conceivable that SR Ca^{2+} release is regulated by more than one mechanism. Channels linked to the T-tubule voltage-sensing molecules could be envisioned to directly open in response to T-tubule depolarization. Released Ca^{2+} could then amplify SR Ca^{2+} release by opening all channels more fully.

An additional requisite would be that there exist a mechanism for the rapid closing of all channels that are and are not linked to the T-tubule particles. This could be achieved if a network of interacting Ca^{2+}-channels exists in the junctional SR membrane. Clearly, although much progress has been made in recent years, our understanding of the mechanisms leading to the release of Ca^{2+} from SR is incomplete and will require further experimentation.

ACKNOWLEDGMENTS.   Research reported here was supported by NIH grants AR18687, HL27430, and HL38835.

REFERENCES

1. Endo, M., 1977, Calcium release from the sarcoplasmic reticulum, *Physiol. Rev.* **57:** 71–108.
2. Somlyo, A. P., 1985, The messenger across the gap, *Nature* **316:** 298–299.
3. Smith, J. S., Coronado, R., and Meissner, G., 1985, Sarcoplasmic reticulum contains adenine nucleotide-activated calcium channels, *Nature* **316:** 446–449.
4. Smith, J. S., Coronado, R., and Meissner, G., 1986, Single channel measurements of the calcium release channel from skeletal muscle sarcoplasmic reticulum: Activation by Ca^{2+} and ATP and modulation by Mg^{2+}, *J. Gen. Physiol.* **88:** 573–588.
5. Rousseau, E., Smith, J. S., Henderson, J. S., and Meissner, G., 1986, Single-channel and $^{45}Ca^{2+}$ flux measurements of the cardiac sarcoplasmic reticulum calcium channel, *Biophys. J.* **50:** 1009–1014.
6. Lai, F. A., Erickson, H. P., Rousseau, E., Liu, Q-Y., and Meissner, G., 1988, Purification and reconstitution of the calcium release channel from skeletal muscle, *Nature* **331:** 315–319.
7. Lai, F. A., Anderson, K. Rousseau, E., Liu, Q. Y., and Meissner, G., 1988, Evidence for a Ca^{2+}-channel within the ryanodine receptor complex from cardiac sarcoplasmic reticulum, *Biochem. Biophys. Res. Commun.* **151:** 441–449.
8. Inui, M., Saito, A., and Fleischer, S., 1987, Isolation of the ryanodine receptor from cardiac sarcoplasmic reticulum and identity with feet structures, *J. Biol. Chem.* **262:** 15637–15642.
9. Imagawa, T., Smith, J. S., Coronado, R., and Campbell, K. P., 1987, Purified ryanodine receptor from skeletal muscle sarcoplasmic reticulum is the Ca^{2+}-permeable pore of the calcium release channel, *J. Biol. Chem.* **34:** 16636–16643.
10. Meissner, G., 1984, Adenine nucleotide stimulation of calcium-induced calcium release in sarcoplasmic reticulum, *J. Biol. Chem.* **259:** 2365–2374.
11. Smith, J. S., Coronado, R., and Meissner, G., 1986, Single-channel calcium and barium currents of large and small conductance from sarcoplasmic reticulum, *Biophys. J.* **50:** 921–928.
12. Meissner, G., 1983, Monovalent ion and calcium ion fluxes in sarcoplasmic reticulum, *Mol. Cell. Biochem.* **55:** 65–82.
13. Nagasaki, K., and Kasai, M., 1983, Fast release of calcium from sarcoplasmic reticulum vesicles monitored by chlortetracycline fluorescence, *J. Biochem. (Tokyo)* **94:** 1101–1109.
14. Ikemoto, N., Antoniu, B., and Meszaros, L. G., 1985, Rapid flow chemical quench studies of calcium release from isolated sarcoplasmic reticulum, *J. Biol. Chem.* **260:** 14096–14100.
15. Meissner, G., Darling, E., and Eveleth, J., 1986, Kinetics of rapid calcium release by sarcoplasmic reticulum: Effects of Ca^{2+}, Mg^{2+}, and adenine nucleotide, *Biochemistry* **25:** 236–244.
16. Meissner, G., and Henderson, J. S., 1987, Rapid calcium release from cardiac sarcoplasmic reticulum vesicles is dependent on Ca^{2+} and is modulated by Mg^{2+}, adenine nucleotide and calmodulin, *J. Biol. Chem.* **262:** 3065–3073.
17. Kushmerick, M. J., 1983, Energetics of muscle contraction, in: *Handbook of Physiology, Section 10: Skeletal Muscle* (L. D. Peachey, R. H. Adrian, and S. R. Geiger, Eds.), pp. 189–236, American Physiological Society, Bethesda, MD.
18. Gupta, R. K., and Moore, R. D., 1980, ^{31}P NMR studies of intracellular free Mg^{2+} in intact frog skeletal muscle, *J. Biol. Chem.* **255:** 3987–3993.
19. Baylor, S. M., Chandler, W. K., and Marshall, M. W., 1982, Optical measurements of intracellular pH and magnesium in frog skeletal muscle, *J. Physiol. (London)* **331:** 105–137.
20. Meissner, G., 1986, Evidence of a role for calmodulin in the regulation of calcium release from skeletal muscle sarcoplasmic reticulum, *Biochemistry* **25:** 244–251.
21. Kim, D. H., and Ikemoto, N., 1986, Involvement of 60-kilodalton phosphoprotein in the regulation of calcium release from skeletal muscle sarcoplasmic reticulum, *J. Biol. Chem.* **261:** 11674–11679.
22. Meissner, G., 1988, Ionic permeability of isolated muscle sarcoplasmic reticulum and liver endoplasmic reticulum vesicles, *Meth. Enzym.* **157:** 417–437.

Chapter 33

Calcium Channels in Sarcoplasmic Reticulum Membranes Isolated from Skeletal Muscle

Benjamin A. Suárez-Isla, Juan José Marengo, Verónica Irribarra, and Ricardo Bull

1. INTRODUCTION

Calcium channels have been detected in sarcoplasmic reticulum (SR) membranes isolated from rabbit[1,2] and frog[3-5] skeletal muscle. Several lines of evidence indicate that these conductances participate in calcium release during excitation–contraction coupling.[1-6] High-conductance channels present in SR isolated from rabbit[1] and frog SR muscle[3-5] are activated by ATP and by calcium applied at the myoplasmic side, and are blocked by magnesium and ruthenium red.[1,6] The channel present in SR membranes from frog is activated by micromolar concentrations of 1,4,5-inositol trisphosphate,[4,5] a postulated internal agonist of excitation–contraction coupling.[7-10] This drug increases fractional open time (P_0) in a concentration-dependent manner without an effect on single-channel conductance. Gating and conductance of the same channel are modified by nanomolar concentrations of ryanodine,[3,11] a plant alkaloid that elicits irreversible muscle contractures.

In addition to the high-conductance calcium channels, a low-conductance channel selective for calcium and barium, and insensitive to calcium and ATP, but activated by caffeine and blocked by dantrolene, is present in native SR membranes from rabbit.[2] A related low-conductance channel insensitive to calcium (above 10^{-5}M), ATP, or IP_3, but

BENJAMIN A. SUÁREZ-ISLA • Departamento de Fisiología y Biofísica, Facultad de Medicina, Universidad de Chile, and Centro de Estudios Científicos de Santiago, Santiago, Chile. JUAN JOSÉ MARENGO, VERÓNICA IRRIBARRA, and RICARDO BULL • Departamento de Fisiología y Biofísica, Facultad de Medicina, Universidad de Chile, Santiago, Chile.

activated by caffeine has been detected in SR from frog.[3] At present the role of low-conductance channels in SR is unclear.

The available evidence reviewed in this chapter favors the view that at least one class of calcium channels in SR are agonist-activated conductances. Endogenous or exogenous agonists activate the channels by increasing P_0 without alteration of single-channel conductance. Channel activation results mainly from the concentration-dependent decrease of closed-time intervals.

2. HIGH-CONDUCTANCE CALCIUM CHANNELS IN SR MEMBRANES

Sarcoplasmic reticulum membrane vesicles isolated from frog skeletal muscle fuse into phospholipid bilayers and display high-conductance calcium channels of ca. 100 pS

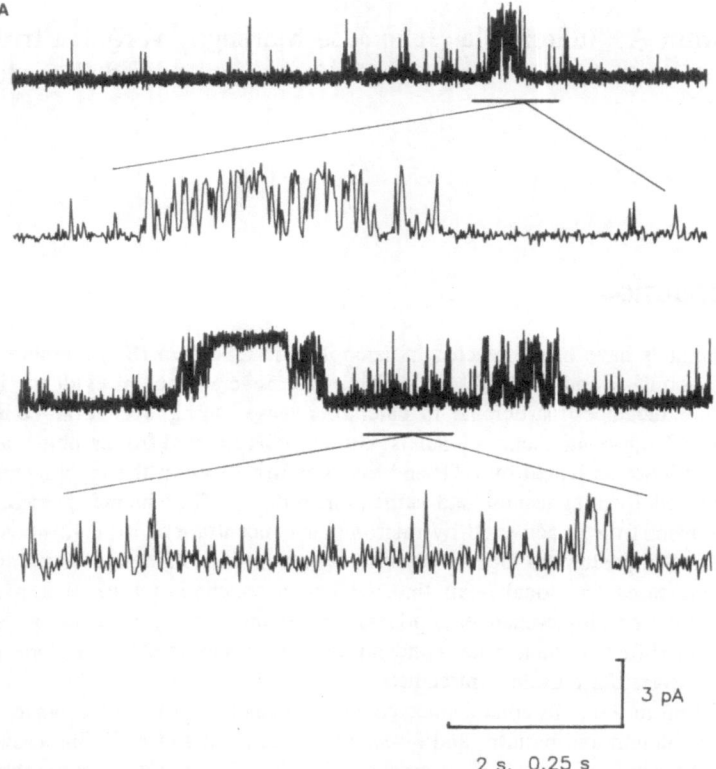

Figure 1. (A) Single-channel current fluctuations at 0 mV in the presence of 1 μM *cis* Ca^{2+}. Note different time calibrations that show intraburst activity. Channel incorporated in charged bilayers (POPE/PS 1 : 1). Channel openings corresponding to cation flux in the *trans-cis* direction are plotted upward. Voltages were applied in the *cis* compartment, while the *trans* chamber was kept at virtual ground.

with 37 mM *trans* Ca or Ba HEPES used as current carrier (Fig. 1A). The *trans* side is assumed to correspond to the intrareticular space. These channels were found to be selective for calcium and barium over Tris ($P_{divalent}/P_{tris}$ = 9.2) (Fig. 1B). As shown in Fig. 1A, records at low time resolution revealed heterogeneous kinetics. Unconditional interval distribution histograms of closed or open times could be approximately described by the sum of two exponential functions, with characteristic time constants of 2 to 6 msec (fast component accounting for ca. 60% of all intervals), and a slow component in the range of 15 to 100 msec. In addition, a variable and small fraction (about 1%) of very long open and closed intervals was observed that exceeded the previous range (more than 100 msec to several seconds). However, these less frequent events could contribute in some cases up to 40% of the calculated fractional open time (P_0).

Fractional open time was not significantly voltage dependent at pH 7.4 in the voltage range from −20 to +20 mV applied in the *cis* compartment (Fig. 2). Open- and closed-time distributions could be described by biexponential functions, and the derived fast and slow time constant did not depend on the voltage applied.

It has recently been reported[12] that the ryanodine receptor channel exhibits a voltage dependence at pH 7.1 but not at pH 7.4. However, the reported voltage sensitivity affects channel gating only at voltages above +60 mV, a value probably not attained at the SR membrane during excitation–contraction coupling.[13]

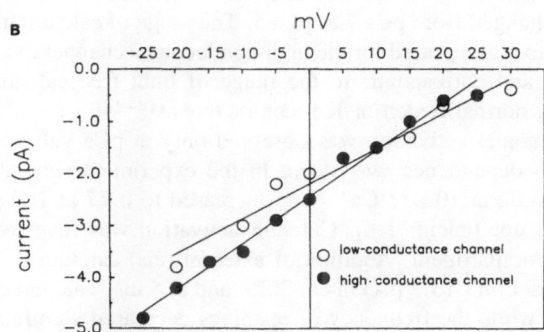

(B) Current–voltage relationship. Open circles: low-conductance Ca^{2+}-channel; 37 mM *trans* Ba^{2+}, conductance 70 pS. Closed circles: high-conductance Ca^{2+}-channel; 37 mM *trans* Ca^{2+}, conductance 100 pS. Channels incorporated in charged bilayers (POPE/PS 1 : 1). Highly purified SR membranes were prepared from the leg muscles of the Chilean frog *Caudiverbera caudiverbera* by homogenization of the tissue followed by sucrose density gradient centrifugation essentially as described.[5,33] Planar phospholipid bilayers were formed from mixtures of palmitoyloleoyl phosphatidylethanolamine (POPE) and phosphatidylcholine (PC) or POPE and phosphatidylserine (PS) obtained from Avanti Polar Lipids, Inc. (Birmingham, AL). Lipids were dissolved in decane at 25 mg/ml. All chemicals used were of reagent grade. Sarcoplasmic reticulum membrane vesicles (50–100 μg protein) were added to the cis side (corresponding to the cytoplasmic side) with stirring. Recording conditions[5] were: *cis* compartment, 225 mM HEPES Tris containing variable concentrations of calcium and drugs; *trans* compartment: 37 mM Ba HEPES or Ca HEPES. All HEPES buffers titrated to pH 7.4. The experiments were carried out at 22–24°C. Data were recorded on tape at 1 kHz (−3dB), postfiltered with an analog Bessel filter or digitally at 200 Hz and digitized at 1 kHz with a microcomputer equipped with an Axolab interface. Data analysis was carried out with versions of Axess or pClamp software (Axon Instruments, Foster City, CA). Fractional open time (P_0) was calculated as the ratio between the sum of all open event durations in a given record and the total record duration.

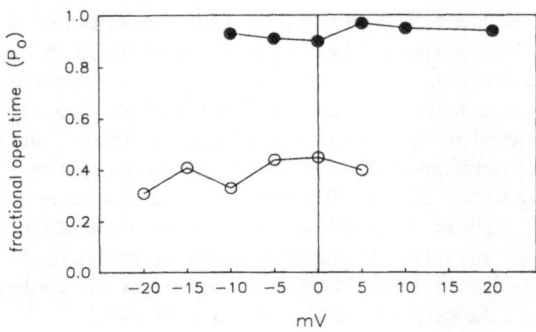

Figure 2. Voltage independence of the high-conductance Ca^{2+}-channel at pH 7.4. Data from two different incorporations carried out as described in the legend to Fig. 1. Filled circles: *cis* chamber contained 150 μM calcium and 1 mM ATP. Open circles: *cis* chamber contained calcium at pCa 4.5. Fractional open time (P_0) was obtained from records of 100 sec containing between 1200 and 2000 fluctuations.

3. THE HIGH-CONDUCTANCE CALCIUM CHANNEL IS NOT ACTIVATED BY PHYSIOLOGICAL CALCIUM CONCENTRATIONS

The effect of *cis* calcium on fractional open time and on closed- and open-time distribution histograms was studied in the presence of HEPES Tris buffers at pH 7.4 with the cis compartment under tightly controlled pCa values in the range 7 to 4. As shown in Fig. 3, fractional open time was not modified when the free calcium concentration in the *cis* chamber was changed from pCa 7 to pCa 5. The range of calcium insensitivity (0.1–10 μM) is similar to that reported for the high-conductance channel found in mammalian SR membranes[6] and corresponds to the range of bulk free calcium concentrations approached during normal twitch or K^+ contractures.[14–16]

Significant channel activation was observed only at pCa values of 4.5 or lower, where the calcium dependence was steep. In the experiment shown in Fig. 3 (open squares), P_0 was 0.02 at 10 μM Ca^{2+} and increased to 0.77 at 150 μM Ca^{2+}; (estimated pseudo-Hill coefficient: 1.6). Calcium activation was reversed by addition of EGTA to the cis compartment. Addition of external (cis) calcium to 1 mM decreased current amplitude at 0 mV (3.2 pA) in ca. 20%, and at 5 mM calcium current amplitude was only 0.8 pA, while the frequency of openings decreased significantly (Fig. 4).

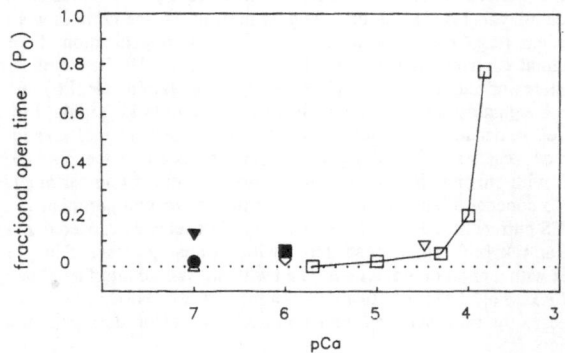

Figure 3. Effect of *cis* $[Ca^{2+}]$ on the fractional open time at 0 mV. Different symbols correspond to seven different incorporations. P_0 was computed as mentioned in the legend to Fig. 1.

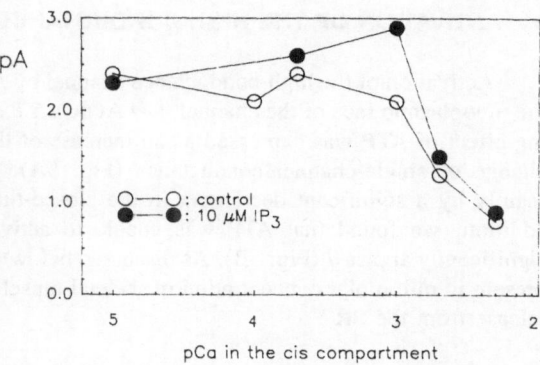

Figure 4. Effect of high *cis* calcium on apparent current amplitude of the high-conductance channel at 0 mV. *Trans* compartment contained 37 mM Ba HEPES. Channel incorporated into POPE:PS (1:1) bilayer. Closed circles: measurements carried out after washing of high calcium and addition of 10 μM IP$_3$ in the same experiment.

4. EFFECT OF MAGNESIUM

Addition of *cis* Mg increases the frequency of brief closures, and above 2 mM Mg^{2+} a decrease in single-channel current amplitude of 0 mV is observed without change of slope conductance (+25 mV to −30 mV). The shift of the current–voltage relationship suggests a significant channel permeability for magnesium and indicates that this ion also exerts a rapid blockade of the channel.[6]

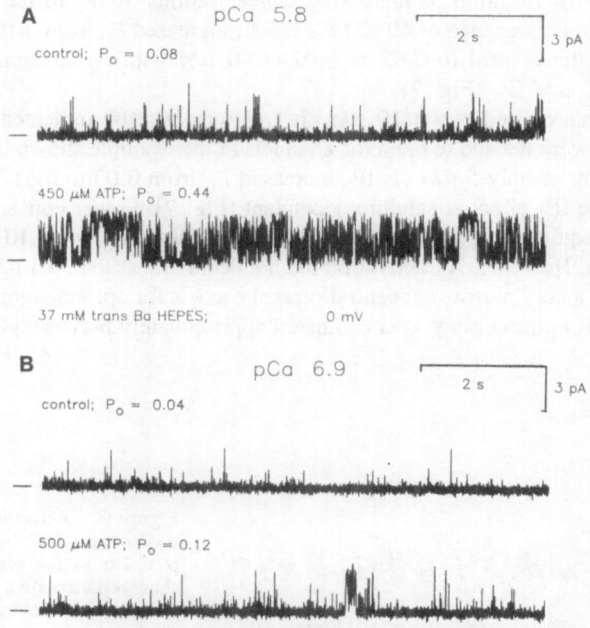

Figure 5. Activation of the high-conductance Ca^{2+}-channel by ATP. (A) ATP addition to the *cis* chamber at pCa 5.8. (B) ATP addition at pCa 6.9, same experiment.

5. ACTIVATION OF THE HIGH-CONDUCTANCE CHANNEL BY ATP

Activation of the high-conductance channel by ATP depends on its concentration at the myoplasmic face of the channel.[4,6] At pCa 5.8 in the *cis* compartment, the activating effect of ATP was expressed as an increase of the fractional open time P_0 without change of single-channel conductance (Fig. 5A). The increment in P_0 was caused mainly by a significant decrement in the closed-time constants (to be published). In addition, we found that ATP was unable to activate the high-conductance channel significantly at pCa 7 (Fig. 5B). As discussed below, this finding may explain why ATP, present in millimolar concentration in skeletal muscle at rest, is unable to elicit calcium release from the SR.

6. THE HIGH-CONDUCTANCE CALCIUM CHANNEL FROM SR IS IP₃-SENSITIVE

IP₃ has been proposed as an agonist of excitation–contraction coupling in skeletal muscle.[7–10,17,18] We found that IP₃ was an activator of the Ca^{2+}-channel present in frog SR membranes and we presented the first quantitative account of direct IP₃ effects on SR Ca^{2+}-channels.[3–5] We observed that micromolar concentrations of IP₃ activated the high-conductance Ca^{2+}-channel present in SR from frog and rabbit, by increasing channel fractional open time (P_0) without effect on the single-channel conductance (Fig. 6). Addition of increasing concentrations of IP₃ to the *cis* side (myoplasmic side) in the presence of 40 μM *cis* Ca^{2+}, increased P_0 from 0.10 ± 0.01 (mean ± S.E.M.) in the control to 0.85 ± 0.02 at 50 μM with an apparent half-maximal activation at 15 μM IP₃ (Fig. 7).

Experiments carried out at 10 μM cis Ca^{2+} or less,[16] indicated that *lower* IP₃ concentrations were needed to open the channel. In the example shown in Fig. 7 (closed circles), addition of only 5 μM cis IP₃ increased P_0 from 0.03 to 0.51. These findings indicate that the IP₃ effect is calcium dependent (Fig. 7). Lower concentrations of this agonist were required to produce a significant increase in P_0 when 10 μM or less *cis* Ca^{2+} was used. However, IP₃ activation did not occur at less than 0.1 μM free calcium, suggesting that a very narrow concentration range exists for optimal agonist effect (to be published). This optimal range was estimated approximately between pCa 7.0 and 6.5.

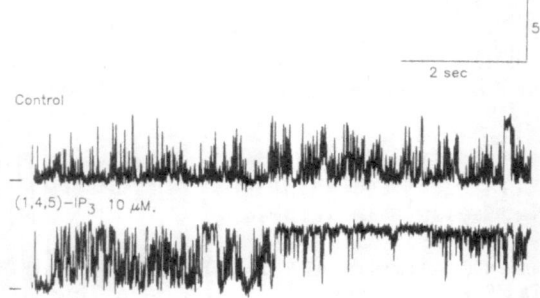

Control

(1,4,5)-IP₃ 10 μM.

5 pA

2 sec

Figure 6. Activation of the high-conductance Ca^{2+}-channel by IP₃ at pCa 7.0 in the *cis* chamber. Traces show representative 8 sec segments of 100 sec records obtained in the absence (top) and presence (bottom) of 10 μM IP₃. Note the appearance of very long (1 sec or more) open intervals.

Figure 7. Activation of the high-conductance calcium channel by IP_3 in the presence of different free calcium concentrations in the *cis* chamber. Symbols correspond to four different experiments where only one channel was incorporated. P_0 was evaluated at 0 mV from single-channel records of 100 to 240 sec digitized at 1 kHz sampling rate and low-pass filtered at 200 Hz ($-3dB$, Bessel filter). P_0 was obtained from consecutive windows of 10 sec, and each point represents the mean P_0 value $\pm$ S.E.M. over 10 to 24 consecutive windows.

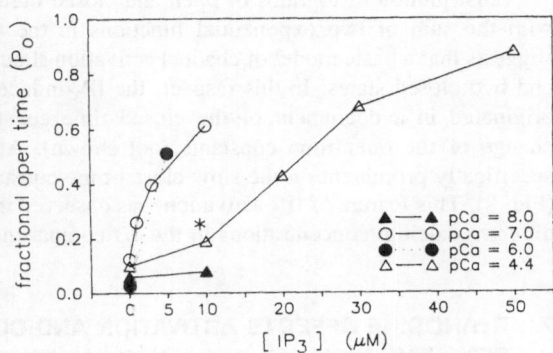

These results suggest that a critical calcium concentration is required to allow for maximal agonist potency. The calcium concentration needed by IP_3 to exert a sizable activation of the high-conductance channel was near to the measured resting free calcium levels.[19] This points out a significant difference with the calcium requirements for ATP, a physiological agent present in millimolar concentration in skeletal muscle at rest that does not activate the high-conductance channel at pCa 7 and that requires pCa 6 or lower to activate the channel.

Calcium dependence of IP_3 effects was also observed when monitoring calcium signals with aequorin from mechanically skinned muscle fibers (see Chapter 29, this volume).

The stimulating effect of IP_3 was also observed after channel incorporation into neutral POPE/PC bilayers. In this case, addition of 13.5 μM IP_3 to the cis side in the presence of 10 μM free cis calcium produced also a significantly greater increase in P_0 than that observed in 40 μM cis Ca^{2+} in charged bilayers from 0.01 to 0.75 (not shown).

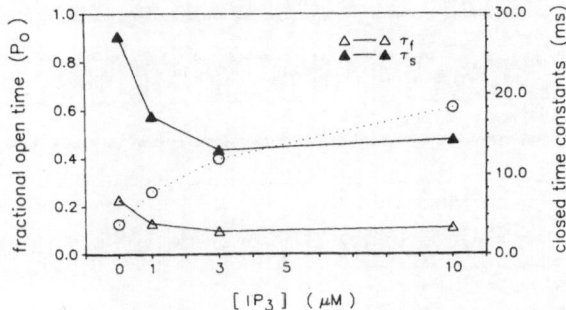

Figure 8. Kinetic changes during IP_3 activation of the high-conductance channel at pCa 7.0. Unconditional open- and closed-interval distributions could be fitted by a biexponential function if all events equal or shorter than 1 msec (limit of time resolution of the recording system) were excluded from the analysis. Maximum interval considered was 400 msec. Open circles: fractional open time (P_0). Closed triangle: slow time constant. Open triangles: fast time constant. Notice that IP_3 effect is maximal between 1 and 3 μM.

Distribution histograms of open and closed durations could be fitted satisfactorily with the sum of two exponential functions in the absence or presence of IP_3. This suggests that a basic model of channel activation should contain a minimum of two open and two closed states. In this respect, the IP_3-induced increase of P_0 was found to be originated in a decrement of the closed time constants (Fig. 8) without significant change of the open-time constants (not shown). At pCa 7.0 this kinetic effect was specifically prominent on the slow closed-time constant and was maximal at 3 μM IP_3 (Fig. 8). This feature of IP_3 activation was observed in all the experiments carried out at different calcium concentrations in the *cis* compartment.

7. RYANODINE AFFECTS ACTIVATION AND CONDUCTANCE OF THE IP_3-SENSITIVE HIGH-CONDUCTANCE CHANNEL

Ryanodine is a plant alkaloid that severely affects cardiac and skeletal muscle function[20] in a concentration- and time-dependent manner. Ryanodine elicits irreversible contractures in skeletal muscle,[21] interacting specifically with the SR. The drug binds with high affinity to SR vesicles isolated from rabbit skeletal muscle (K_D: 5 to 50 nM).[22-24] The ryanodine receptor isolated and purified from SR terminal cisternae[24,25] seems to correspond to a part of the "feet" structures found in the triadic junction.[26,27] After incorporation of the receptor into planar bilayers,[24,28] the macromolecule displays electrophysiological properties similar to those of the high-conductance Ca^{2+}-channel of heavy SR from rabbit[1] and frog muscle.[5]

Two types of binding sites for ryanodine were found in heavy SR membranes

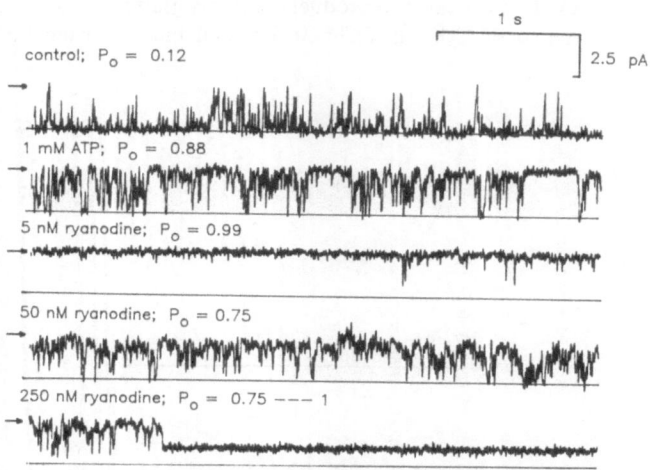

Figure 9. Ryanodine effects on single-channel current fluctuations at 0 mV. Drug concentrations and fractional open times (P_0) calculated from 100 sec records are displayed to the top left of each trace. For display purposes records were filtered at 100 Hz and digitized at 500 Hz. Solid line indicates zero current level; arrows show the open current level. Current carrier was 37 mM Ca^{2+} *trans*. Notice subconductance open level in the bottom trace.

isolated from frog skeletal muscle,[11] one of high affinity (K_D = 1.3 nM, B_{max} = 3.3 pmol per mg) and another of lower affinity (K_D = 90 nM, B_{max} = 7.0 pmol per mg). In addition, ryanodine produced complex concentration-dependent effects on the IP_3-sensitive high-conductance channel of SR from frog (Fig. 9). Low concentrations of ryanodine (5–10 nM) activated the high-conductance channel after a delay of 5 to 10 min, increasing the open probability of the channel without change in conductance. This agonist effect could be elicited from conditions of high or low open probability, and was due to an increase of open-time constants with a concomitant decrement of closed-time constants. Under conditions of low open probability (< 10%), long closed interburst periods (>800 msec) were not affected by ryanodine. Higher concentrations (250 nM) of the drug locked the channel into an open state of ca. ⅓ conductance that displayed very infrequent closures to zero current level, as has been reported in other cases at higher ryanodine concentrations.[25,29]

These findings suggest that the agonist effect of ryanodine may be a result of drug interaction with high-affinity sites, whereas binding to low-affinity sites may be associated with the expression of the subconductance state.

8. BLOCKERS OF THE HIGH-CONDUCTANCE Ca²⁺ CHANNEL

Drugs or ions that are known to block calcium release from isolated SR vesicles were found to exert blockade effects on this channel.[3,4] Ruthenium red increased the frequency of rapid closures at 0.1 μM and blocked channel activity at 0.2 μM. In addition, the channel was inversibly blocked by 0.5 mM lanthanum in the cis chamber.[5] Dantrolene inhibited channel activity at 25 μM after a delay of several seconds (to be published).

9. LOW-CONDUCTANCE Ca²⁺ CHANNELS ARE FOUND IN SR MEMBRANES

The presence of a low-conductance cation channel selective for Ca and Ba over tris (69 pS in 37 mM trans Ca or Ba), was demonstrated in vesicles from frog SR.[3] This channel was observed in ca. 1 out of 10 incorporation attempts and, at variance with the high conductance channel, it was not activated by cis Ca in the range tested (> 10 μM), ATP, or IP_3. However, this conductance was activated by millimolar concentrations of caffeine. This channel may correspond to the very low conductance calcium channel that was detected in native membranes of SR from rabbit muscle,[2] using the tip-dip method. Calcium channels of lower conductance have been also reported in SR membranes isolated from rabbit muscle.[29]

10. THE HIGH-CONDUCTANCE Ca²⁺ CHANNEL IN SR IS AN AGONIST-OPERATED CHANNEL

One of the central questions put forward about the mechanism of excitation–contraction coupling is how the Ca²⁺-channels present in the SR are gated as a consequence of T-tubule depolarization. Activation of the high-conductance Ca²⁺-channel from frog SR membranes is produced by several endogenous and exogenous agents after channel incorporation into planar bilayers. The receptor–channel complex recognizes calcium, ATP, other adenine nucleotides, and IP_3. In general, these agonists activate the channel by increasing fractional open time in a concentration-dependent manner, without significantly affecting ion conduction properties. This conductance is not gated by voltage under *in vitro* conditions, and, thus, it can be characterized as an agonist-operated channel displaying receptor sites for agonists and/or modulators on the myoplasmic moiety of the macromolecule. Our results suggest that the key signal leading to rapid channel opening cannot be given by the preexisting calcium or ATP at rest and, as a consequence, a transient physicochemical process is required to account for the rapid transduction of the depolarization of the T-tubule into a large increase in calcium permeability in the SR.

11. DISCUSSION

Two basic models of excitation–contraction coupling have been put forward to explain how the depolarization of the T-tubule membrane triggers calcium release from the SR. One of them is the *"mechanical model"*[31,32] (also see Chapter 25, this volume). According to this model a sort of intimate configurational interaction ("allosteric modification") would occur between macromolecules situated in the T-tubule membrane and in the SR membranes associated with the electron dense "feet" present at the T-SR junction; depolarization of the T-tubule membrane would lead to opening of a SR calcium channel via a transmitted conformational change between the T-tubule and the SR membranes. The asymmetric charge movement observed during voltage clamp pulses in the absence of ionic currents might report the conformational change.[31,32]

A different proposal is the *"chemical messenger"* model[7–10,17,18] (also see Chapter 27, this volume). According to this model depolarization of the T-tubule membrane would elicit rapid production of IP_3 that would function as an internal agonist, eliciting calcium release from the SR by opening calcium-selective channels.

Our results address directly the fundamental question of how the SR calcium channels are gated and lend support to the notion of a chemical messenger that would initiate the calcium release process. The reported findings[4,5] demonstrate for the first time the existence of IP_3-sensitive calcium channels in the SR and give direct support to one of the key assumptions of the chemical transmission hypothesis. Our studies also demonstrate[3–5,16] that the IP_3-sensitive high-conductance channel is the same channel affected by ryanodine.

In this regard, we think that the recent discussion on the "mechanical" as opposed to the "chemical" hypothesis has been taking place on grounds of increasing consensus. Some of the facts about which agreement exists are:

1. The T-SR junction is the specialized organelle at which transduction occurs.
2. The time course and voltage dependence of calcium signals and suprathreshold charge movement are similar. Both can be abolished by lowering external calcium.

3. The ryanodine receptor is part of the foot protein and is activated by micromolar concentrations of IP_3 as is the native SR Ca^{2+}-channel.
4. Depolarization increases IP_3 sensitivity in skinned muscle fibers.[30]

However, views diverge as to the interpretation of some components of charge movement and the role of the hypothetical voltage sensors generating the charge movement. Proponents of the "mechanical" model would assign to the voltage sensors the role of gating the SR Ca^{2+}-channels by allosteric modification and transmission of a conformational change through the backbones of neighboring macromolecules in the T-SR junction. From the side of "chemical transmission" it would be argued that charge movement would be reporting voltage-dependent activation of an enzyme or a group of them that would be responsible for the breakdown of PIP_2 into IP_3 and diacylglycerol: the voltage sensor is then part of an enzyme.

Knowledge of the activation and modulation profiles of the high-conductance Ca^{2+}-channel cannot solve this fundamental question per se. However, we think that the available evidence (see Chapters 27, 29, and 30) does give substantial support to the notion of a chemically mediated transduction mechanism where IP_3 may be the direct agonist of the SR channel. The necessary experiments to help solving this problem should monitor simultaneously calcium transients induced by IP_3 and the effects of voltage on these signals.

ACKNOWLEDGMENTS.   This work was supported by NIH grant GM-35981, by a grant from the Tinker Foundation to the Centro de Estudios Científicos de Santiago, Muscular Dystrophy Association, FONDECYT 902, 598 and 1340, and Universidad de Chile DIB grants No. 2123 and 2149. Additional support was provided by NSF International Collaborative grant INT86-13052 and NIH AM-25201 to Dr. J. Vergara at UCLA.

REFERENCES

1. Smith, J. S., Coronado, R., and Meissner, G., 1985, Sarcoplasmic reticulum contains adenine nucleotide-activated calcium channels, *Nature* **316**: 446–449.
2. Suárez-Isla, B. A., Orozco, C., Heller, P. F., and Froehlich, J. P., 1986, Single calcium channels in native sarcoplasmic reticulum membranes from skeletal muscle, *Proc. Natl. Acad. Sci. USA* **83**: 7741–7745.
3. Irribarra, V., Bull, R., Oberhauser, A., Marengo, J. J., and Suárez-Isla, B. A., 1988, Two types of calcium channels in frog sarcoplasmic reticulum (SR) membranes, *Biophys. J.* **53**: 609a.
4. Suárez-Isla, B. A., Irribarra, V., Oberhauser, A., Bull, R., Larralde, L., Jaimovich, E., and Hidalgo, C., 1988, Inositol 1,4,5-trisphosphate activates a calcium channel in isolated sarcoplasmic reticulum (SR) membranes, *Biophys. J.* **53**: 467a.
5. Suárez-Isla, B. A., Irribarra, V., Oberhauser, A., Bull, R., Larralde, L., Hidalgo, C., and Jaimovich, E., 1988, Inositol 1,4,5-trisphosphate activates a calcium channel in isolated sarcoplasmic reticulum (SR) membranes, *Biophys. J.* **54**: 737–741.
6. Smith, J. S., Coronado, R., and Meissner, G., 1986, Single-channel measurements of the calcium release channel from skeletal muscle sarcoplasmic reticulum: Activation by Ca^{2+} and ATP and modulation by mg^{2+}, *J. Gen. Physiol.* **88**: 573–588.
7. Vergara, J., Tsien, R. Y., and Delay, M., 1985, Inositol 1,4,5-trisphosphate: A possible chemical link in excitation–contraction coupling in muscle, *Proc. Natl. Acad. Sci. USA* **82**: 6352–6356.

8. Vergara, J., and Delay, M., 1986, The measurement of a transmission delay and the effect of temperature at the coupling process of the triadic junction in skeletal muscle fibers, *Proc. Roy. Soc. Lond.* **229:** 97–110.

9. Vergara, J., Asotra, K., and Delay, M., 1987, A chemical link in excitation–contraction coupling in skeletal muscle, in: *Cell Calcium and the Control of Membrane Transport* (L. Mandel and D. C. Eaton, Eds.), pp. 133–151. Rockefeller University Press, New York.

10. Vergara, J., and Asotra, K., 1987, The chemical transmission mechanism of excitation–contraction coupling in skeletal muscle, *News Physiol. Sci.* **2:** 182–186.

11. Bull, R., Marengo, J. J., Suárez-Isla, B. A., Donoso, P., Sutko, J. L., and Hidalgo, C., 1989, Activation of calcium channels in sarcoplasmic reticulum from frog muscle by nanomolar concentrations of ryanodine, *Biophys. J.* **56:** 749–756.

12. Ma, J., Fill, M., Knudson, M., Campbell, K. P., and Coronado, R., 1988, Ryanodine receptor of skeletal muscle is a gap junction-type channel, *Science* **242:** 99–102.

13. Oetiker, H., 1982, An appraisal of the evidence for a sarcoplasmic reticulum membrane potential and its relation to calcium release in skeletal muscle, *J. Muscle Res. Cell Mot.* **3:** 247–272.

14. Maylie, J., Irving, M., Sizto, N. L., Boyarsky, G., and Chandler, W. K., 1987, Calcium signals recorded from cut frog twitch fibers containing tetramethyl-murexide, *J. Gen. Physiol.* **89:** 145–176.

15. Miledi, R., Parker, I., and Zhu, P. H., 1982, Calcium transients evoked by action potentials in frog twitch muscle fibers, *J. Physiol.* **333:** 655–679.

16. Eusebi, F., Miledi, R., and Takahashi, T., 1983, Aequorin–calcium transients in frog twitch muscle fibers, *J. Physiol.* **340:** 91–106.

17. Volpe, P., Salviati, G., Di Virgilio, F., and Pozzan, T., 1985, Inositol 1,4,5-trisphosphate induces calcium release from sarcoplasmic reticulum of skeletal muscle, *Nature* **316:** 347–349.

18. Volpe, P., Di Virgilio, F., Pozzan, T., and Salviati, G., 1986, Role of inositol 1,4,5-trisphosphate in excitation–contraction coupling in skeletal muscle, *FEBS Lett.* **197:** 1–4.

19. Tsien, R. Y., and Rink, T. J., 1980, neutral carrier ion-selective microelectrodes for measurement of intracellular free calcium, *Biochim Biophys. Acta* **599:** 623–638.

20. Sutko, J. L., Ito, K., and Kenyon, J. L., 1985, Ryanodine: A modifier of sarcoplasmic reticulum Ca^{2+} release in striated muscle, *Federation Proc.* **44:** 2984–2988.

21. Katz, N. C., Ingenito, A., and Procita, L., 1970, Ryanodine-induced contractile failure of skeletal muscle, *J. Pharmacol. Exp. Ther.* **171:** 242–248.

22. Inui, M., Saito, A., and Fleisher, S., 1987, Isolation of the ryanodine receptor from cardiac sarcoplasmic reticulum and identity with the feet structures, *J. Biol. Chem.* **262:** 15637–15642.

23. Lattanzio, F. A., Schlatterer, R. G., Nicar, M., Campbell, K. P., and Sutko, J. L., 1987, The effects of ryanodine on passive calcium fluxes across sarcoplasmic reticulum membranes, *J. Biol. Chem.* **262:** 2711–2718.

24. Lai, F. A., Erickson, H. P., Rousseau, E., Liu, Q. Y., and Meissner, G., 1988, Purification and reconstitution of the calcium release channel from skeletal muscle, *Nature* **331:** 315–319.

25. Campbell, K. P., Knudson, C. M., Imagawa, T., Leung, A. T., Sutko, J. L, Kahi, S. D., Reynolds, C. R., and Madson, T., 1987, Identification and characterization of the high-affinity [³H]ryanodine receptor of the junctional sarcoplasmic reticulum Ca^{2+} release channel, *J. Biol. Chem.* **262:** 6460–6463.

26. Kawamoto, R. M., Brunschwig, J. P., Kim, K. C., and Caswell, A. H., 1986, Isolation characterization and localization of the spanning protein from skeletal muscle triads, *J. Cell Biol.* **103:** 1405–1414.

27. Franzini-Armstrong, C., 1970, Studies of the triad. I. Structure of the junction in frog twitch fibers, *J. Cell Biol.* **47:** 488–499.

28. Imagawa, T., Smith, J. S., Coronado, R., and Campbell, K. P., 1988, Purified ryanodine receptor from skeletal muscle sarcoplasmic reticulum is the Ca^{2+}-permeable pore of the calcium release channel, *J. Biol. Chem.* **262:** 16636–16643.

29. Smith, J. S., Coronado, R., and Meissner, G., 1986, Single-channel calcium and barium currents of large and small conductance from sarcoplasmic reticulum, *Biophys. J.* **50:** 921–928.

30. Donaldson, S. K., Goldberg, N. D., Walseth, T. F., and Huetteman, D. F., 1988, Transverse tubule voltage control of inositol trisphosphate-induced Ca^{2+} release in peeled skinned muscle fibers, *Biophys. J.* **53:** 468a.

31. Schneider, M. F., and Chandler, W. K., 1973, Voltage-dependent charge movement in skeletal muscle: A possible step in excitation–contraction coupling, *Nature* **242:** 244–246.

32. Rios, E., and Brum, G., 1987, A possible role of dihydropyridine receptor molecules in excitation–contraction coupling, *Nature* **325:** 717–720.

33. Hidalgo, C., Parra, C., Riquelme, G., and Jaimovich, E., 1986, Transverse tubule from frog skeletal muscle. Purification and properties of vesicles sealed with the inside-out orientation, *Biochim. Biophys. Acta* **855:** 79–88.

Index